S0-EKT-362

FOUNDATION CONSTRUCTION

FOUNDATION CONSTRUCTION

A. Brinton Carson, P.E.

Chief of the Inspection Division, Bureau of Engineering and Construction, Commonwealth of Pennsylvania. Member: American Society of Civil Engineers; Society of American Military Engineers; National Society of Professional Engineers. Formerly: Adjunct Assistant Professor, Evening College, Drexel Institute of Technology; Commander, Civil Engineer Corps, U.S. Naval Reserve (Retired).

McGRAW-HILL BOOK COMPANY

New York Toronto London Sydney

FOUNDATION CONSTRUCTION

 Library of Congress Catalog Card Number 64-18400

10167

PREFACE

The first decision that must be made when a set of contract documents is prepared for a construction project is whether they shall include specific instructions as to the construction methods to be employed or whether only the desired result shall be stated, leaving the methods to be used in obtaining that result to the discretion of the contractor. The choice of approach has been the subject of considerable controversy among those charged with the preparation of such contract documents, but for the contractor preparing a bid, or even executing the resulting contract, the choice is not as wide as would appear, since his selection of method is frequently circumscribed by the mere inclusion of specific materials.

Take the case of pre-stressed, pre-cast concrete cylinder piling discussed in Chapter 5. The mere fact that this material is to be used indicates how it shall be manufactured, handled, driven, spliced, and cut-off just as surely as if each of these operations was exhaustively detailed in the specifications. For this reason, the author has been at some pains to present the various materials commonly used for piling in sufficient detail to guide the contractor in purchasing, and then to proceed to develop the methods appropriate to the placing of each kind.

Materials are not the only things which dictate the method of construction. Open caissons, to which Chapter 8 is devoted, are designed with the construction method clearly in view and an alternate method of construction would definitely alter the design. Open caisson construction, once limited in use to underwater bridge piers, in recent years has been extended to providing support for other types of structures built in open water, and even to the construction of building foundations on dry land.

Chapter 7, General Foundation Construction, covers a wide range of foundations, often differentiated only by a variation in excavation methods or by whether they are dug-in, drilled-in or punched-in. Equipment foundations, also discussed, frequently involve only nominal excavation but the newer techniques, for example, pre-stressing, differentiate one type from another and alter construction methods.

But there are many problems facing the foundation contractor to which no reference can be found in the contract documents. Wind and weather, the personalities of the individuals involved, the selection and storage of tools and equipment, and the effects of landslides, earthquakes, and other catastrophes will dictate or alter construction methods, as will procedures as to payment for unexpected job conditions. These matters are discussed in Chapter 1, at the beginning, since they too must be considered in bidding.

Where construction operations are to be performed under water it is occasionally necessary to employ compressed air to hold back this pervading medium. Methods of working under water as well as under compressed air are discussed in Chapter 2, where the subject of dredging, or underwater excavation, is also treated. Dredging, once a method used chiefly for the maintenance of harbors, has now become a desirable method for the general contractor involved in foundation construction.

Ground conditions have always played an important role in foundation construction. In Chapter 3, in addition to a discussion of foundations placed on rock, sand, and clay, the treatment of unusual soil conditions involving loess and expansive clays has been included. Soil stabilization is often involved in foundation preparation and a full discussion of the various types of grouting forms a portion of this chapter.

Chapter 4 covers the step-by-step procedures used in under-pinning and Chapter 6 treats of the construction of cofferdams, both indispensable requirements of the foundation contractor and two processes seldom to be found in a set of specifications, no matter what approach was used in their preparation.

It has long been the author's opinion that photographs do not adequately convey sufficient detail to properly delineate construction processes, and in this volume all illustrative material has been limited to line drawings, an approach more commonly used in Europe than in the United States. Many sources have been drawn upon in preparing the more than 160 pages of illustrations in this volume, but each page has been designed solely for the purpose of illustrating the page of text facing it, thus forming a composite unity for easy assimilation by the reader.

Approximately 25 pages are devoted to tabular material, not buried in an appendix, as is so customary, but facing the text where the subject is discussed, so that here again the subject can be comprehended as a whole. (Perhaps this approach is the result of the author's endeavor, over a period of ten years, to convince undergraduate students that the material in the appendix is also pertinent.)

To augment the author's personal experiences in the field of foundation construction many sources of information have been tapped and these have been listed by chapter and section in the bibliography. Only those sources that have been useful to the author have been listed; there are, of course, a vast number of additional articles, pamphlets, and volumes that could be added. No references earlier than 1950 have been included since we are concerned only with foundation construction currently being performed.

Only two or three especially pithy direct quotations have been employed, the author preferring to flavor the subject under discussion with his own word patterns and, hopefully, to add a touch of humor to the gray profundities of foundation construction.

A. Brinton Carson

CONTENTS

FOUNDATION CONSTRUCTION

1 Pre-bid Factors

The process of preparing a bid is a feat of the imagination, particularly if it be a bid concerned, to any degree, with subsurface construction. There are, first of all, the uncertainties which excavation into the earth's surface will disclose. No matter how much factual material in test borings or geological investigations is available, these must be converted into a series of probabilities, into a concept with logic and consistency. Secondly, once the site conditions have been fully conceived, construction methods must be devised to cope with this conception. In a word, the estimator must build the entire project in his mind before he ever puts a figure on paper.

Experience is highly praised as a background for this pre-bidding feat of the imagination. How does this project compare with a similar project once conceived and executed, with due allowance for discovered variations between conception and execution? But experience, once the province solely of the individual, has now become the domain of the many. There is first the "formal education" designed to impart the accumulated wisdoms distilled from the minds of an individual's predecessors. Beyond this is the accumulated experience of specialists, devoting an entire career to the execution of a single series of functions, a role assumed more and more by the manufacturers of particular kinds of equipment and the purveyors of specialized services. Experience, once an assortment of loose ends tossed into the top drawer of the individual's mind, now becomes an intricate filing system, carefully cross-indexed.

The contractor in the role of broker pulling all the many specialists together into a unified pattern has never found unqualified acceptance by the engineer dealing with the construction industry. The reason for this is a little obscure, but in any case such a course has not been too frequently essayed in subsurface construction contracts and need not

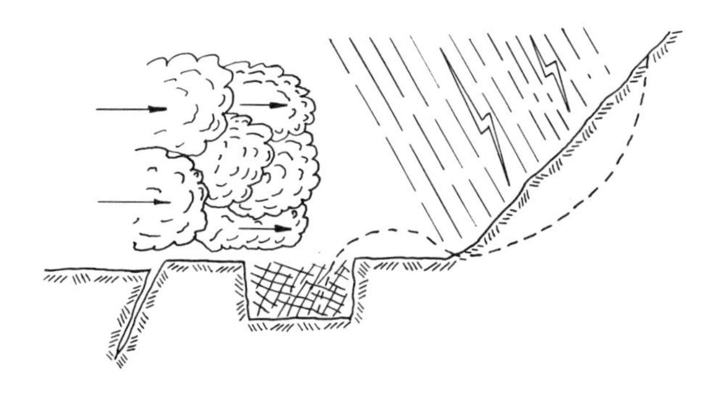

concern us here. The fact remains that whether these specialists are assembled as employees or as sub-contractors, their selection and conceived functioning must go into bid preparation.

This is not a volume concerned primarily with estimating or with the preparation of bids, but rather with construction methods. But, in fact, an estimate is simply a tabulation of the costs of each item, each step involved in the construction methods to be employed. The role of the estimator is that of breaking down a contract into the number of items of work in a selected construction method and of then establishing the costs of each item.

There are numerous volumes dealing with the ordinarily encountered items involved in routine construction methods. There are, however, a number of aspects newly recognized that are not commonly found in printed texts relating to construction. Frequently considered to be factors influencing only design, they are either ignored or unknown to the general contractor.

The items discussed here as pre-bid factors are related primarily to subsurface construction and are only such items as are frequently overlooked in pre-bidding analyses. They are selected and discussed as a stimulant to the estimator's imagination. Only those aspects having or appearing to have some bearing on his concept of construction methods have been introduced. Many of them deserve, and have had, a full volume devoted to them in all their aspects, including those which influence the design of structures.

For those who regularly deal in these construction intangibles, the discussion in this chapter will seem too basic to be of interest. For those encountering these aspects for the first time, further thought and study from other sources may be indicated.

1.1 PROJECT EVALUATION

When a construction project swims into the ken of a contractor, it does so in any one of a number of ways. It may be through the medium of a printed service to which he subscribes, such as *Dodge Reports;* it may be through the medium of a contractor's association of which he is a member; it may be through notices circulated by the originating agency such as a state highway department or the consulting engineer; it may be through advertisements in newspapers or other publications such as the *Engineering News-Record;* it may be by invitation from the owner, in the case of industrial construction. When knowledge reaches him by word of mouth, he must still revert to one of the above sources for further information.

The first things the contractor will want to know are the kind and size of project. If the project is of a type with which he has had experience directly or has knowledge related thereto, if it is within a size range comparable to his ability to finance and for which he has the correct plant or can procure it, he will then want to know where plans and specifications can be obtained, what deposit, if any, is required to obtain them, and whether his deposit will be returned if he decides not to bid on the project.

Once the plans and specifications (the bidding documents) are in his possession, the contractor is faced with the primary decision of "to bid or not to bid." This decision will be based on four appraisals.

1. From a careful summary of the plans and specifications, a final appraisal of the size and nature of the project must be made in relation to the contractor's experience, finances, and plant. If the project is wholly outside his experience or involves broad facets which appear, superficially, to offer difficulties, he may reject it. Experience, or the lack of it, is frequently "no bar to these wild spirits, from the four corners of the earth they come to woo fair" fortune or misfortune.

Financial ability and plant are closely allied. Plant may be acquired with sufficient finances, although the reverse is not quite so simple an operation. A substantial plant will ordinarily be required for subsurface construction. The possession of an adequate plant may facilitate financing by a contractor whose finances are otherwise meager, but construction is, alas, more difficult to finance than other kinds of endeavor.

2. The second appraisal is that conducted at the site. Some aspects of this will be discussed in subsequent sections, but decisions at this point come mostly from the reactions of the contractor to the milieu existing at the site of the project. It is not uncommon for a contractor, trailing a cloud of aides behind him, to trot briefly over the site of a subsurface project and to decide summarily not to bid without ever opening the massive roll of prints that some small character in the rear is laboriously toting.

True, this decision may be actuated by off-site factors such as the volume of work currently under contract or likely to be available for future bidding, but just as frequently it will be an educated feeling for the site in terms, perhaps, of a previous contract. "This looks like the site of a project where I lost a pile of dough." Or, conversely, it may remind him of a very profitable venture. The progress being made on current contracts, or the difficulties encountered, may enter this thinking as well. Once a decision has been made at this level, subsequent appraisals are less likely to reverse it.

3. The next appraisal is made in the office and is concerned with the detailed nature of the contract documents, particularly of the specifications and of their general and special conditions. Not only do the terms of payment as a whole now enter into the thinking, but such matters as provisions for payment of subsurface contingencies.

4. An appraisal based on a detailed examination of the plans and the technical portion of the specifications can now proceed, but this appraisal will seldom alter previously made decisions.

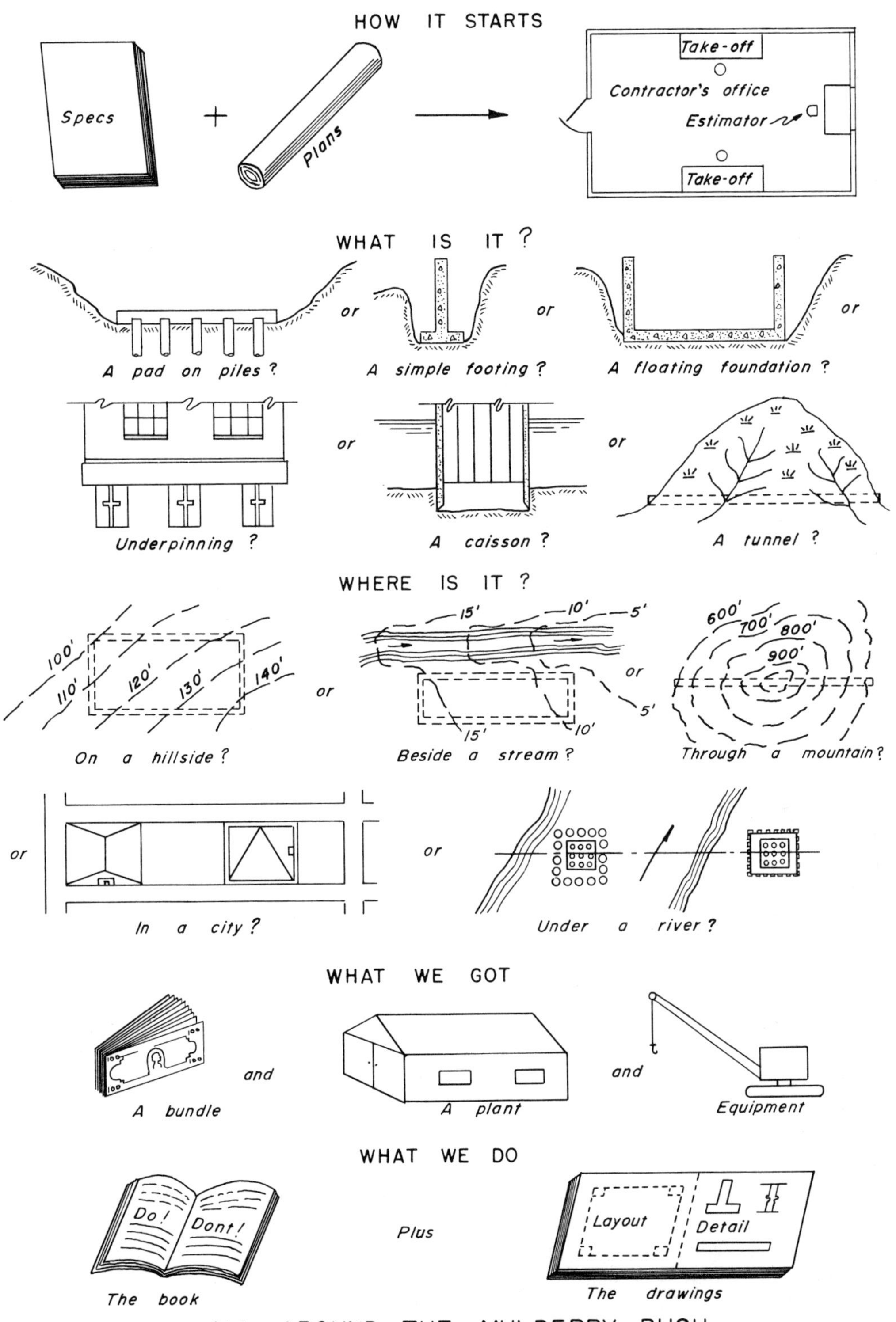

ALL AROUND THE MULBERRY BUSH

1.2 GENERAL CONSTRUCTION PLANT

The term "construction plant" is widely used but seldom precisely defined. From an accounting standpoint it may include everything used by a contractor to prosecute his business. From a more limited point of view it may be everything used by a contractor to complete a particular project or contract. To limit the scope still further and for a more precise cost control of a construction project, "construction plant" as discussed here will cover only those items which cannot be exactly allocated to a specific construction operation or method.

A list of items which may be considered general construction plant is shown. Generally speaking, they constitute items whose costs will affect the entire cost of the project but will not be directly applicable to any single operation. Many of them may be lumped together and added to a bid as "overhead" simply because they cannot be distributed. Some items may, as a matter of preference, be charged to a single item of method.

Concrete-mixing plants certainly apply to all items of concrete, and ordinarily concrete will be a major item in subsurface construction. There are two reasons for pulling this item out of direct concrete costs. First, it is often necessary to make a decision as to whether to buy concrete mixed off site or to mix it as a contract operation. This naturally segregates it in the early planning stages. In the second place, types of concrete, as well as the costs of each type, may vary widely on a single subsurface contract. Good cost control would indicate that 4000-lb concrete placed in forms should be separated from 2500-lb mass concrete. The costs of placing alone will be substantially different.

A crane, similarly, may be used for a number of purposes such as demolition, excavation, concrete placing, and rigging. This usage is sometimes distributed later, as a cost, from the records of the time any particular attachment such as the headache ball, the clamshell, the concrete bucket, or the hook was used; sometimes not.

Shop items are viewed variously. The yard for reinforcing steel fabrication has clearly a limited use, and so, generally, will the carpenter shop (since on subsurface work it is generally confined to formwork); but equipment shops such as lubrication, repair, and wash racks will have general application.

In the final analysis, not only the particularities of each job but the peculiarities of the contractor and his system of cost accounting will determine whether an item belongs to general plant or to a specific operation.

Many contractors operate for years without ever encountering any very special problems of manpower importation or housing; and material or equipment parts procurement problems are minor. Adjacent to a city such as New York City, none of these present any great problem beyond locating the various sources of supply in advance of the need for them. In the African bush the situation is quite otherwise, and everything may need to be considered, developed as a cost, preplanned, and provided.

The questions involving small tools furnish some other aspects of the problem. How small are small tools? A spud wrench is a small tool, but is an impact wrench? Perhaps portability should be the criterion; can they be carried to a shed and locked up? The proper tool for the proper job is a beautiful creed, but how many of each will be needed at one time, how many will wear out and at what rate, how many will be lost in the mud, how many will be employed for purposes not intended and will not be available when needed, and, last but not least, how many will be so portable that they can be carried off the job for a do-it-yourself project?

To cover all these factors can only be done with a percentage based on the size and intricacy of the contract, graduated from perhaps 2% on a $100,000 job to one-half of 1% on a million-dollar operation.

The contingency item is to the contractor what the factor of safety is to the engineer. As any contractor with money in the bank will tell you; he knows all the answers; what he really needs are the right questions.

Check List of Construction Plant Questions

MANPOWER

1. Is local labor available? ________ % skilled? ________ % unskilled? ________
2. Is housing required for local labor? ________ A commissary? ________ Transportation? ________
3. Is local housing available? ________ For how many? ________ Adequate? ________
4. What special facilities will be needed for local labor? ________
5. Must labor be imported? ________ How many skilled? ________ Unskilled? ________
6. Is a camp site required? ________ For how many? ________
7. Will facilities be for single men? ________ Families? ________
8. Location for camp site? ________ No. of buildings? ________ Miles of road? ________
9. Feeding facilities? ________ Food supply? ________
10. Is water supply needed? ________ Power? ________ Sanitation? ________
11. Medical facilities? ________ School facilities? ________
12. Will fire protection be needed? ________ Police protection? ________
13. Mail service? ________ Entertainment facilities? ________

MATERIAL

14. Area available to contractor? ________ Area to be acquired? ________
15. Working area needed? ________ Storage area needed? ________
16. Access roads to be built? ________ Rights of way required? ________
17. Transportation to site: Rail? ________ Ship? ________ Truck? ________
18. Transhipment facilities: Rail to truck? ________ Ship to truck? ________
19. Freight-handling facilities needed at site? ________ Off site? ________
20. Source of supply of fine aggregates? ________
21. Source of supply of coarse aggregates? ________
22. Source of supply of cement? ________
23. Aggregate plant requirements? ________
24. Source of water supply? ________ Type? ________
25. Water distribution materials? ________
26. Temporary power facilities? ________
27. Power plant equipment? ________

EQUIPMENT

28. Heavy equipment: Cranes? ________ Attachments? ________
    Tractors? ________ Cable ways? ________ Concrete plant? ________ Trucks? ________
29. General equipment: Compressors? ________ Pumps? ________ Small tools? ________
30. Floating equipment: Dredges? ________ Barges? ________ Whirleys? ________
31. Plant structures: Offices? ________ Labor shanties? ________
    Storage sheds? ________ Trestles? ________ Ramps? ________
32. Maintenance shops: Fuel storage? ________ Fueling site? ________
    Lubrication? ________ Repair? ________ Steam cleaning? ________
33. Fabrication shops: Carpenter? ________ Blacksmith? ________
    Rebar yard? ________ Testing facilities? ________

1.3 PAYMENT FOR SUBSURFACE CONTINGENCIES

In no field of construction does the problem of responsibility for unknown or unanticipated problems or conditions arise so frequently as in subsurface construction. The facts that adequate test borings are now more often the rule than the exception and that developments in the science of soil mechanics have led to better interpretation of test-boring results have not significantly decreased the number of lawsuits or bankruptcies arising from the execution of subsurface work.

Modern practice, for the most part, establishes a contractual relation between the owner and the contractor in which the architect or, more generally on subsurface work, the engineer is simply an agent for the owner. The designer, be he engineer or architect, has created a design and prepared plans and specifications based on *his* idea of subsurface conditions gleaned from test borings and other geological investigations. In conveying the intent and extent of the design to the contractor by means of plans, specifications, and contract, the designer may cover the uncertainties of subsurface behavior in any one of three ways.

1. The contract may clearly assign the responsibility for subsurface conditions to the contractor, not only as they existed prior to beginning the contract but as they may develop during the course of construction. This is the more usual approach.

2. The contract may state that the owner will assume the liability for underground conditions, paying the additional costs of unexpected hazards and securing the savings if subsurface conditions are routine. Such a clause, though seldom seen *per se,* is frequently implied by the use of unit prices for various facets of the work.

3. The contract may be set up for prospective lawsuits by "straddling the fence," by deftly referring one clause to another and, in contradictory language, leaving the subject so ill-defined that each contractor puts a widely varying interpretation on his responsibility.

In the first case, where the contractor is invited to take full responsibility, the contract documents may include well-detailed plans with borings or other data to substantiate them, the whole of which are fully intelligible to the contractor, or they may not. In either event, the contractor is further invited, nay exhorted, to visit the site, to take additional borings, to dig test pits, to put divers over the side where water is involved, and, in general, to make his own investigations. Such shenanigans, if pursued conscientiously by a large group of bidders, would clutter up the site, involve unconscionable expense, and be too time-consuming in the face of a fixed bid date. In lieu of further exhaustive investigations the contractor generally sets up a contingency item or a "risk fund" covering uncertainties and adds it to his price bid.

In the second case, where the owner assumes responsibility, he is paying precisely for the amount of work performed. Where conditions turn out to be as represented, the savings accrue to the owner. If unanticipated conditions arise, they are paid for by the owner, and properly so, since to him all benefits in the completed structure ensue.

Where payment by the owner is contemplated, it is now accepted practice to insert a clause in the specifications stating specifically the procedure to be followed in adjusting the contract documents.

In the third case of poor plans, incomplete data, or obscure specifications, the possibilities for trouble are so immense that, while a few contractors will carefully double their contingency item, many more will simply walk away from the situation, while they can, and return the plans unbid.

On the opposite page a few cases of lawsuits involving subsurface conditions are cited. They do not necessarily involve inadequate contract documents, nor do they pretend to cover all the infinite range of excess-cost possibilities inherent in the construction of subsurface structures.

Subsurface Contract Problems

1. *Problem:* Responsibility for damage to contiguous structures incident to subsurface construction.

Discussion: Contracts for major works in developed areas, such as subways in cities, generally contain precisely designed provisions for the shoring, underpinning, or protection of contiguous structures. This will often be true where new buildings are being constructed adjacent to existing structures, even where the new foundations are somewhat shallow.

In some instances such precautions may be covered, not by specific design provisions but by general clauses, placing the responsibility on the contractor. Where such structures are visibly apparent and immediately adjacent to the new subsurface structure, some allowance, if nothing more precise, can be provided in the contractor's bid. In some cases, however, site conditions deemed wholly outside the area of construction at the time of bidding may still be subject to damage.

Dewatering operations, for instance, can produce settlement over a wide area underlain by saturated soils. Even unexceptional blasts can induce vibrations that can travel along rock strata with sufficient intensity to cause damage. Changes in underground stability, incident to excavations, can alter bearing capacities at points remote from the construction site, particularly in homogeneous soil masses such as old fills, whether natural or man-made in origin.

Property-damage insurance, generally a requirement in contracts, especially those involving underground work, is not a final solution. Frequently the insurance company will let the case go to court, where it becomes necessary to prove that the contractor was negligent. Unfortunately, in jury trials, contractors are frequently presumed negligent, simply from the nature of their calling. In any case such proceedings are time-consuming and costly.

2. *Problem:* Insufficiency of geological data.

Discussion: At least one case, reaching the Supreme Court, has found the owner liable for damages to adjacent structures through his failure to have had test borings made prior to construction. However, where the contract between owner and contractor places responsibility on the contractor for subsurface conditions, the owner's liability can be transferred to the contractor.

3. *Problem:* Quantities of work.

Discussion: Although many contracts involve a lump-sum payment for the whole construction project, quantities are involved in some cases.

a. The contract is lump sum, but a tabulation of quantities is given for information. Where these run counter to the contractor's take-off, they may still influence his bidding, by causing confusion in the estimator's mind.

b. On a unit-price job, where a wide discrepancy in the quantities of an item is encountered, how shall it be bid? The cost of execution for one small unit may be quite different from that for 100 of the same unit. Here also the question of unbalanced bid items enters the picture. If the contractor bids the item as he sees it, he risks the accusation of presenting an unbalanced bid.

c. Unit prices to be quoted upon for additions to, or deletions from, the contract. Here the possibilities are very broad indeed, since, as one individual phrased it, "You can't crawl down a man's belly to find out what he's thinking."

One case in point involved a unit price for additional wood piling, per foot. The number of pilings was shown clearly, and the specifications stated that the bid was to be based on each being 40 ft long. A unit price for additions or deductions in length was requested, on a nominal quantity of 2500 lineal feet. The contractor bid (some years ago) a price of 25 cents per ft, the bare cost of the pile delivered.

The owner, with unexampled cupidity, decided to redesign some of the structures, placing them on 30,000 ft of additional piling which he then insisted the contractor place at 25 cents per ft, although such additional piling was worth about 75 cents per ft. Because the attempt was so flagrant, the matter was finally settled without a lawsuit, but only after the contractor had installed the additional piling, without current payments, in order to complete the contract and place himself in a position to sue.

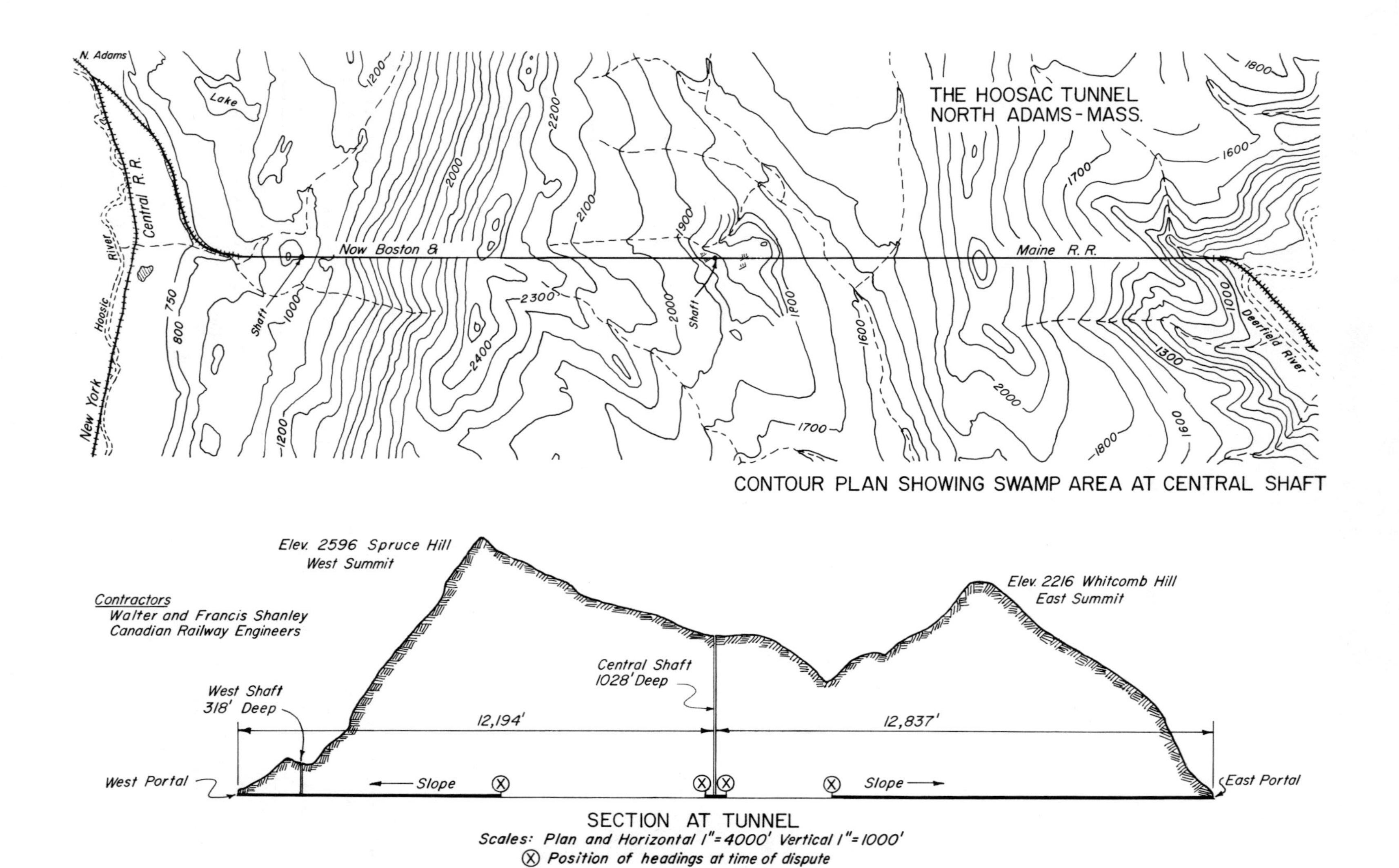
N. Adams
Lake
Central R.R.
Hoosic River
New York
Now Boston &
Maine R.R.
Shaft
Shaft
Deerfield River
THE HOOSAC TUNNEL
NORTH ADAMS - MASS.
750
800
1000
1200
1300
1600
1700
1800
1900
2000
2100
2200
2300
2400
CONTOUR PLAN SHOWING SWAMP AREA AT CENTRAL SHAFT
Elev. 2596 Spruce Hill
West Summit
Elev. 2216 Whitcomb Hill
East Summit
Contractors
Walter and Francis Shanley
Canadian Railway Engineers
West Shaft
318' Deep
Central Shaft
1028' Deep
12,194'
12,837'
West Portal
Slope
Slope
East Portal
SECTION AT TUNNEL
Scales: Plan and Horizontal 1"=4000' Vertical 1"=1000'
Ⓧ Position of headings at time of dispute

1.4 "HOSTILE SUPERVISION"

Another aspect of the situation involving the contractor and the engineer (or possibly the architect) is that concerning decisions as to how a project shall be performed. In many cases of subsurface construction, the design of the structure is based on the method to be employed, requiring that the method of construction be specified in detail. Even where methods of construction are dropped squarely into the contractor's fist, many joint decisions remain. It is a wise engineer who can renounce his favorite brain child.

The idea of a railway tunnel through the Housatonic Mountains in northern Massachusetts was born in 1848 with the chartering of the Troy and Greenfield Railway to connect the Connecticut River westward to Williamstown. Five miles of this run was to be in tunnel under Hoosac Mountain. Tunneling was begun in 1851 from both ends. Since the work was jobbed out to a number of contractors, progress was slow and 4 years later the project was abandoned. The state of Massachusetts had loaned the railroad $750,000, and the mortgage covering this was foreclosed in 1862. The construction of the tunnel now became a matter of state politics.

The state engineer decided that a shaft sunk near the center from which headings could be opened in two directions would speed up construction. The position selected for the 1000-ft-deep shaft lay in a depression between the peaks which collected all the surface drainage from the surrounding area. Indeed, it was in the middle of a swamp. Five hundred feet of the shaft had been sunk in 1867 when a fire in the buildings containing the pumping and hoisting equipment terminated operations and trapped a number of men at the bottom of the shaft.

The following year the state of Massachusetts asked for bids for completing the entire tunnel, stipulating that this central shaft was to be completed by the successful bidder. Walter and Francis Shanley, two railway engineers from Toronto, Canada, who had formed a company for the purpose, were finally awarded the contract in 1868 for the sum of $4,598,268.

The state, as per contract, pumped out the shaft and removed the 13 drowned bodies earlier trapped in it; and the Shanleys, in October of 1870, completed the shaft to final grade.

Headings from the shaft were begun both eastward and westward. In the first 200 ft of the westerly heading, however, a great volume of water (in terms of pumping equipment available at that time) was encountered and they suspended operations here, pushing on the easterly heading. They felt that once the tunnel had been completed to the east, water from the west heading could be drained off by gravity.

The state's engineer felt that the solution lay simply in providing more pumps. The Shanleys pointed out that these would so fill up the shaft as to make it impossible to remove the excavated rock. Suspension of payments was threatened in November of 1870, and the Shanleys poured $200,000 into the fruitless endeavor before, by tacit agreement, the task was abandoned.

A number of claims involving not only the shaft but the patching up of work done by previous contractors were contained in a petition to the state's Committee on Claims presented in January, 1875, following the Shanleys' successful completion of the tunnel. In it they said in part:

> . . . It was sought to be shown that in our way of prosecuting our work, we paid but scant heed to the stipulations of the contract in respect of rates of progress. . . . That is quite true. We did not: and we readily and cheerfully enter a plea of guilty to that count in the indictment; adding that whenever and wherever it came to be an issue between carrying *out* the specification and carrying *on* the work, we never for a moment paused in that course we had laid down for ourselves, or hesitated to treat that document—the specification—so sacred in the eyes of State Engineers, as other than a meaningless and obstructive encumbrance, where too literally construed. But as a consequence . . . we were ever afterwards made to feel that we had to carry out our engagements . . . under a hostile supervision.

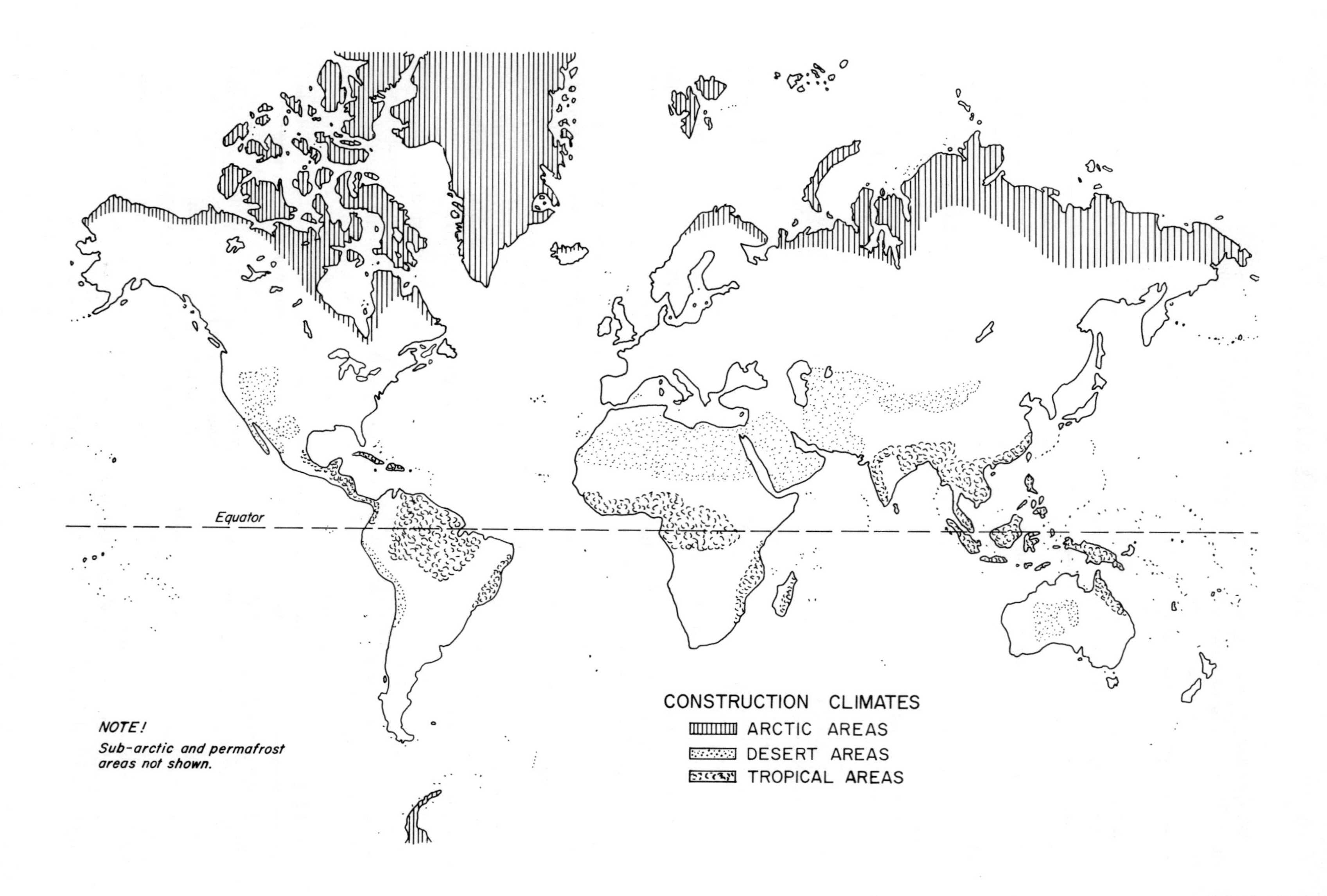
Equator
NOTE!
Sub-arctic and permafrost areas not shown.
CONSTRUCTION CLIMATES
ARCTIC AREAS
DESERT AREAS
TROPICAL AREAS

1.5 PREVAILING WEATHER CONDITIONS

Weather conditions prevailing at the site of a construction project are an important factor which must be considered in evaluating a contract prior to bidding. Not too many decades back, the contractor seldom strayed far from a specific milieu in which he was accustomed to operate. He knew what the weather conditions were and what to do about them regardless of their variations. Today, the scope of many contractors has so broadened that they are apt to be working under various types of weather conditions, many of them adverse, that exist anywhere on the surface of the earth.

Construction techniques as practiced today have largely developed in what we know as the Temperate Zone of the earth's surface, where the extremes of temperature and precipitation vary within relatively narrow limits. There may be exceptions that jump the margins, but their occurrence is rare and of short duration. The rare event may indeed create havoc with construction, but its incidence has been studied and some provision can be made for it.

As construction moves away from the Temperate Zone, it encounters new conditions where temperatures may be continuously higher or continuously colder than those it is accustomed to work in and, similarly, finds precipitation that is greater or considerably less in amount. These are not emergency conditions but standard working conditions with their own range of variations. They cannot successfully be covered by some contingency item, but must be considered as an integral part of the project and provided for accordingly.

Many of the factors due to extremes of temperature or precipitation do not seriously interfere with subsurface construction as they might with buildings or structures rising above the ground surface, but their effect cannot be ignored.

Tropics and *sub-tropics* are characterized by high average temperatures, with a consistent variation between 75 and 100°F. Rainfall may average out anywhere between 80 and 160 in. per year, with as much as 30 in. falling in a month and more than 2 in. in a single hour. Humidity will stand at about 80. Normally the winds will be light, but tropical storms, typhoons, or hurricanes can be a recurring danger.

Desert areas, though generally considered as hot and dry, can be subject to great extremes of temperature, in some parts of the world, leaving only their characteristic aridity as a common denominator. For the most part their temperatures are high, ranging from 120°F at noon to half that figure at night. Rainfall is low, often less than 10 in. per year, and this may be jammed into a short season. Because of the lack of vegetation and without the cooling effect of moisture, the surface soils, not necessarily sands, may attain temperatures of 150°F. Humidity is negligible. Winds range from 15 to 20 mph, but even at these velocities may kick up dust storms of impenetrable density.

Arctic regions or *sub-arctic* regions have sometimes been divided into those which vary between a frozen and a melting state and those where the temperature does not rise above freezing and is generally much below it. Areas near the poles, around the Arctic Sea, in Greenland, in Siberia, and in the Antarctic present such special problems due to their thick layers of ice and their prevailing high winds that special techniques, rather than the adaptation of customary techniques of construction, are indicated.

In what may be termed *sub-arctic* conditions, wide temperature variations, from −94°F to +100°F, exist. Wind velocities cover an equally wide range, as does precipitation, generally falling in the form of snow. During the spring and fall seasons heavy fog may blanket large areas for days at a time.

The worst extremes of climatic conditions are not found in the United States, but the illustrated locations of the various weather areas will indicate their prevalence around the world. As the extremes vary, their effects on working conditions, on materials, on equipment, and on the health and productivity of men become major factors in preparing bids.

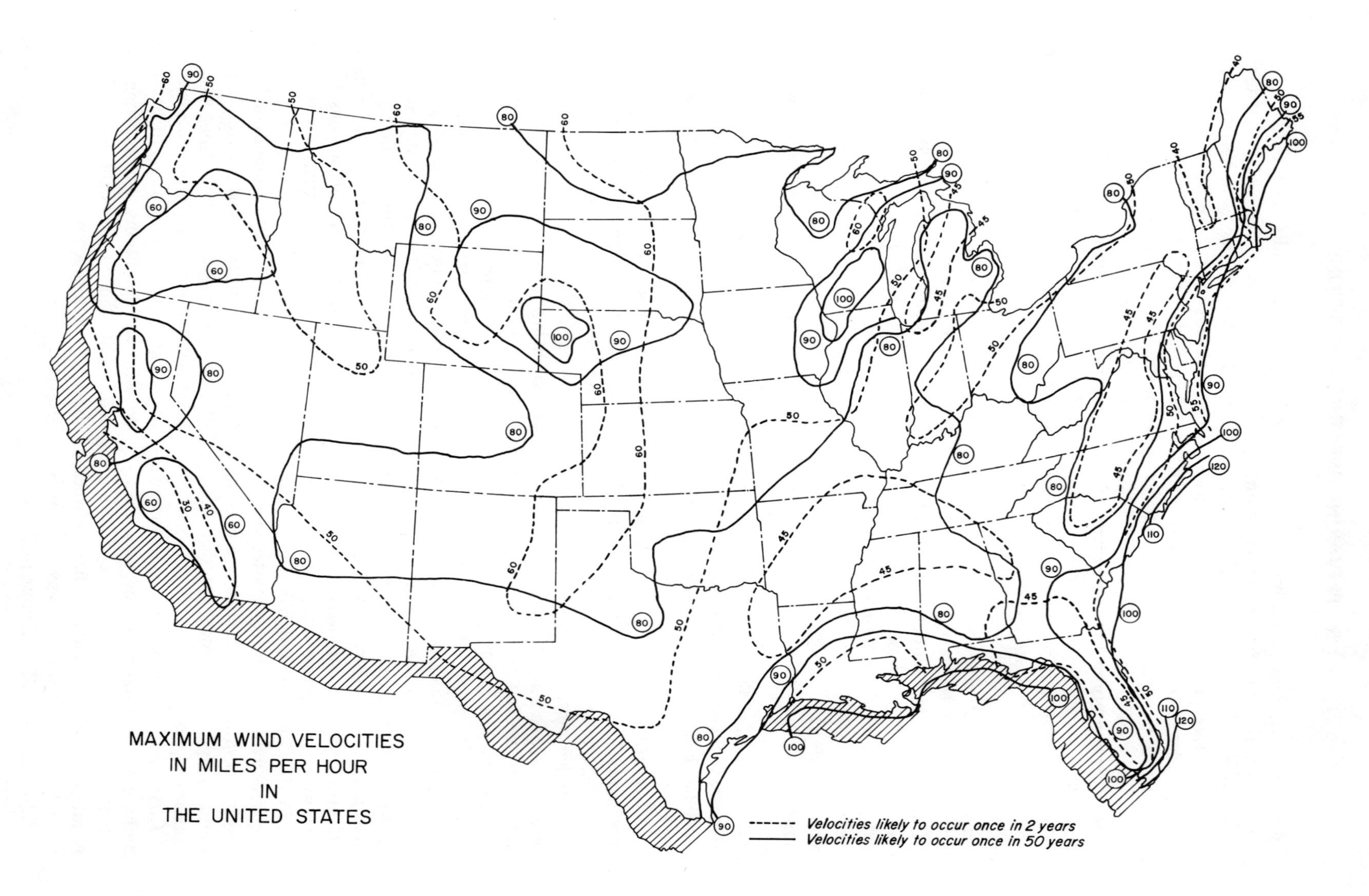
MAXIMUM WIND VELOCITIES
IN MILES PER HOUR
IN
THE UNITED STATES
Velocities likely to occur once in 2 years
Velocities likely to occur once in 50 years

1.6 THE WIND AND THE RAIN

In Sec. 1.5 it was pointed out that, as we move from the Temperate Zone into other zones, prevailing weather conditions change. Yet common to all zones is the occurrence of winds of high velocity, which cannot be classed as prevailing but are still a factor to be reckoned with. Although there are areas of the earth where hot dry winds blow at certain seasons of the year without other storm manifestations, it is more common to think of high winds in connection with heavy rainfalls or blizzards and with the combined effect of these storm factors on open bodies of water. Heavy rainfalls accompanying high winds can flood any low-lying depression in the earth's surface, but on rivers or lakes of wider extent the wind effect is added to the rainfall to produce wave action. This same result occurs on the shores of ocean bodies, where its effect is magnified by tidal ranges.

It would appear that subsurface construction lying, by definition, below the surface of the earth would be unaffected by high winds with their storm connotations. This is not necessarily the case, nor is the effect limited to those conditions due to floods or abnormal tides. In the case of simple foundations, for example, the contractor must consider the effect of storms on the soil on which foundations are to be constructed, particularly if these soils are clays or silts. Soils which tend to absorb water under prolonged rainfall may change their characteristics so as to require secondary excavation or grading, involving more work than merely "cleaning off the mud." Temporary enclosures are quite practical for rain falling vertically, but rain combined with wind requires substantial temporary construction.

Although choosing a season when a minimum of rainfall can be expected and storms might also be at their lowest frequency seems like an obvious solution, it may seldom be possible to so arrange the construction schedule.

Many of the types of subsurface construction discussed here can be seriously affected by storms, whether accompanied by flood conditions or not. Piling, which would seem to be least affected, is, nevertheless, more difficult to handle and drive in a high wind and is, of course, impossible in cyclones or thunderstorms. Many types of subsurface construction involve large bodies of water on which caisson or tunnel sections are floated. These structures are particularly susceptible to damage from high winds and consequent wave action, without the necessary accompaniment of rainfall or flood conditions.

Contractors have generally become accustomed to obtaining rainfall information for an area in which they are to work, using intensity and duration as a means of setting up an allowance for possible lost time during construction. Similar information is available for maximum wind velocities for the United States and Europe, and to limited extents for some other parts of the world. Like rainfall, the frequency of high-velocity winds will vary with the area as well as with the length of time being considered. Thus for a certain area, there is the likelihood of a 60-mph wind every 2 years, whereas a 100-mph wind may occur only once in 50 years and higher winds only once in 100 years.

Wind velocities vary with height above ground surface, increasing generally up to heights of from 1000 to 2000 ft. At ground surface there is a frictional drag slowing down the velocity. This is independent of terrain conditions such as forests, hills, and cities, which would further affect not only the velocity but the direction of the wind's path.

The illustration shows the distribution of maximum wind velocities in the United States which might be expected to occur with frequencies of 2 or 50 years. Methods of obtaining this information have been standardized. A wind-measuring device (an anemometer) is set at a site and provides a relatively uniform surface friction condition for some 25 miles in front of it and at a height of 30 ft above ground surface.

1.7 TROPICAL CONSTRUCTION

The principal characteristics of tropical areas have already been mentioned: the high temperatures, the extensive rainfall, the high humidity, and the incidence of storms. Also typical of these areas are tropical rain forests or jungle, varying from place to place in their assembly of species but always lush and impenetrable. Although not all portions of the tropical belts are necessarily covered with jungle, the portion that is combines all the worst features of the tropics, and this discussion will be limited to such areas.

Although tropical rain forests are generally thought of as lying primarily on the flat coastal plains, they will often be found extending up to ground elevations of 9000 ft. There will be some fundamental differences with rise in elevation; the more widely spaced large trunks will give place to more closely spaced boles of a smaller size, for instance, yet the general density will be the same.

Despite the cover provided by the branches of the main trees at heights of perhaps 100 ft, the growth at ground level and up to this cover is very heavy, unlike that in Temperate Zone forests of equal tree density.

The problem that immediately presents itself is that of how much of the jungle shall be cleared away. Where construction operations are to be performed there can be no doubt, but the clearing of adjoining areas is a question of removing cover that may be valuable. Although the generally hot, humid climate cannot be controlled, the presence of trees to serve as an umbrella during the day and to provide perhaps a slight amount of cooling at night may be desirable.

General policy indicates the procurement of construction labor from as close to the site as possible. In tropical climates the quantity and the ability of even common labor are apt to be limited, and a substantial portion of the work force must be brought to the site and housed. Regardless of what native labor is used, there will always be a cadre of experienced men for whom the best possible living conditions must be provided.

The three rules for productive survival in any hot climate: (1) drink plenty of water, (2) consume extra salt, and (3) keep away from alcohol—are too well known to need any amplification here.

In most areas of tropical rain forest there will be plenty of water, but the problem will be to get it sufficiently treated to be potable. The very factors that lead to the lush jungle growth lead to a lush bacterial growth in the surface and even underground waters. Treatment facilities which might be quite adequate in any other part of the world, filtration and chlorination, will frequently not provide a safe drinking water in the jungle. A chlorine dosage which will be adequate for ordinary bacteria may not be enough for certain organisms found in the jungle, such as amoebic cysts. Heavier dosage is required of the water supply, and this can definitely produce a taste.

The temperature of the water supply, which can get up well above 100°F, is another element of potability. Although ice water, as such, is dangerous, some means of reducing the temperature of drinking water will be required.

Insects are another problem most difficult to control in the jungle. These fall into two classes: those which, sweeping down in hordes, are annoying and distracting to the workmen and those which are disease carriers. Extensive mosquito-control operations may be required, such as drainage and ditching, the general spraying of land areas, or the oiling of water surfaces. It may be advantageous to establish campsites at an area remote from low-lying pondage, even when this involves hauling the workmen to and from the construction site.

Other types of insects such as flies can be controlled by proper sanitary facilities with adequate disposal of human and garbage wastes.

In addition to flies and mosquitoes, many other insects such as armies of ants can be annoying although not actually dangerous, and fungus growths on the human body, particularly in the moist crevices of the groin, toes, armpits, and even on the scalp, while not seriously disabling, can be very uncomfortable.

Problems of Tropical Construction

MATERIALS

1. *Problem:* Storage in closed places accelerates the growth of mold.

Solution: Provide open shed storage with tight roof but open sides to permit circulation of air. Screen sides to prevent attacks by rodents.

2. *Problem:* Can local lumber be used?

Solution: A careful survey of local timber should be made. Lumber will generally resemble hard, dense ironwood or soft coconut palm. Neither is suitable for formwork. Hardwoods may be usable for poles or piling.

3. *Problem:* Handling of imported form lumbers.

Solution: Store as noted in item 1 rather than in open piles. Combination of high humidity and direct heat will warp and distort dimension lumber rapidly. Plan shipments to avoid storage of excess quantities.

4. *Problem:* Special provisions for handling concrete.

Solution: High temperatures of aggregates and water will produce quick-setting characteristics, especially under high ambient temperatures. Additional water for curing will be needed, to prevent surface cracking and to maintain workability. Retardants may be required to increase setting time.

EQUIPMENT

5. *Problem:* General protective measures.

Solution: Do not cover with tarpaulins or plastic covers. Overnight condensation on the underside can provide the precise milieu needed for growths of mold. Seat covers or other fabrics should be carefully chosen. Some synthetic fibers such as rayon or nylon resist mold, whereas others do not.

6. *Problem:* Protection of metal parts from corrosion.

Solution: Non-moving parts should be coated with shellac or varnish. Moving parts must be frequently oiled with a clean oil likely to be free of air and moisture.

7. *Problem:* Protection of rubber.

Solution: Rubber is readily rotted in a hot, damp climate. Some synthetic rubbers are highly resistant to the action of sunlight; natural rubbers are not. Rubber hose and accessories should be painted. A heavy coat of paint on tire sidewalls helps to resist the action of sunlight.

8. *Problem:* General maintenance.

Solution: Rust forms quickly on parts operating in oil. Equipment should be operated daily even where it is not used productively. Tool or electrical boxes should have holes drilled in the bottom to permit condensation to drain off.

PERSONNEL

9. *Problem:* General health control.

Solution: There are really two parts to this problem. There must be provision for adequate medical facilities, with personnel experienced in tropical illnesses and disease, to treat imported labor. The same facilities, although adequate for, may not be precisely suited to treatment of native labor, and special facilities may be required.

10. *Problem:* Control of disease-breeding organisms.

Solution: Careful site sanitation and proper disposal of wastes, both human and garbage. Spraying of swamps may or may not be practical, depending upon their prevalence. Provide proper screening of housing facilities.

11. *Problem:* Control of productivity.

Solution: Fungus growths on the human body are very prevalent in the tropics. Personal sanitation offers some control, but medical treatment is always desirable. These growths are not disabling, but they can reduce productivity. Swarms of insects, although not disease-bearing, can slow production. Control as indicated in item 10. The heat and the humidity may require a change in working hours. Work may have to be halted during the middle of the day. Reduction in working hours may be necessary.

SITE

12. *Problem:* Effects of heavy and frequent rainfalls.

Solution: Provide adequate drainage for site roads and elevate these if practical. Provide protection for erodible slopes.

13. *Problem:* Temperatures in underground structures.

Solution: Foundation areas in which no air circulates may become unbearably hot and unworkable. In some instances provision for blowers may improve working conditions. Work in tunnels may require the cooling of the air circulated in ventilation systems.

1.8 DESERT CONSTRUCTION

Although there are, of course, many similarities between the tropics and the deserts and in some areas the one merges imperceptibly into the other, it is the differences that are important and will be discussed here.

A few decades ago the contemplation of construction in the middle of the desert would have been considered fantastic, but the oil industry has changed all that. Water, the one missing element in the desert, can now be obtained by deep-well drilling and pumping or be brought in overland in pipelines parallel to those carrying the oil out. Once in the possession of water, in sufficient quantity, the desert can be made to blossom.

For desert construction not only the amount of water needed for the working force, a minimum of 30 gal per man per day, but that needed for construction, particularly for that indispensable foundation material, concrete, must be carefully estimated. Because of the abnormally high temperatures, the normal allowance for water in relation to yards of concrete must be doubled because of surface evaporation. Where water is not available in quantity, washdown techniques at the end of a pour may have to be tightly controlled if the concrete-placing equipment is to be properly maintained. Curing compounds may have to replace water, but these must be carefully tested beforehand to be certain they will not disintegrate prematurely under the 160°F heat that can occur at the concrete surface.

Since the ultimate use of a construction project will require water, the ultimate sources of water will not generally concern the contractor. If, under temporary emergency, he has to resort to local water sources such as oases for drinking water, he can assume extreme contamination and should provide complete treatment of it. Even where it is hauled in, temporarily, there is the problem of temperature, as in the tropics. Drinking water at temperatures in excess of 70°F not only is unpalatable but can actually produce nausea and vomiting. Since cooling processes also demand water, "desert" water bags, which permit slight percolation through their canvas and which cool by evaporation, may have to be used.

Deserts lie generally within 15 and 25 deg of latitude, but their temperatures have wide variations. In the central Sahara Desert, temperatures range from 120°F during the day to an average at night of 77°F; but in the higher grounds, the variation during the winter months is from 70 to 50°. In the Gobi Desert, whose altitude varies between 3000 and 5000 ft, temperature extremes are greater, falling below freezing at night during the winter months.

The "sands" of the desert vary from fine silts to coarse gravels, not only from desert to desert but within the area of a single desert. As a result, the processing of the soils to produce concrete aggregates varies. In many areas it is simply a matter of screening to produce the gradations required, while in other areas proper aggregates may be virtually unobtainable.

All desert construction operations will be conditioned by sandstorms. These occur with not quite predictable frequencies driven by winds of 15 to 20 mph which twirl dust to heights of 10,000 ft. In the Sahara these storms occur generally in the late afternoon, while in other deserts they may occur during the heat of the day or with less predictable regularity.

Wherever such storms occur, although there may be means of protecting the workmen on the project, lost time and often damage to work under way will be involved, and sand may be piled up against partly completed structures and require removal. Damage may be due to wind alone but will more often be due to the sandblasting effect of the blown particles. A sandstorm, for instance, could readily scrape off the curing compound on a concrete surface, leaving it once again exposed to high temperatures.

Sanitary problems are just as severe in the desert as in the tropics, although basically different. Native labor is more apt to be usable here than in the tropics, and for brick and mud-wall construction it is perfectly suitable. However, natives contribute their own disadvantages.

Problems of Desert Construction

MATERIALS

1. *Problem:* Storage.

Solution: Effects of high heat would indicate open-sided sheds, but the sandstorm potential makes this impossible. The use of native mud huts with deep-set windows and doors that can be hermetically sealed against blown or drifting fine sands is indicated. Ventilate during midday when winds are quiescent.

2. *Problem:* Local material.

Solution: The only local materials likely to be available are concrete aggregates, sands, and gravels. Gravels may lie buried, as may sands. Surface sands may be too fine for satisfactory concrete.

3. *Problem:* Form lumber.

Solution: All lumber for forms will be imported. Quantities in excess of normal use will be required. The dry, high heat and the abrasiveness of blowing sand, added to the normal effect of drying concrete, will reduce the number of uses. Form faces, often coated with oil, will here require a plastic coating.

4. *Problem:* Special provisions for handling concrete.

Solution: See item 4 under Problems of Tropical Construction. For large surfaces requiring finishing, the effect of blowing sand must be added to the difficulties due to the dry heat. Substantial shelters may be required. High shrinkage rate and shortage of water may demand a heavy curing compound.

EQUIPMENT

5. *Problem:* General protective measures.

Solution: Desert conditions are particularly hard on construction equipment. Sand works into the bearings and through the finest filters. When it becomes mixed into a lubricant it is particularly deadly. More maintenance will be required. Equipment should be torn down frequently, cleaned, and re-lubricated. All openings should be closed whenever equipment is shut down at the end of a shift.

6. *Problem:* Control of equipment efficiency.

Solution: The only solution here is provision for higher operating costs. Engine operation in terms of fuel consumption is always lower at higher temperatures. Less productive work can be expected from mobile equipment working constantly on soft footings. Crawler tread life may be reduced from 1500 to 700 or 800 hr. Cable, on cable-operated devices, will also have less than normal life.

7. *Problem:* Rubber-tire use in the desert.

Solution: Traction is always difficult. The high temperatures may increase inflation pressures by as much as 40%. To maintain adequate traction, low inflation pressures must be used. These pressures and the abrasiveness of the sands will reduce tire life. Investigate special tires designed for this duty that operate on pressures of 5 to 10 lb.

PERSONNEL

8. *Problem:* Handling native labor.

Solution: Native labor adaptable to construction needs may be more available adjacent to deserts than in the jungle. Hire through native leaders and use native foremen, transmitting all orders through them and dealing through them in all matters. Productivity will vary widely, depending on area. Check local pay rates.

9. *Problem:* General health control.

Solution: This problem will be somewhat more complex where native labor is used since control of sanitation, even on site, is difficult. Imported labor will generally be set up in a separate campsite, where the precautions noted in Sec. 1.8 will be used.

10. *Problem:* Protection from sand or dust.

Solution: Since in some desert areas dust storms occur with predictability at certain times of the day, work schedules may have to be adjusted to avoid these periods. Clothing that will provide protection from blowing dust should be furnished, and generally, some sort of goggles for eye protection are advisable. Because of the heat, a major difficulty will be to prevail on the labor forces to accept these confining safeguards.

SITE

11. *Problem:* Effect of sand on project.

Solution: Although windstorms bearing sand or dust cannot be controlled, continuous light winds can cause a drifting of soil. Once the direction of prevailing winds has been determined, throw up embankments across this side of the project site to provide drift protection—the same principle as the snow fence. In selecting campsites, if possible, place them behind some natural protection.

12. *Problem:* Local conditions.

Solution: Heat as characteristic of the desert has been stressed here, but many deserts have sharp drops in temperature at night. During the day, sands may be too hot to walk on, and tools may become too hot to handle, requiring them to be kept in a bucket of water. At night, some modest heating facilities may be required.

1.9 ARCTIC CONSTRUCTION

In Sec. 1.5 the terms arctic and sub-arctic were used. The arctic is defined as that area where the mean temperature for the warmest summer month is less than 50°F, the sub-arctic having temperatures higher than this. The tree line more or less defines the limit between arctic and sub-arctic, but there is no specific limit to the sub-arctic although it extends well beyond the limit of permafrost (see Sec. 3.14).

Not many decades ago construction operations, even in parts of the Temperate Zones, halted abruptly with the onslaught of the first snow in December and did not again start until March or later. This is still true, to some extent, since the costs of winter construction are excessive. The military exigencies of World War II, the political pressures of the cold war, and the enfoldment of Alaska into the Union as a state have lengthened the construction time span per year not only in the Temperate Zone but in the sub-arctic and arctic areas.

Most of the construction techniques described here have been used in the arctic and sub-arctic, though sometimes to a limited extent and in much modified form. Most of these modifications arise from the nature of permafrost, but general weather conditions also have a profound influence.

Even several degrees south of the Arctic Circle the construction season extends only from the middle of June to the middle of September, during which time temperatures average 40° above freezing. The period during which 12 hr or more of sunlight can be expected begins in April and extends to the end of September. The arctic night descends completely from about the middle of November until the end of January, requiring artificial light for all operations.

This short favorable season affects every aspect of the construction plant. During the latter part of this period, water may be drawn from the surface of lakes or rivers, but during as much as 9 months of the year provisions must be made for drawing it from under 4 to 5 ft of ice. Generally only routine chlorination is required.

Precipitation in the arctic regions is almost as low as in the deserts, with a few exceptions, ranging from as little as 6 in. at the Arctic Circle to 19 in. in Labrador. Some of this, very little, falls as rain, most of it arriving as snow. Arctic regions have the fewest heavy storms of any part of the world; but, again like the deserts, relatively light winds, averaging 12 mph, pick up the light snow and migrate it into drifts.

Although transportation by air is feasible during most of the year, it is obvious that this is too expensive for the bulk shipments of construction materials required, and transportation of these items must be limited to those months when movement over water is possible. Special freight-hauling self-propelled vehicles have been developed, moving on large tires with 5 to 10 lb inflation pressures for use over the snow as over the desert. These have extended the transportation season but have not materially reduced costs.

The water-borne sanitary sewage systems common in most parts of the world, involving cesspools, septic tanks, and tile fields, cannot be used in the arctic since the frozen ground does not permit seepage of the liquids. More primitive methods of waste disposal are required, although experiments with heating these waste facilities until they can be unloaded into large bodies of water are under way.

Living or working structures must be tightly constructed, as in the desert, to prevent snow from drifting through openings. Structural insulation must be doubled not only to provide comfortable conditions but to cut down on fuel consumption, for this again is a matter of transport.

Fuel in the arctic, like water in the desert, affects every operation. Quite aside from equipment and power requirements, the need for heat is paramount. Except in a very limited season, tools must be heated for use, even with gloves, and nails or other assembly items often require heating.

Concreting can be successfully accomplished, provided heating facilities are available for each step.

Problems of Arctic Construction

MATERIALS

1. *Problem:* Storage.

Solution: Storage will be of two kinds: that requiring some control of temperature and that for products not subject to deterioration in low temperatures. Storage buildings where heat will be used to any extent must be set above ground to permit blow-through under them. (See sections on permafrost construction.) Storage facilities where no heat is required must be clearly marked by poles to provide access to them under drifting snow. As in the desert, construction must be tight to prevent entrance of snow.

2. *Problem:* Local lumber.

Solution: In the arctic proper, no lumber will be available. In the sub-arctic, lumber may be had, but transportation problems must be considered.

3. *Problem:* Concrete aggregates.

Solution: Gravel is available in some areas, although size and grading will cover a wide range. Effective processing and stockpiling is limited to a few months. Sand is very scarce and, if stockpiled, will require some heating except during the very short summer period.

EQUIPMENT

4. *Problem:* General protective measures.

Solution: In general, heat is the solution to most problems with equipment. Specially designed equipment, providing cabs or other protection for operators, with heating facilities, is essential. Batteries provide only 50% of their power at zero temperature and should be removed to a heated place when equipment is not in use.

5. *Problem:* Fuels and lubricant selection.

Solution: Diesel-powered equipment should be used since, volume-wise, only two-thirds as much diesel is required as gasoline, and transport requirements are thus reduced. Special lubricants are required since standard lubricants stiffen in low temperatures. Since lubricants will vary with function, consultation with supplier as well as manufacturer of equipment is essential.

6. *Problem:* Start-up of equipment.

Solution: Pre-heaters will always be required at low temperatures. These are generally space heaters with integral blowers. Heat is applied to the intake manifold, crankcase, oil pan, and carburetor. Open-flame blow torches are dangerous and should not be used.

7. *Problem:* Repair and spare parts.

Solution: Low temperatures adversely affect most metals, increasing brittleness and consequently their tendency to crack. The same thing is true of synthetic rubber, glass, and other materials, especially where sudden temperature changes are involved. A higher rate of failure, coupled with transportation problems, requires an increase of 30 to 40% in the number of spare parts required.

8. *Problem:* General maintenance.

Solution: Clearing drift snow from engine areas will be required before start-up. Fuel tanks must be serviced daily to prevent accumulations of condensed moisture. Careful cleaning of all accessible parts should be undertaken after each period of operation.

PERSONNEL

9. *Problem:* General health control.

Solution: Adequate protective clothing designed to keep the body warm and yet provide sufficient movement for work operations is the prime requisite. Frostbite with its danger of gangrene can strike any portion of the body exposed for a considerable time. Protection of the eyes by the use of colored glasses is important whenever snow lies on the ground.

10. *Problem:* Insect control.

Solution: It has been said that there are ten times as many mosquitoes in tundra areas, during the summer months, as there are in the tropics. They do not carry malaria or yellow fever, but will attack in force in the late afternoon on warm, quiet days. Black flies and midges which can get inside the clothing are blood-sucking insects whose bites become painful with time. The use of smudges and sprays and even head nets is often required.

11. *Problem:* Control of productivity.

Solution: Production is necessarily reduced, the greater the protective clothing required. June, July, and August are the three months during which protection from cold is minimal, and in some areas insects are no problem. December, January, and February will generally be wholly non-productive. Production during the intervening 2- or 3-month periods will vary, depending on location. Careful scheduling of operations is the only solution.

SITE

12. *Problem:* Effects of snow.

Solution: Construction during periods of snowfall is expensive and, except in emergencies, should be avoided. Transportation over snow is generally more satisfactory than over the summer-thawed tundra. Tractors with cleated tracks are usable over snow if track plates are perforated to permit extrusion of packed snow. For distance hauling, low-pressure tires are preferable.

13. *Problem:* Temperature control.

Solution: The only practical temperature control is under cover during at least half of the year. Even in summer, temperatures may vary from 15 to 60°F within 24 hr and some protection of concrete must be anticipated.

1.10 OFF-SITE FACTORS

Closely allied with factors involved at the site of the project are those which connect the project to its sources of supply. Generally these factors involve transportation.

Most subsurface construction requires the use of concrete in considerable quantities. In the developed areas of the United States, commercial concrete-mixing plants are not uncommon. If the capacities of these plants are adequate, hauls of 20 or 30 miles in transit mix trucks present no problem. Where the capacities of these plants are inadequate, where the haul becomes excessive, or where the economics are adverse, the contractor will set up his own mixing plant. Sources of water, cement, and coarse and fine aggregate must be investigated and the transportation to the site planned.

Water can present a considerable problem at all stages of concrete construction. Water used in mixing, curing, and washdown operations is one requirement. Where coarse or fine aggregates are to be manufactured or processed by the contractor, additional substantial volumes of water will be needed. At fresh-water sites—rivers, lakes, or similar bodies of water—construction water is a minor problem. Further inland or in salt-water areas the problems are not so simple. Wells, intakes, storage facilities, pumping equipment, and pipe installations are the simplest of the facilities that may temporarily be needed. Overland pipelines involving rights-of-way, intermediate lift stations, or, in more extreme cases, truck tankage may be required.

Although special cement plants have been set up for large construction projects (Boulder Dam, for instance), they are not always warranted economically. Coarse aggregates, on the other hand, are frequently prepared by the contractor. This may involve dredging operations, where gravel lies in watercourses, or quarrying operations. Suitable stones for coarse aggregates are readily available in most parts of the world, although their manufactured costs may vary considerably. Where operating quarries exist, a pre-bidding comparison of delivered prices versus self-manufactured costs is warranted.

Suitable sands are less readily come by and, where there are no natural beds within a reasonable distance, may require manufacture. This is a somewhat more difficult operation than in the case of coarse aggregates since, in the first place, many rocks will not crush to a suitably graded fine aggregate and even where they do will produce vast quantities of fines requiring extensive washing operations. Generally the high costs of manufactured sands will warrant a much longer haul. Moreover, since sand gradation has considerable bearing on the quality of the concrete, manufactured sand is often the subject of disputes.

Within the United States, shipping problems of purchased materials required for subsurface construction are generally left in the hands of the material supplier. At the project site, however, the contractor must decide whether he wants to receive shipments by rail, truck, or water. This decision will depend on facilities available, storage space, and unloading conditions.

Outside the continental limits other factors enter the picture. Choices between materials available locally and those imported are not always simply a matter of first cost. In the case of cement, for instance, chemical variations in the content may determine the time of set and may influence stripping operations; a greater quantity per yard may be required to produce a satisfactory concrete.

Materials shipped overseas are generally shipped by the supplier FAS (free at ship). The contractor must pay the overseas freight and the stevedoring costs at both ends as well as the movement from the port of destination.

Because the subsurface contractor is generally the pioneer in starting a job remotely located, he will be the one to solve these problems initially, leaving the superstructure contractors to follow the path he has hacked out.

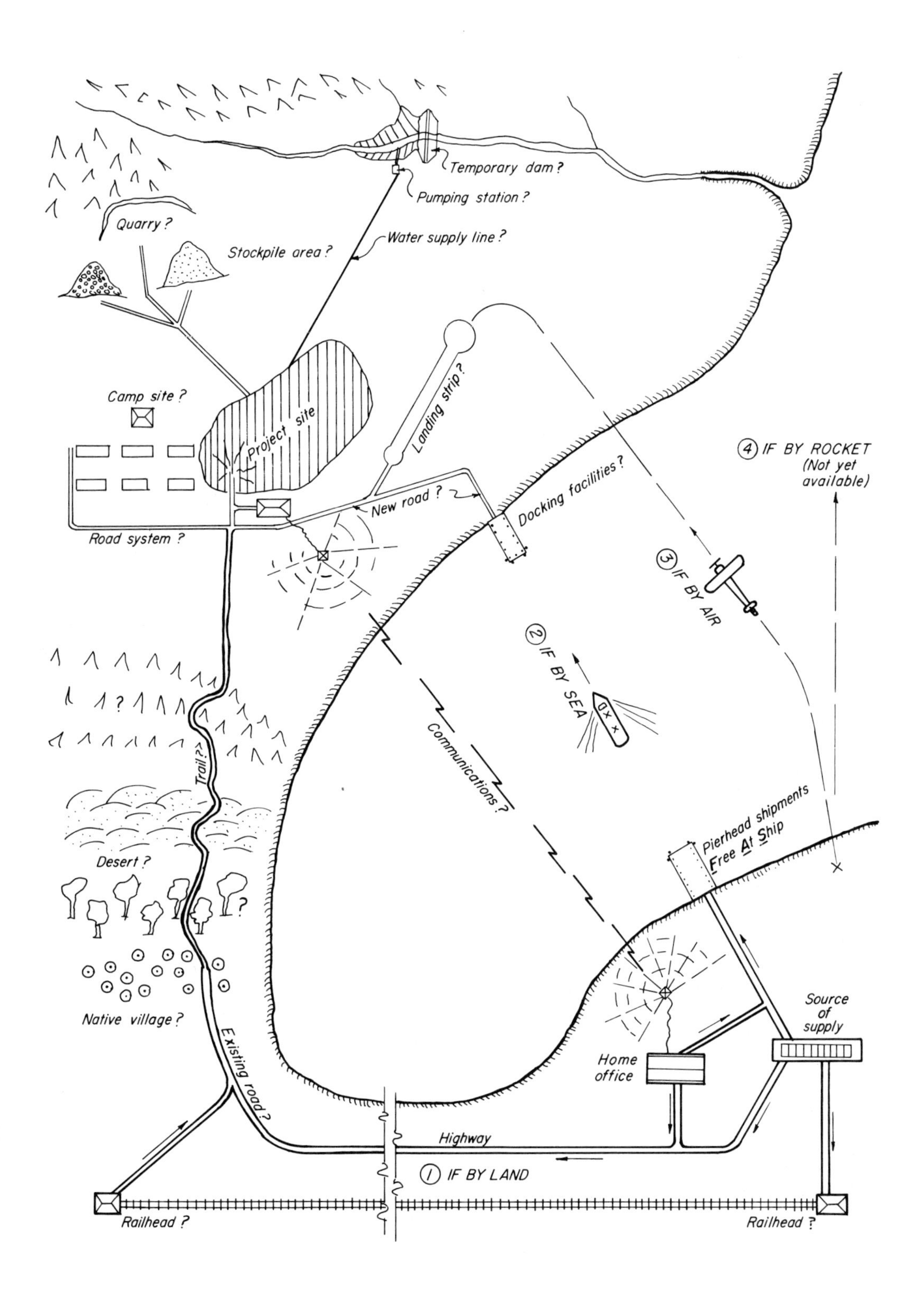

Temporary dam?
Pumping station?
Quarry?
Water supply line?
Stockpile area?
Camp site?
Project site
Landing strip?
New road?
Docking facilities?
Road system?
④ IF BY ROCKET
(Not yet available)
③ IF BY AIR
② IF BY SEA
Communications?
Trail?
Desert?
Native village?
Pierhead shipments
Free At Ship
Source of supply
Home office
Existing road?
Highway
① IF BY LAND
Railhead?
Railhead?

1.11 BLASTS AND EARTHQUAKES

The earth quakes at the rate of 250 shocks per day, most of them minor and passing unnoticed except by the delicately balanced seismograph. There are enough shocks of importance, however, to have a disturbing effect on partially completed construction. Direct damage to building foundations depends on their proximity to the quake site, but the secondary effects of earthquakes, which produce landslides and tidal waves, are also devastating.

In the San Francisco earthquake of 1906 a railroad tunnel in the Santa Cruz mountains cut across the San Andreas fault (the source of the quake). The shock and slippage produced a 5-ft deflection in the line of the tunnel in the vicinity of the fault and affected the alignment of 5000 of its 6200 ft length. In the Kern County earthquake of 1952, another tunnel, that of the Southern Pacific Railroad, had one wall pushed in sufficiently to cover portions of the wildly twisted track. Such shocks could, of course, be concurrent with the construction period.

The earth resembles a not-yet-dried-out litchi nut (once available in Chinese laundries) with its hard, dense core mantled in pulp and contained in a thin, dry, wrinkled crust. The great density of the earth's core is surmised, but its constituents are unknown. The earth's mantle is almost equally unknown except that, far from being a pulp, its materials appear denser than those found in the earth's crust and may, to some extent, act as a plastic mass. Only the crust is hard and brittle.

Tracking the shock waves induced by earthquakes with seismic equipment, A. Mohorovicic in 1909 noted a difference in the refraction of these waves at a depth of about 60 km (37 miles). Named after the discoverer, this variation in effect has come to be known as the M discontinuity and, by inference, is assumed to be the bottom of the earth's crust. More recent studies (1960) by the University of Wisconsin and others, using the more precisely timable shock waves induced by blasts of dynamite (from ¼ to 12 tons), taken together with other pertinent data, have established the depth of crust to the M discontinuity beneath the United States (*b*).

Earthquakes have been classed as *volcanic, tectonic,* or *plutonic* in origin, but very little real information is known concerning them. Earthquakes, sometimes severe, generally accompany volcanic eruptions; whether this is due to the bursting of the crust in the volcano's crater or due to the same pressures that force the lava to the surface is unknown. Whether the lava is contained solely in the earth's crust or is forced upward through a vent from the mantle is equally unknown.

Tectonic earthquakes spring from structural alteration of the earth's crust, and this would appear to be the principal source since at the epicenter of the numerically largest number of earthquakes the source is less than 60 km below the earth's surface. The best-known aspect is that produced by the shifting of the crust along a fractured surface known as a fault, but other aspects such as folding or massive subsidence, as in the Assam earthquake in 1897, also occur.

Seismic records indicate that at the epicenter of a substantial number of earthquakes the disturbance originates at depths of up to 700 km, that is, within the mantle. Since the true consistency of the mantle is unknown, the precise cause of these plutonic earthquakes is conjectural.

The contractor working in earthquake-prone areas of the earth's surface must provide himself with some protection from this possibility. Purchased insurance is by no means always available, and a contingency item may be required. Certainly such a contingency must be higher adjacent to a fault zone than remote from it, but by no means are all fault zones known.

In 1961 seismographs throughout large portions of the world recorded what was thought to be an earthquake somewhere in Russia which shortly turned out to be an atomic blast of 50 megatons. Only limited information on the earthquaking effects of such blasts has been published, and almost nothing has been said on the possibility of their triggering tensions already built up in the earth's taut crust.

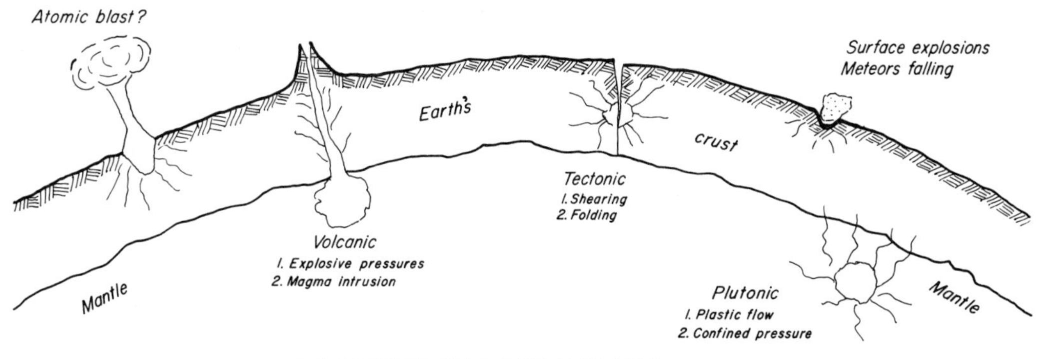

(a) ORIGINS OF EARTHQUAKES

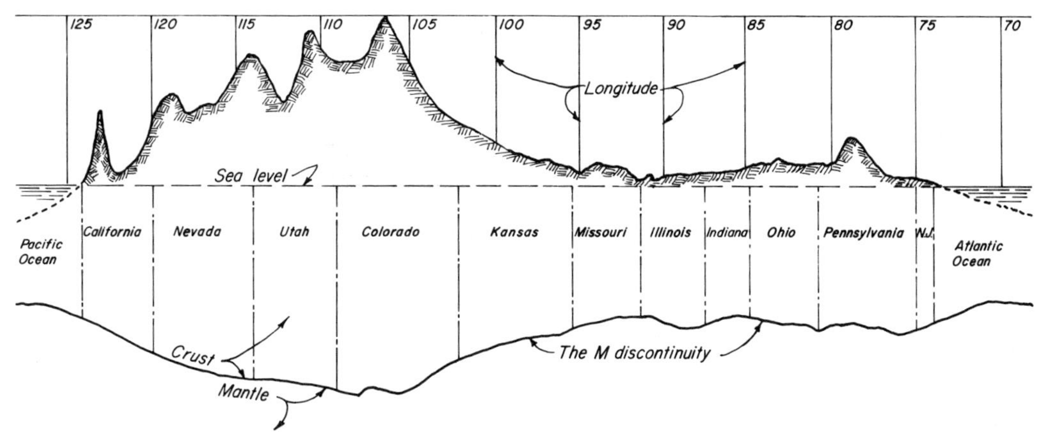

(b) CRUSTAL THICKNESS UNDER THE UNITED STATES ALONG THE FORTIETH PARALLEL

Horizontal Scale 1" = 300 Miles

Vertical Scale Below Sea Level 1" = 15 Miles *Vertical Scale Above Sea Level 1" = 1.18 Miles*

Atlantic Coast	*1.0*
Atlantic Coast (50 Miles Inland)	*0.5*
Mississippi Valley	*2.0*
Mississippi Valley (Outside Alluvial)	*1.0*
Mid-Continent	*1.0*
Rocky Mountains	*3.0*
Pacific Coast (Wash. & Oregon)	*6.0*
Pacific Coast (California)	*10.0*
Pacific Coast (500 Miles Inland)	*4.0*
Imperial Valley (California)	*10.0*

(c) RELATIVE VALUE OF INSURANCE PREMIUMS FOR EARTHQUAKE PROTECTION

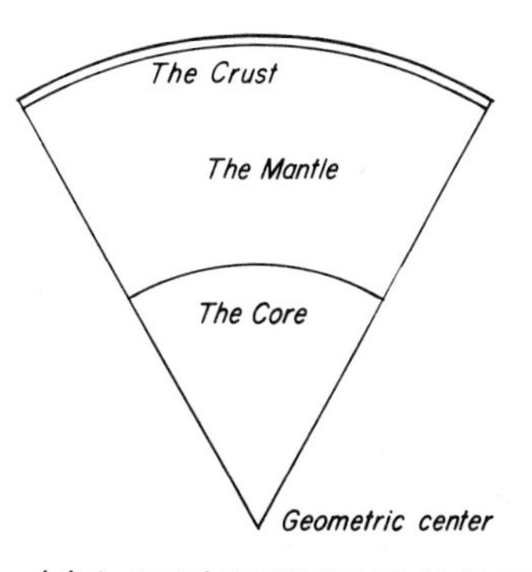

(d) A SEGMENT OF THE EARTH

1.12 LANDSLIDES

A powerhouse was under construction below a steep hillside in a narrow gorge of the Cowlitz River near Tacoma, Washington, when, in 1959, it was struck by a landslide. The movement followed heavy rains which also brought the river to flood stage. The sliding earth and roaring water buried under a sea of mud the partially completed foundations being constructed in a cofferdam. The contractor spent 2 months and $100,000 cleaning up the mess.

The same year an earthquake in Yellowstone Park triggered a slide of one wall of Madison Canyon on the Madison River in Montana that filled the valley with 43 million yards of rock, creating a dam 300 to 400 ft high (*c*). There were nine known dead, and eighteen others were believed buried under the dam, which formed a lake quickly dubbed Quake Lake. No contractors were involved initially, but they were brought in to move some million yards of rock to stabilize this natural dam and prevent further catastrophe.

In the years prior to 1956 a residential area was developed at Portuguese Bend on the Palo Verdes Peninsula some 25 miles south of Los Angeles, California. In the summer of that year, the whole area began to slip toward the ocean, and continued to do so at various daily rates, a total distance of from 71 to 100 ft in 4 years. One hundred and fifty-six dwellings and a country club were either severely damaged or demolished and abandoned. Lawsuits proliferated: the development company (the contractors?) sued Los Angeles County, as did the insurance companies; the county sued the developers; the homeowners sued everybody.

Closer to the city of Los Angeles and to the north of it, landslides have intermittently blocked the Pacific Coast Highway, in as many as 10 different locations, since its construction. Satisfactory remedial measures are fantastically expensive.

Landslides may involve any type of soil or rock and result from a number of causes (see Sec. 1.13). Their disastrous effects have been noted through the ages, but it was Marshal Vauban, the great French fortifications builder, who seems to have published the first treatise on them in 1699. C. A. Coulomb's studies, which first appeared in 1773, dealt with the sloping bank of soil generally. The first systematic study of landslides appears to have been undertaken by Alexandre Collin, an engineer with the French Corps Royaux des Ponts et Chaussées.

Collin, the son of a building contractor, was assigned as engineer on an extensive canal-building program then under way in France and, over a period of 7 years, made careful notes on all landslides he encountered. Because canal embankments were constructed of clay, his studies were limited to that soil and were published in 1846 as "Experimental Research on Spontaneous Slides in Clay Soils." (A translation by W. R. Schriever was published in 1956 under the title "Landslides in Clay.") As shown in (*a*), he determined that sliding took place along a cycloidal surface. (A cycloid is a curve generated by a point on a circle rolling on a straight line.)

Little more was done in the field of landslides until in 1916 some hundreds of feet of quay wall slid into Göteborg Harbor in Sweden. A geotechnical commission which had been appointed 3 years previously began a study which culminated 10 years later in the publication of a mathematical analysis by W. Fellenius which has since come to be known as the Swedish cylindrical-surface method (*b*).

Actually the Swedish cylindrical-surface method for determining the potential extent of a landslide is concerned only with clays in the special case where the toe is submerged below water surface as at Göteborg. Approximately the same approach has been proposed by Karl Terzaghi for determining the extent of an earthquake-induced landslide.

The majority of landslides involve areas and shapes which cannot be predicted—as in the Madison Canyon slide (*c*).

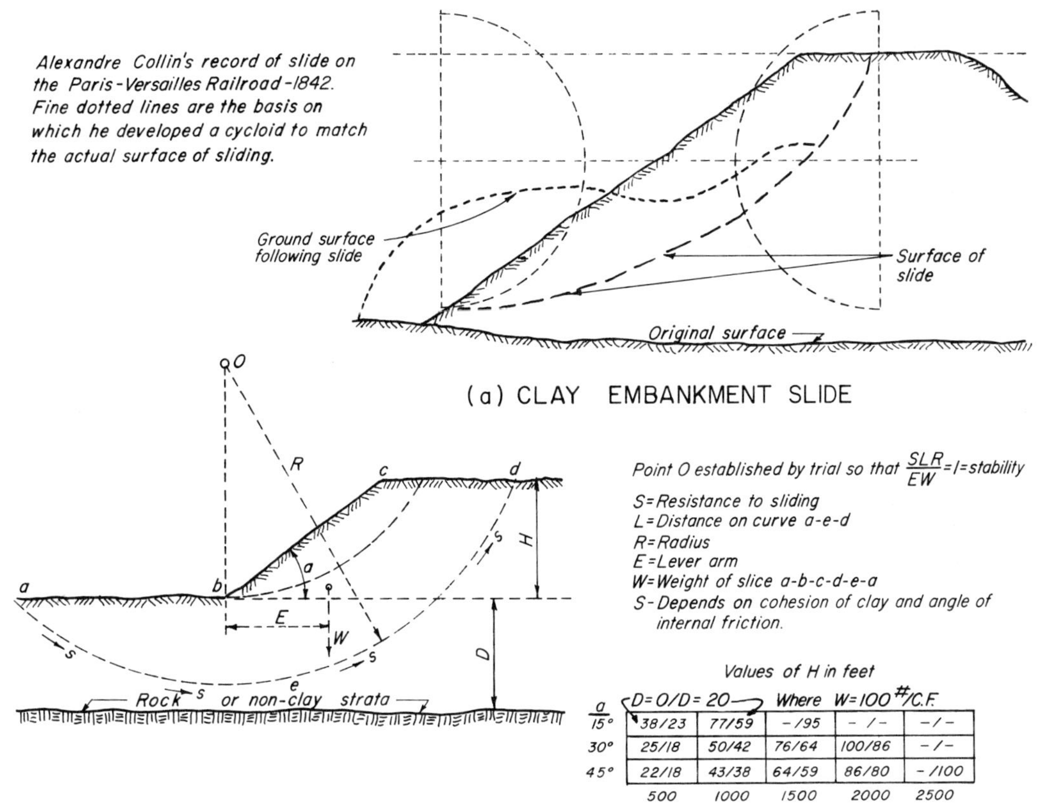

(a) CLAY EMBANKMENT SLIDE

Values of H in feet

D = O / D = 20   Where W = 100 #/C.F.

a	500	1000	1500	2000	2500
15°	38/23	77/59	–/95	–/–	–/–
30°	25/18	50/42	76/64	100/86	–/–
45°	22/18	43/38	64/59	86/80	–/100

Bearing capacity lbs./sq. ft.

(b) SWEDISH CYLINDRICAL SURFACE METHOD FOR CLAY ONLY

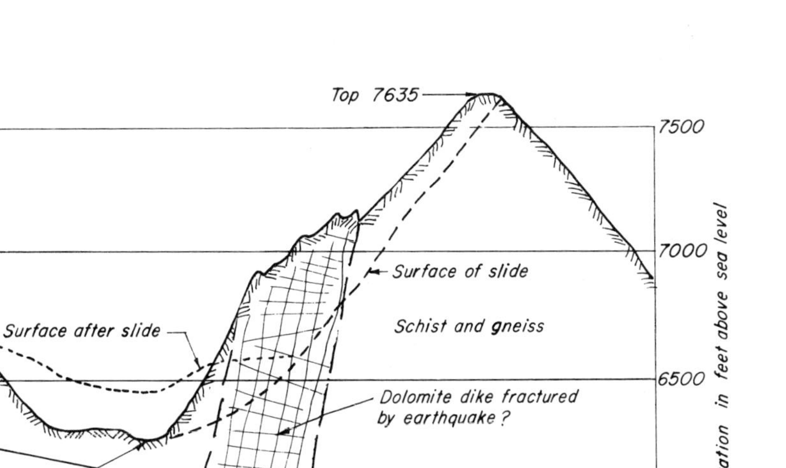

(c) MADISON CANYON LANDSLIDE IN ROCK

1.13 LANDSLIDE-PRODUCING CONDITIONS

Landslides in which the contractor can become involved may be due to so-called acts of God; they may be inherent in the design he is endeavoring to execute; they may be due solely to his own actions—or lack of them. In the majority of instances, landslides occur following heavy or prolonged rainstorms; but this is not always true, and in some cases the action may be so long delayed that any connection appears tenuous. Nine basic causes of landslides can be identified.

1. The undercutting of the toe of a slope, as in the case, for instance, of excavating a highway through a hillside, can produce a slide of the earth above the cut. This action may be accelerated by stripping off the natural ground cover on the slope. The contractor may produce such a slide by cutting away the toe of a spoil bank or stock-pile.

2. Overloading the top edge of an embankment will produce slides, particularly if the embankment material is saturated (*b*). Construction fills are carefully controlled to avoid this condition, but often less care is used on spoil banks, stockpiles, and waste dumps. Sometimes the overloading material, moving down the slope, will seal off the natural drainage through the surface.

3. An increase in the quantity of water trapped in a soil mass can produce a landslide. Rainfall is the most common source of water producing this effect, but dredging operations or the super-positioning of other wet wastes will have the same result (*c*). Saturation of soils may result when either the quantity of underground flow or its height is altered by construction methods.

4. Clays or fine silts may be saturated as noted in item 3. Coarser-grained soils may tend to slide if the groundwater table is raised. Here also either rainfall or construction procedures can produce the result. Careless location of pump discharges can affect large areas often remote from the actual point of discharge (*d*).

5. Landslides have been produced by the rapid drawdown of a body of water, the lowering of a lake or reservoir. The voids in the adjacent soils are water-filled. If the body of water is lowered slowly, the water in the soil drains off more or less horizontally with it. However, if the open water is removed much more rapidly than the water in the soil can flow out, a steep drawdown curve ensues which carries the soil down in the facsimile of a landslide (*e*).

6. Changes in the inherent nature of the soil due to leaching or chemical action can produce landslides. Soils containing binders such as calcium carbonate, subject to gradual dissolution by water, are an example. Such soils may have been overlain and protected by an impervious layer until construction excavation exposed them to weathering action.

7. What has been termed *spontaneous liquefaction* occurs in fine sands or coarse silts "heaped" in an unstable arrangement of soil particles immersed in water. Like a "poof" at a house of cards, the vibration due to blasting or pile driving can shatter the unstable arrangement. The presence of water retards conversion to a stable condition, and a thick viscous liquid results with virtually no angle of repose.

8. Structural flaws such as faults, fissures, or canted strata can contribute to the creation of landslides, especially when combined with any one of factors 1 to 6. In the Turtle Mountain slide in Alberta in 1903 (*f*) the slide could have been due to the fault near the base of the slope or to the initiation of coal mining in the seam at a lower level, in effect undercutting the toe.

9. Earthquakes are precipitating agents for the most part, contributing the vibration needed to set a mass in motion. Yet in some cases their role is not clear. At the Madison Canyon slide no one had certainly gone looking for a fault in the dolomite dike before the slide, and the earthquake may very well have shattered the stone, creating the fault. In the earthquakes and landslides that occurred in the Kansu Province of China in 1920, water was not a factor. The soil was loess and "bore the appearance of having shaken loose, clod from clod and grain from grain, and then cascaded like water," the shattering result of the earthquake.

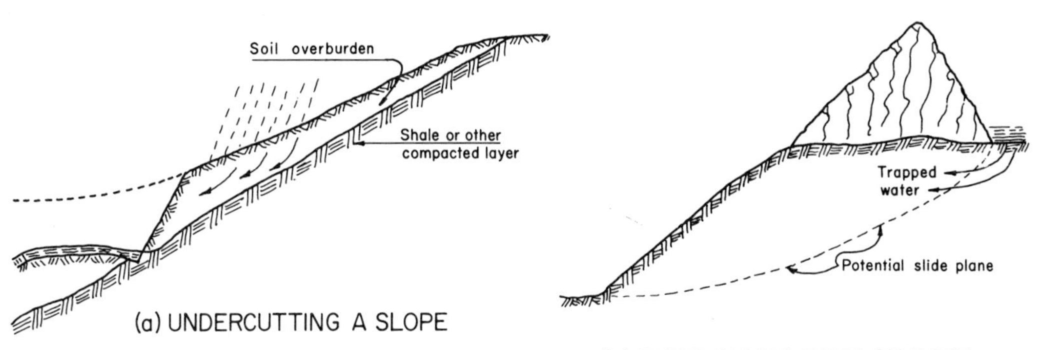

(a) UNDERCUTTING A SLOPE

(b) OVERLOADING EDGE OF BANK

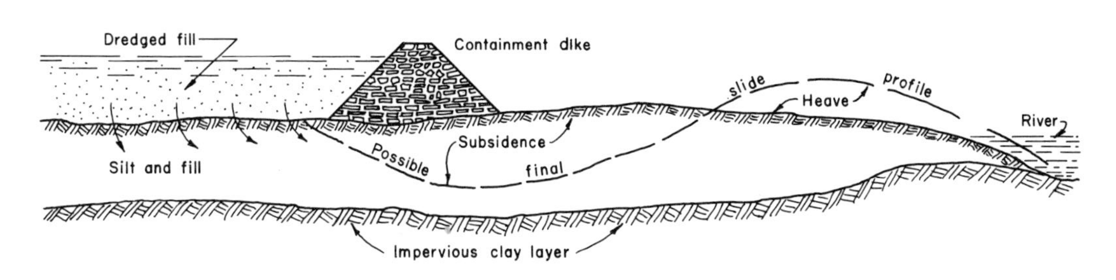

(c) SATURATION OF SUBSOILS

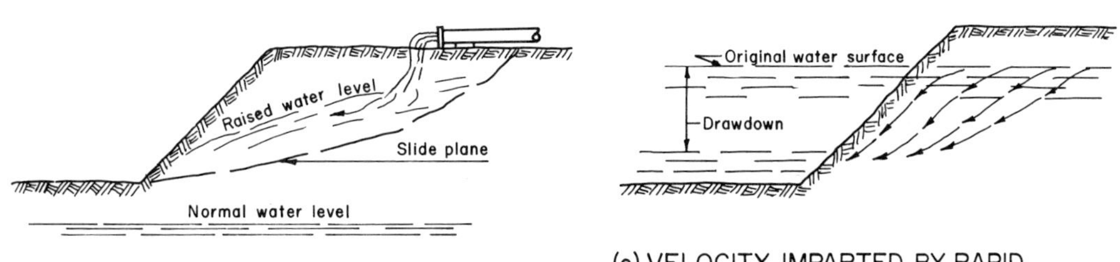

(d) SATURATION OF PERVIOUS SOILS

(e) VELOCITY IMPARTED BY RAPID DRAWDOWN

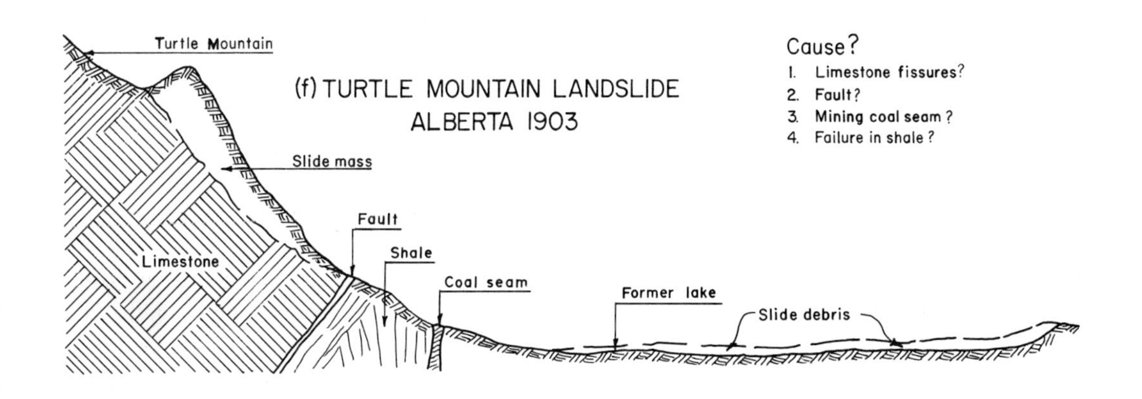

(f) TURTLE MOUNTAIN LANDSLIDE ALBERTA 1903

1.14 LANDSLIDE INVESTIGATIONS

Landslides, unless due to earthquakes or spontaneous liquefaction, begin with a slow creep of the earth mass not often apparent to the eye, but evidential for all that. In 1806 the landslide at Goldau in Switzerland (Sec. 1.15, Fig. *a*) took the village by surprise, wiping it out and killing 457 persons, but the horses and cattle had been restless for several hours before the slide started and "bees deserted their hives." A slide occurred during the construction of the German superhighway between Munich and Salzburg in 1935. The laborers working on the project had been insisting for a week before the slide that "the slope becomes alive."

Wherever there seems any likelihood of a slide's occurring in connection with construction operations, a series of observation wells should be established over the area. These can be used for observation of the groundwater level, as well as any change in the pressure (pore-water pressure) within the soil mass. They can also be used to house a newly developed device known as a slope indicator which will record any horizontal movements in the soil.

The slope indicator consists of a tiltmeter housed in a watertight brass cylinder about 2.5 in. o.d. and about 15 in. long. Within it, the tip of a pendulum contacts a precision-wound resistance coil forming one half of a Wheatstone bridge. The other half of the bridge, together with a potentiometer, resistors, switches, and batteries, is contained in a control box. The tiltmeter, which is lowered into the observation well, is connected by cable to the control box (*f*). When the circuit is in balance, the component of inclination in a plane defined by any two of four small wheels on the tiltmeter is proportional to the dial reading of the potentiometer.

Observation wells must be adapted to the use of the tiltmeter by lining them with a special plastic casing of 2⅞ in. i.d. prepared in 5-ft lengths. Four longitudinal slots are machined into the interior of the casing to accommodate the wheels on the tiltmeter. Lengths of casing are joined by aluminum couplings cemented and crimped in place after careful alignment of the slots. After the casing is positioned in the well with one pair of slots in line with the slope, the remainder of the space in the well is packed with sand or a clay-cement grout.

The tiltmeter is lowered into the casing with the wheels engaged in one pair of slots, and readings are taken at intervals of 2 ft. The same procedure follows in the slots at 90 deg to the original pair, readings being observed at the same depths in both instances.

Measurements are usually taken daily at first and then with extended time intervals as the pattern develops. The inclination can be determined within 3 deg of arc; but, of course, where large movements occur, the casing will be deflected to such an extent as to make further readings impossible. When the movements reach this magnitude, however, the contractor has already learned what he needed to know.

Figure *e* shows one set of a pair of readings with the corresponding deflection in inches, taken in connection with the slide indicated in Fig. *a*. A series of such readings from a number of wells provides not only the area of possible sliding but also the prospective cycloidal or cylindrical surface along which the slip will occur, if in clay.

Continuing observations of the water level in a series of wells will provide some idea of the slope of the water table if there is one. In finer soils, pore pressure readings can be taken with piezometers. Increases or decreases in the level or pressure of the water below the surface may give warning of a change in the slope's stability.

Essentially these observations are steps to be taken "until the doctor comes." The interpretation of any of these data must depend to a very great extent upon knowledge of the geological formations underlying the site; this requires the services of an experienced geologist. The geologist cannot stop the slide, but he can suggest methods of preventing it. Even where a slide cannot be prevented, advance notice will permit the contractor to minimize damage and loss of life.

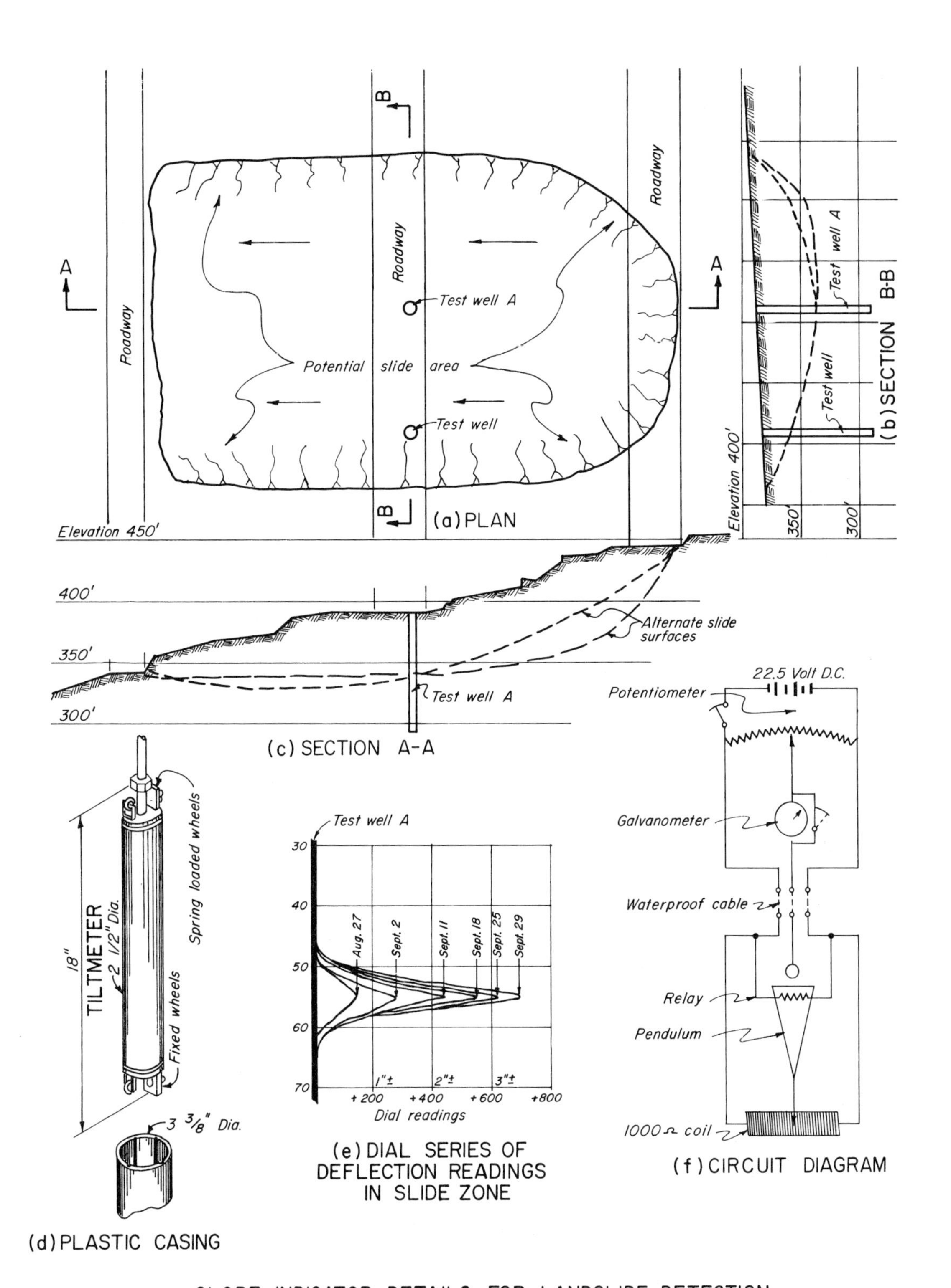

SLOPE INDICATOR DETAILS FOR LANDSLIDE DETECTION

1.15 LANDSLIDE CONTROL

Before discussing methods of controlling landslides it is necessary to know that a potential landslide exists. It would be foolish to gaze up at a hillside, assume that it will slide, and take measures to prevent it. It is just as absurd to assume that since a hillside has never moved, it never will. It is apparent that there is no type of control for landslides induced by earthquakes or blasts since the occurrence of these cannot be predicted. (Blasts from the contractor's operation are not involved here.) Where other factors noted in Sec. 1.13 exist, they should be treated for control regardless of secondary factors which may trigger them.

The presence of faults and fissures must be approached cautiously. Their mere existence proves nothing; they may have stood by, immobile, during uncounted centuries. On the other hand, they cannot be ignored.

Canted or tilted strata, particularly of beds of shale, have often led to slides along the natural clay seams which they may contain. Control of the sliding of one mass over a lower bed can be difficult (*a*), since any excavation is apt to introduce water that may accelerate the movement.

Most landslides are associated with water either from rainfall or from underground sources; consequently, once there is dimensional evidence of an impending slide as discussed in Sec. 1.14, drainage appears to be the most satisfactory solution.

Drainage along the toe of the impending slide mass, consisting of a pipe laid with open joints in a bed of crushed stone, can be effective (*b*). The pipe should be of substantial size, not because of the prospective flow but because there will be a tendency for it to silt up because of percolation of fines through the stone. Where any knowledge of the surface of sliding at the toe can be gleaned, the drain should be set to straddle and parallel it.

Any known sources of water entering the slope, whether surface or underground, should be tapped and diverted. In some cases a drain similar to the toe drain may be run along the top of the potential slide area as indicated by surface cracks, to intercept water entering the slide mass (*c*).

The suggestion that the slope be covered with a layer of pervious gravel or crushed stone to intercept rainwater and lead it to the toe drain is not always practical (*d*). The alternative suggestion that the surface be covered with an impervious blanket, presumably of rolled clay, is in a like category (*d*). The presumption that it is the rain falling directly on the surface may be far from true, and in many cases the depth of the surface of sliding may be such as to make such palliatives ineffective. Any loading of the surface may actually precipitate a slide.

These objections may be overcome, however, with horizontal drains (*e*). These can be drilled into the surface with standard drill rigs, angled as required, with a pipe casing (generally 2 to 4 in. in diameter) installed in the hole and projecting from the surface. This method is particularly effective in rock masses. Angled pipe extensions should convey the flow to the toe drain.

In soils of low permeability (silts and clays) the most satisfactory recourse is to decrease the slope angle by excavation or to erect barrier walls or rows of piling to thwart the path of the slide. Such obstacles should penetrate well below the surface of sliding to be effective. However, in the Palo Verdes slide noted in Sec. 1.12, some 25 drilled-in caissons were sunk near the toe into an apparently stable layer and were no more effective in stemming the land flow than was King Canute in holding back the sea.

Many of the preventive measures within the contractor's cognizance are purely negative. Rapid drawdown against a slope can be slowed up (although in this case a layer of gravel on the slope may prove helpful). The careful drainage of water from saturated fills produced by dredging or tunnel mucking operations serves to prevent their soaking into deeper, susceptible layers.

Slopes on fills or waste piles should be controlled to conform to the angle of repose of the soil being deposited. Where this would require too much space, layering of these piles with drier or coarser soils or rocks will prevent initial movement.

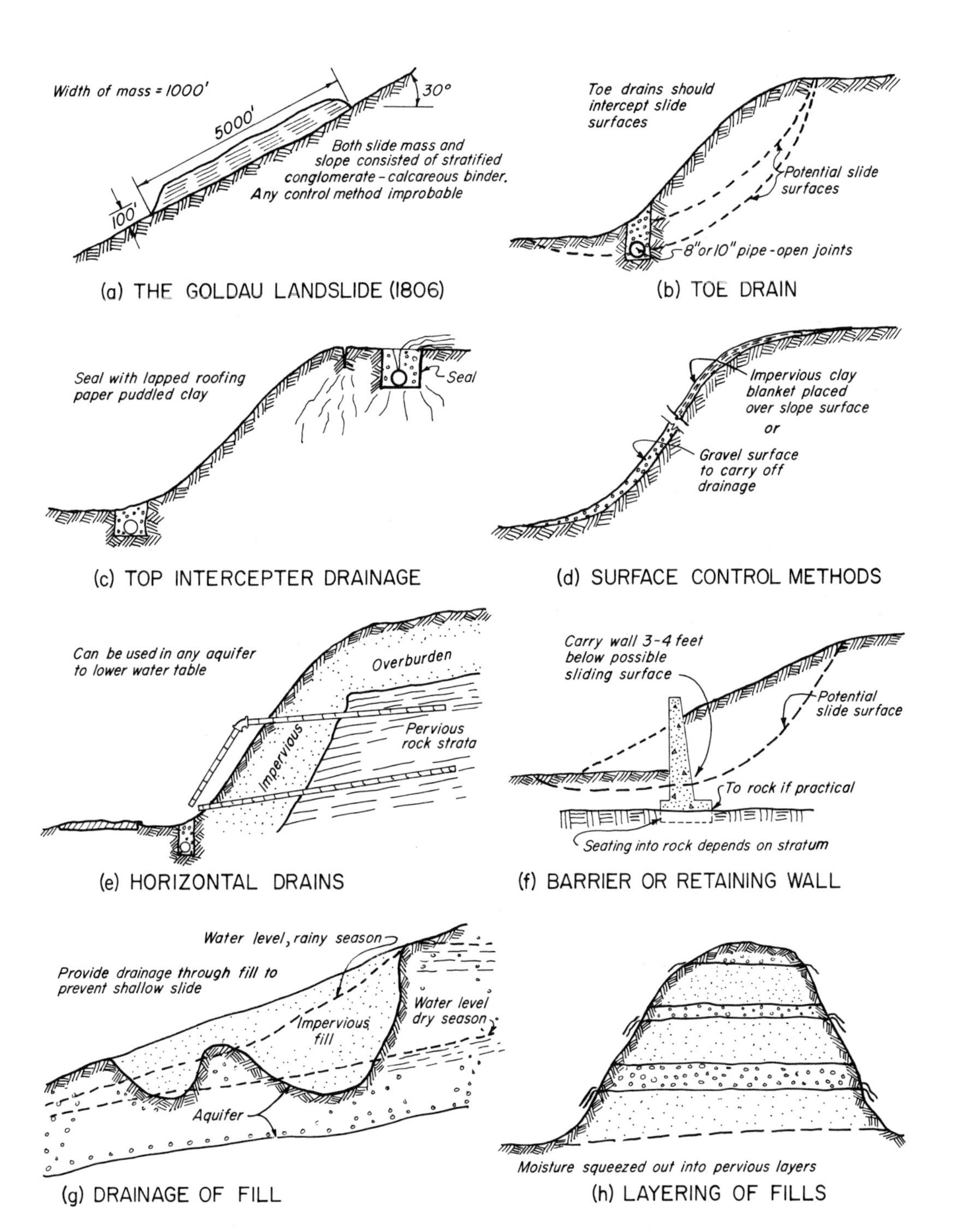
Width of mass = 1000'
5000'
30°
100'
Both slide mass and slope consisted of stratified conglomerate - calcareous binder. Any control method improbable
(a) THE GOLDAU LANDSLIDE (1806)
Toe drains should intercept slide surfaces
Potential slide surfaces
8" or 10" pipe - open joints
(b) TOE DRAIN
Seal with lapped roofing paper puddled clay
Seal
(c) TOP INTERCEPTER DRAINAGE
Impervious clay blanket placed over slope surface
or
Gravel surface to carry off drainage
(d) SURFACE CONTROL METHODS
Can be used in any aquifer to lower water table
Overburden
Pervious rock strata
Impervious
(e) HORIZONTAL DRAINS
Carry wall 3-4 feet below possible sliding surface
Potential slide surface
To rock if practical
Seating into rock depends on stratum
(f) BARRIER OR RETAINING WALL
Water level, rainy season
Provide drainage through fill to prevent shallow slide
Water level dry season
Impervious fill
Aquifer
(g) DRAINAGE OF FILL
Moisture squeezed out into pervious layers
(h) LAYERING OF FILLS

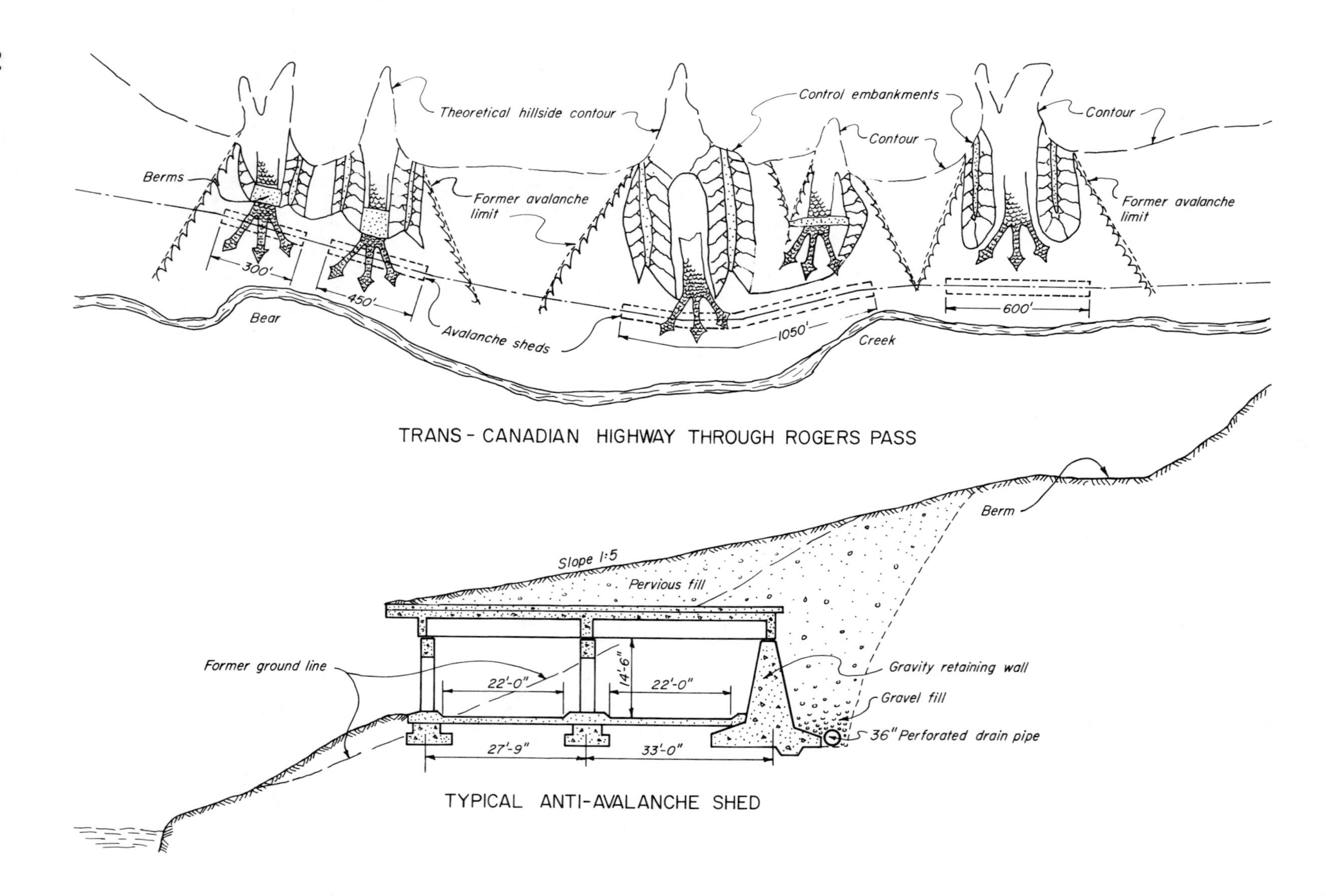

TRANS - CANADIAN HIGHWAY THROUGH ROGERS PASS

TYPICAL ANTI-AVALANCHE SHED

1.16 AVALANCHE OR LANDSLIDE?

In January, 1962, in the mountains of Peru, high in the valley above the village of Raurahirca, a great block of ice broke loose from Glacier 511 and became an avalanche. Traveling from a height not many thousands of feet below the 22,000-ft peak of Peru's highest mountain, it swept down the valley at the rate of a mile and a half per minute to bury the village 40 ft deep under a mass of ice, snow, soil, boulders, and debris. Spreading out as a blanket, a mile and a half wide, it trapped and buried over 3000 people before it came to rest. Twenty years earlier in a valley only 30 miles from Raurahirca another avalanche had destroyed the village of Huarás, killing 5000 people.

Although the death toll in Peru was, perhaps, higher than in most cases, avalanches are not uncommon in many parts of the world, and scarcely a year goes by without recording at least one substantial fall in the Alps or in some of the other chains of mountains.

Above the snow line, growing and compacting heaps of snow, sometimes in the form of glaciers, exceed their evaporation losses and create towering masses. Spring or summer rains or partial thawing attack the soils or rocks underlying these masses, disturbing them or removing their support. When this occurs, the mass of snow or ice breaks off and starts down the precipitous slope to become an avalanche. In the Swiss Alps it is said that thunder, or a good big yodel, can start the mass moving, but it can spring alive just as readily in absolute quiet.

An avalanche is a violent form of landslide to which the element of snow has been added. Many huge accumulations of snow never result in avalanches or landslides because they rest on primary rocks which provide continuous support. In other words, the same factors must exist for an avalanche as for the landslides previously discussed. The catastrophic aspect of avalanches arises from the steepness of the slopes down which they are hurled.

Because avalanches are in some cases more predictable than landslides, an experiment with controlling them in the Canadian Rockies is of interest. Certain aspects might be of use in controlling landslides of other sorts in highway cuts through mountainous terrain.

A portion of the trans-Canadian Highway was being cut through Rogers Pass in the Canadian Rockies in 1960 to 1962. In the pass area, avalanches were known to be an annual occurrence. In order to keep the highway open a system of snowsheds was devised, along with other remedial measures, to pass the landslide falls over the highway. The details are shown in the illustration.

In this approach, where snowfall is not the prime mover of the landslide, certain basic factors must be considered. The first of these is whether the landslides recur with sufficiently predictable frequency to warrant the cost, which is substantial. Is the relative depth of road cut sufficient, and is the hillside steep enough to pass the major portion of the slide over the shed? Although the weight of wet snow can be as great as certain soil masses, it will in time melt away, whereas soils or rocks will remain, altering the angle of slope and presenting an obstacle to later slides.

Another factor of importance is that, in this case, heavy timber growths limited the probable avalanche area to predictable gulleys and valleys, permitting dikes, berms, and dams to channel the slide over the portion of highway where the snowsheds were constructed.

In other respects, this is the standard preventive treatment for landslides of any sort. Toe drains are provided along the base of the cut on the high side, coupled with a retaining wall of substantial proportions. The retaining wall is further buttressed by the remaining columns forming the shed. In addition to adequate approach slopes continuing over the shed, a clear and continuous drop must be available on the low side of the shed so that the bulk of the avalanche will pass over into the lower valley paralleling the highway.

2 Techniques under Air and Water

The concept of the construction estimator in the creative role of juggling the elements of a construction project nimbly in his mind before converting them into staid columns of dollars and cents was briefly mentioned in the introduction to Chap. 1. Where this feat is successfully performed, the contractor receives the award of a construction contract. (Let us ignore some of the facts of life that intervene, such as that the contractor receives the award because the estimator made a not-so-successful error or that he does not receive the award because some other contractor's estimator made an equally unsuccessful mistake.) Once an award has been made and the contract has been executed, management then takes over.

Construction management's prime function is that of marshaling the men, materials, equipment, and ideas necessary for the execution of the contract and, once they are assembled, the keeping of these elements moving in pre-determined directions. The project must now be broken down into a new set of component parts consisting of a series of operations to be performed. Each of these operations involves one or more techniques. The estimator's "500 tons of reinforcing steel at $200.00 per ton, in place, equals $100,000.00" now becomes a series of items: detailing; secondary design; tons of reinforcing steel by sizes; cutting; bending; identifying; delivery; placing; tying. Reinforcing steel consists of a relatively narrow range of operations and techniques compared with concrete, and concrete is much less involved than the estimator's item of excavation.

This was all rather simple—once. In the days when the great Gothic edifices were being built, the owner called in a master mason who assembled his associates, ranging from stone setters to sculptors, collected a group of strong-backed local boys to dig foundations and hustle mortar, had in a few workers in wood to put on a roof and hang a few doors; glaziers sealed it, and the thing was done. Time was not of the essence and money could be had—in time—for the asking; a hot euchre party was organized or a new tax levied, and the work went on.

Such work was conducted close to ground

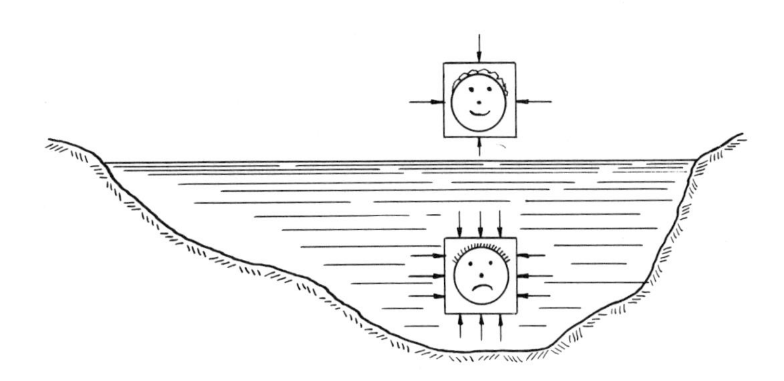

level at sites where underground conditions involved no special problems. Other kinds of work at or adjacent to rivers, harbors, or waterfronts did involve some other problems dealing with water, but these were not severe. Bridges were limited in span and location to suitable sites; tunnels, except those used as an adjunct in mining, were not needed and not built. Ships were light, with shallow draft, and waterfront structures required very little but piling to make them satisfactory.

This picture of things began to change in the early nineteenth century. Bridge spans of greater extent, at sites selected not for ease of construction but for efficient railroad operation, were needed. Also, it was found that railroads could not efficiently go over or around the mountain; a straight line was the shortest distance between two points even where it meant going through the mountain. One thing led to another; iron ships developed greater draft and, empty, needed higher clearance under bridges and deeper channels. Means of deepening ancient seaports were needed first, and later, means of going under a river rather than over it.

All these developing techniques had one thing in common, the control of water, for as bridges became larger, so did their foundations become not only larger but deeper. Moreover, the site for a bridge could not be determined by the ideal foundation conditions but had to be selected regardless of such conditions. Tunnels were in a like category.

Control of water can be approached in two basic ways: It can be pushed aside or removed so that construction can be done in the dry, or work can be conducted within or "under" it. Earlier methods had been employed to permit work in the dry, but to only very limited depths and within narrowly limited areas. The nineteenth century developed the techniques of working under water with divers and dredges and of pushing it aside with compressed air. These newer techniques have now been firmly added to construction management's already bulging bag of tricks.

2.1 COMPRESSED AIR

Compressed air is so widely used in building operations today that the concept of modern construction without it is impossible. Although its characteristics were first studied 300 years ago, its application to construction problems is not much over a century in age, and its use as motive power for innumerable tools is somewhat more recent.

Air is a molecular mixture of two gases containing about 78% nitrogen and about 21% oxygen with a few other inert gases totaling about 1%. As a mixture of gases, air behaves characteristically as a perfect gas in three ways: (1) there is complete diffusion of the oxygen in the nitrogen and consequently the mixture is uniform in composition; (2) the molecules, of which the mixture is composed, are in constant motion, having relatively vast distances between them; (3) the impact of the molecules striking any surface produces the pressure exerted on that surface by the gas.

Because of complete diffusion (1), any law that applies to a perfect gas applies to air. Because of the spacing of the molecules (2), any quantity of air can be reduced in volume. Because the rate at which the same number of molecules strike a surface (3) is increased when the gas volume is decreased, the pressure on that surface is increased.

Man has been dependent upon air for his very life since his ancestors emerged from the primeval ooze, but his curiosity regarding its nature was only aroused about 1660 when a number of individuals, including particularly Robert Boyle, began experimenting with air and other gases. Among the many observations made by Robert Boyle was one basic premise now known as Boyle's law: that the pressure of a gas multiplied by its volume is always the same so long as the temperature does not change.

$$P_1V_1 = P_2V_2$$

Further observations indicated that the pressure of the air mantle under which man survived varied with the distance above the surface of the ground. And while it was established that the pressure of the air at "sea level" was 14.7 psi, by inference it was deduced that at some distance from the earth's surface there would be no air at all, a fact that did not assume any importance for several centuries.

It was a hundred years after Boyle before a French physicist, J. A. C. Charles, developed the law which bears his name: that the pressure of a gas increases as the temperature increases so long as the volume remains the same.

$$P_1T_2 = P_2T_1$$

By the inference so beloved of the philosophers who flourished mightily in the eighteenth century it might have been assumed that at some extremely low temperature there would be no pressure at all, that the frozen molecules would cut their cavortings. But it was nearly a hundred years before Lord Kelvin, about 1848, established in fact that there was an absolute zero, 273.2° below zero on the centigrade scale (460° below zero on the Fahrenheit scale).

All this can be assembled into a single equation $P_1V_1/T_1 = P_2V_2/T_2$, in which P is the absolute pressure in pounds per square inch (remember that pressure at sea level is 14.7 psi), T is the temperature in degrees above *absolute* zero, V is the volume in cubic feet. In the metric system, occasionally used, P would be given in kilograms per square centimeter; temperature would be converted into degrees centigrade; volume would be in liters or thousands of cubic centimeters.

For practical purposes, when dealing with quantities of compressed air in cubic feet per minute, these figures are given in terms of "free air," that is, the cubic feet per minute of air at atmospheric pressure at sea level (or 14.7 psi) and at a temperature of 60° F (see calculations). Generally ignored is the fact that atmospheric air, having a variable moisture content, has a higher density (i.e., the molecules in a cubic foot are increased in number by the number of molecules of water vapor), slightly varying the pressure.

DEFINITIONS—CALCULATIONS—CONVERSIONS

GAGE PRESSURE

The ordinary pressure gage used on air lines indicates only the difference between the pressure being measured and the pressure of the atmosphere surrounding it. Thus a gage reading 100 psi at sea level would actually be indicating 114.7 psi absolute pressure. For calculation purposes the conversion to absolute pressure must be made, but where differences in pressure only are concerned, no conversion is necessary. Where both absolute and gage pressures are being used, gage pressure is indicated as psig.

A. *Boyle's Law:* $P_1V_1 = P_2V_2$

EXAMPLE: A compressor is rated at 100 cfm of free air at sea level and 60°F. How much air can be delivered by the compressor at 100 psig?

$P_1 = 0$ psig + 14.7 = 14.7 psi $P_2 = 100$ psig + 14.7 = 114.7 psi
$V_1 = 100$ cu ft (per min) $V_2 = ?$
14.7 psi × 100 cu ft = 114.7 psi × V_2 $V_2 = 12.8$ cu ft (per min)

B. *Charles' Law:* $P_1T_2 = P_2T_1$

EXAMPLE: A compressor compresses free air at atmospheric pressure, raising the temperature from 60°F to 120°F. What would be the final pressure?

$P_1 = 0$ psig + 14.7 = 14.7 psi $P_2 = ?$
$T_2 = 120°F + 460° = 580°$ abs $T_1 = 60°F + 460° = 520°$ abs
14.7 psi × 580° abs = 520° abs × P_2 $P_2 = 16.4$ psi

C. *General Gas Law* (combining *A* and *B*): $\dfrac{P_1V_1}{T_1} = \dfrac{P_2V_2}{T_2}$

EXAMPLE: Solve for volume of air, using data in *A* and *B*.

$P_1 = 0$ psig + 14.7 = 14.7 psi $P_2 = 100$ psig + 14.7 + 114.7 psi
$V_1 = 100$ cu ft (per min) $V_2 = ?$
$T_1 = 60°F + 460° = 520$ abs $T_2 = 120°F + 460° = 580°$ abs

$$\frac{14.7 \text{ psi} \times 100 \text{ cu ft}}{520° \text{ abs}} = \frac{114.7 \text{ psi} \times V_2}{580° \text{ abs}} \qquad V_2 = 14.7 \text{ cu ft (per min)}$$

ATMOSPHERES

The pressure of the air in the atmosphere at sea level and 60°F is termed 1 atmosphere = 14.7 psi. For most practical purposes 15 psi is sufficiently accurate. In diving, atmospheres are frequently used instead of psi to denote depth.

1 atmosphere of air pressure = a depth of 33.9 (34) ft in fresh water
= a depth of 33 ft in salt water

WATER PRESSURE

Water pressure is dependent on the weight of water.
1 cu ft of water weighs 62.4 lb (salt water = 64 lb).
Pressure is the weight (in psi) of a column of water 1 sq in. in area and 1 ft high.
62.4 lb/144 sq in. = 0.433 psi per foot of depth
50 psi is the legal limit of air pressure under which men can be worked.
50 psi/0.433 psi = 115 ft, the equivalent depth of water for compressed-air operations.

2.2 AIR COMPRESSORS

Air is widely used in the construction and other industries for a variety of purposes, requiring from 1 cfm (in some automotive tools) to as much as 50,000 cfm (in tunnel construction) and at pressures ranging from 10 psi (in caisson construction) to 500 psi (in deep-well drilling). Because of this wide range in demands, compressors are generally manufactured, or adapted, to meet the particular range of values required for a specific operation.

All compressors were at one time of the reciprocating type, compressing air by means of a piston moving in a cylinder. The rotary compressor was introduced in 1950. The rotary compresses air by means of vanes sliding centrifugally from a shaft revolving eccentrically in a cylinder. In 1960 another variant, the screw rotary, with fixed but twin spiraled vanes set on a concentrically revolving shaft was brought out. About 80% of portable construction compressors in the range from 125 to 900 cfm are now rotaries. Smaller compressors, some portables, and larger non-portable sizes (500 to 2000 cfm) are still manufactured as reciprocating units.

Air may be compressed in either one or two stages. Although single-stage compression is generally used for pressures up to 80 psi, many compressors use single stage for the construction range of pressures from 80 to 125 psi.

Compressors may be air-cooled, oil-cooled, or water-cooled, referring to the reduction of heat generated in the compression cycle. Reciprocating compressors are often air-cooled (by air blown back by the flywheel fan); rotary compressors are oil-cooled by an injection of atomized oil into the incoming airstream; fixed, non-portable compressor installations (such as those used in tunneling) are generally water-cooled by circulating cold water in a jacket surrounding the compression cylinder. Water cooling is the most efficient method but is not practical for portable units and requires an adequate water supply.

Water and oil in compressed air are undesirable and require removal. Not only is clean, dry air required where men are breathing it, but water in the air feeding to pneumatic tools will wash away the lubricant, resulting in rapid wear and excessive air consumption. The removal of water and oil vapor from compressed air before it enters transmission pipelines requires condensation of these vapors. Condensation is accomplished by additional cooling of the air in aftercoolers. The compressed air flows through a nest of pipes enclosed in a jacket through which water is pumped, producing cooling to within 15° of the temperature of the incoming water. The liquid water and oil drain to separators, where they can be drawn off either manually or automatically.

From the aftercooler the air goes to a receiver whose prime purpose is to maintain a pre-determined range of pressure. (In some instances, as with rotary rigs, the receiver also contains an oil filter for the removal and recirculation of oil.)

Compressor capacity is rated at the intake side in terms of cubic feet per minute of "free air." The quantity of air that can be drawn from the compressor is dependent on the pressure maintained in the air receiver. Compressor engines, whether gasoline, diesel, or electric, operate at variable speeds, depending upon demand at the receiver. This may vary between 10 and 15 lb of pressure, or may be set within much narrower limits. In all cases some type of "unloading" device must be provided so that engines do not start against a load of partially compressed air. At maximum pressure in the receiver, the pilot device unloads the compressor, either by holding the inlet valves open or by closing the compressor intake line. When the pressure drops to its minimum setting, the pilot device permits air to re-enter the compression cylinder.

Stationary set-ups for tunnel or caisson work generally use independent air receivers even where portable compressors supply the air. This not only provides closer control of variations in pressure on the inlet side but provides closer control of the pressures and quality of the air fed to the working face.

COST OF COMPRESSED AIR

1. *Operating Cost*

   (*a*) Power: Diesel or gasoline fuel cost per 100 cfm (see below) ______________

   (*b*) Labor: Operator's time

$$\frac{\text{Hourly rate}}{\text{Cfm/per hr}} \times \frac{100}{60}$$ ______________

   (*c*) Maintenance, repairs, overhaul, lubrication

$$\frac{\text{Increment of operating time}}{\text{Cfm/increment}} \times 100$$ ______________

2. *Depreciation*

$$\frac{\text{First cost of compressor}}{\text{Life in operating hours}} \times \frac{1}{\text{Rate in cfm/hr}} \times \frac{100}{60}$$ ______________

3. *Interest on investment* (interest on non-depreciated portion of first cost converted to cfm as in #2) ______________

Total cost (cfm) ______________

FUEL COST OF COMPRESSED AIR

(In cents per 100 cfm)

Engine brake horsepower required to deliver 100 cfm of free air	Diesel fuel, cents per gal					Gasoline, cents per gal					
	8	10	12	14	16	10	12	14	16	18	20
16	0.133	0.167	0.200	0.234	0.266	0.333	0.400	0.467	0.534	0.600	0.666
18	0.150	0.189	0.226	0.262	0.300	0.375	0.450	0.525	0.600	0.676	0.750
20	0.167	0.210	0.250	0.292	0.334	0.416	0.500	0.584	0.666	0.750	0.832
22	0.183	0.230	0.276	0.320	0.366	0.458	0.550	0.642	0.734	0.826	0.916
24	0.200	0.251	0.300	0.350	0.400	0.500	0.600	0.700	0.800	0.900	1.00
26	0.217	0.272	0.326	0.380	0.434	0.542	0.650	0.759	0.866	0.976	1.08
28	0.223	0.293	0.350	0.404	0.466	0.584	0.700	0.816	0.934	1.05	1.17
30	0.250	0.314	0.376	0.438	0.500	0.626	0.750	0.875	1.00	1.13	1.25

TEMPERATURE EFFECT ON DELIVERY OF AIR

(Intake temp. = 60°F)

Deg F	Deg abs	Relative delivery
−20	440	1.18
0	460	1.13
20	480	1.083
32	492	1.058
40	500	1.040
60	520	1.000
80	540	0.961
100	560	0.928
120	580	0.896
140	600	0.866
160	620	0.838

ALTITUDE EFFECT

(Relative brake horsepower required to deliver 100 cfm of free air)

Altitude, ft	Single-stage, at delivered pressure			Two-stage, at delivered pressure			
	60 psi	80 psi	100 psi	60 psi	80 psi	100 psi	125 psi
Sea level	16.3	19.5	22.1	14.7	17.1	19.1	21.3
1,000	16.1	19.2	21.7	14.5	16.8	18.7	20.9
2,000	15.9	18.9	21.3	14.3	16.5	18.4	20.5
3,000	15.7	18.6	20.9	14.0	16.1	18.0	20.0
4,000	15.4	18.2	20.6	13.8	15.8	17.7	19.6
5,000	15.2	17.9	20.3	13.5	15.5	17.3	19.2
6,000	15.0	17.6	20.0	13.3	15.2	17.0	18.8
7,000	14.7	17.3	19.6	13.0	14.9	16.6	18.4
8,000	14.5	17.1	19.3	12.7	14.6	16.2	18.0
9,000	14.3	16.8	18.9	12.5	14.3	15.9	17.6
10,000	14.1	16.5	18.6	12.3	14.1	15.6	15.7

Air is widely used to operate construction tools, but the economics involved in the use of air must be borne in mind. Where a single small tool such as a concrete vibrator, an impact wrench, or a small air pump is to be used from a portable air compressor, the overall cost may not be justified. Although compressors for construction use are available down to 65 cfm capacity, some small tools may require only a third or a quarter of this capacity. Moreover, in some areas, operators are required by union rule to stand by compressors of any size (although in other cases this applies only to capacities in excess of 65 cfm). For these small tools, therefore, other types of motive power may be desirable. For underwater work the economics must be ignored since no other source of motive power has been found to be practical.

Air requirements for pneumatic tools vary widely with the type of tool, its size, the service for which used, the condition of the tool, and the pressure of the air supply. Although many types of tools use the piston principle, others use a rotary type of air motor. Tools used intermittently require less air than those used continuously. As parts become worn, more air will be required to produce the same amount of work. Similarly, where only a limited pressure is required, higher air pressures produce no more work and, in effect, waste air. For these reasons, although a new tool can be said to require so many cubic feet per minute of air under certain working conditions and air pressures, the practical method is to average the variables and list the numbers of a particular type of tool which can operate on standard construction sizes of air compressors at specific receiver pressures (see table opposite).

All air tools require oil lubrication: as a sealant between the close-fitting moving parts; as a coolant to dissipate the heat developed; as a means of absorbing moisture (by forming an emulsion) which may have passed through the primary devices. Standard, portable, construction compressors are provided with both aftercoolers and receivers designed to provide clean, dry air. Lubrication devices must therefore be provided additionally.

Some air motors employ built-in lubricating wells or wicks, but piston-actuated equipment is lubricated by means of the compressed air passing through the cylinders. A line oiler or lubricator is placed in the line as close to the tool as practical. These are designed to provide a minute continuous supply of oil.

In working air tools in a milieu of compressed air or under water, consideration must be given to the back pressure. If a rock drill is designed to work efficiently using air at 80 psi under atmospheric conditions, when it is operated in a chamber having a 20-psi air pressure (gage), the effective pressure in the tool will be only 60 psi, insufficient for efficient operation. At pressures of 50 psi, therefore, a compressor furnishing air at 130 psi would be required. This factor must be considered in setting up compressor requirements in tunnel and caisson work.

An additional consideration, where a number of air tools are worked simultaneously in a chamber, is the effect of the released air on the pressure inside the chamber. Although a considerable amount of the initial pressure of the air is lost in passing through the tool, sufficient pressure remains to vary the chamber pressure. (If the chamber pressure were higher than the exit pressure, the tool would not work at all.)

As noted, air tools require continuous lubrication, some of which escapes with the spent air as a spray. Although recirculation of the air supply is mandatory in tunnel or caisson work, working a number of tools may produce dense clouds of oil particles, introducing respiratory hazards and lowering sight distances.

Working under water, air released from air tools simply rises to the surface and, unless it becomes trapped en route, presents no special hazard. Here too the back pressure required to release the spent air must be greater than the water pressure at the depth of water being worked.

Characteristics of Air Equipment

Description	Weight, lb	Hose size, in.	Approx. air for each, cfm	85		125		250		365		600		900	
Compressor size →				85		125		250		365		600		900	
Air pressure, lb →				80	90	80	90	80	90	80	90	80	90	80	90
Impact wrench:															
1¼″ heavy duty	20	½	30	2	2	4	3	9	8						
1½″ heavy duty	28	¾	45	2	1	3	3	7	6						
1¾″ heavy duty	70	1	80	1	...	1	1	3	3						
Wood borer:															
1″ drill	15	½	30	2	2	4	3	9	8						
2″ drill	32	½	65	1	1	1	1	5	4						
3″ drill	57	1	80	1	1	1	1	2	2						
Back-fill tamper:															
Single	42	½	35	2	2	4	3	10	9						
Triplex	170	¾	80	...	...	1	1	3	3						
Clay spade:															
Light	15	¾	35	4	3	6	5	9	8						
Heavy	27	¾	45	4	3	6	5	9	8						
Paving breaker:															
Light	30	¾	45	...	...	4	3	7	6						
Medium	55	¾	50	...	...	3	2	6	5						
Heavy	85	¾	75	...	...	2	2	4	3	8	7	12	10		
Sheeting driver	82	¾	60	...	...	2	2	4	3	8	7	12	10		
Rock drill or jackhammer	28	¾	60	...	...	2	2	4	3	7	6				
	40	¾	85	...	...	1	1	2	2	5	4	10	9	15	14
	50	¾	95	...	...	1	1	2	2	5	4	...	...	14	13
	62	¾	100	...	...	1	1	3	2	5	4	8	7	12	11
Concrete vibrator:															
2½″ dia.	30	¾	40	...	...	5	4	11	9						
3″ dia.	45	¾	50	...	...	3	3	8	7						
4¼″ dia.	55	1	70	...	...	2	2	4	3						
5½″ dia.	83	1	85	...	...	1	1	3	3						
Sump pumps:															
220 gpm @ 60′ head	58	1	100	...	...	1	1	2	2	3	3	5	4		
295 gpm @ 60′ head	130	1	100	...	...	...	...	1	1	1	1	2	3		
Sludge pump, 50 gpm @ 75′ head	97	1	90	...	...	2	1	4	4	6	5				
Drifters:															
3″ cylinder bore	166	1	200	...	...	...	...	1	1	2	2	4	3	6	5
3½″ cylinder bore	195	1	250	...	...	...	...	1	...	1	1	3	3	5	4
4″ cylinder bore	275	1	300	...	...	...	...	...	...	1	1	2	2	4	3

2.4 COMPRESSED AIR AS A PHYSIOLOGIC MEDIUM

Contemplating the possibility of life on other planets, man is made more acutely aware of the fact that he is a product and sometimes the victim of his own environment. Without his blanket of air, with its exact existing proportion of gases, which envelops him and in which he functions, he would be a vastly different creature. Something over 100 years ago when the first efforts were made to work men in an atmosphere of compressed air, the physiological aspects of this change of environment were little appreciated. Air was air was air, was it not?

The realities of water pressure, that is, the increase of pressure with depth, had been appreciated for centuries, but it was not until the beginning of the nineteenth century that the suggestion was made that air under pressure be used to counteract the pressure of deep water. The first actual use of compressed air in construction began toward the middle of the nineteenth century for sinking caissons, but compressed air had developed a vogue before this as a cure-all for many diseases.

Following Robert Boyle's basic experiments with air in the years before 1660, and lasting well into the twentieth century, "air baths" in which patients were placed in cylinders under slightly compressed air were quite fashionable and were alleged as cures for all manner of ailments. Under the very limited pressures used it did occasionally provide a momentary relief. At the higher pressures used in caisson construction the effects were not so fortunate, and a malady developed which came to be called "the bends" (from the fact that it developed intense pains in the joints which caused the individual to contort grotesquely in seeking relief) and, simply, "caisson disease."

The phenomenon was not properly understood until, in 1878, the French physiologist Paul Bert published the results of his investigations in "La Pression Barométrique." (An English translation was prepared in 1943 for the use of the United States Army Air Force.) His experiments indicated that it was the effect of the nitrogen component of air that induced the bends.

Under a pressure of 2 atm (equivalent to an underwater depth of 33 ft) a man working in compressed air breathes twice the amount of air he would breathe under atmospheric pressure and consequently twice as much nitrogen. At pressures higher than atmospheric, the nitrogen does not pass off in exhalation but goes into solution in the blood and tissues of the body. The greater the pressure, the greater the absorption of nitrogen. When the pressures decrease, the nitrogen comes out of solution in the form of minute bubbles. As the pressure further decreases, the bubbles continue to expand, tending to block off capillaries, veins, and arteries. This can damage the nervous system and affect the spinal cord. The term "black froth," to describe the one aspect of the symptoms, arises from these bubbles.

Bert had suggested the solution to preventing caisson disease, that of decreasing the pressure slowly so that nitrogen passed off gradually. Roebling's Dr. Andrew Smith had already used this method, in 1873, on sandhogs coming out into atmospheric pressure too rapidly from the caissons of the Brooklyn Bridge. But it was J. S. Haldane, an English physiologist, who, working with deep-sea divers (afflicted with the same problem), developed the decompression pattern now in use. He found that the bends could be avoided by raising divers in stages, each stage being half of the previous pressure, holding them at that pressure for a specific time, and then raising them to a lower stage of pressure. He also discovered that different types of individuals reacted differently during stage decompression.

Since that time, decompression rates for workmen in pneumatic caissons and air-driven tunnels have been developed adequately to cover all types of persons who might be involved. Other factors are also generally specified, such as the length of time work should be performed under specific pressures, the number of times per 24-hr day, the number of shifts a man should work under pressure, and the length of rest periods between shifts.

Decompression Rates

Working pressure	First shift						Second shift						Hours in 24-hour period		
	1st Stage		2nd Stage		3rd Stage		1st Stage		2nd Stage		3rd Stage		Max. total working time	Max. shift in comp. air	Min. rest interval
	Drop Press.	Min.	Drop Press.	Min.	Drop Press.	Min.	Drop Press.	Min.	Drop Press.	Min.	Drop Press.	Min.			
6 psi	6–3	½	3–1½	1	1½–0	1½	6–3	½	3–1½	1	1½–0	3	6	3	2½
10 psi	10–5	1	5–2½	1½	2½–0	2½	10–5	1	5–2½	1½	2½–0	5			
15 psi	15–7½	1½	7½–3½	2	3½–0	3½	15–7½	1½	7½–3½	2	3½–0	7			
20 psi	20–10	2	10–5	2½	5–0	5	20–10	2	10–5	2½	5–0	10			
25 psi	25–12½	2½	12½–6	3	6–0	12	25–12½	2½	12½–6	3	6–0	24	4	2	3½
30 psi	30–15	3	15–7½	4	7½–0	15	30–15	3	15–7½	4	7½–0	30			
35 psi	35–17½	3½	17½–8½	4½	8½–0	17	35–17½	3½	17½–8½	4½	8½–0	34	3	1½	4
40 psi	40–20	4	20–10	5	10–0	20	40–20	4	20–10	5	10–0	40	2	1	4½
45 psi	45–22½	4½	22½–11	5½	11–0	22	45–22½	4½	22½–11	5½	11–0	44	1½	¾	4¾
50 psi	50–25	5	25–12½	6½	12½–0	25	50–25	5	25–12½	6½	12½–0	50	1	½	5

Use of Tables:

First shift comes from tunnel or caisson after 2 hr work at 30-psi pressure:

In man lock drop pressure to 15 psi; hold for 3 min.

Then drop pressure to 7½ psi; hold for 4 min.

Then drop pressure to 0 psi; hold for 15 min.

Men rest 3½ hr before starting 2d shift; after second shift:

In man lock drop pressure to 15 psi; hold for 3 min.

Then drop pressure to 7½ psi; hold for 4 min.

Then drop pressure to 0 psi; hold for 30 min.

Minimum total elapsed time 8½ hr; actual working time less than 4 hr.

INSTRUCTIONS FOR COMPRESSED-AIR WORKERS

1. Never go on shift with an empty stomach.
2. Avoid all alcoholic liquors.
3. Eat moderately.
4. Sleep at least 7 hr daily.
5. Keep the bowels regular.
6. Take extra outer clothing to avoid becoming chilled during decompression.
7. Move limbs freely during decompression to stimulate circulation.
8. Drink hot coffee; take warm shower; have brisk rubdown after each shift.
9. Do not give intoxicating liquor to men suffering from compressed-air illness.
10. Report for medical checkup for any ailment, no matter how slight.
11. Wear identification badge prominently displayed at all times on or off the job. This will show physician to be notified if you become ill because of delay reaction from working under compressed air.
12. Stay at job site for a least ½ hr after locking out.
13. All employees must be medically examined periodically. When first employed a re-examination is required after first half-day period.
14. Do not work more than two shifts or periods in any 24 hr.

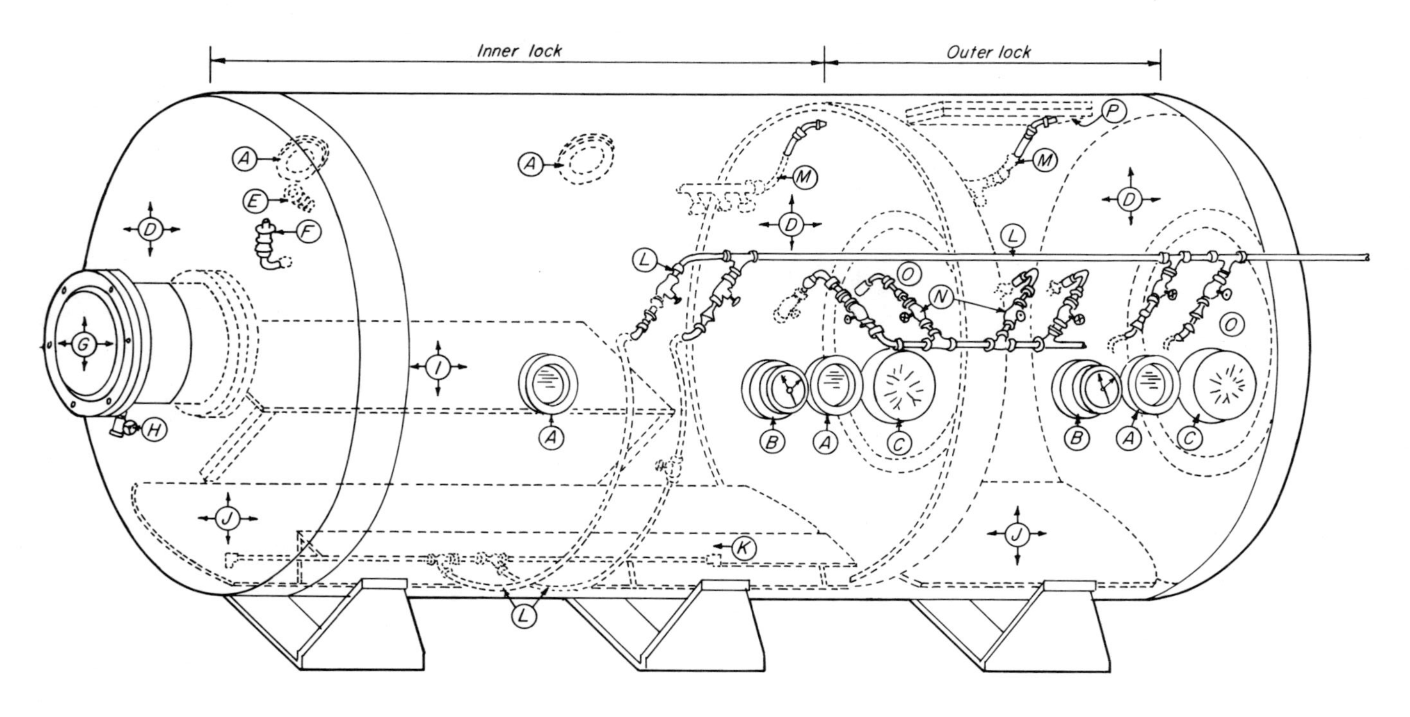

MEDICAL LOCK OR RE-COMPRESSION CHAMBER

A *Observation port*
B *Pressure gauge*
C *Caisson gauge*
D *Bulkheads*
E *Light*
F *Relief valve*
G *Medical supply lock*
H *Equalizing valve*
I *Cot or bunk*
J *Floor board*
K *Bench*
L *Air supply lines*
M *Oxygen supply and manifolds*
N *Exhaust control lines*
O *Access ports*
P *Heating element*

2.5 AIR LOCKS

An air lock is a chamber used in transferring men and materials from atmospheric pressure to the higher pressures used in caisson and tunnel work. For most projects a man lock and a muck lock are provided as separate units since the problems of operations are, or can be, considered separately. In addition to the man lock and muck lock, either of which may be in duplicate, an emergency man lock is provided where space permits, as well as a medical lock which is usually entirely separate, but should be maintained within a short distance of the man locks (within a radius of 100 yd).

Air locks today are generally designed for the specific operation, although otherwise suitable units may be adapted from site to site. Essentially they are steel tanks manufactured in conformance to the A.S.M.E. Unfired Pressure Code and adapted with airtight access doors as required. (For tunnels these will usually be in the ends; but with caissons, where the tank is upended, they may be on the side.) Additional adaptations for air and pressure are required.

All types of locks, except medical locks, are faced with the necessity of maintaining pressure within the working space or chamber; any variation here is dangerous not only to personnel but to the entire operation being performed. However, since decompression is not a factor in the operation of muck locks, they can generally be set up on a more rapidly operating basis than man locks and therefore constitute an entirely separate operation.

Requirements for man locks are generally specified by regulations. The number of men that can be passed through at one time will be limited by the cubic feet of air per man available within the lock (generally 30 cu ft per man). A minimum headroom is generally specified as 5 ft, and, since the men will spend some time in the chamber during decompression, adequate seating facilities are usually provided. Electric lighting is required within the lock as well as telephone communication with the outside. Ventilation can be maintained by cracking the air-outlet valve as well as the intake valve, to provide air passage without loss of pressure.

As was noted in Sec. 2.1, air during compression acquires an increase in temperature, and one of the problems involved in working under compressed air is control of temperature in the working area or chamber since air is supplied from the compressors directly to this point. Aside from regulations which may stipulate that the temperature be maintained at 80°F, or below 100°F, the debilitating effects of working at high temperatures and the consequent loss of productivity will dictate control of temperature. However, decompression will generally involve a corresponding *decrease* in temperature, sometimes of as much as 25°F, during the initial stage, so that heating for the man lock should be provided.

Although many operations can be automatically controlled, manual pressure-control valves should be available both inside and outside the lock. A shatterproof observation port should be provided in both ends of the lock. In addition to a pressure gage and an electric clock inside the lock, a recording pressure gage should be installed so that a continuous record of pressure variations is later available. A lock tender should be maintained on the atmospheric side of the lock, one of whose principal duties is to log in and out every person passing through the air lock. This information is necessary for proper treatment of individuals who may later develop symptoms of the bends.

Medical locks are essentially the same in construction as other types of man locks except that they are divided into two separate compartments by a bulkhead, permitting entrance to the inner lock during decompression treatment (see illustration). In addition to the gage and control conditions already cited and the usual facilities for first aid or medical treatment, they will require a separate supply source of compressed air of sufficient capacity to raise the pressure to 75 psi within 5 min, at which pressure the temperature shall be controlled so as not to exceed 90°F. Because of its use in treatment of the bends, pure-oxygen inhalation equipment is usually additionally provided.

2.6 OPERATIONS UNDER COMPRESSED AIR

The quantities of air required for caisson sinking, or tunnel construction, under compressed air are widely variable and depend upon an assortment of factors. Industrial codes usually specify the minimum quantity of free air which must be supplied, per man, to the working face. This varies from 1800 cfh (30 cfm) in the New York State Code to 900 cfh (15 cfm) in England. The quantity of air to be compressed will vary in accordance with the working pressures required. The number of men that can be worked may depend on the size of the working chamber, the operation being performed, or the type of material being excavated.

The quantity of air required per man would be the quantity to be exhausted from the chamber if there were no other losses, but there is considerable absorption of air in the working face, the quantity depending upon the nature of the soil. In dense clays the loss may be very small, although seemingly minor tubes left in the clay by the disintegration of ancient root systems may account for substantial losses.

Losses in silts will depend upon their moisture content, but since the presumption, where air is being used, is that they are saturated, the losses may vary from quite small amounts to substantial quantities and may increase progressively as the face, exposed to air, dries out.

Where the formations range from fine sands to coarse gravels the air losses will be the greatest, although here the degree of saturation will have an important bearing. Where saturation is not complete, that is, where the formation is not under direct hydrostatic head, an air bubble extending into the voids will form, protruding beyond the working face. The surface of this bubble may extend into an unsuspected porous stratum, thus bleeding off considerable air.

With caissons, the losses between the locks and the working chamber are minor; but in tunnel construction, losses through the tunnel lining must be considered. These losses will depend not only on the length of tunnel being worked but also on whether the secondary lining has been placed. Primary linings often permit large losses of air.

There is a loss of air in passing through either the muck lock or the man lock, but this is readily calculated by estimating the number of passages to be made and the quantity of air required to raise the pressure at each passage.

Ignoring such unforeseen events as "blows," a comprehensive empirical figure of 20 cfm per sq ft of working face has sometimes been used. In some formations such as dense clays this is extravagant, while in several sub-aqueous tunnels an allowance of 30 cfm per sq ft of face has proved inadequate. In the case of the caisson discussed in Sec. 8.22, working under 50 psi of air pressure, 3000 cfm was consumed. The bottom area was about 1500 sq ft, and the exposed vertical face beneath the cutting edge was another 1500 sq ft. The air required in this case was therefore only 1 cfm per sq ft of face.

Air supplied to the working chamber will be low-pressure air, ranging from a few pounds per square inch above atmospheric to the legal limit of 50 psi. Although the standard compressor can provide this air through the use of pressure-reducing valves, compressors specially adapted for the lower pressure ranges will be more economical to operate.

For most caisson and tunnel work, semi-permanent installations of fixed compressors driven by electric motors will be used.

Power may be generated on the job by using diesel-driven generators, but more generally public power will be brought into the site. In this case separate feeders will be required, each feeder having sufficient capacity for the entire load and normal overload. Dual feeders are generally each run to the site by diverse routes and enter the compressor plant at separate locations. This duplication of facilities is carried through in the plant by the use of separate bus-bar connections and automatic switch-over of power in case of failure. Diesel standby engines are often provided where feeder lines traverse considerable distances or precarious terrains.

UNDERGROUND COMMUNICATION BASED ON NEW YORK STATE INDUSTRIAL CODE

RULE NO. 22—OCTOBER 15, 1960

"A telephone intercommunication system ready for use at all times shall be maintained between the working chamber, the power house, the source of compressed air, the place of compressed air control, the first aid room and the superintendent's office."

IN ADDITION

"Effective and reliable signalling devices shall be maintained at all times for communication between the bottom and top of hoisting shafts. Every person transmitting or receiving signals shall be able to speak and read the English language. Signal codes shall be kept posted in a conspicuous place at the entrance to shafts, in the working chamber and in the change house so as to be readily seen by persons affected thereby."

BELL SIGNALS FOR HOISTING

1 Bell—Stop if in motion or hoist if not in motion
2 Bells—Lower
3 Bells—Run slowly and carefully

NOTE

The above refers to all hoisting operations whether in tunnel shafts, caisson shafts, or in shafts operated above ground. The use of electrically operated bells is assumed. An additional signalling system employing manual whistle blasts or hammer blows is necessary for use in caissons and can be used in tunnel headings to transmit information into or out of locks, especially in emergencies due to current interruption. In the case of "blows," slides, cave-ins, or other catastrophic events, hammer blows may be the only means of communication between trapped men and the "surface." Although of primary importance in caisson construction, it has an equal use in tunnel headings.

SIGNAL CODE

"The following code shall be used to transmit whistle signals from the working chamber in caissons. Where hammer blows are used to transmit signals to or from the working chamber the same code shall be used, substituting a single rap of the hammer for a whistle blast. The word 'rattle' as used in the following code means a rapid repetition of hammer raps or whistle blasts."

1 whistle .... Hoist
Rattle and 1 whistle .... Hoist slowly
2 whistles .... Stop hoisting
3 whistles .... Lower
4 whistles and 1 whistle .... Open high pressure air
4 whistles and 2 whistles .... Shut off high pressure air
5 whistles .... Bucket is out; lock in
5 whistles and 1 whistle .... Increase low pressure air
5 whistles and 2 whistles .... Lower low pressure air
6 whistles .... Lights are out
7 whistles .... Lights are OK
Rattle and 4 whistles and 1 whistle .... Open blow pipe
Rattle and 4 whistles and 2 whistles .... Shut blow pipe
2 whistles and 2 whistles and 1 whistle .... Call foreman
4 whistles .... Drop caisson
Rattle and 5 whistles .... Call gang out
6 whistles and 2 whistles .... Turn off lights
8 whistles .... "DANGER!" (Get out of working chamber)

"In all cases except 'DANGER!' a reply signal, repeating the original signal, shall be made before proceeding. Additional signals or modifications of the above code to meet local conditions may be adopted."

2.7 DIVING

Many of the construction operations to be discussed here are either wholly or intermittently performed under water and require the use of divers. Although diving gear and techniques have improved rapidly since World War II and certain types of diving have become extremely popular as a sport, the contractor will use regular commercial divers, trained and certified by the United States Navy Diving School. This brief discussion is intended merely to indicate the possibilities and the limitations.

Construction diving will generally be shallow-water diving, that is, up to a maximum depth of 130 ft. This is the limit of diving with compressed air; and though, of course, greater depths are possible with other techniques, they will seldom involve the contractor. There are two types of shallow-water diving, "scuba" diving and diving with lifeline and air hose.

Scuba (Self-Contained Underwater Breathing Apparatus) diving is done with compressed-air flasks fastened to the diver's back, a mouthpiece breathing device, and an air-pressure regulator at the air supply, connected by flexible rubber hose to the mouthpiece. In warm climates the balance of the outfit may include only a swimming suit, canvas or rubber shoes, and a double-edged knife worn sheathed in a belt. In addition, a quick-release weighted belt may be worn. In colder waters or where submersion will be of extended duration, an insulated or non-insulated rubber suit designed to retain body heat will be worn.

The scuba diver is used, if at all, for underwater reconnaissance and for minor repair work—any function that requires mobility rather than localized, sustained operations. For more extensive operations, shallow-water diving should be done with a lifeline and air hose.

Diving with a lifeline and air hose involves the use of a face mask, generally the so-called "hard-hat" type with usually a rubber suit and additional weighting. The lifeline is secured around the diver's waist by using a spring-clip connection. The air hose is fastened to the lifeline at 6-ft intervals so that the line takes any strain on the air hose. The air hose is coupled to an air-control valve on the diver's mask and leads up to a filter on the pressure tank topside.

The scope of operations performed by the diver using lifeline and air hose is extended to include not only underwater reconnaissance but limited repair work on piers and pilings, underwater cutting with either an oxyhydrogen or an arc-oxygen torch, or placing explosive charges.

The advantage of a surface supply of air is that greater diving time is available. The average scuba flask holds 70 cu ft of air, sufficient for about 70 min of work near the surface. At a depth of 33 ft only 35 min of working time is available, and at a depth of 66 ft the available diving time is down to 23 min.

From a diving standpoint alone, with a continuous air supply (lifeline and hose) 200 min is available at 33 ft and 50 min at 66 ft. Since part of this time is used in descending and ascending, working time at the bottom can be quite limited even here.

For more extended operations, or where strong underwater currents exist, the deep-sea or suited-helmet diver should be used. This gear consists of a helmet (hard hat also) sealed to an inflatable suit and includes the weighted belt as well as weighted shoes. The complete gear weighs 196 lb, sufficient to hold the diver in position at the bottom for the length of time considered safe on other scores. His movements and buoyancy can be controlled by the degree of inflation of his suit.

Deep-sea diving gear is necessary where air tools such as rock drills and paving breakers are to be used or where underwater welding is required. The principal disadvantage is the time required to prepare the diver both before and after the dive. For lengthy, relatively deep, dives additional decompression time on the ascent will be required.

Other diving methods now more or less in the experimental stage, such as helium-oxygen diving, may extend the depth to which dives are possible, but at the present time are seldom used on construction work.

U.S. Navy Standard Air Decompression Table

Depth, ft	Bottom time, min	To first stop, min	Min at decompression stops				Total ascent time, min
			40 ft	30 ft	20 ft	10 ft	
40	200					0	0.7
	210	0.5				2	2.5
	230	0.5				7	7.5
	250	0.5				11	11.5
	270	0.5				15	15.5
	300	0.5				19	19.5
50	100					0	0.8
	110	0.7				3	3.7
	120	0.7				5	5.7
	140	0.7				10	10.7
	160	0.7				21	21.7
	180	0.7				29	29.7
	200	0.7				35	35.7
	220	0.7				40	40.7
60	60					0	1.0
	70	0.8				2	2.8
	80	0.8				7	7.8
	100	0.8				14	14.8
	120	0.8				26	26.8
	140	0.8				39	39.8
	160	0.8				48	48.8
	180	0.8				56	56.8
70	50					0	1.2
	60	1.0				8	9.0
	70	1.0				14	15.0
	80	1.0				18	19.0
	90	1.0				23	24.0
	100	1.0				33	34.0
	110	0.8			2	41	43.8
	120	0.8			4	47	51.8
	130	0.8			6	52	58.8
	140	0.8			8	56	64.8
	150	0.8			9	61	70.8
80	40					0	1.3
	50	1.2				10	11.2
	60	1.2				17	18.2
	70	1.2				23	24.2
	80	1.0			2	31	34.0
	90	1.0			7	39	47.0
	100	1.0			11	46	58.0
	110	1.0			13	53	67.0
	120	1.0			17	56	74.0

Depth, ft	Bottom time, min	To first stop, min	Min at decompression stops				Total ascent time, min
			40 ft	30 ft	20 ft	10 ft	
90	30					0	1.5
	40	1.3				7	8.3
	50	1.3				18	19.3
	60	1.3				25	26.3
	70	1.2			7	30	38.2
	80	1.2			13	40	54.2
	90	1.2			18	48	67.2
	100	1.2			21	54	76.2
	110	1.2			24	61	86.2
	120	1.2			32	68	101.2
100	30	1.5				3	4.5
	40	1.5				15	16.5
	50	1.3			2	24	27.3
	60	1.3			9	28	38.3
	70	1.3			17	39	57.3
	80	1.3			23	48	72.3
	90	1.2		3	23	57	84.2
	100	1.2		7	23	66	97.2
110	20					0	1.8
	30	1.7				7	8.7
	40	1.5			2	21	24.5
	50	1.5			8	26	35.5
	60	1.5			18	36	55.5
	70	1.3		1	23	48	73.3
	80	1.3		7	23	57	88.3
	90	1.3		12	30	64	107.3
120	20	1.8				2	3.8
	30	1.8				14	15.8
	40	1.7			5	25	31.7
	50	1.7			15	31	47.7
	60	1.5		2	22	45	70.5
	70	1.5		9	23	55	88.5
	80	1.5		15	27	63	106.5
	90	1.5		19	37	74	131.5
130	10					0	2.2
	20					4	6.0
	30				3	18	22.8
	40				10	25	36.8
	50			3	21	37	62.7
	60			9	23	52	85.7
	70			16	24	61	102.7
	80		3	19	35	72	130.5

Table shows times required for dives of varying depth. *Bottom time* is time between leaving surface in descent and leaving bottom in ascent. All times shown are in minutes. *Decompression stops* are in feet below water surface. Rate of ascent = 60 fpm.

Example: A diver working for 90 min at a depth of 100 ft will be raised at 60 fpm to 30 ft below the surface and held there for 3 min; then raised to 20 ft and held for 23 min; then raised 10 ft and held for 57 min before being surfaced.

The decompression time indicated is safe, but does not, in fact remove *all* the nitrogen from the blood stream. For repetitive dives, residual nitrogen time as well as a surface interval must be allowed. See U.S. Navy Diving Manual, Navships 250–538 for more information.

2.8 UNDERWATER CUTTING AND WELDING

Underwater cutting and welding should be done only by an experienced underwater diver who has been trained in this technique in a United States Navy school, since the field has been opened up largely in salvage operations. However, the contractor should have some knowledge of what is involved in the process.

The familiar oxy-acetylene torch is not suitable for underwater cutting. At pressures over 15 psi the acetylene becomes unstable and tends to break down into its carbon and hydrogen components, resulting in an explosive mixture. Although it has been used to depths of 20 ft, it is safer to depend on other methods. Those most commonly used are the oxyhydrogen and the arc-oxygen methods.

The oxyhydrogen method can be used for cutting any thickness of metal at any depth. Its operation is based on the combustion of 2 parts of oxygen and 1 part of hydrogen. One part of the required oxygen is delivered to the torch under proper control from the oxygen cylinder. The other part of oxygen comes from compressed air which is used to surround the flame at the end of the tip. The discharge end of the air nozzle must be designed to supply the compressed air without creating turbulence and without contaminating the purity of the oxygen delivered through the central orifice of the tip. Turbulence will interfere with pre-heating, while the contamination of the oxygen will reduce the cutting speed and may even make cutting impossible.

The first practical oxyhydrogen torch was developed by Capt. Edward Ellsberg, United States Navy, for use in raising the sunken submarine S-51 in 1926; and the torches now available from a number of sources are based on his original design. Controls consist of compressed air, hydrogen, and cutting oxygen valves as well as an oxygen preheating valve and an oxygen cutting lever. The tip has an adjustable air nozzle surrounding it.

An underwater torch is subject to more rigorous working conditions than one used on the surface, and must be constructed so as to permit easy and direct access to all valve seats.

Underwater gas torches are generally ignited on the surface. The hydrogen valve is opened one-half to two-thirds of a turn and ignited; then the oxygen needle valve is opened and the flame adjusted. The compressed-air valve is now opened to a predetermined setting based on depth of use (see table) and passed down to the diver, who makes the final adjustments. The compressed air should provide a bubble 3 in. long for proper operation.

The oxyhydrogen torch can be ignited under water, but the process is time-consuming and is resorted to only where it is impractical to pass the torch down.

The arc-oxygen method may be used for both cutting and welding under water and is similar to electric welding on the surface except that, for cutting, the electrode is tubular and hollow. The method combines the use of the electric arc as a source of heat with pure oxygen under pressure as a means of rapidly oxidizing the molten metal.

Although several other types have been used, tubular steel electrodes are now generally preferred, since they permit the changing of electrodes under water. These electrodes are coated and waterproofed and are designed for a special insulated torch with an oxygen-valve lever.

The welding machine used should be a 300-amp d-c machine designed for tubular steel electrodes. (Other types of electrodes such as ceramic or cored carbon electrodes will require higher capacities.)

For welding operations, a special type of electrode holder is required that is completely insulated. Standard electrodes 5/32 or 3/16 in. in diameter are used after dipping them in a waterproof solution such as Ucilon or Selac or a cellulose-acetone solution.

Direct-current requirements for underwater welding are about 30% in excess of those used for surface welding, and the strength developed by underwater welds is only about 50% of that developed on the surface; consequently the method should be used only in emergencies.

Methods of Cutting and Welding Under Water

GENERAL PRECAUTIONS

1. Only a qualified diver assisted by an experienced tender and a trained torchman should use underwater equipment.
2. Before starting any cutting, be sure there are no combustible gases, liquids, or solids adjacent to the point of cutting or within a radius of 30 ft, particularly overhead.
3. The diving gear should be in good condition and should be equipped with a loudspeaker telephone designed for uninterrupted two-way communication. A trained torch attendant should be on duty continuously on the surface at the oxygen and gas regulators.
4. No work of any kind should ever be permitted on the surface over the space in which the diver is working or within a radius less than the depth of operations.

SPECIAL PRECAUTIONS—OXYHYDROGEN METHOD

1. Because of the difficult footing and poor visibility under water, the diver should handle the torch with care, staying completely clear of the torch hose and avoiding excessive slack in the lines.
2. The oxygen regulators used should be adequate in size for the delivery of the needed volume to avoid "freezing."
3. The orifices in the cutting tips should be kept clean, but care should be used in cleaning them so as not to distort or enlarge them.
4. At the end of the work the torch should be cleaned thoroughly and dried, inside and out, including the oxygen high-pressure valve, seat, and spring.

OXYHYDROGEN SUPPLY REQUIREMENTS

Thickness of metal → Size of tip →		Up to 1 in. No. 1	Up to 2 in. No. 2	Up to 3 in. No. 3
Depth below water surface, ft	Air pressure, psi	Oxygen pressure required, psi		
10	20	50	60	70
20	25	55	65	75
30	30	59	69	79
40	35	63	75	84
50	42	66	80	90
60	48	71	84	94
70	55	76	87	98
80	62	82	95	104
90	70	85	98	108
100	75	89	103	115
Hydrogen pressure required, psi		20	25	30

SPECIAL PRECAUTIONS—ARC-OXYGEN METHOD

1. The cable and torch insulation and all joints in the circuit should be carefully checked for current leakage at frequent intervals.
2. The diver should not permit any part of his body or gear to become a part of the electric circuit.
3. While a-c current may be used, if the diver's body or gear inadvertently "enters the circuit" the resulting "shock" is more pronounced than with d-c current.
4. The diver should inspect his helmet and other metallic parts of his gear regularly for deterioration resulting from electrolysis and place his ground connection at points, with reference to his position while cutting, that will reduce electrolysis to a minimum.
5. Always wear diver's rubber gloves or mittens, telephone the surface to shut off current before changing electrodes, and keep it shut off except when actually cutting. Be sure to close the hinged frame holding the welding lens on the outside of the helmet before striking the arc, and remove the electrode before taking the torch under water or returning it to the surface.

2.9 UNDERWATER BLASTING

The contractor engaged in subsurface construction will occasionally find it necessary to blast under water either for demolition purposes or to remove rock or other compacted materials, as, for instance, in the sinking of open caissons. Here again is a method that can be executed only by experienced divers under the direction of an expert in this type of blasting. Most of the explosive manufacturers can furnish such experts to direct operations. Here only a general discussion of the operations involved can be undertaken.

The drilling and blasting discussed here might almost be called incidental, that is, the removal of incidental masses of rock or debris which may appear in sinking a caisson or in driving piling.

Two factors make underwater blasting more difficult than ordinary surface blasting; one of these sums up all the difficulties of working under water, and the other is the weight or pressure of the water itself. At no very great depth the water acts like a giant mudcap, and blasting can be done by using the water as a surface restraint. Since this is not an economical procedure, it is seldom used.

Because of the pressure, holes must be drilled shallower and closer together than for surface blasting. Since we are dealing here with blasting of isolated or limited masses, spacing has little meaning since it will be determined by the conditions. Boreholes will range in size from 2½ to 4½ in. in diameter, and the size of dynamite stick used will be appropriate to the size of the hole.

For general use a 60% gelatin dynamite will give satisfactory results; however, as depths increase, this should be increased to a 70% or even an 80% gelatin. For shattering compacted soils, shales, or weathered rocks, strengths as low as 40 or 50% are practical.

The quantity of dynamite used will be greater than that for surface work by two or three times, depending, of course, on depths and conditions. Whereas in soft shales at shallow depths effective blasting can be done with 1 lb of dynamite per cubic yard, this will increase with depth and conditions up to 5 lb or more per cubic yard. On the surface, of course, 2 lb per cu yd is a very high rate for dynamite. Dynamite for underwater blasting should be packed in special heavy paper shells known as "submarine packing."

Special water-resistant blasting caps are available and, where any depth of water is involved, should be used two to a hole. All wire splices should be carefully taped to prevent bare wire from touching not only the water but any exposed metal topside.

Where general blasting of an area is contemplated, loading of boreholes should be done through a loading tube, a sheet-metal shell of proper diameter to slide easily to the bottom of the borehole. The tube is loaded topside and lowered into the borehole, where the plunger pushes the charge out as the shell is raised.

In the blasting under consideration here, where direct access to the borehole from the surface is impractical, as, for instance, near the cutting edge of an open caisson, the charge can be loaded into a sheet-metal tube which is then lowered to the bottom and dropped into the borehole intact. The length of this tube should be only the length of the charge, and a new tube would be required, of course, for each blast.

The difficulties of loading a borehole under water are largely due to difficulties in handling the dynamite, as well as difficulty in seeing what is being done due to fines clouding the water. Although the weight of dynamite will vary with its density, generally speaking it is only about one-third heavier than water, and handling it under water may be difficult. With a heavy paper wrapping it may not displace much more than the weight of the water.

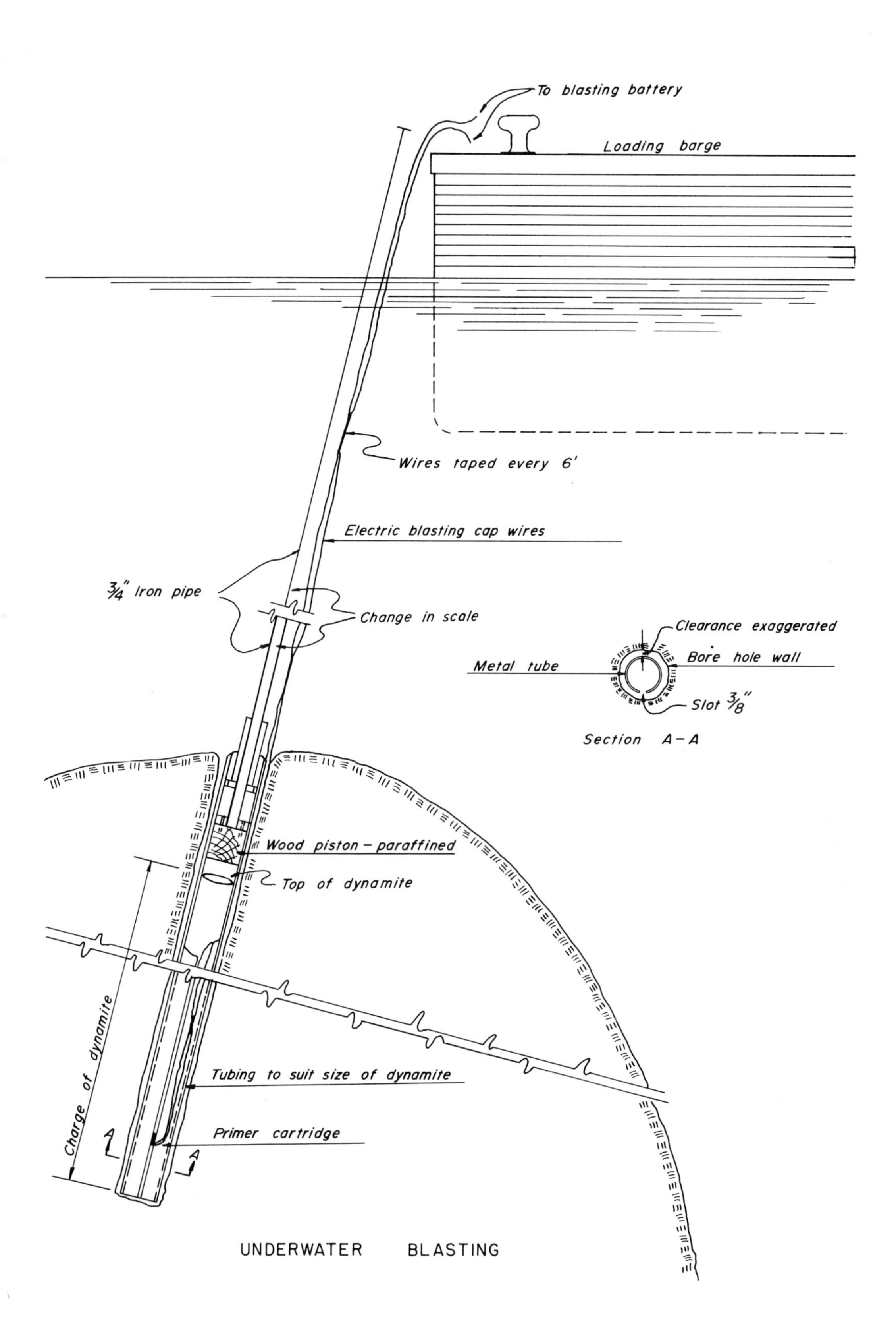

UNDERWATER BLASTING

2.10 DREDGING

Dredging, the process of excavating under water, by a variety of means, can be a very useful technique for the subsurface contractor and, in some cases, an indispensable one.

In ancient times cities often developed where the mouth of a river, emptying into the sea, provided harbor facilities and, in turn, disappeared when floodwaters dumped silt into the shipping lanes and made them impassable. The first crude dredging devices were employed in the maintenance of harbors and the deepening of shipping channels.

The earliest types of dredges—the dipper, grapple and bucket, or ladder, dredges employing the principles respectively of the power shovel, the clamshell, and the trencher—are still in use, particularly in mining operations and in the processing of sands and gravels for concrete aggregates. However, much dredging today employs the hydraulic principle by which the soils, mixed with water, are pumped from the bottom of the waterway.

In the early 1800s the Corps of Engineers, United States Army, was charged with surveillance and maintenance of the navigable waters of the United States. One of the first projects involved the removal of sandbars from the upper portions of the Mississippi River. The process has been continuous, not only on inland rivers but in the harbors and river mouths of the coastal areas, not only for removing deposits laid down by the rivers but for deepening channels to meet the increasing size of shipping. The original 6-ft channel in the Mississippi was only fully converted to a 9-ft channel in the 1950s. Currently, only one-third of the main channel depths in the harbors of the United States will accommodate fully loaded vessels larger than 38,000 d.w.t., and some super tankers now under construction run from 40,000 to 130,000 d.w.t.

Hydraulic dredging was begun in the United States in the years immediately following the Civil War; and although its principal use has been in maintenance dredging, one of the earliest units was developed by Eads for use in connection with the caissons of his Mississippi River Bridge. Although the Corps of Engineers continues to be the single largest owner and operator of dredging equipment, many privately owned dredging companies are in operation in the United States.

The most familiar type of dredge is that which conveys its dredged material through a pipeline laid on floats over the water to shore, and thence overland by continuing pipeline to the point of discharge. In cases where pipelines are not practical, the same dredge may discharge into barges from which the water is drained off, and the finally loaded barge is then towed to a dumping site. Another type of dredge is the hopper dredge, provided with integral storage hoppers filled by dredging, in which the dredge itself transports its load to the disposal area.

All hydraulic dredges employ the suction principle, drawing the mixture of water and soil up through centrifugal pumps for discharge, but there are a number of suction heads used. Hopper dredges employ a screened drag head which is drawn over the underwater surface. Dustpan dredges have suction heads that are pushed over the underwater ground much as a dustpan would be. About 8 in. high, the dustpan may be from 20 to 40 ft long and is supplied with jets along its face to stir up the bottom surface.

Both the hopper dredge and the dustpan dredge are principally used in maintenance and in many cases participate only in what is known as agitation dredging. This consists in pumping the discharge directly into the sea and using the tide to carry the fines to deeper-water areas. Agitation dredging is employed only during ebb tide in tidal estuaries having swift tidal flows that will disperse the accumulations of silt. It is of little interest to the subsurface contractor.

The cutterhead dredge is provided with a rotating cutter ahead of the suction pipe to cut into the waterway bottom and reduce the soil to transportable size. Most dredges used in land reclamation are cutterhead dredges since any type of soil or soft rocks can thus be handled.

Dredging Equipment

As typical of the wide range of types and capacities of dredging equipment in a limited area, the following table of Available Major Dredging Equipment in the Great Lakes Area is given. It was compiled in 1957 by the Detroit office of the U.S. Corps of Engineers as part of a program of planning for the deepening of 168 miles of channels to a depth of 27 ft required by the completion of the St. Lawrence Seaway Project in 1959 and the connecting channels in 1962.

Note: This is a list of *major* equipment and does not include the many smaller dredges available, which would more than double the list and increase the volume. The Great Lakes represent only a small portion of the waterways requiring dredging within and surrounding the United States.

HYDRAULIC DREDGES (16 IN. OR OVER)

Dredge	Owner	Size, in.	Horsepower	Cu yd per month
Illinois	Great Lakes Dredge and Dock Co.	30	3500	250,000
Niagara	Duluth–Superior	22	2400	225,000
Three Brothers	Price Bros.–McClung	18	1800	200,000
Alice Loraine	Control State Dredging	18	1800	150,000
Ohio	Great Lakes Dredge and Dock Co.	16	500	130,000

DIPPER AND DRAGLINE DREDGES (4 CU YD OR OVER)

Dredge	Owner	Dipper, cu yd	Cu yd rock per month	Cu yd other per month
Sampson	Marine Operators (Dragline)	12	90,000	160,000
Mogul	Great Lakes Dredge and Dock Co.	12	80,000	150,000
Old Hickory	Duluth–Superior	11	60,000	102,000
No. 9	Great Lakes Dredge and Dock Co.	10	60,000	120,000
Hellgate	Merritt–Chapman–Scott	10	60,000	120,000
M. Sullivan	Merritt–Chapman–Scott	9	60,000	120,000
No. 7	Great Lakes Dredge and Dock Co.	8	55,000	100,000
War Horse	Great Lakes Dredge and Dock Co.	8	55,000	100,000
Chicago	Great Lakes Dredge and Dock Co.	8	70,000	120,000
No. 6	Fitzsimmons & Connell	8	55,000	85,000
Omadon	Dunbar–Sullivan	7	50,000	72,000
Buffalo	Great Lakes Dredge and Dock Co.	5.5	40,000	60,000
No. 27	Zenith Dredging	5	35,000	50,000
Tipperary Bay	Dunbar–Sullivan	4	35,000	50,000
Ring Coal	L. A. Wells	4	35,000	50,000

CLAMSHELL AND BUCKET DREDGES

Dredge	Owner	Bucket, cu yd	Cu yd per month
Nos. 53, 55, 56	Great Lakes Dredge and Dock Co.	5–12 each	64,000 each
Gotham	Merritt–Chapman–Scott	6	55,000
Four Spot	Merritt–Chapman–Scott	4½	50,000
Ojibway	Fitzsimmons & Connell	4	50,000
Handy Andy	Dunbar–Sullivan	4	50,000
Wauhassee	L. A. Wells	4	50,000
Wellston	L. A. Wells	4	50,000
Five Spot	L. A. Wells	4	50,000

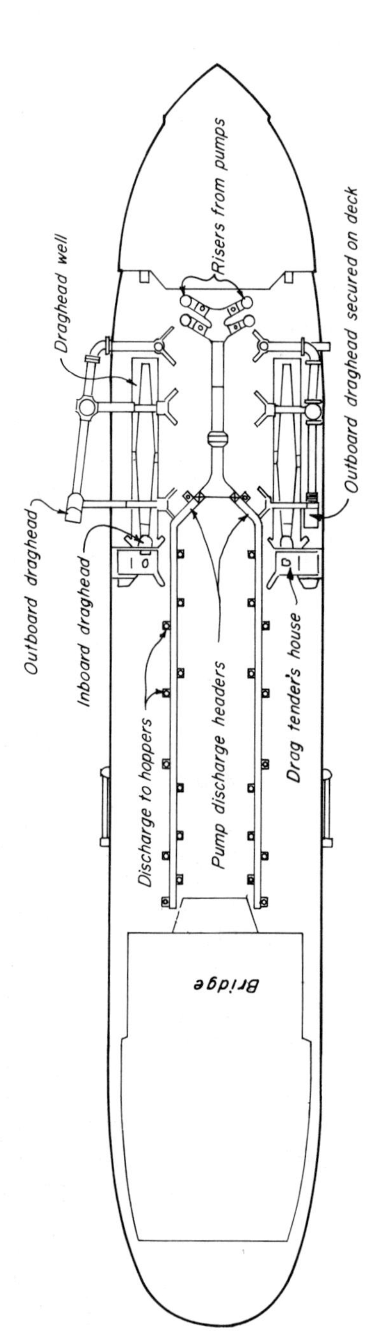

TOP DECK - HOPPER DREDGE

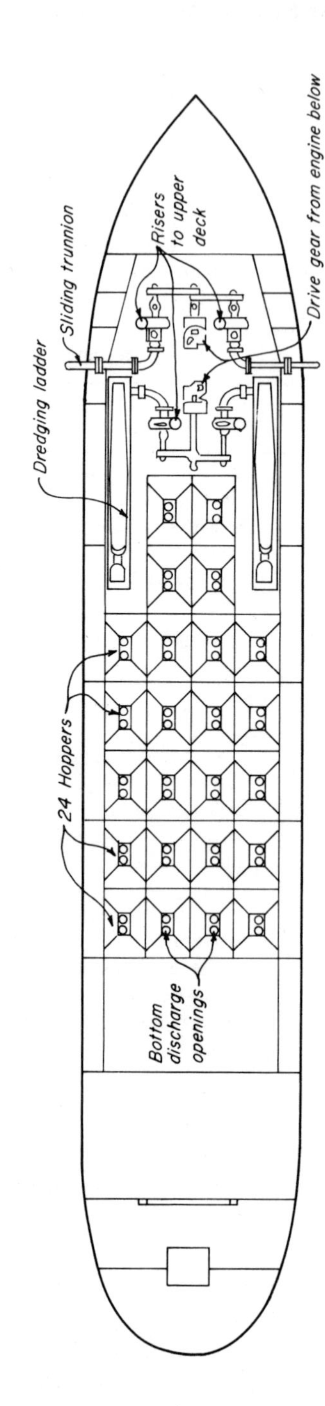

LOWER DECK - HOPPER DREDGE

2.11 HOPPER DREDGING

The hopper dredge is a seagoing ship designed for dredging but additionally provided with hoppers into which the dredged material is discharged. It is used when dredging must be done in open harbors where a line of floating pipe would interfere with harbor traffic; in bodies of water with swift-moving currents where floating pipe would require anchorage; where there is no fill area within economic discharge range of the dredging area. Loading dredged material onto barges is also generally impractical under these circumstances.

The techniques and the vessels used in hopper dredging have been developed largely by the Hopper Dredge Board of the United States Corps of Engineers, whose fleet of hopper dredges handles some 60 million cu yd of soils annually at a cost of under 30 cents per yd. There are also a number of privately owned hopper dredges, of similar design, available for use by the general contractor.

Hopper dredges are generally over 300 ft in length and have 60-ft beams. They are distinguishable from cargo ships chiefly by the presence of hanging gear supporting the suction drags to port and starboard.

Suction lines 20 to 30 in. in diameter are raised and lowered on swivel joints by dual systems of block and tackle. The lower end is fitted with a drag head resembling a vacuum-cleaner intake tilted on its long edge. The drag head is about 6 ft to 8 ft long and about 18 in. high. The drag-head opening is screened to limit the size of object that can enter. The drag head is equipped with scarifier bars, but these reduce dredging speed and are seldom used. When in operation, the drag heads rest on the underwater bottom and are dragged over it by the forward motion of the vessel.

Drag tenders, housed just above the drags, control the density and character of the dredged material. Formerly this was done by observation of a small pilot stream coming from the discharge of the pumps, but nuclear density gages have replaced the unaided eye. The SOFRIAC (Solids Flow Rate Automatic Indicating Computer) uses a radiation device to automatically indicate and record the density. With this is used the LIMAR (Light Mixture Automatic Rejection) to automatically cast unsuitable mixtures back into the sea, by the operation of motorized cone valves. Drag tenders, however, are still necessary to raise and lower the drag heads.

The hoppers, in multiple units, have a total capacity of between 2500 and 3000 cu yd when full. Dredged material is run into them at a rate which will permit the solids to settle and the clearer water to run off their surface. In practice, they are seldom filled since the detention time of the entering mixture is gradually reduced as the soil accumulates.

Hopper-dredging cycle time consists of the dredging time, the running time to and from the disposal area, and the dumping time. Generally the dredging time is the biggest variable. This is controlled by taking soundings in the hoppers more or less continuously to determine yardage accumulation. The total cycle time is then computed and divided into the stored yardage to determine the rate of accumulation in cubic yards per minute. When the rate of filling starts to decrease, the economic load will have been reached and further dredging is stopped.

In channel maintenance work the dredged material is usually transported to underwater disposal areas offshore. For this the hoppers are provided with bottom dump doors opened and closed (on the latest models) by geared mechanisms. However, for other types of service it may be necessary or desirable to discharge the load on shore. For this purpose a suction line is provided in each hopper together with a jetting system. The jets break up the soil mass and provide the water necessary for re-pumping. With this method of discharge, mixes containing 30% solids can be obtained. (Initial dredging provides mixtures containing between 5 and 15% solids, depending on underwater soil conditions.)

2.12 HOPPER-DREDGE USE

Privately owned hopper dredges are now employed in procuring offshore deposits of sand and gravel for fine and coarse concrete aggregates. The subsurface contractor also will frequently require sand fill for cofferdams and sand islands incidental to caisson construction and for other similar purposes. Hopper dredging to provide fill can be a fairly expensive method, and other means or other types of dredging may be more economical. Hopper dredges, because of their size and complexity, carry crews of from 30 to 50 men. The first cost of such dredges will run between 3 and 4.5 million dollars. A case where this method was considered to be the least expensive is illustrated.

The northern approach to the Throgs Neck Bridge, completed in 1961, connecting the boroughs of Bronx and Queens of the City of New York, needed 1 million cu yd of fill. Some of this material was required for raising the grade of the approach area, some of it to replace areas of swampland muck as much as 4 ft deep. Suitable materials to provide the specified compaction were not available within any reasonable hauling distance. Moreover, hauling over the intricate pattern of streets in the Bronx would be time-consuming and therefore costly.

The offshore material in areas adjacent to the site consisted generally of heavy deposits of organic silt wholly unsuitable for fill. The solution was to bring in suitable sands and gravels from some other point where they were available to hopper dredging.

The hopper dredge used here was a vessel regularly employed in obtaining construction aggregates from sand and gravel deposits in the Atlantic Ocean just outside New York Harbor off the tip of Long Island. Here was a vast shoal area of suitable material known as East Bank, much of it in water less than 10 ft deep and continuously restocked by the tides and the river. Adjacent to Ambrose Channel, a dredged entrance to New York Harbor, removal operations would always be beneficial, since maintenance dredging was continuously required on the channel in any case.

The route to Throgs Neck, by water, lay through lower and upper New York Bay and thence up the East River to its junction with Long Island Sound, a distance of 30 miles.

No docking facilities for the seagoing hopper dredge were available near the site, and to provide these a 90-ft scow and a car float were moored to three 30-pile dolphins, driven for this purpose. The hopper dredge, similar to those described in Sec. 2.11, had insufficient power to push the heavily loaded mixture the full distance to the point of discharge, which began a mile from the dredge and extended over a mile and a half of roadbed. An auxiliary dredge, a barge-mounted cutterhead (see Sec. 2.14), was therefore anchored between the hopper dredge and the shoreline. A short floating discharge line was run from the hopper dredge to the cutterhead dredge, whose 1600-hp pumps picked it up and pushed it to the discharge point ashore.

Initial discharge of the 20-in. line began about 5000 ft from the shore and was gradually extended to a final point of discharge 12,000 ft away. A portion of the line was buried under streets in order to prevent interference with traffic along the route.

The nominal capacity of the hopper dredge was 3000 cu yd, but as noted in Sec. 2.11, this capacity was seldom obtained. With about 2500 cu yd per trip, loaded in 1½ hr, with a 2½-hr trip upstream and a 2-hr trip downstream, 6 hr would be consumed. Unloading time, including docking, hook-up, and cast-off, would not be less than 1 hr. On this basis not more than two loads could be handled per day. With a total of 1,500,000 gal with a 30% density to be handled per load, this would be equivalent to 100,000 to 120,000 yd per month.

Sand and gravel placed thus provide the 95% compaction required. Although delivered costs may run $1 per yd and the piping and drainage facilities (see Sec. 2.15) may add another 50 cents per yd, this still represents a lower figure than overland hauling and compaction costs in many instances.

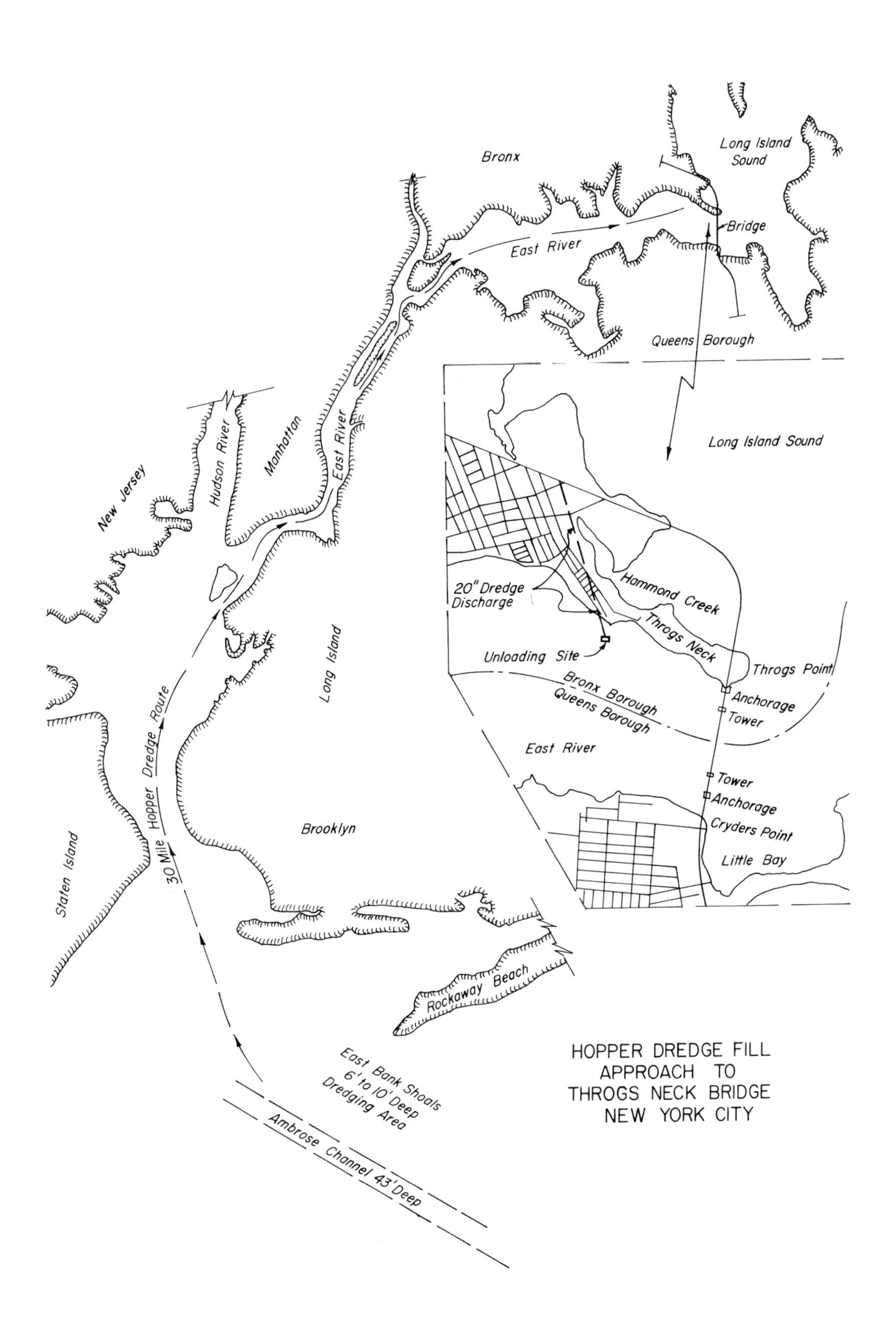

HOPPER DREDGE FILL
APPROACH TO
THROGS NECK BRIDGE
NEW YORK CITY

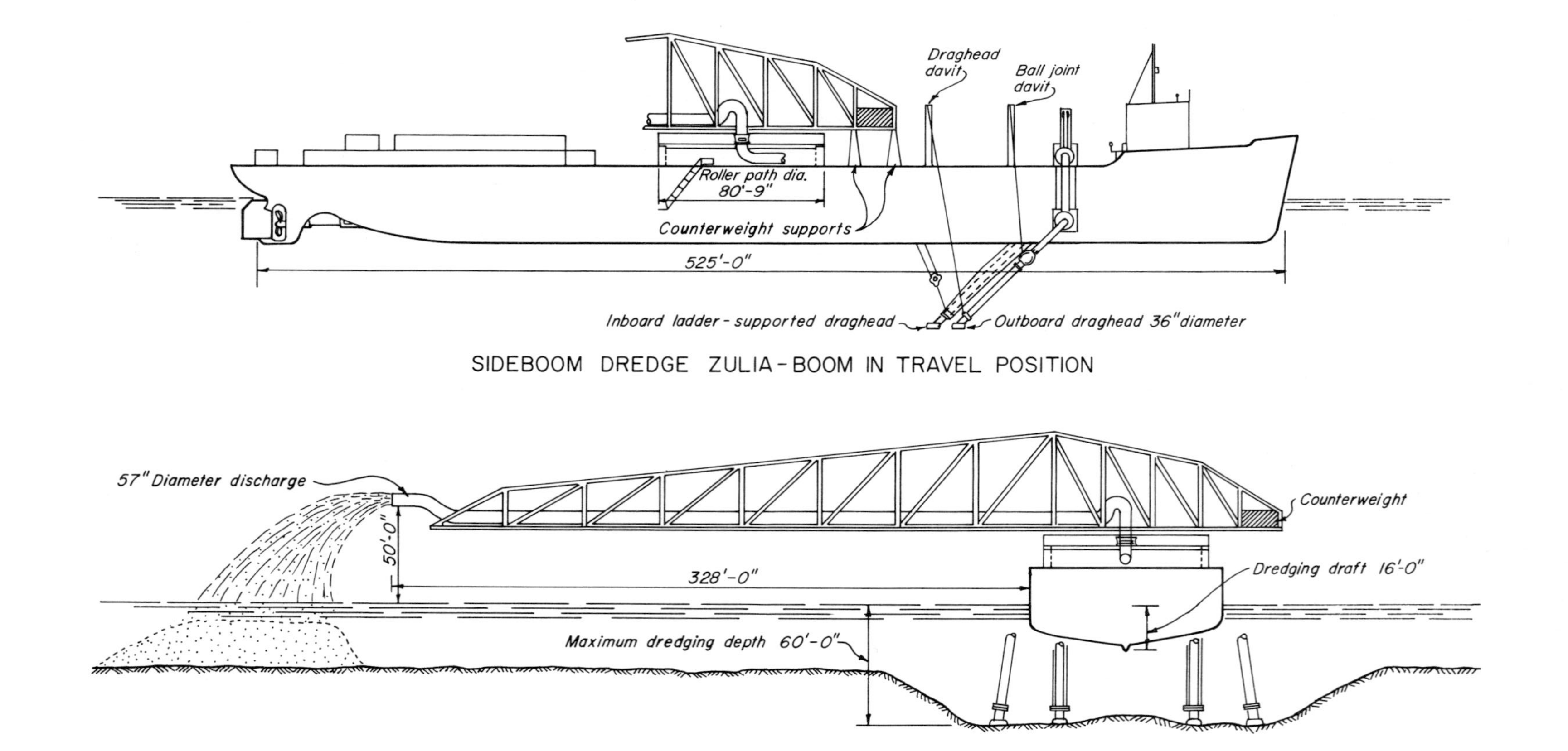

SIDEBOOM DREDGE ZULIA-BOOM IN TRAVEL POSITION

CUTTING A CHANNEL WITH SIDEBOOM DREDGE

2.13 SIDE-BOOM DREDGING

Side-boom dredging is a very recent development born of necessity and principally applicable, at present, to maintenance dredging. A discussion of it is included here since it would appear to be extremely valuable as a technique in preparing waterway bottoms for pre-fabricated sub-aqueous tunnels or open caissons, for the reconstitution of eroded ocean beach fronts, and possibly for some other construction uses for which the hopper dredge is currently the only other solution.

The side-boom dredge is essentially similar to the hopper dredge except that the discharge, instead of going into hoppers or directly back into the sea, is carried in a discharge pipe hung from a boom, a distance of from 200 to 500 ft directly to port or starboard of the vessel, and there discharged into the atmosphere, dropping vertically from a height of about 50 ft onto the surface of the sea. The drag heads of the dredge provide a channel, and the excavated soil is spread over a wide shoal area on either side, without the necessity of hauling it to sea.

The necessity out of which the side-boom dredge was born was that of moving 60,000-ton ore vessels coming down the Orinoco River in Venezuela across the 25-mile shoal area, only 12 ft deep in many places, at the river mouth known as the Boca Grande. This age-old deposit of mud and silt required constant dredging to maintain any sort of suitable channel for the fully loaded ore ships, which, with hopper dredges, would have been enormously expensive.

The first side-boom dredge was a converted T-2 tanker fitted out similarly to a hopper dredge except that no hoppers were provided. The boom was a trussed section fixed in position forward on the starboard side, hung from a second boom and gantry placed athwart the top deck. The boom, carrying two 25-in. rubber-lined discharge pipes, extended 250 ft from the side of the ship, with the pipes themselves forming the upper and lower chords of the boom. The vessel was equipped with two 32-in. dredge pumps, each discharging through one of the dual pipelines.

The results achieved by this first side-boom dredge (named the *Sealane*) were so spectacular that a specially designed side-boom dredge named the *Zulia* (after a state in Venezuela) was commissioned and placed in operation in the Orinoco in 1960. In this instance, the boom was of tubular steel construction, 425 ft in overall length, mounted on an 80-ft-diameter trunnion so that it was revolving, and provided with 1000 tons of counterweighting. This boom carried a 57-in.-diam discharge pipe 50 ft above water level.

Hoppers were built into the *Zulia,* and four 32-in. dredge pumps were manifolded to provide hopper loading, unwatering, and jetting, as well as boom discharge.

The vessel, 525 ft long and with a draft of 26 ft 6 in., is manned by a crew of 96 men and can be said to be the world's largest earth-moving machine. In its first year of operation it removed 57 million cu yd of soils at a cost of about 15 cents per cu yd. This compares to the 60,500,000 cu yd handled by the 15-unit hopper-dredge fleet of the Corps of Engineers, United States Army, at a cost of 28 cents per cu yd.

Average pumped discharge of the *Zulia* is about 200,000 gpm at a velocity of about 25 fps. The quantity of solids pumped is determined by a SOFRIAC installation (see Sec. 2.11), and velocity in the length of the boom is checked by pneumatically injecting a dye capsule into the shipboard end of the boom and timing its travel to the exit.

The use of side-boom dredges, if and when they become available in more limited sizes, for projects such as cutting a channel for a sub-aqueous tunnel, will depend on the depth required for such a tunnel below underwater ground level and the drag depth available, as well as the initial depth of water in relation to the draft of the dredging vessel. It will also depend on the density of the waterway bottom, since drag heads have little capacity for loosening compacted soils; and unless an optimum density of solids (1150 g per liter) can be maintained, the efficiency of such dredging is no improvement over other methods.

2.14 CUTTERHEAD DREDGES

Cutterhead pipeline dredges are the type with which the general contractor is most familiar and for which many contractors are finding a wide variety of uses.

Cutterhead dredges consist of a shallow draft barge on which the pumping and other machinery are mounted. A "ladder" is hung from the forward end which supports the suction pipe. The suction action is augmented by a rotating ball-shaped cutter operating just ahead of the suction intake. The cutterhead, powered by a hydraulic motor hung on the ladder just above it, cuts into and loosens compacted soils and soft rocks such as shales. It also increases dredge capacity by channeling the soils into the suction pipe. The ladder is raised and lowered to control the pressure of the cutterhead into the underwater soil.

Cutterhead pipeline dredges are designated by the size of their discharge pipe and are available in sizes from 6 to 36 in. Dredges in the size ranges from 6 to 12 in. are available as portable units which can be disassembled and hauled overland. Dredges in these sizes are not self-propelled but are limited to "walking" (see below), although outboard motors can be attached when necessary. Larger dredges are generally self-propelled and have deeper draft hulls.

Dredge pumps are heavy-duty pumps with chrome-carbide or manganese steel liners and impellers. In silts or rounded sand grains their life is often a matter of months, but where sharp-grained sands or large gravel sizes are being handled, casing and impeller lives may be figured in hours. On a project along the coast of Newfoundland in 1952 the dredged material was 70% sharp gravel up to 5 in. in size, 28% sharp sand and 2% boulders from 5 to 10 in. in size. Here, a new pump lasted 18 days, its shell worn down from a 3 in. thickness to ¾ in. A rubber-lined pump lasted 27 hr before the rubber wore away. The sixth and final pump lasted 507 hr and handled 300,000 cu yd of soil. The life expectancy of dredge pumps and piping can only be approximated since soil conditions will seldom be uniform.

Two hydraulic-ram-operated spuds are suspended from the after end of the barge, and these are used to control dredging operations. One spud is lowered and, in fact, driven into the bottom. Anchors controlled by winches are cast off to port and starboard at the bow. By hauling in on one anchor and slacking off on the other the cutterhead is swept through an arc. At the outer limit of the arc, the digging spud is raised and the alternate or "walking" spud is lowered. The dredge is now swung back to the opposite side, where the digging spud is again lowered. The cycle or "sweep" is resumed by raising the walking spud.

Portable dredges currently available have a separate hull, about 10 by 36 by 5 ft deep for a 12 in. dredge, in which all the machinery except the ladder is contained. The 10-ft width is suitable for most limited hauling but is too narrow for stable operation, and four flotation tanks are provided, 5 ft in diameter, which not only increase the width by 10 ft but, also increase length. They are fastened in saddles welded to the sides of the hull and not only serve for flotation but, by compartmentation, provide fuel storage. Ten-inch dredges start with an eight-foot-wide hull more suitable perhaps in tight hauling conditions.

The width of cut will depend on the depth and the angle of swing used in walking, and the operating depth will depend on the length of ladder, usually 36 ft long with a 12-in. dredge and 32 ft with a 10-in. dredge. These can work at angles up to 60 deg from the horizontal. A 32-ft ladder operating at 60 deg will cut at a depth of 25 ft. If a full 90-deg swing were used, it would sweep a width of about 60 ft, but angles less than 90 deg are customarily used.

For uniform excavation of the bottom, the area to be dredged must be laid out in strips or channels of the width of the proposed sweep. Range markers are then driven along the center of the strip so that at least two are always in front of the dredge operator and can be used for his alignment. Sometimes the edges of each strip are laid out with a line of stakes so that the width of sweep can also be gaged.

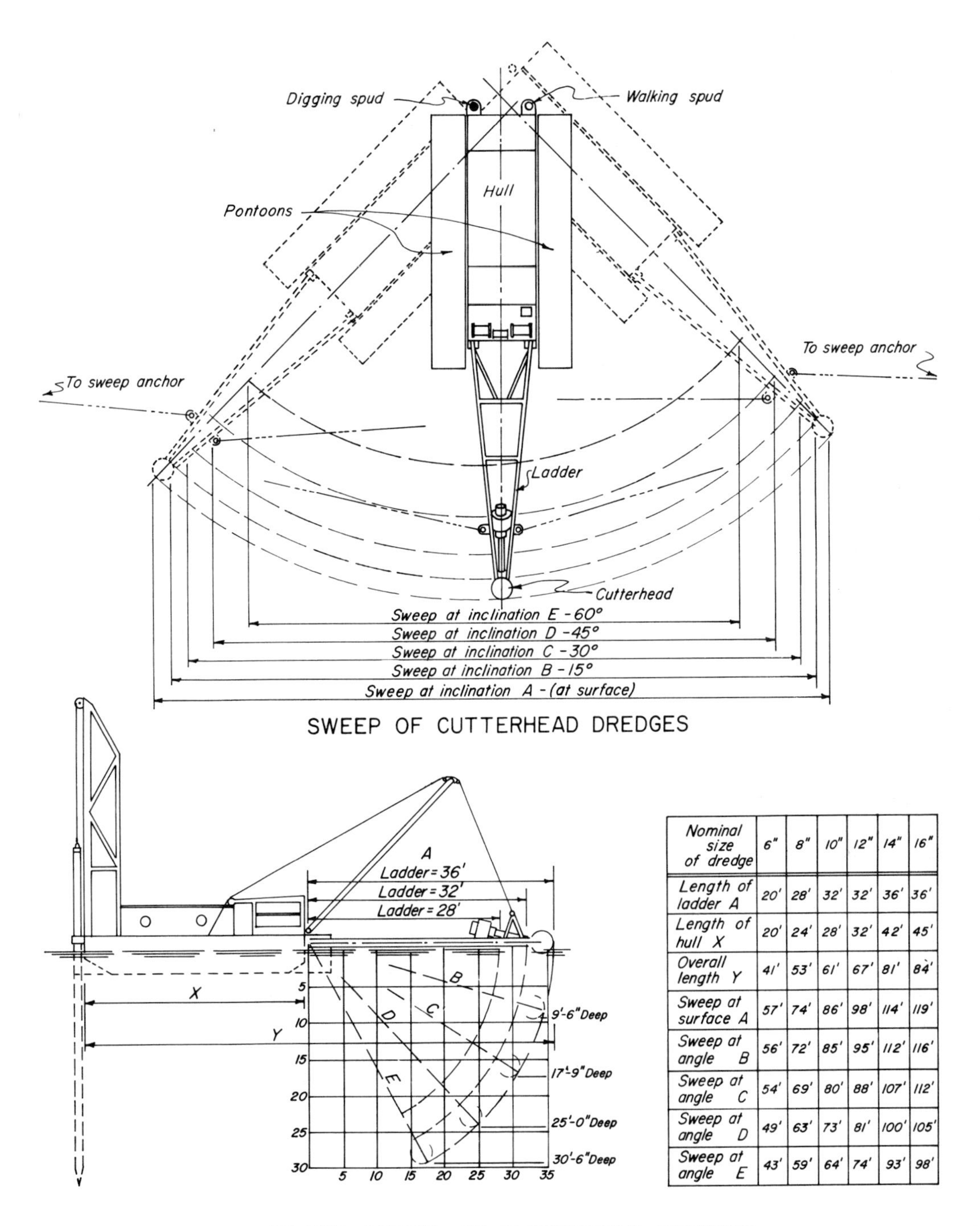

SWEEP OF CUTTERHEAD DREDGES

Nominal size of dredge	6"	8"	10"	12"	14"	16"
Length of ladder A	20'	28'	32'	32'	36'	36'
Length of hull X	20'	24'	28'	32'	42'	45'
Overall length Y	41'	53'	61'	67'	81'	84'
Sweep at surface A	57'	74'	86'	98'	114'	119'
Sweep at angle B	56'	72'	85'	95'	112'	116'
Sweep at angle C	54'	69'	80'	88'	107'	112'
Sweep at angle D	49'	63'	73'	81'	100'	105'
Sweep at angle E	43'	59'	64'	74'	93'	98'

DIGGING DEPTHS OF CUTTERHEAD DREDGES

Data shown is based on dredges manufactured by the Dixie Dredge Corporation of Miami, Florida. It is typical for all dredges in the portable class.

2.15 DREDGE DISCHARGE

Portable dredges are sold as a unit, generally complete, without the discharge piping and its accessories. Prices of dredges are quite variable, and their location of manufacture (many of them in Florida) will determine shipping costs. In 1962, 8-in. portable dredges were available at about $50,000; 10-in. dredges at about $60,000. Weights also vary widely: one 8-in. unit weighs 22 tons, a 10-in. unit 32 tons, and a 12-in. unit 52 tons, while another manufacturer's product runs 24 tons for an 8-in. model and 26 tons for a 10-in. model.

Discharge piping is often considered an accessory item and some manufacturers include 400 ft of "float" pipe and 600 ft of "shore" pipe in a package of accessory working equipment. Such accessory items (described below) will amount to about $8000 in addition to the cost of the basic 10-in. dredge.

Pipelines laid over the water are supported on pontoons or floats whose size will vary with the pipe size being carried. Two floats are provided per 40-ft length of pipe with saddles welded to them to which the pipe is bolted. Floating pipelines must be flexible and, to avoid continuous adjustment, are laid out in long loops or are "snaked" across the water. Various thicknesses and types of end connections are available, but a flexible joint must be used over water. Grooved-end pipe with rubber sleeve connections fastened on with banded couplings, is recommended. A special swivel elbow must be used to connect the floating pipeline to the dredge, and a special pipe connector must be used where the floating section connects to the shore piping.

Shore piping requires no flexibility but does require adjustment for length so that a quickly assembled pipeline is provided. Generally this will consist of a tapered-end pipe which slides into a grooved end as described above. Two hooks or lugs are welded on opposite sides, a few inches back of each joint, and connections across the joint, between hooks, are made with a short length of chain. Other available types of joints such as wedge lock, victaulic, or dresser couplings are no faster and somewhat more expensive.

Fittings are generally avoided in shore piping since abrasive soil particles, striking them rapidly, cut them away. However, it is necessary to provide means of continuous dredge discharge, and this is done by inserting a wye in the line, with a valve, so that as one area is filled up, dredged material can be diverted to another area while pipelines are being extended.

When dredging must be stopped to extend float pipe or to change position, clear water must be pumped through for a time before shutdown to prevent settlement of solids in the discharge piping as the velocity of flow decreases.

Dredging, once begun, is generally performed around the clock; and provisions for reaching the dredge, as each shift reports, must be provided. The accessory list referred to will include a workboat or dredge tender (equipped for adding an outboard motor). Not only must crews be hauled to and from the dredge and the fuel supply maintained, but range markers must be set; and frequently it may be necessary for the contractor to run soundings ahead of or behind his dredge, in order to estimate progress and quantities of work.

Small dredges are not generally equipped with devices for judging the density of flow, and in any case this should be checked ashore. Shortwave walkie-talkies should be provided for contact between shore and dredge. Shutdowns are often needed in a hurry where abrasion has completely pierced a pipe shell or where a joint has broken through. Too heavy discharges of organic silt may have to be halted to prevent too great a concentration at a single discharge point.

Concentration of solids in the dredge flow runs between 10 and 20% and will depend not only upon the nature of the soil and the depth of penetration of the cutterhead and its effectiveness but upon the length of pipeline and the height above water level at the discharge point. Capacities of the smaller-size dredges vary widely.

Check List For Cutterhead Dredge Selection

Type of Project
(*a*) Cutting a new canal? __________
(*b*) Widening an existing canal? __________
(*c*) Deepening a river channel? __________
(*d*) Shaping an existing lake? __________
(*e*) Providing hydraulic fill? __________
(*f*) Correcting beach erosion? __________

Materials to Be Dredged
1. Light, soft soils? ____% (*a*) Silt? ____% (*b*) Muck? ____% (*c*) Sediment? ____%
2. Heavy, abrasive soils? ____% (*a*) Sand? ____% (*b*) Gravel? ____% (*c*) Shell? ____%
3. Tight, hard soils? ____% (*a*) Clay? ____% (*b*) Shale? ____% (*c*) Rock? ____%

Dredging Conditions
(*a*) Fresh water? __________ (*b*) Salt water? __________
(*c*) Roots? __________ (*d*) Debris? __________
(*e*) River current (mph)? __________ (*f*) Tidal range (ft)? __________

Scope of Dredging
(*a*) Width of area to be dredged? __________
(*b*) Length of area to be dredged? __________
(*c*) Average depth to be dredged? __________
(*d*) Maximum depth of dredging? __________
(*e*) Estimated total yardage in place? __________
(*f*) Gallons of dredged material to be pumped? __________

Production Required (cu yd)
(*a*) Per hour? __________ (*b*) Per day? __________ (*c*) Per month? __________

Operating Conditions (Working Hours)
(*a*) Per day? __________ (*b*) Per week? __________ (*c*) Per month? __________

Discharge Conditions
Length of floating pipe line: (*a*) Maximum? __________ (*b*) Minimum? __________
Length of shore piping: (*c*) Maximum? __________ (*d*) Minimum? __________
(*e*) Total maximum length of discharge piping? __________
(*f*) Maximum height of dredge pipe above water level? __________

Access Limitations
1. Delivery method
   (*a*) Rail? __________ (*b*) Truck? __________
   (*c*) Shipboard? __________ (*d*) Floating (towed)? __________
2. Size limitations
   (*a*) Maximum width of hull? __________ (*b*) Maximum length of hull? __________
   (*c*) Maximum draft of hull? __________
   (*d*) Maximum clearance height of dredge above water line? __________

Payment for dredging is customarily made in terms of the quantity of yardage laid down as fill, rather than in terms of yardage removed from the waterway. For one thing, it is easier to measure; and if any large percentage of it is sand, as it usually is, hydraulic fills lay down at a uniform density usually in excess of 95% Proctor (or AASHO). See Sec. 3.1 for definition. In order to obtain full payment, therefore, certain precautions must be observed.

The first consideration in laying down hydraulic fills is the effect of the loading of water and soil on the ground beneath the fill. In most cases, particularly in swamp areas being filled by hydraulic means, settlement will occur. Where payment is by depth of fill placed, settlement platforms, as illustrated, should be provided. These will settle with the underlying soil and still provide adequate depth measurements. The frequency of such platforms will depend on the uniformity of the surface being filled (at 100-ft intervals on level swamp areas; closer on rolling terrain).

Another effect of hydraulic fills is the tendency, in saturating a soil mass, to create landslides, as discussed in Sec. 1.13. Some knowledge of the underlying structure of the fill area may be useful in these instances, but more probably control of the rate of fill will be required. Control of the rate of fill can be accomplished by spreading the fill over wide areas in successive layers rather than by piling the material deeply in a single location. This technique also prevents landslides in the material itself. The Fort Peck Dam slide was of the hydraulic fill, not of the underlying strata.

Drainage systems may depend very largely on the type of soil being dredged. Although the dredging of clay deposits is not recommended, it can be done. Clay, being finely divided, can remain in suspension for long periods of time, and drainage by direct run-off of water from this dredging product cannot be used. Even after it has been pooled and left standing, substantial quantities of colloidal particles remain in suspension. Direct draining of this water may create objectionable conditions, and separate settling facilities may be required before returning it to a main body of water. (Such a condition could exist where dredge water discharge occurred above a water-supply intake on a river.)

Although the inclusion of clay in dredging operations cannot be avoided, it is desirable to layer this with substantial seams of sand to provide the necessary drainage. Laying down a sand layer over suspended clay helps in settling and consolidation and provides drainage within the mass. Fine silts often resemble clays in this respect and should be treated similarly.

A system of dikes is generally used to contain hydraulic fills, but here again the size of soil particle being handled will have a bearing on such systems. The distance that soil particles can be transported by flowing water will depend on the size of the particle and the velocity of flow. As dredged material is discharged from the pipeline, the velocity of flow decreases sharply, and particles from gravel down to intermediate sand sizes quickly settle out. Finer sands, silts, and clays may travel some distance, and sufficient settlement may take hours.

For fine-grained materials substantial dikes must be built, sufficient to retain the water the length of time required to prevent the uncontrolled overflow of dredge effluent over adjoining lands and back into the waterways. In swamp areas such dikes are built of marsh sod stacked in pyramids, but where this is not available any soil may be used for the embankment. Timber sluice gates are generally provided in the dikes, which are closed during dredging but are opened to release the water after sufficient settlement takes place.

Interior drainage of the area is provided to channel the run-off to the sluice gates. Such channels often result from the operation of stacking the dike; but in many locations, adjoining developed areas, special drainage provisions are required. In the case cited in Sec. 2.11 an elaborate system of ditches, spillways, and settling basins was required to control the flow-off of water.

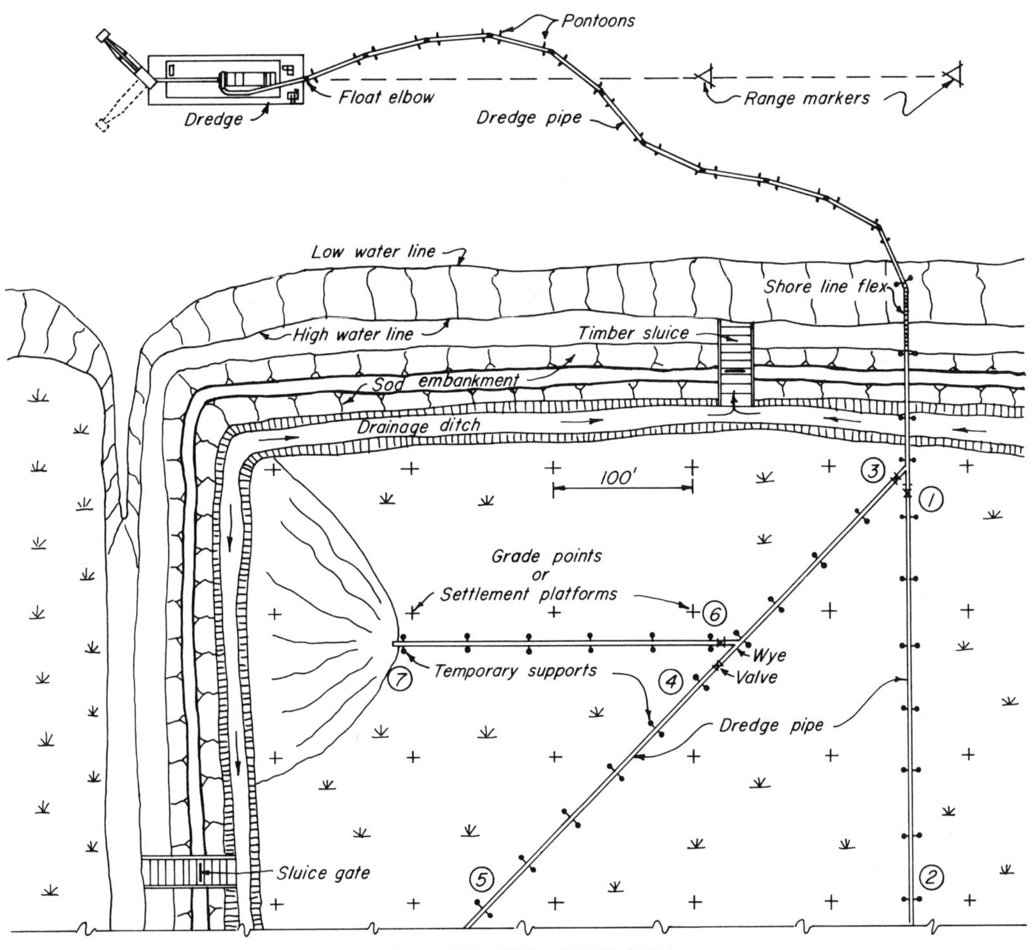

THE DREDGING OPERATION
(1) to (7) Indicate relative points of discharge

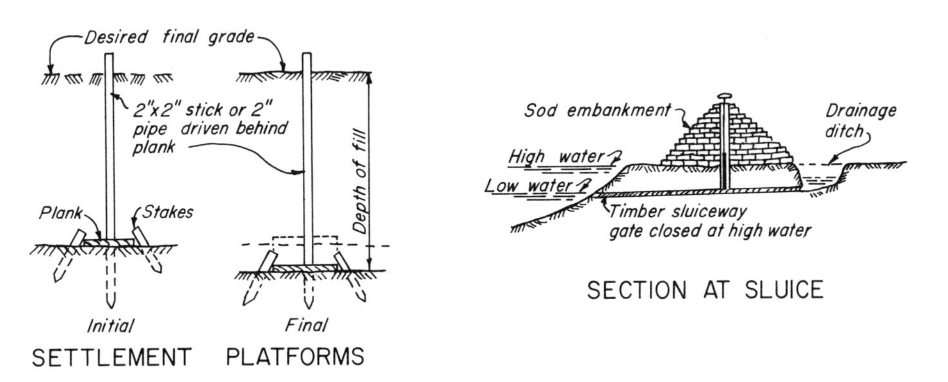

SETTLEMENT PLATFORMS

SECTION AT SLUICE

DREDGED FILLS

3 Subsurface Conditions

Any volume dealing with underground construction must give prime consideration to the problems incident to the ground within which, or upon which, these structures will be placed. Despite the relative youth of the science of soil mechanics, the literature on the subject is extensive, particularly that relating the foundation structure to the commonly encountered soil or rock formation upon which it will be built. To cover this field fully here would be, in effect, to insert a secondary book within the primary volume. There are, however, some special subsurface conditions not commonly encountered on which information is scant and scattered but which may be of inestimable use to the contractor in an emergency. The discussion in this chapter will therefore limit itself to these special cases.

The desirability of a broad knowledge of special soil conditions is well illustrated by an occurrence incident to a dam construction in the Walla Walla area of the state of Washington. The dam was built in a loess (Sec. 3.14) formation with whose characteristics the designers were familiar. In addition to other special precautions, an internal drainage blanket was provided to control underseepage. Neither the inspector nor the contractor understood its function, and major segregation was permitted to occur in placing it. The result was a progressive erosion of the loess which produced "piping" and seriously impaired the designed function of the structure to retain stored water.

In the beginning, the surface of the earth was a rock—slowly cooling. During this process, internal forces broke it into great masses that heaved and tilted or, over considerable areas, shattered it into fragments. Then external forces came into play: rains drenched it; turbulent salt seas beat upon it; ice masses—glaciers—crawled over its surface, ripping and rolling its fragments. These and yet other forces remolded it, producing new rock formations and casting aside mountains of broken fragments. Some of these remnants were large (boulders) but time and the tides and the rolling reduced them to smooth and rounded gravel—or to sand, or later, much later, to silt. One asks finally—how small can a rock fragment get?

In the grinding process more and more faces of the rock were exposed to water,

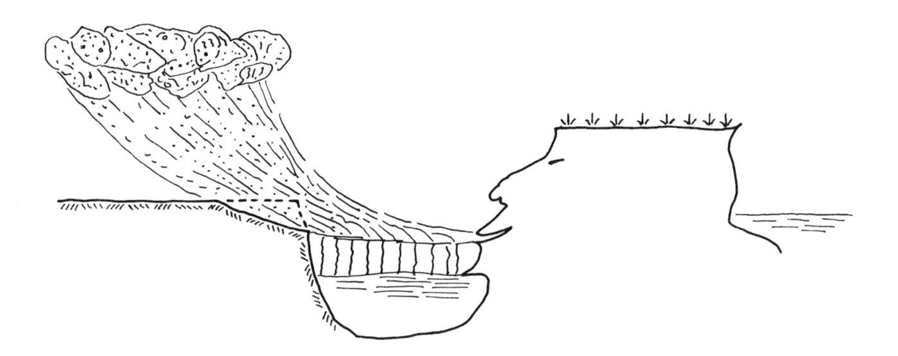

often saline, and those elements which dissolved in water were leached away; first the alkaline earths, then the calcium and magnesium carbonates. The removal of these alkaline elements left a slightly acid residue which tended to oxidize, further breaking down the rock structure. What remained, finally, were crystalline sheets of atoms bound together variously in thin layers. These are the clays.

At the clay stage we are dealing with various atomic aggregations subject to electrolytic and chemical forces, rather than mechanical, so that any further breakdown leads simply to new formations on the same scale. The rock has lost its identity, but matter, of course, is not so simply destroyed.

There are several classifications of rock residues—or soils—by screen sizes. The A.S.T.M. classification places the particle size of sand as being between 2.0 and 0.05 mm; that of silt, between 0.05 and 0.005 mm. Clays are particles less than 0.005 mm in size. Clays are further distinguished as particles ranging from 0.005 to 0.001 mm, the sizes below this being known as colloids.

Because the process of rock disintegration is a random thing, no natural deposit of soils will lie wholly within a particular grain size but will consist of agglomerations. References to silty sands and silty clays as well as mixtures of clay, sand, and gravel are frequent, for instance. Separation of these mixtures by mechanical screening is possible down to a 200-size screen, equivalent to about 0.08 mm. Below this, the relative size of particles can be determined only by sedimentation, that is, by observation of the rate at which they settle out.

But natural forces have often provided segregation by sedimentation. Swift-flowing waters will pick up a mass of assorted particles, dropping out first the coarser and then the progressively finer particles as the velocity of flow decreases. Since this is not a completely selective process, silts may fall with sands or be laid down in a bed of clay. Yesterday's swift-flowing waters may today grow sluggish and drop a layer of clay over yesterday's coarse gravel. And to compound the confusion, today's water may be free of contaminants and tomorrow's may be highly saline, tending to produce chemical changes in yesterday's clays.

3.1 SOIL DENSITIES AND MOISTURE CONTENT

The density of a soil, and consequently its ability to support a load, often depends upon its moisture content. The maximum density of a soil is achieved, in these pertinent cases, at what is termed the optimum moisture content. The basic tests for relating the dry density of a soil to its moisture content were first propounded by R. R. Proctor in 1933. Extensively used in evaluating the compaction of highway sub-grades, the Proctor test was later modified into the AASHO (American Association of State Highway Officials) test reflecting more advanced methods of compaction.

Either test procedure consists in compacting a soil sample under specific conditions with varying amounts of moisture until the moisture content at which maximum compaction occurs is ascertained. The *dry* density under this condition is determined, and then the moisture content as a percentage of this dry density is established. These figures are then used as a criterion for comparable tests made in the field.

The field test is conducted by removing a sample of soil from a small hole. It is immediately weighed. Later, in the laboratory, the sample is completely dried and again weighed. The difference, of course, represents the weight of the water. Volume is determined by filling the hole from which the sample was taken with a calibrated quantity of a uniform sand. The dry weight divided by this volume gives the dry density in pounds per cubic foot. The determined weight of the water content is then readily expressed as a percentage of the dry weight.

These test procedures have been standardized, are well known, and are frequently specified. Less familiar are procedures employing nuclear devices for determining dry density and moisture content. The nuclear method, although not yet widely accepted, would appear to offer some distinct advantages, among them being non-disturbance of the soil, results obtained in one-fifth the time, and improved accuracies reducing the possible error variation from 10% to 5%. An additional advantage is that testing can be conducted to greater depths, a special advantage where structures are involved. The chief disadvantages are that the nuclear method, employing complicated electronic gear, is much less rugged than conventional methods and is considerably more costly.

Most of the units available for measuring density or moisture use a radium-beryllium radioactive isotope (for which no Atomic Energy Commission license is needed) as a source of rays. This material may be contained in a tube used as a probe for insertion into the ground or may be sealed in a case which rests on the ground surface.

Density determinations depend upon measuring either the absorption or the backscattering of the emitted gamma rays impinging on the soil particles. The extent of backscattering (the most common method) can be related directly to soil density by the use of calibration curves and is detected, per unit of time, with a suitable counter.

Moisture determinations are made using the same source material, except that here the density of the thermal neutron cloud around the radioactive source of fast neutrons is proportional to the number of hydrogen atoms in the vicinity of the source (that is, in the moisture in the soil). From the concentration of the hydrogen atoms, the moisture is readily established.

In either case, the impulses picked up by the probe, after transistor pre-amplification, are sent by cable to a "scaler," or counter. This includes a rate meter as well as a precision timer with the necessary transistorized circuits. A converter steps up the power source (of wet or dry cell batteries) to high-range voltages of from 1100 to 1500 volts. The operator merely holds down a switch for 60 sec and either the gamma ray or the slow neutron count is dial-recorded.

The count thus obtained does not directly reflect either the density or the moisture content. The scaler must be co-related to a standard count, and the readings are compared with known densities in similar types of soil. Soil type has less influence on the count for moisture content; but here also, calibration and relation to a standard are required.

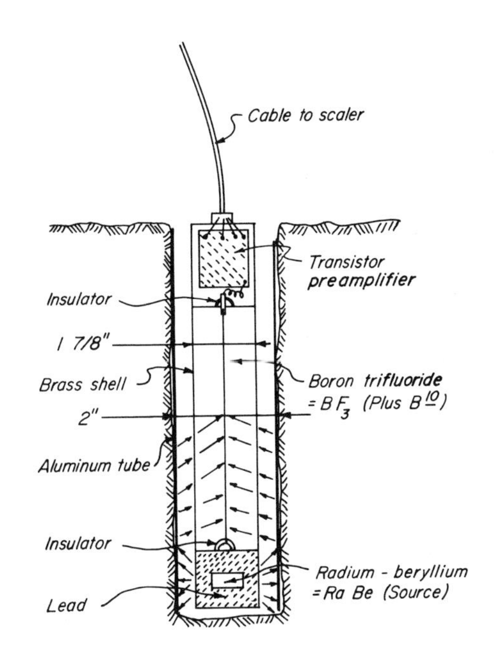

(a) MOISTURE PROBE

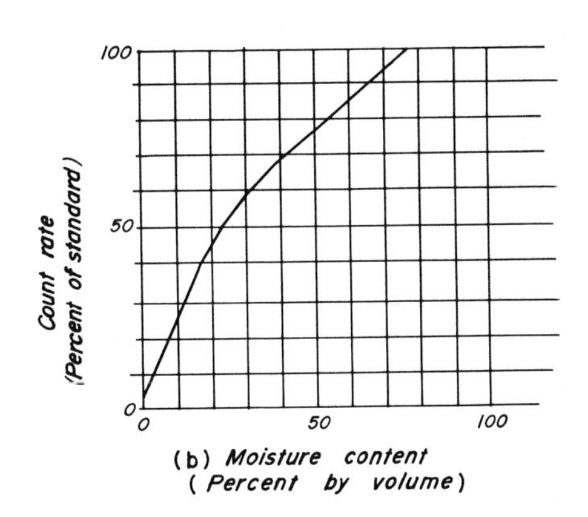

(b) Moisture content (Percent by volume)

ALUMINUM TUBE DRIVEN OR PUSHED INTO SOIL AND CLEANED OUT WITH SCREW AUGER. PROBE IS CHECKED WITH FIELD STANDARD BEFORE INSERTION AND AFTER WITHDRAWAL AND AVERAGE PERCENTAGE OF VARIATION FROM STANDARD REFERRED TO CHART ABOVE TO FIND MOISTURE CONTENT.

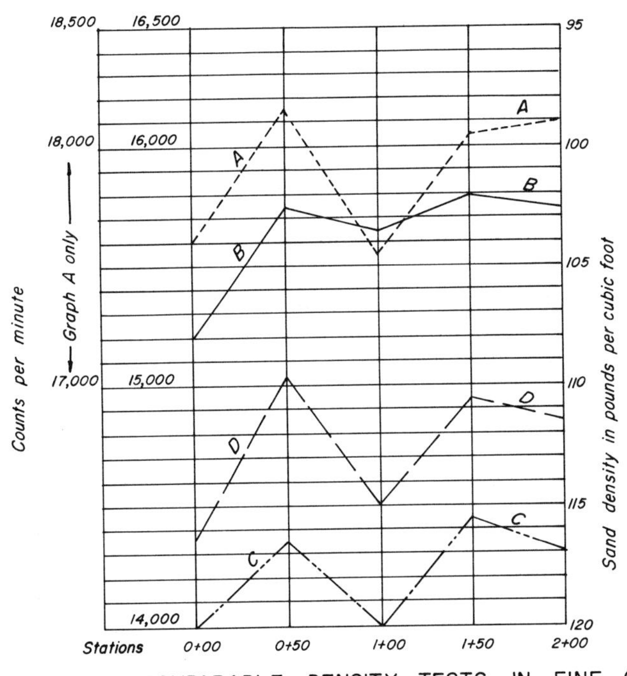

(c) COMPARABLE DENSITY TESTS IN FINE GRAY SAND

PLOTTED POINTS TAKEN AT 50 FOOT INTERVALS ALONG A HIGHWAY SUB-GRADE. EACH POINT REPRESENTS THE AVERAGE OF THREE TESTS TAKEN IN A TRIANGLE.

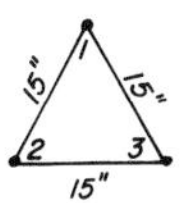

Graph A - Nuclear device A
Graph B - Nuclear device B
Graph C - Nuclear device C
Graph D - Standard proctor test

NUCLEAR MOISTURE AND DENSITY DEVICES

3.2 ROCK

Rock is often deemed the ideal foundation material, being presumably solid and homogeneous. In fact it may not necessarily be either, and these discontinuities and variabilities have frequently a greater influence on the cost of the contractor's operations than is the case in solid rock excavation. Many of these eccentricities will not be apparent from core borings or other pre-excavation examinations and must therefore be dealt with after excavation is in progress or when it is virtually complete.

It was the *weathering* of the ancient rocks that produced the soil which so extensively coats the earth's surface today. Other natural forces, such as heat and pressure, converted some of these soils back into rock, different in nature and texture, but hard enough for all that. These re-converted rocks also continue to weather, changing back again to soils. Excavation often exposes rock formations to the weathering actions of sunlight, air, and water from which it has been protected for centuries. Certain shales, for instance, requiring blasting for initial removal, disintegrate to their basic clays under the influence of air and water. These actions are often so rapid (particularly if a chemical change or leaching is involved) that there is not time to place a foundation before the nature of its bedding has been changed. But even where weathering has occurred over long spans of time, the results, being not everywhere uniform, can produce uneven settlements under concentrated loads.

The term *bedrock* is much abused, being sometimes offered as the solution to the problems encountered due to weathering. "Take the foundations to bedrock!" But real bedrock can lie a long way down, and the intervening rocks can present a variety of aspects that involve special consideration.

Rock frequently lies in strata or layers, particularly if it was reconstituted, and these are seldom horizontal. They may dip and strike, that is, may slope in two directions (Fig. *a*). The sloping surfaces resulting must generally be leveled or stepped in benches in order to apply the structural loading vertically (Fig. *b*). Obtaining these level or benched surfaces, particularly in hard strata, can result in extensive breakback on the one hand or, where all layers are not equally hard, can extend blast results and shatter rock layers otherwise adequately leveled. To control this requires careful blasting and often diagonal drilling to limit blast effects. The costs of this special care in excavation as well as the added volumes to be handled can be considerable. Figures *c* and *d* show some methods of control.

Stratified rocks in which the rock strata are separated by erodible soils have been mentioned in connection with landslides and, to a smaller degree, are not uncommon in excavation. Undetected clay seams in rock have led to several notable and disastrous structural failures. Such erodible strata, underlying a whole foundation site, may not be detected until the partly completed structure begins to move (Fig. *e*).

The heaving and sinking of the earth's surface during earlier molten states or during the period of cooling thereafter has produced various types of distortion and cracking. Stratified rock may be cracked, the cracks in each stratum running in different directions (Fig. *f*). The heaving may have been up as in an anticline (*g*) or down as in a syncline (*h*), and these in particular will show cracked strata. Generally, cracked rock strata will have substantial bearing power; but here again, once they are disturbed by blasting or even by shovel or backhoe excavation, the effect, not only on contiguous rock in the same layer but on layers immediately below, can be considerable. The solidity of the foundation becomes questionable, and extra excavation may be required.

Faults were mentioned in discussing earthquakes, but many minor faults (Figs. *i* and *j*) occur. Faults are often classified as "active" or "passive," that is, liable to further movement or not. There is no sure way of determining which type any particular fault will be. Earthquakes can convert passive to active faults, but, as noted, various types of blasting can produce the same effect.

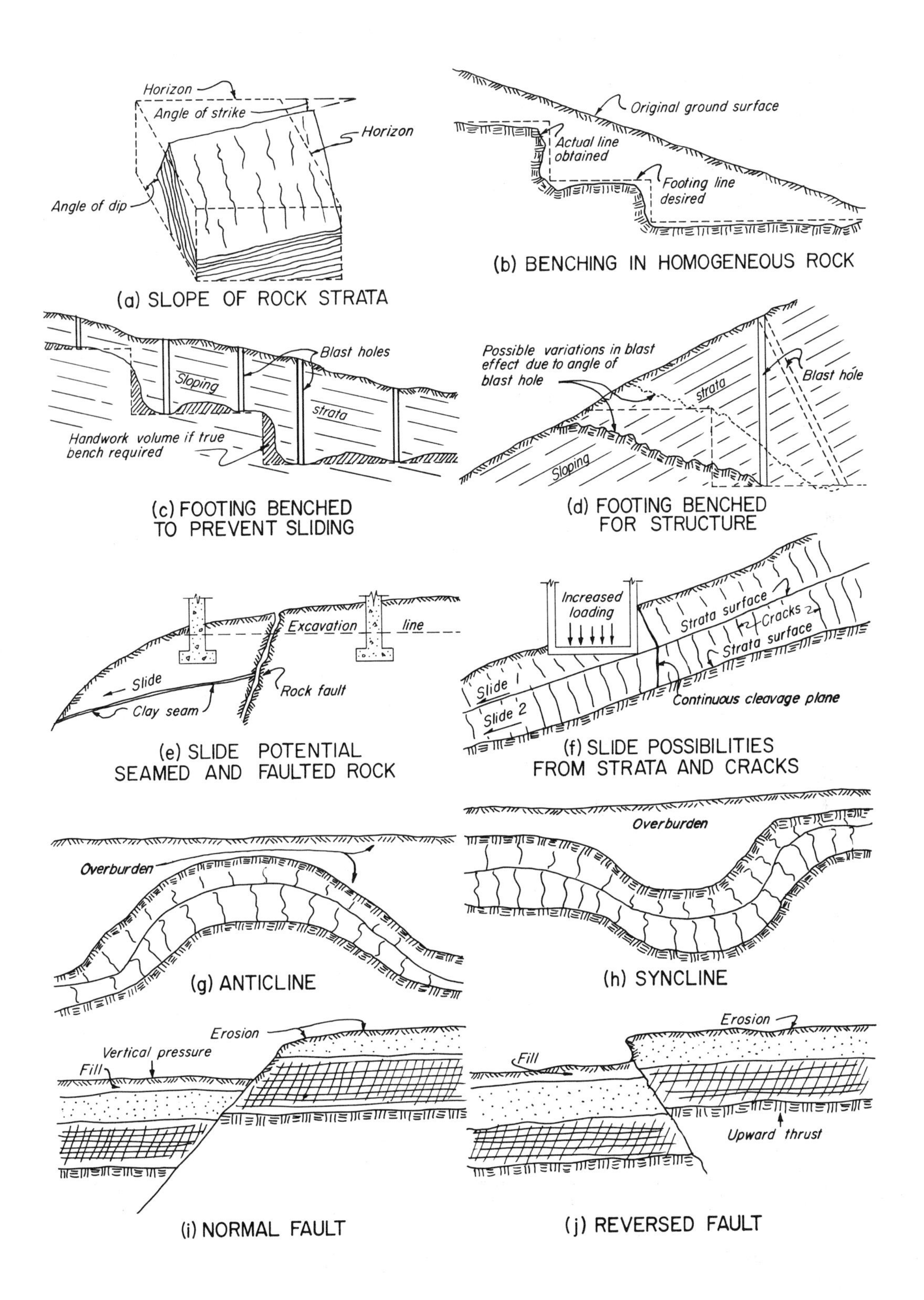
Horizon
Angle of strike
Horizon
Angle of dip
(a) SLOPE OF ROCK STRATA
Original ground surface
Actual line obtained
Footing line desired
(b) BENCHING IN HOMOGENEOUS ROCK
Blast holes
Sloping
strata
Handwork volume if true bench required
(c) FOOTING BENCHED TO PREVENT SLIDING
Possible variations in blast effect due to angle of blast hole
strata
Blast hole
Sloping
(d) FOOTING BENCHED FOR STRUCTURE
Excavation line
Slide
Clay seam
Rock fault
(e) SLIDE POTENTIAL SEAMED AND FAULTED ROCK
Increased loading
Strata surface
Cracks
Strata surface
Slide 1
Slide 2
Continuous cleavage plane
(f) SLIDE POSSIBILITIES FROM STRATA AND CRACKS
Overburden
(g) ANTICLINE
Overburden
(h) SYNCLINE
Erosion
Vertical pressure
Fill
(i) NORMAL FAULT
Erosion
Fill
Upward thrust
(j) REVERSED FAULT

3.3 PAYMENT PROBLEMS ON ROCK SURFACES

A few of the major but special conditions that can occur in rock excavation were described in Sec. 3.1 without exhausting the possibilities. In most cases remedial action involves re-consideration of design or alterations in site location, for which the contractor can receive payment. In many quite usual rock formations, however, the rock surface to be prepared for a foundation lies wholly within the province of the contractor.

In the case of a hard or relatively hard rock of uniform consistency, standard blasting methods call for "shooting out the toe." This is done by drilling 2 to 3 ft below the desired grade and placing at least part of the dynamite at the bottom of the drill hole (Fig. *a*). The result is a surface which may be from 0 to 3 ft below grade. The question now arises whether this shall be filled with concrete or whether some other means is satisfactory.

The first decision is the contractor's. Is it cheaper to blow the bottom out and refill or to keep major blasts to or above grade and be faced either with secondary blasting or trimming the surface with breakers, a sometimes very difficult procedure in basalts or granites. Although specifications will occasionally recognize this dilemma and provide a unit price for this type of fill, engineers are understandably leery of this approach since they have no control over blasting techniques. Where no unit price is provided, the contractor must provide a contingency for this type of overbreak in his estimate.

The second decision rests generally with the designing engineer. Does the structure for which this is the foundation require the full bearing capacity of the original rock, or will a somewhat lower loading be suitable?

In some instances, where the sub-grade will be exposed to substantial fluctuations of temperature, pouring this void as an integral part of the base slab may not be desirable, especially if the rock surface is deeply pitted. Variations in the coefficient of expansion between the concrete and the rock can produce stresses that can crack the base slab. Where the full bearing of the slab is required, a weaker fill can be used, substituting perhaps a 1500-lb concrete to bring the sub-grade up to the bottom of the slab, instead of the 4000-lb concrete required in the slab itself.

Where substitution of mixes seems inadvisable, the use of cyclopean concrete may be acceptable. The bottom is brought up to slab sub-grade by pouring concrete into which sound whole fragments of the excavated rock, in one-man or two-man sizes, are *hurled* (for embedment) and the surface roughly screeded to grade. This provides a well-graded surface on which reinforcing steel can be placed and held. (This should not be considered as masonry. The attempt to lay out the stone in the pockets and run grout into their interstices does not produce a sound bearing.)

Although the primary concern in the above has been with overshooting the bottom in homogeneous formations, the same problems and solution may arise where local and extensive faulting occurs, where cracked and shattered rock formations exist, or where natural rock surfaces are deeply pitted or have extensive clay pockets.

There are many special cases of which the contractor must be wary. The bottom illustration shows a case where a 250-ft stack foundation sub-grade was specified at a fixed elevation based on a test-boring elevation for rock. One hundred and fifty feet away the test-boring elevation indicated a sound, hard surface perfectly suitable for its function. Under the stack, however, although the test boring correctly indicated rock surface, on excavation, it turned out that the area was a mica schist formation, tilted almost vertically, with deep pockets of weathered rock having no more bearing capacity than loose sand. A suitable rock surface was from 5 to 7 ft below that originally indicated.

Many foundations are designed to permit minor settlements provided they occur uniformly under the whole foundation. In cases of tall structures and in particular such units as the stack just noted, or elevated water tanks, very slightly non-uniform settlements can produce dangerous tilts.

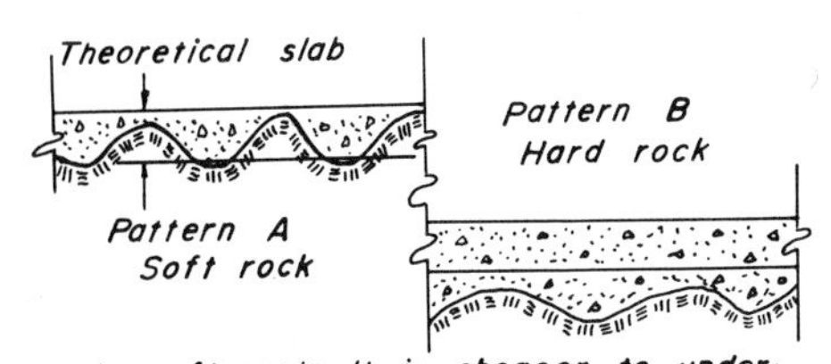

In soft rock: It is cheaper to undershoot, knocking off humps by paving breaker.
In hard rock: Overshoot the bottom and level with concrete or cyclopean masonry.

(a) PAYMENT LEVEL OF ROCK EXCAVATION HOMOGENEOUS FORMATIONS

Theoretical slab thickness

Dotted lines indicate possible secondary cracking, loosening the section but not freeing it completely. Grade to slab bottom with paving breakers and grout surface.

(b) PAYMENT LEVEL OF ROCK EXCAVATION WITH VERTICAL STRATIFICATION

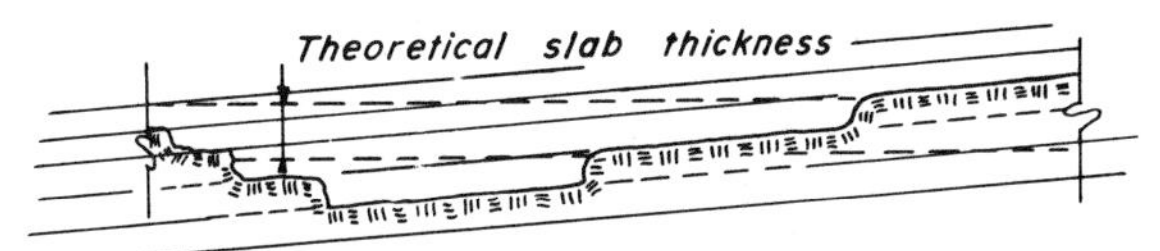

Dotted lines along strata indicate possible loosening without complete separation of slab. Test by prying with point of pick.

(c) PAYMENT LEVEL OF ROCK EXCAVATION WITH HORIZONTAL STRATIFICATION

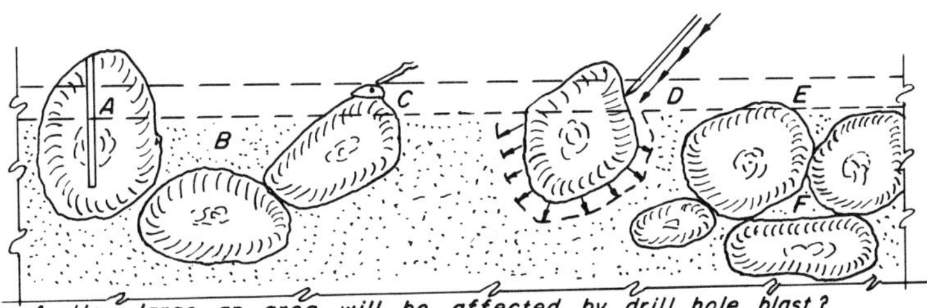

A. How large an area will be affected by drill hole blast?
B. Is clay lying over boulders adequate for bearing?
C. Will mud capping shatter the top or crack it vertically?
D. Will breaker chipping rock loosen or compact clay?
E. How much projection into slab is permissible?
F. What may be the effect of complete voids in the mass?

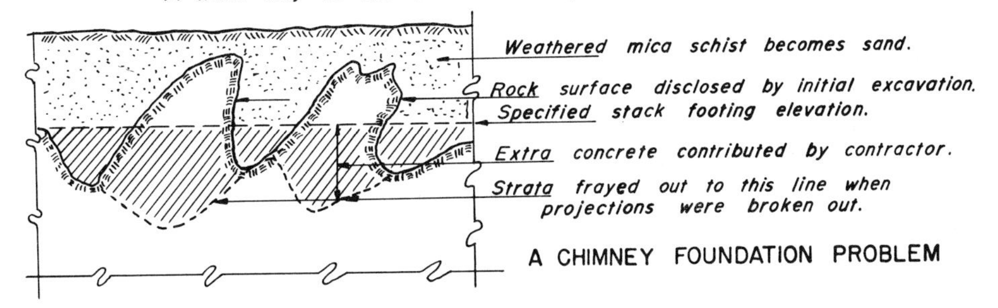

A CHIMNEY FOUNDATION PROBLEM

PAYMENT PROBLEMS ON ROCK SURFACES

3.4 ROCK ANCHORING OF VERTICAL WALLS

Rock anchoring may be divided into two aspects: that used on vertical rock faces and that used beneath a foundation as a tie-down. Rock anchoring on vertical faces can be performed for either of two reasons: to tie a concrete wall such as a spillway wall, or retaining wall, to a rock face or to tie together an exposed rock face which, because of faults, seams, or vertical strata, might otherwise have a tendency to slide, spall, or collapse.

Two substantial failures in the early 1960s highlighted this subject, even though it is doubtful whether rock anchoring would have been contemplated for use in either case or whether, if it had, it would have prevented the failures. Reference is made to the catastrophic failure of the Malpasset Dam in France and the less damaging but quite costly failure of one wall of the Wheeler Lock on the Tennessee River near Athens, Alabama. Both were attributed to undetected clay seams faulting the rock and along which slip occurred.

The primary problem in rock anchoring of vertical faces is that of drilling the holes for the anchors. Drilling horizontally into the rock, up to the limitations of drill rigs, presents no problem. This height can be extended by mounting elevated timber drill platforms on top of a Euclid-type dump truck or other vehicle, or by setting such platforms on hydraulically operated material lifts, front-end loaders, or even fork lifts. Patent or wooden scaffolding can be used where diagonal bracing can be added to take the thrust of the drill tools if the wall is truly vertical; however, in most cases the rock face is either rough or tilted backward, making it difficult to maintain suitable platforms close enough to the working face.

At greater heights of rock face, various expedients have been used. The simplest of these is that of lowering men in boatswain's chairs from the top, equipped with separately supported jack hammers. Support and movement may be provided by winches if suitable anchorage for them can be had at the top of the cliff, or they can be suspended from the end of a boom on a crane traveling well behind the edge. A platform, similarly suspended, can be used. In both instances the weight of the driller against the jack hammer is relied on for thrust. The rope suspensions (steel cable is not flexible enough) must hang without major contact with jagged rock edges which could quickly fray or cut it. Depths of drilling would be limited to the 10-ft capacity of jack hammers.

Special rigs have been devised like the one shown, supported on a boom from a crane traveling on the surface below the wall. Here, an elevator tower is suspended from the boom point pin. A cage moving on the tower furnishes the drilling platform. Drilling is done by drifters mounted on shafts which are breasted from the face of the wall. These compensate for the various angles at which the tower must lean against the rough rock face.

Drilling patterns will vary with the condition of the rocks. Spacing of holes may be on 4- to 5-ft centers both horizontally and vertically or may be erratically spotted to suit the contour conditions of the face. Depth of anchorage may vary from 10 to 40 ft, again depending on conditions. Diameter of holes will vary from 1½ to 3½ in.

Bolts used may have the usual square head, either thrust down to the bottom of the hole or, where wedging devices are being used, may be placed on the outside. Generally rods with both ends threaded will be more satisfactory. Double nutting can secure the plate on both ends; and against the face, if the thread is short, it can be extended with a "nut-runner." Bolts are pressure-grouted in place with an expandable grout (Sec. 3:22). Steel plates 8 by 8 by ¼ or ⅜ in. are set on the outer end of the bolt for concrete anchorage 6 to 8 in. from the face of the rock. Where rock anchorage is involved, these are drawn up hard against the rock face on a leveling pad of grout.

Such installations will generally specify pull tests up to 20,000 psi conducted after the grout has attained its full set. Calibrated torque wrenches with extended handles can be used in these cases.

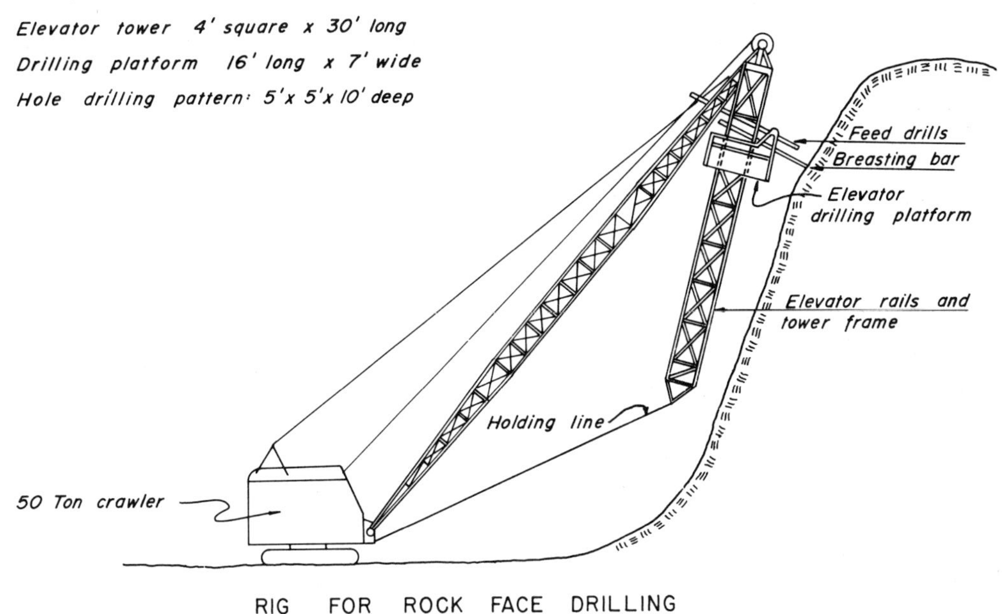

RIG FOR ROCK FACE DRILLING

FIRST STEP

Drill holes 5' o. c. vertically and 10' o.c. horizontally - drilling from platform mounted on truck.

Existing 24' wide road

Desired slope 1½ : 1

105' ±

Drill holes

24'

100' ±

Walker Lake

SECOND STEP

40' Deep slope line drilling @ 6' o. c.

Area shot loose and removed

48'

THIRD STEP

Projection shot loose and removed.

But - faces above and below could first be stabilized by rock bolting if necessary.

100' ±

SPLITTING A VERTICAL SLOPE INTO SEGMENTS

Used here for rock removal, the same approach can be used for rock bolt drilling.

3.5 ROCK ANCHORING OF STRUCTURES

The rock anchoring discussed in Sec. 3.4 was principally concerned with tying rock masses together for consolidation purposes and, in the process, often enough, tying the structure to the rock mass so consolidated. Here we are concerned with a somewhat different type of anchorage provided to prevent uplift, to hold down a structure subject to periodic flotation, or to counteract other natural forces, such as heave due to frost. Here, as in other cases, the basic rock structure and its dip, strike, and faulting will influence the decision to use this method and may modify the specific approach noted here.

A television antenna tower was planned for Mt. Wilson, which rises 5600 ft above the surrounding terrain. High wind velocities at the site together with the tower height itself indicated the need for substantial design to prevent overturning. A heavy mass-concrete footing was considered, but the prospect of hauling the large quantity of concrete up the sheer sides of the mountain was deemed a costly and time-consuming project. It was estimated that rock anchorage might be as much as 56% cheaper, and this method was used.

The primary rock at the site was a solid homogeneous mass. Footing pits were excavated in this to a depth of 5 ft below grade. To obtain a working surface after the rock excavation was complete, a 5-in.-thick pad of concrete was placed and screeded level over the rock.

Eight 7-in.-diameter holes were drilled diagonally through the 5-in. slab to a depth of 25 ft. Into these were dropped 1¾-in. high-tensile-strength rods capped on the lower end with a 6½-in. plate washer held by a nut. (Such close tolerances of a 6½-in. plate in a 7-in. hole are possible only in solid rock formations, diamond-drilled.)

The lower 10 ft of the positioned rods was pressure-grouted with an expandable grout. After the grout had obtained its ultimate strength, each bolt was tested to 60 tons by using a pre-stressing jack mounted on a special steel frame (see detail). At this loading a 45-psi friction existed between the grout and the drill-hole wall. With the stressing load still on the jack, the balance of the hole was grouted.

In this case, the uplift resisted by the whole foundation amounted to 530,000 lb, providing a factor of safety of 3. To have provided a safety factor of only 1.5 would have required another 150 cu yd of mass concrete.

The anchoring of concrete tanks (or steel tanks tied to a concrete slab) to prevent flotation is a somewhat simpler operation. The uplift forces, though substantial, are distributed over a wide area. Holes of 1½ in. diameter drilled into the basic rock on 4- or 5-ft centers and 3 ft deep will generally provide the doweling required. Number 6, 7, or 8 reinforcing rods are grouted into these holes with their tops hooked over the steel in the upper half of the slab.

A special case involving rock anchoring to prevent uplift due to frost is indicated. The foundation shown was one of a number provided to carry a log conveyor system at a log mill in northern Ontario. Although the site was below the permafrost belt, substantial deep-ground freezing occurred. Rock lay at a variable depth below the muskeg at ground surface, and its upper layer was extensively shattered.

A 10-in.-pipe pile was driven to rock surface and cleaned out with air. A 6-in.-diameter hole was then drilled 10 ft into the rock through the pipe. A 3-in.-diameter steel bar, deformed with welded lugs, extended the full depth and was grouted in under pressure.

This installation was tested to 120 tons of uplift pressure using a collar fastened to the 3-in. dowel, with no indication of movement. However, as a further precaution, the faces of the piers were coated with grease and covered with impregnated paper to permit the freezing muskeg to slide upward along their surfaces. (Muskeg is an organic silt which retains water and expands greatly as it freezes.)

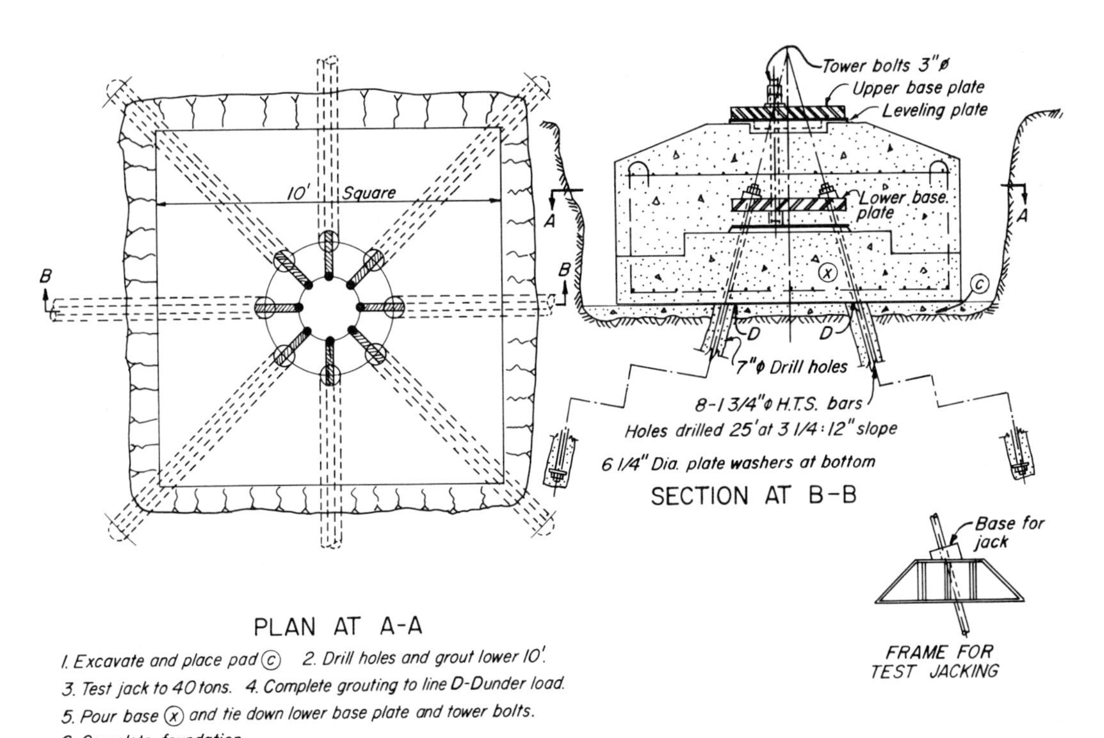

AN ANCHORED FOUNDATION

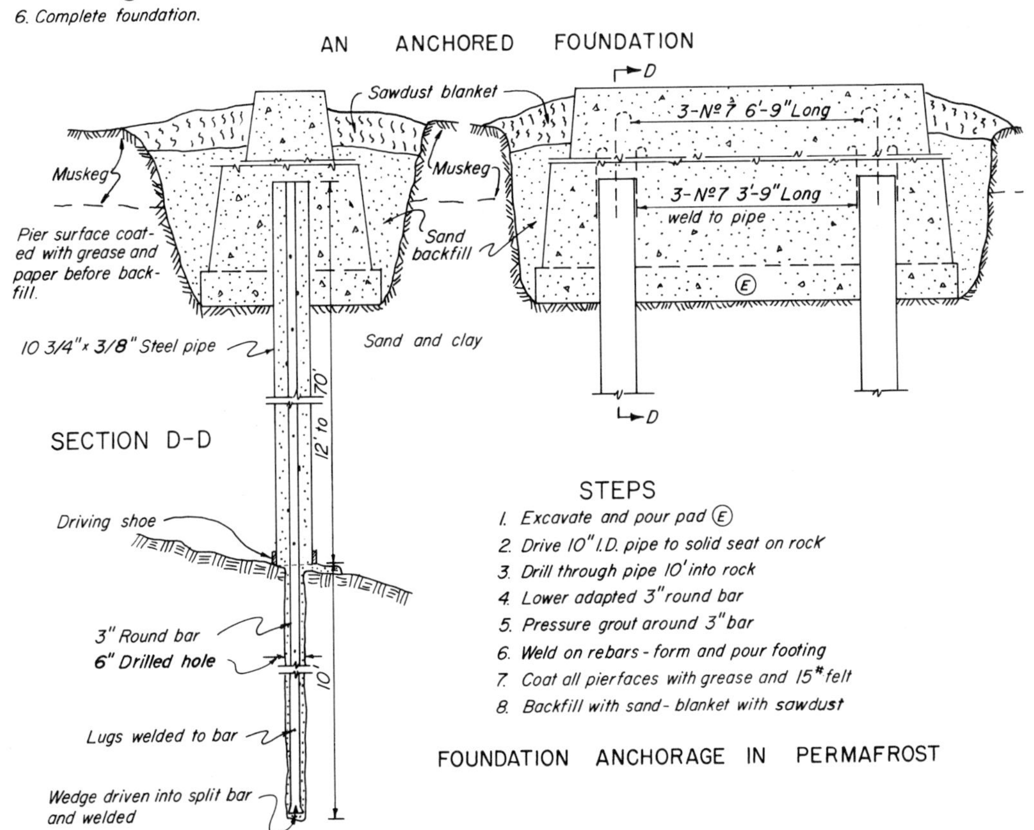

FOUNDATION ANCHORAGE IN PERMAFROST

3.6 SAND

Literature abounds with uncomplimentary references to sand as a foundation material. There are, first, the several Biblical references warning the unwary against building their houses on sand. The phrase "the shifting sands of the desert" has become a cliché. Even Lewis Carroll wanted none of the stuff:

> The Walrus and the Carpenter
> Were walking close at hand:
> They wept like anything to see
> Such quantities of sand:
> 'If this were only cleared away,'
> They said, 'it *would* be grand!'
> 'If seven maids with seven mops
> Swept it for half a year,
> Do you suppose,' the Walrus said,
> 'That they could get it clear?'
> 'I doubt it,' said the Carpenter,
> And shed a bitter tear.

(Someone, rather unkindly, has said that the Walrus was, in fact, a contractor inspecting the site of a building project and has further alleged that the Carpenter was not shedding tears over the sand but over his boss's threat to bring in non-union labor since it is well known that the International Sand-Sweepers is an all-male union.)

As the scope of construction has widened, as civilization has encroached more and more on the desert, it has become less and less practical to consider "sweeping it all away." Greater study has gone into methods of treating sand *in situ,* and such studies have begun with the facts known to every surf bather.

Sand is a cohesionless material, an aggregation of unbound grains, lacking the plasticity of clay or the stronger bonds of rock, yet the salt waves pounding up the beach can pack it into a relatively dense mass. As the tide recedes and the sun beats down, the surface of the sand dries out, and walking in it again becomes difficult as it fluffs up and loses its temporary cohesion. The well-traveled bather knows that this effect varies from beach to beach without, perhaps, realizing that the difference is due to the size and condition of the sand particles involved.

Ancient beaches whose sand particles have been long rolled against each other consist of fine grains, well rounded; but in large subterranean areas, sand deposits exist essentially in the state in which they were chipped from the mother rock, with large grain sizes predominating and with their sharp chipped edges still retained. Between these extremes are many variations of sand deposits, while beyond these extremes of "pure" sand lie deposits heavily laden with clay, silt, or gravel. Dune sand or "blow" sand is a fine-grained sand with well-rounded particles, which has been accumulated by the winds from more mixed deposits and blown into shifting heaps.

Many methods of stabilizing sand have been tried. The introduction of clay in sufficient quantity to fill the voids is practical only where adequate mixing of the two soils is possible. Where clay can be added, gravel as well will help to form what in its natural state is called a hardpan; but these methods are costly and chiefly employed on surface sands. Other surface treatments have employed portland cement (6%) with water (10%) to stabilize sand, using a traveling mixer which picks up the loose sand, mixes the cement and water into it, and respreads the mixture to a few inches of depth behind it.

The use of water, as on sea beaches, will provide temporary compaction alone, but this moisture content cannot permanently be maintained. The vibration of wet sands by rolling equipment can produce surface compaction and for road sub-grades or backfill may be satisfactory, but there is no control of compaction unless the sand is placed in layers, seldom practical for deep foundations.

To get deep compaction, blasting in drill holes has been used, particularly in very fine sands with a high water content. In these cases it is the excess water that causes the difficulty, providing a high pore pressure between fine sand particles and producing a liquefied mass. In this case the removal of the accumulated pore water is the crux of the matter, and the blast-induced vibration provides additional compaction.

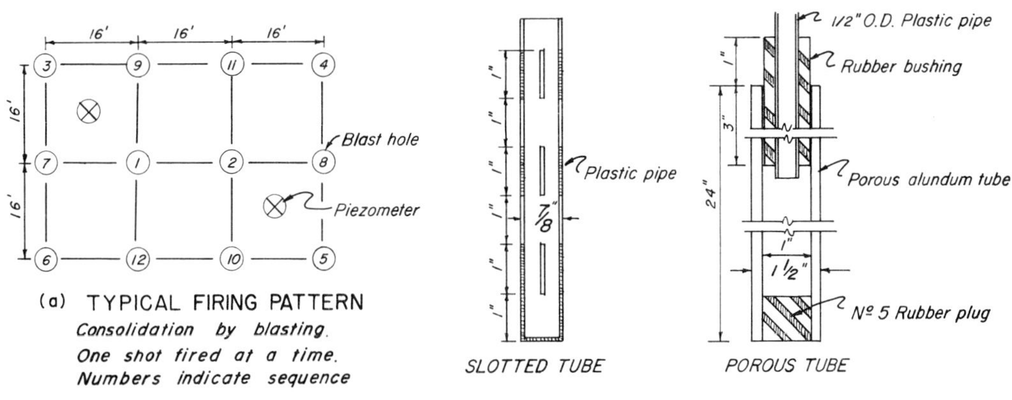

(a) TYPICAL FIRING PATTERN
Consolidation by blasting.
One shot fired at a time.
Numbers indicate sequence

(b) TYPES OF PIEZOMETERS

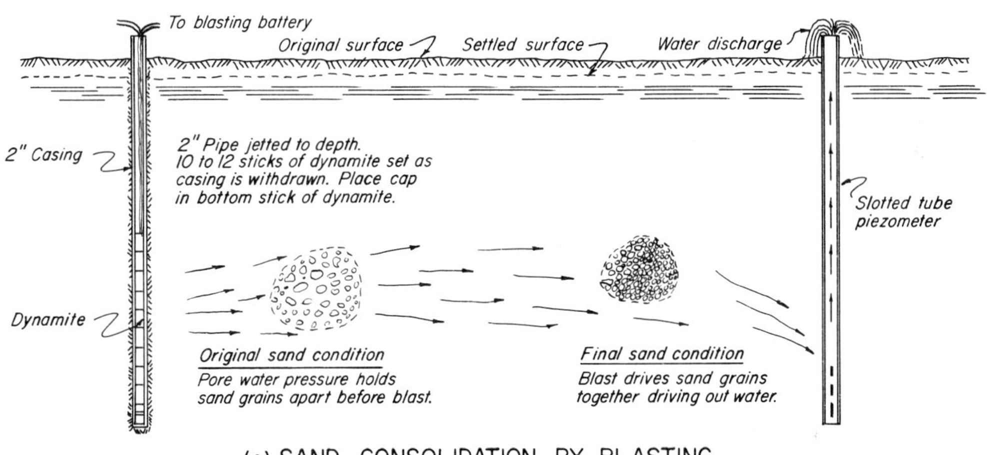

(c) SAND CONSOLIDATION BY BLASTING

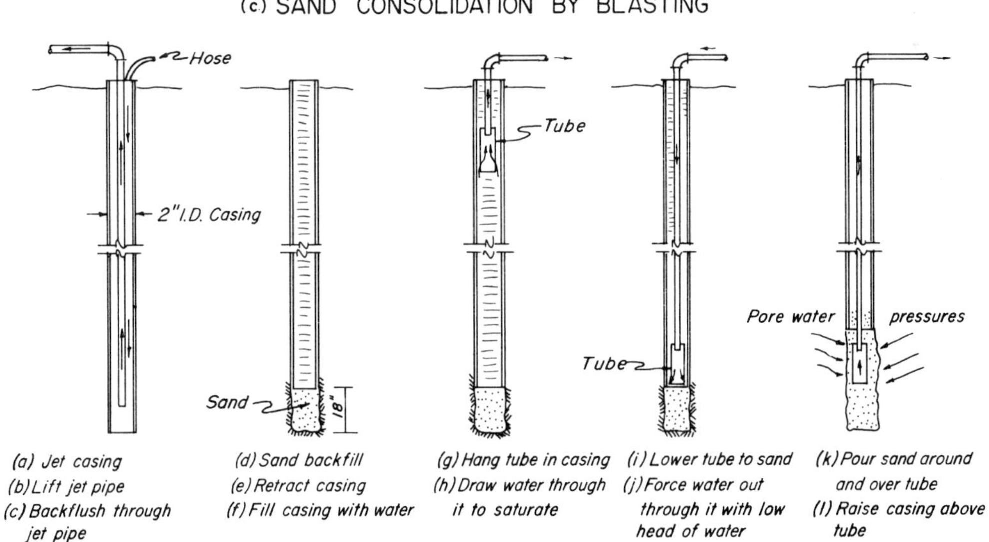

(d) INSTALLING A POROUS TUBE PIEZOMETER

3.7 VIBROFLOTATION OF SAND

The principles mentioned in Sec. 3.6 for compacting sands for foundation purposes were pulled together into a single process involving saturation and vibration in Europe in the 1930s but were not developed and applied to a substantial project in the United States until 1952, when the process was used to stabilize the sand foundation for a phosphate plant at Bartow, Florida. Following this, the Vibroflotation Foundation Company was formed, registering the name Vibroflotation and developing a device known as the Vibroflot, used as a trademark, for performing the operation.

The Vibroflot consists of a 30-hp electric vibrator head, 15 in. in diameter and about 6 ft long, coupled to a follow-up pipe of the same diameter. The length of the follow-up pipe is adapted to the depth required. The bottom end of the head is an open jet supplied by water from within the shell of the vibrator. A second set of jet openings occurs at the bottom of the follow-up pipe and is fed by a separate water supply.

The Vibroflot, generally in excess of 12 ft long, is hung from a crane boom from which also hangs a set of pile leads. The leads provide a guide for the Vibroflot, keeping it vertical as it penetrates the soil.

In operation the Vibroflot is placed on the sand surface and the lower jet is turned on. The jetting produces a "quick" condition in the sand due to the upward velocity of the escaping water. Under this condition the Vibroflot sinks under its own weight.

When the tip of the Vibroflot has reached the desired penetration, the bottom jet is cut off and a flow of water is now directed through the upper jets at a reduced pressure. The flow here is sufficient to produce turbulence without producing buoyancy in the sand, and the sand, still being vibrated, settles around the vibrator. During this process the Vibroflot is slowly withdrawn, a foot at a time, the upper jets forcing sand down into the cavity left. The ensuing consolidation of the sand in the column surrounding the Vibroflot produces settling, and additional sand is added as required.

The Vibroflotation process can be used in any sand as defined by A.S.T.M. or in a sand-and-gravel formation which does not contain in excess of 5% clay or 25% silt.

Depths of 20 ft are customary although compaction has been carried as deep as 60 ft. The loading requirements and stratification of the soil, if any, will determine the final depth. The effective width of influence will depend somewhat on the grading of the sand. The zone of influence is roughly a circle; consequently for complete area compaction these zones of influence must overlap. Generally a circle 2 ft in radius from the center of the Vibroflot is the safest assumption, but zones up to 4 ft in radius have produced satisfactory densities (see illustration).

Density is determined by standard laboratory procedures either by taking samples in a test pit, generally the best procedure, or by using some standard sampling device such as a Shelby tube. The densities used in defining the compaction produced by Vibroflotation are known as "relative densities" and are calculated from the formula shown. On this basis the relative density of loose sand would vary from 0 to 33%; of medium sand, from 33 to 66%; and of dense sand, from 66 to 100%. Relative densities in the vicinity of 70% are obtained by Vibroflotation. (The Proctor tests are not, of course, applicable to sands, and a special AASHO test for sands which relates the in-place weight directly to maximum weight provides a different figure.)

As previously noted, the sand column produced by Vibroflotation does not eliminate the hazards of building on sand. So long as the initial condition of containment prevails, however, a more satisfactory foundation bearing is obtained less subject to vibrations induced at the surface above it and less subject to the possibility of drying out deeply.

Costs of Vibroflotation will vary widely, depending upon conditions, but will generally run somewhat higher than $3 per cu yd stabilized. The process requires substantial quantities of water.

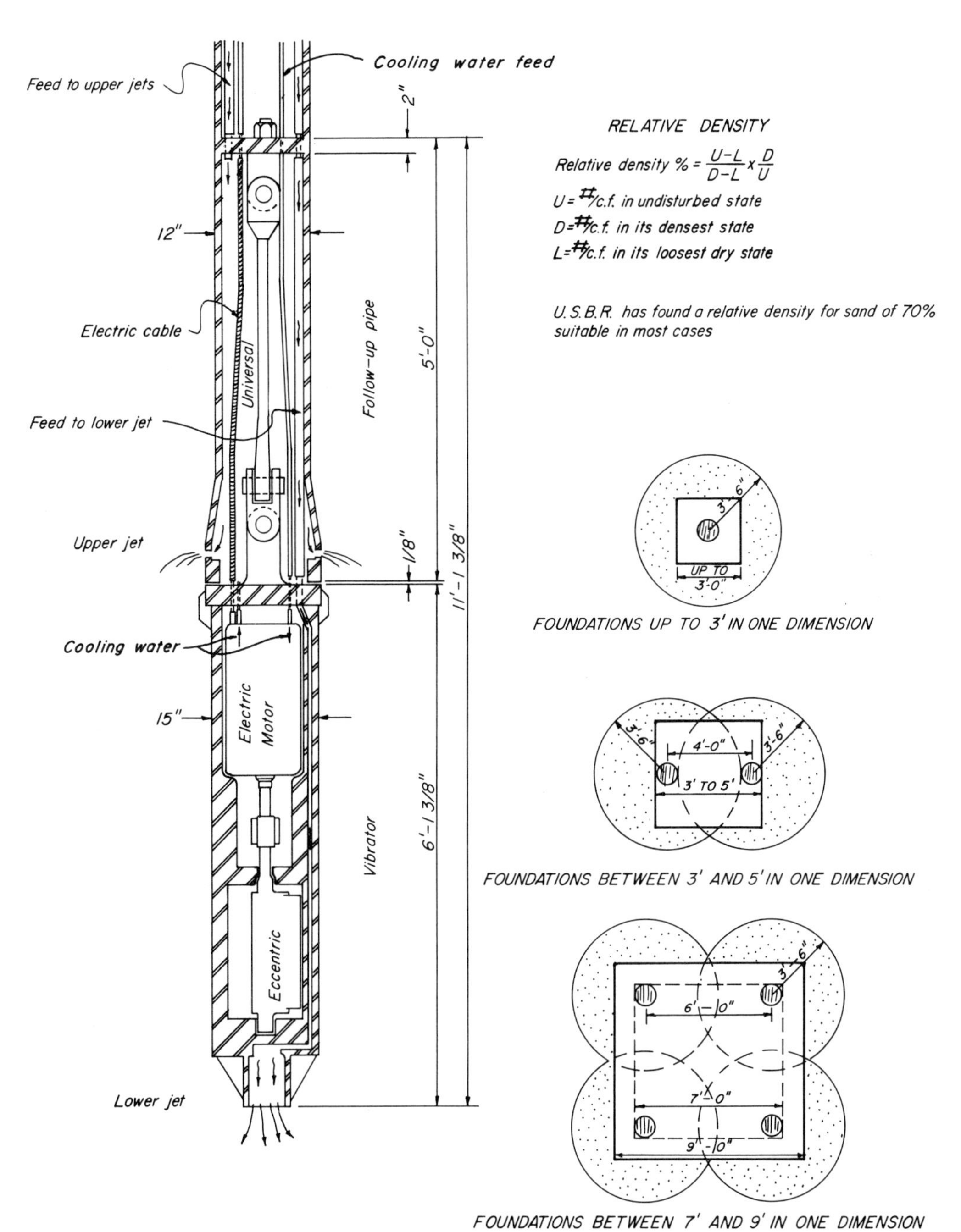

THE VIBROFLOT

VIBROFLOT PATTERNS

Position of vibroflot

Influence radius 3'-6" assumed

3.8 CLAY

Clays are atomic assemblages formed as crystalline flakes and composed of atoms of alumina, silica, oxygen, hydrogen, and hydroxyls. The arrangement of these elements in the crystal results in three different types of clay: kaolinite, illite, and montmorillonite, each with different physical characteristics. (There are several additional types of limited occurrence.)

Kaolinite occurs as hexagonal flakes composed of alternate sheets of silica and alumina atoms bound together in a tight structure.

Montmorillonite consists of an alumina sheet of atoms between two silica sheets, loosely bound, so that water molecules can enter between them and cause the flake to expand, producing what is termed an expansive clay (see Sec. 3.10).

Illite resembles montmorillonite, but has a tight structure similar to that of kaolinite due to the inclusion of additional atoms of potassium, iron, or manganese.

Clays, because of their rather basic atomic structure, cover the smallest of the grain sizes found in the earth's crust, and at one time a clay was simply any fine-grained soil. Current A.S.T.M. specifications define a clay simply as a particle of 0.005 mm in diameter or less, and colloids, or colloidal clay, as less than 0.001 mm in size.

Particles of colloidal clay all carry a negative charge and mutually repel each other in water. The movement thus initiated is accelerated by their resulting impact with water molecules. Because of this continuous movement, clay colloids do not readily settle out in water and they cannot, therefore, be obtained by sedimentation. When larger clay particles are introduced having occasional positive charges on their faces, the colloids will attach themselves and settle with these.

Clay deposits are widespread throughout the world, but very rarely are they of uniform consistency. Not only can a single deposit have any combination of atomic structures, but the size ranges at a single site may cover the whole gamut from 0.005 mm down, and mechanical screening cannot suitably separate or designate the gradations.

Because mechanical screening will not satisfactorily delineate the nature of clay, methods requiring laboratory facilities must be used. This volume is not concerned with laboratory test procedures or their application to the design of structures resting on clays; however, a brief definition of test terms used and their function may be helpful in evaluating construction problems.

Void ratio is the ratio of the volume of the voids in a soil mass to the volume of the solids. A similar term, *porosity,* defines the voids as a percentage of the volume of the mass, but the volume of the mass can be changed by pressure.

Water content is the percentage of water, by weight, of the total solids in a soil mass. In clays this can vary from 12 to over 200%.

The *Atterberg limits* relate to the consistency of a clay. Clay in the earliest stage of sedimentation is essentially *liquid.* As the volume of voids decreases, the clay becomes more compact and is then fairly soft and *plastic.* A deposit may lose this characteristic of plasticity under compression or drying, and it is then spoken of as *semi-solid.* A final stage is reached where the clay mass will stop shrinking, and this is the *solid state.*

The *plasticity index* defines the range of water content (as a percentage of the weight of the solids) between the point where it ceases to be liquid and becomes plastic and the point where it ceases to be plastic and becomes semi-solid. These limits define the properties of a clay more satisfactorily than grain-size determination. Within the plastic range, clay has *cohesion* which requires a certain force to break. This force is termed the *shearing resistance.*

The *liquid limit,* that is, that water content at which the soil mass loses its characteristics as a liquid and becomes plastic, is affected by the fineness of the clay particles or by their shape. Very flat clay particles will have a high liquid limit. When sands or silts are added to the clay particles, the liquid limit will be lowered, that is, the quantity of water required for plasticity will be less.

Summary of Test Procedures For Clay Soils

(Adapted from A.S.T.M. Designations)

LIQUID LIMIT

Definition: The liquid limit of a soil is that moisture content, expressed as a percentage of the weight of the oven-dried soil, at which the soil will just begin to flow when lightly jarred 25 times.

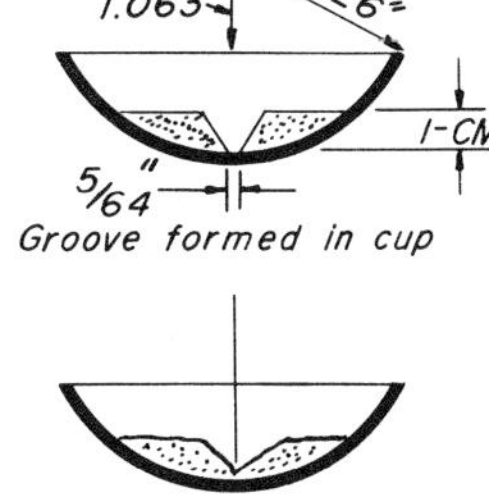

Groove formed in cup

Groove closed

1. Obtain a representative pint sample of soil.
2. Air dry.
3. Hand sieve using a No. 40 sieve. Discard retained material.
4. Mix 100-gram sample with measured quantity of water.
5. Place portion of sample in brass cup to a depth of 1 cm.
6. Level surface and divide with special grooving tool to leave a 5/64-in. opening between the halves.
7. The cup is tilted on line of groove and allowed to drop back on a hard surface until the two sides of the groove meet.
8. Water content is determined from portion around groove.
9. Moisture content and number of blows are recorded.
10. Repeat with variable quantities of water to give readings above and below 25 blows.
11. A curve is plotted with percentage of moisture as ordinate and the number of blows as abscissa (logarithmic scale).
12. The percentage of moisture at 25 blows is picked off the curve and is the value of the liquid limit.

PLASTIC LIMIT

Definition: The plastic limit of a soil is the lowest water content, expressed as a percentage of the weight of the oven-dried soil, at which the soil can be rolled into threads 1/8 in. in diameter without the threads breaking into pieces.

1. Obtain a representative sample of soil.
2. Air dry.
3. Hand sieve using a No. 40 sieve. Discard retained material.
4. Mix 15-gram sample with water until mass becomes sufficiently plastic to be easily shaped into a ball.
5. Roll the ball between the palm of the hand and the surface of a glass plate with sufficient pressure to form the mass into a thread.
6. When diameter of thread is 1/8 in., knead the soil together and again roll out.
7. Continue this process until the soil crumbles when the thread becomes 1/8 in.
8. Gather the portions of the crumbled soil and determine the water content.
9. This water content, expressed as a percentage of the oven-dry weight of the soil, is the plastic limit.

PLASTICITY INDEX

Liquid limit minus plastic limit = plasticity index.

3.9 CONTROL OF FOUNDATIONS ON CLAY

The occurrence of clay as a foundation material is very widespread, its chief prevalence being in relatively low-lying areas adjacent to bodies of water at precisely those locations where, in times past, cities were most apt to be founded.

The city of Chicago is underlain almost completely by a clay bed as much as 200 ft thick, prohibiting economically the pushing of all but the heaviest foundations down through it. Denver has from 20 to 600 ft of clay depth over bedrock, and Detroit has from 60 to 90 ft. Although New York City's subsurface conditions are more variegated, clay deposits are not missing, and the same thing is true in Philadelphia.

Clays, excavated and re-worked, using moisture control and compaction, have been made into virtually impervious masses for use as linings for canals and reservoirs and as impervious cut-offs in earth dams, but practical considerations seldom permit the re-working of clays on which structural foundations are to be placed. Extensive studies of re-worked or "compacted" clays during the past several decades have contributed greatly to our knowledge and control of clay as a building material, but very little of this research has been concerned with undisturbed clays on which foundations are to be built. Yet the findings from research on re-worked clays indicate that proper control of foundation clays is far from adequate.

The determination of the bearing value of a clay bed in its undisturbed state is not a difficult field procedure and, used with a reasonable factor of safety, and with the pre-vision that some limited settlement may occur, is satisfactory within its limitations. The limitations are that such tests be carefully conducted so that the clay is truly undisturbed and that it is protected from water and drying which may change its characteristics. But once excavation is begun, large expanses of the clay bed are exposed alternately to the elements of rainfalls and hot suns, and these alone can alter the bearing capacity of the clay. The clay is then massively loaded as the structure rises upon it and, loaded, is then subjected to wetting or drying or to changes in temperature whose effect has never been studied because it never previously existed.

It was noted in Sec. 3.8 that clays are flakes strung together from crystals of atomic size; in certain ways they resemble a box of soap chips. As the soap chips in the box reach us, they are "all shook up" and the flakes lie at all angles between the vertical and the horizontal. If they are put into a container under pressure the volume is readily reduced, principally by breaking the flakes up into smaller pieces. If the flakes could be oriented so that they would all be in a horizontal position and in contact, face to face, the same reduction in volume would occur, the breakage would be limited, and the supporting strength, in either case, would approach the strength of a homogeneous bar of soap.

These soap chips are dry, but if we introduce a limited amount of water the conditions become quite different, and the temperature of the water will have considerable effect on the result. The analogy cannot be carried too far since the basic characteristics of a soap flake and a crystalline clay flake are quite different, and such factors as the di-electric properties of clays are not present in soap; but the re-orientation of flakes is important and parallel, as are the quantity and temperature of water that may reach the mass.

Once our foundation is in place, applying pressure, can water reach the clay under it, re-orienting the crystals, changing the electrolytic concentration and so decreasing the strength? Will this water contain chemicals that, over a period of time, may produce chemical changes in the clay and so change its structure, or can it, with time, leach away important elements in the mass and so change its characteristics? Will the clay "creep" under continuously applied pressures, or will the initial containment change so that the clay is squeezed out through an aperture like toothpaste?

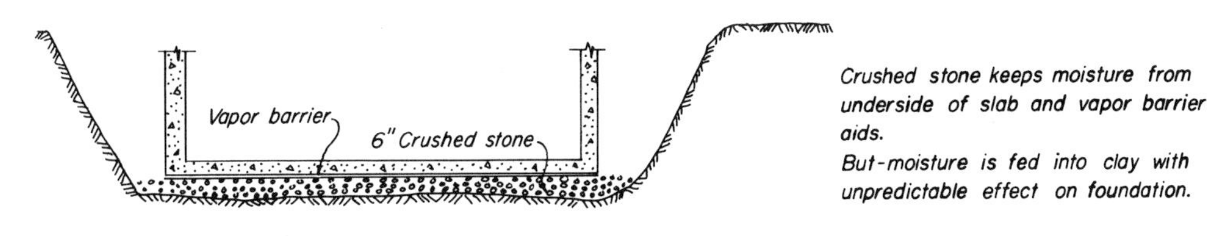

(a) A CUSTOMARY FOUNDATION PROCEDURE

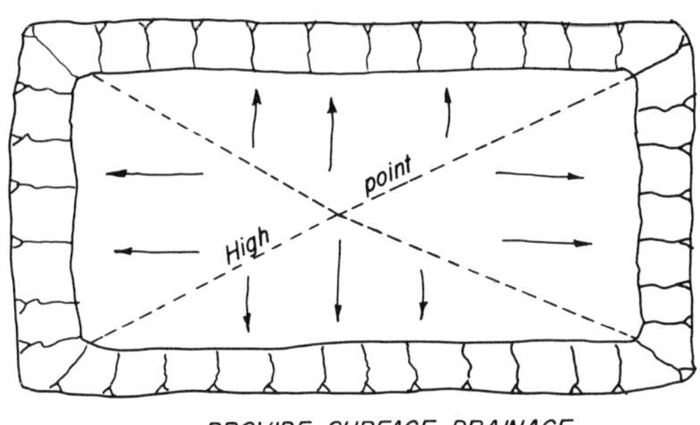

PROVIDE SURFACE DRAINAGE

Moisture barrier

SPRAY SURFACE WITH SEALANT. USE ASPHALT OR A PLASTIC COMPOUND. COVER WITH 2" OR 3" CONCRETE SLAB IF NECESSARY.
WHERE SURFACE IS SLOPED, SIDE DRAINS AND SUMPS MAY BE REQUIRED.

(b) PRECAUTIONS DURING THE CONSTRUCTION PERIOD

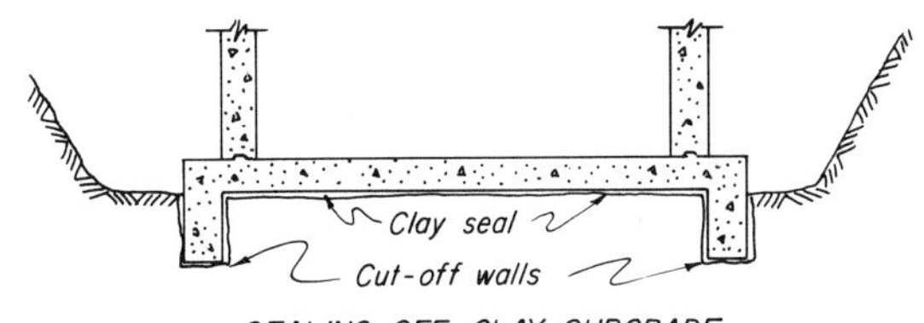

SEALING-OFF CLAY SUBGRADE

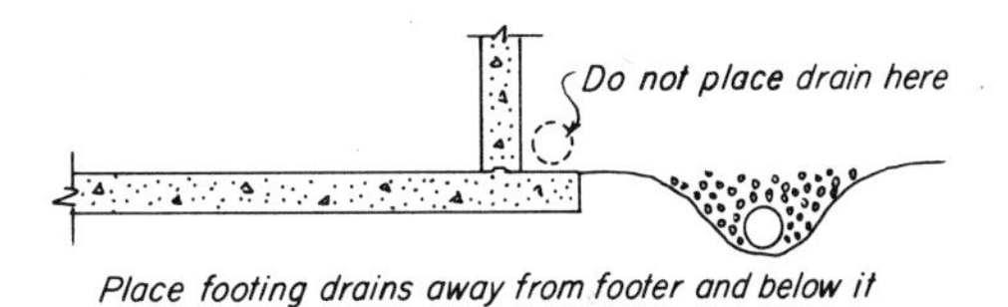

PROPER UNDERDRAINAGE

(c) POSSIBLE CLAY FOUNDATION CONTROL REFINEMENTS

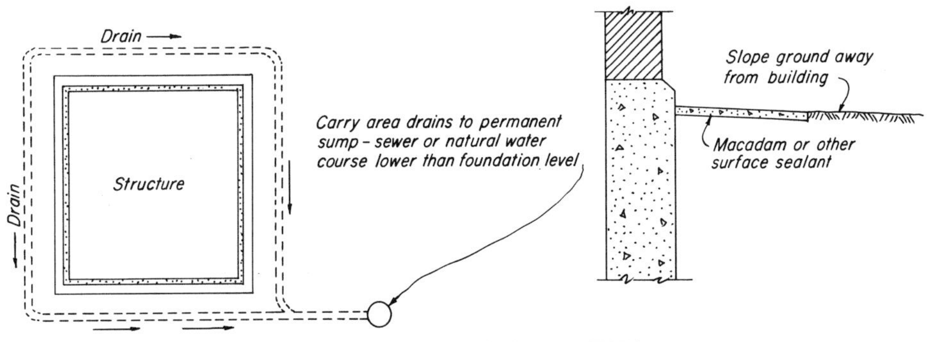

(d) FINAL FOUNDATION CONTROL FACTORS

CONTROL OF FOUNDATIONS ON CLAY

3.10 EXPANSIVE CLAYS

A man waters his lawn in Peru, and plaster falls from the ceilings of his home as the structure slowly rises in the air and ¼-in. cracks develop in the brick masonry walls. The same thing happens in a residential area near Denver, Colorado. The bottom of a canal heaves in California and shatters its concrete lining, while a water-supply siphon of 80-in.-diameter welded steel pipe, over the Malheur River in Oregon, twists out of shape. On "The Road to Mandalay," in Burma, the office building of the Mandalay Race Club is so badly cracked that it must be torn down, and all bets are off. These are some of the destructive results of water on expansive clays.

Expansive clays contain substantial amounts of montmorillonite (Sec. 3.8), their tendency to expand depending very largely upon the percentage of this clay which they contain. There is no field test by which this characteristic of a clay can be determined; only carefully controlled laboratory analysis can suggest the possibility.

Montmorillonite clays may be encountered on the surface, or near the surface of the natural ground, but those clays with which we are here concerned are those which occur at foundation levels or below. Quite frequently these layers of clay have been protected by an impervious layer, perhaps of a non-expansive clay, and have not previously been exposed to water. In these cases the restoration of the ground surface around the structure should be aimed at preventing later access of water to the lower level.

It may be possible to place an impervious fill and seal off moisture penetration. This has been done with canals cut through expansive clays. A compacted earth lining has been used to prevent penetration of moisture where materials for this purpose are within economic hauling distance. But in structures of other types, surface ditching or drainage must be provided to intercept water and carry it away. In the case of the siphon noted above, it was found that irrigation ditches were feeding water to the expansive-clay layer, and some adjacent lands had to be taken out of cultivation.

Removal of the expansive clay and its replacement with well-compacted non-expansive materials is always a possible solution; but if the bed of clay is too deep or too extensive, not only can the cost be excessive, but the problem of finding and placing suitable alternative materials may present itself.

It has been found that the volume change in expansive clays can be controlled by compacting the soil under substantial pressures at high moisture contents. This preconsolidation technique must first be established in the laboratory to develop the effective moisture content and the required load. However, attempts to translate laboratory conditions into field conditions present serious difficulties.

A soil which expanded 10% upon wetting without load still showed a 5% expansion when loaded to 40 psi or about 2.8 tons per sq ft; whereas the same soil, first loaded to 50 psi (3.6 tons per sq ft) and then wetted showed expansion to 7% only when the load was completely removed. Since the "wetting" referred to consisted in increasing the moisture content to about 25% (of the dry weight), the difficulties for field use are at once apparent. To procure complete wetting, the soil must be thoroughly worked. The resulting plastic mass, unless closely confined, will not sustain the loading required for consolidation.

The extension of footings to an inactive soil zone below the expansive-clay level by means of piling or caissons may offer the best solution, but even here some caution is indicated. If the expansive layer is thick, relatively confined, and periodically subjected to saturation, uplift problems can present themselves. Certain types of pilings and caissons are more susceptible to uplift than others; but even where the foundations themselves are not affected, because of their total loading, the basement slabs between them can be cracked by the pressures exerted.

EXPANSIVE CLAYS

1. Qualitative laboratory tests can determine the amount and types of minerals present. The quantity of montmorillonite and the kind and amount of exchangeable bases will indicate the degree of expansiveness. The tests include:

(*a*) Petrographic microscope	(*b*) X-ray diffraction
(*c*) Differential thermal analysis	(*d*) Colloid content
(*e*) Plasticity index	(*f*) Shrinkage limit

2. Quantitative laboratory tests reproducing expected field conditions in a consolidometer:

   (*a*) Saturate undisturbed soil specimens under anticipated loading.
   (*b*) Develop load—expansion curves from a variety of load conditions.
   (*c*) Translate into field operating conditions.

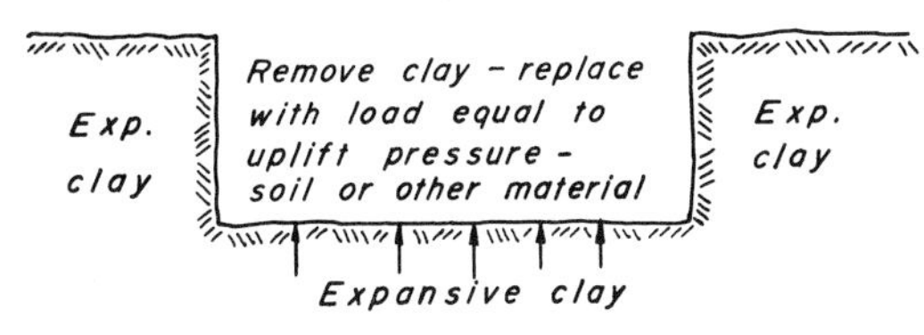

This depends on accurate judgment as to future quantity of water reaching clay. Consider sealing clay - see (c)

(*a*) SUBSTITUTION ON LOADING

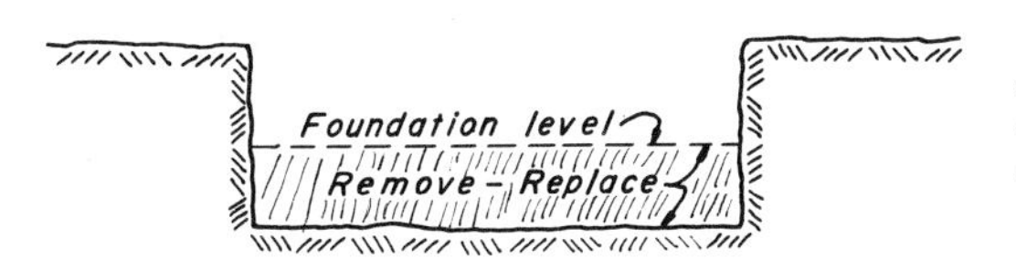

Remove volume indicated. Remold with water to increase moisture content. Replace under controlled loading

(*b*) REWORKING FOUNDATION SUBGRADE

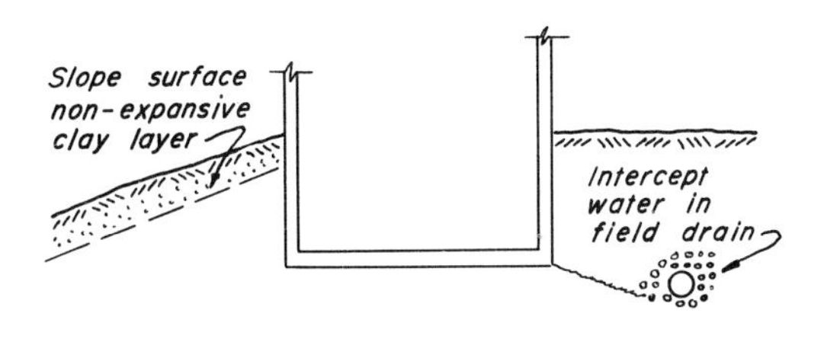

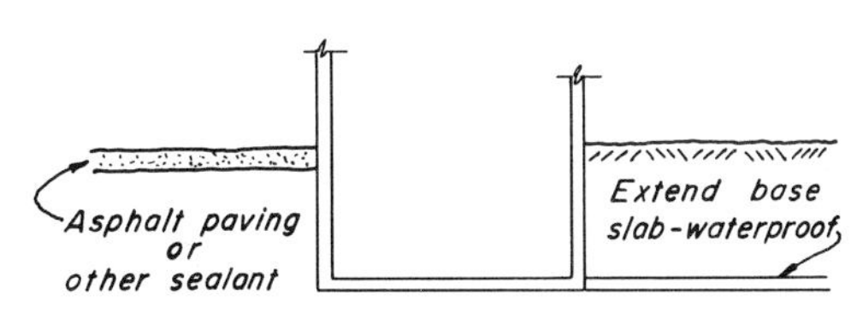

(*c*) CONTROL OF MOISTURE CONTENT

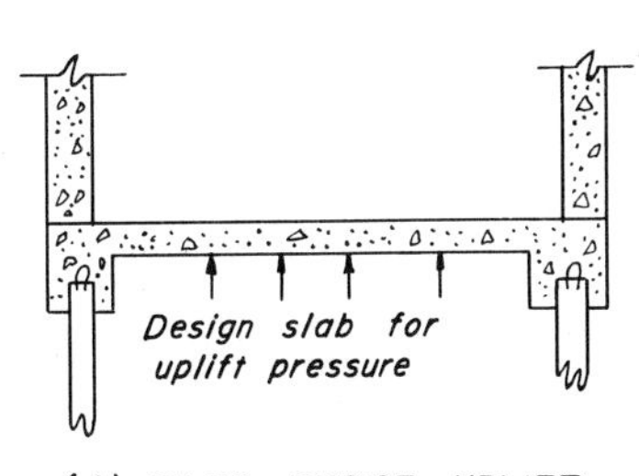

(*d*) PILES RESIST UPLIFT

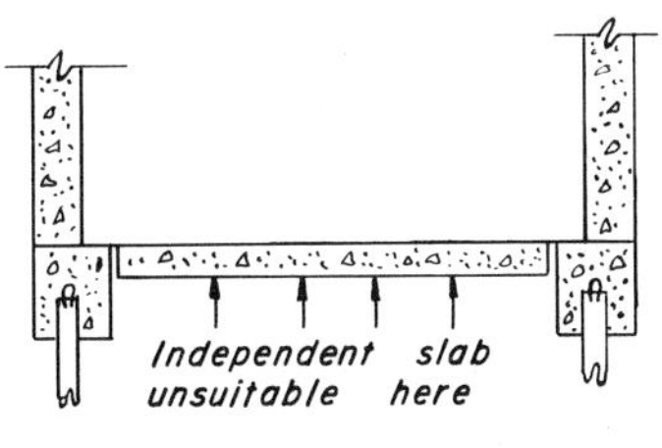

(*e*) PILES RESIST UPLIFT

Some special cases occur in excavations for foundations which cannot be attributed to the factors so far discussed and which involve clays that are not expansive or at least are not expanded because of the action of water. These cases involve the movement of clays by heaving when excavation is performed within them due, basically, to a redistribution of pressures within the mass directly resulting from the excavation. Such heaving occurs in clays of considerable sensitivity.

Sensitivity of clay has been defined (by Terzaghi and others) as the ratio of the unconfined compressive strength in the undisturbed state to the same strength after it has been remolded. The determination of this ratio is obviously a laboratory procedure; and the resulting value, which may vary from 1 to 8 or 10, does not necessarily indicate a value at which heaving will take place. This is very likely to be determined by the nature of the operation.

The illustration (*a*) shows the general condition. As excavation is performed, the confining effect of the superimposed strata is removed from the clay layer below, producing a loading on the clay surrounding the pit. Three things can occur where we are dealing with a sensitive clay: the area surrounding the pit can subside; the side walls of the pit can deflect as indicated, although the effect, in a cohesive soil, may be insufficient to produce sliding; and the bottom of the pit can heave upward.

Superficially it would appear that in an area where the subsidence was of no importance and the pit walls did not collapse, the solution would be simply that of removing the extra soil heaved up at the center. This solution, however, solves nothing, since the loading to be placed by the structure in the area of heave will almost certainly be greater than the original upper strata loading, causing heave outward at a later date. In lieu of this, the sequence of excavation and foundation construction must progress so as to avoid any initial heave at all.

Factors determining the extent of heaving depend on the density of the clay, the shearing value of the clay, and, as has been noted, its sensitivity, factors which require laboratory procedures. Additional factors include the depth of the stratum of sensitive clay as well as the size and dimensions of the pit. Where depth to a stable layer is not far below foundation level, sheathing driven to stable soil can inhibit heaving. Sheathing may or may not be effective in deeper deposits of clay.

The length of a restrained (sheathed) excavation has a bearing on the potential heaving, since the longer it is in relation to its width, the greater the likelihood of heave near the center of its length. This can be controlled by placing a center slab of limited width athwart the center, equal in weight to the removed overburden. Additional slabs can then be poured alternately on either side of it until the entire bottom has been concreted.

A similar operation was found to be necessary in the construction of the generator foundation pits for the Oahe Dam. The site is a soft shale with unconsolidated clay layers of some sensitivity. The shale is cracked and fissured and, at least to a considerable depth, has no stability. The whole mass acts like a sensitive clay, and the sensitivity is increased by wetting (rainfall). Tests indicated that at the 37-ft depth, heave could amount to 2 ft.

The pits required had side slopes of 1:1 so that slope paving was practical. A 7-ft depth of pit was first excavated, as indicated, and a working slab laid down. Through this, anchor bolts were drilled. When these had been grouted, the combined effect of the slab and the partial anchoring was sufficient to hold the sides and bottom while a second 7-ft layer was excavated.

A structural slab was placed over the whole area when all the anchors had been set and grouted.

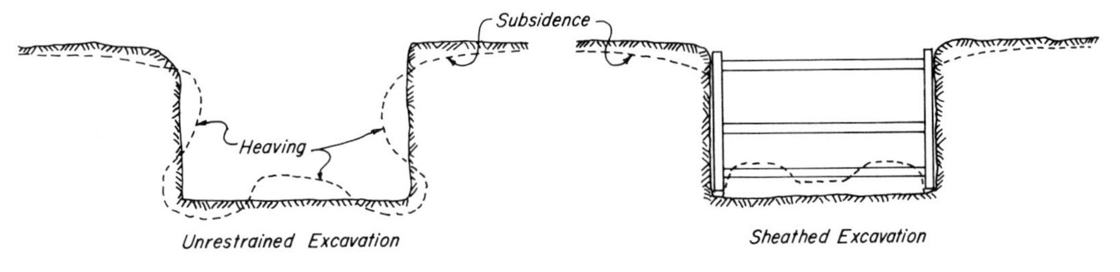

(a) HEAVING IN SENSITIVE CLAYS

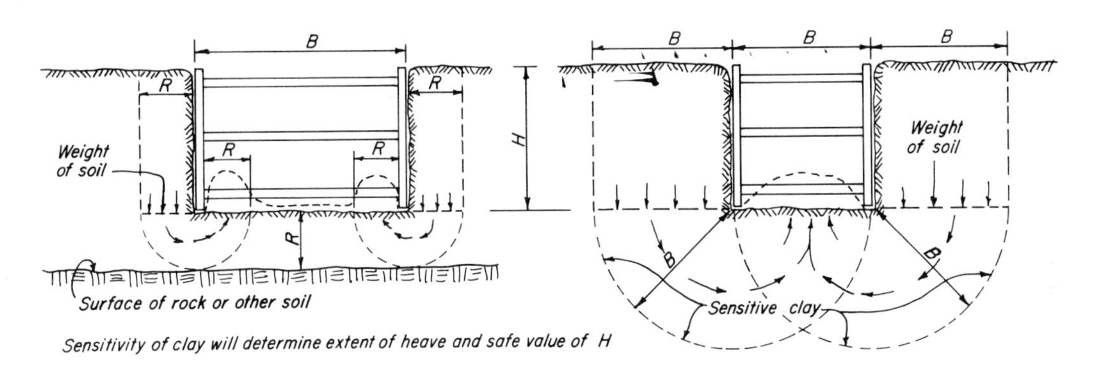

(b) EFFECT OF SUBSOIL CONDITIONS IN PRODUCING HEAVING

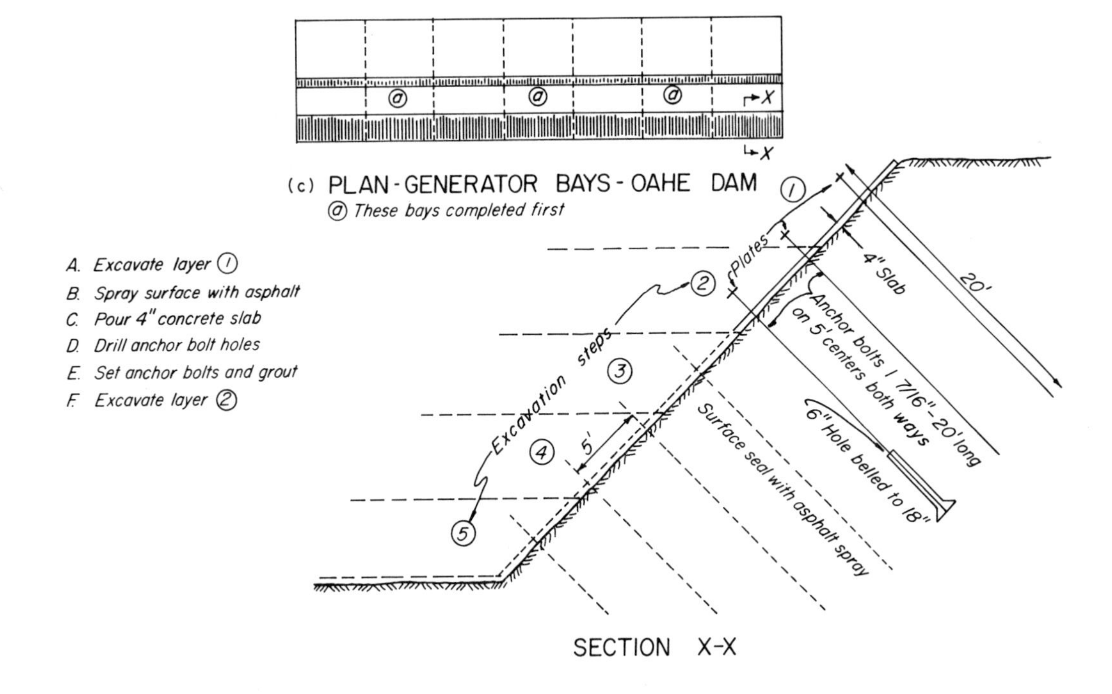

(c) PLAN-GENERATOR BAYS-OAHE DAM

ⓐ These bays completed first

3.12 ORGANIC SILT

Silt consists of soil particles between 0.05 and 0.005 mm in size, as classified by A.S.T.M. In these sizes they are below the range where their grading can be determined by sieving, and their relative sizes must be determined by sedimentation. Since they exist in the intermediate grain sizes between sands and clays, logic, as herein induced, would place them in a position following sand, as we go down the scale. However, on examination, it will be found that silt consists of more finely divided particles of sand but also of oversize clay particles built up by the chemical and electrolytic forces already discussed in connection with clays. Consequently, a proper understanding of silt follows that of clays.

The term "organic silt" has been frequently used in the field. To the contractor encountering it, no doubt remains as to the meaning of the term. It is a dark grey to black, is damp, and has an odor of decay. When it dries out, the color turns to light grey and the odor largely disappears. Microscopic examination reveals the presence of sand fragments of quartz and feldspar, minute mica flakes and chips, clay aggregates, and a variety of micro-organisms feasting on organic debris.

Organic silt is not truly a silt by definition of grain size; for not only are true clays encountered but fine sands as well. It deserves its own consideration, however, since it is very prevalent over extensive areas around the mouths of rivers, particularly along the east coast of the United States. It is frequently encountered in subsurface work incident to bridges, tunnels, and waterfront structures. True silts do occur in other locations, generally as an admixture to soils predominantly classifiable as sands or clays and consequently subject to treatment as the one or the other.

The "organic" element in organic silts more nearly defines their characteristics than any other facet of their existence. As encountered, these silts may comprise a layer as deep as 50 ft on the bottom of a river bed in the lower reaches devoid of swift velocities. Vast areas termed marshes, tidal flats, or swamps are overlain with organic silts. For the most part such soil conditions are geologically new and are even now in the process of formation. They seem generally to be the result of sedimentary deposits of fine soils laid down repeatedly over organic growths.

These silts, because of the organic content, tend to retain water, making them soft and unstable, and are subject to progressive alteration under loading as their organic content changes. They cannot be used as foundation material.

Piling, driven through organic silts to layers of other materials below, is the most usual treatment, but it can have certain disadvantages. Vibrations due to driving may cause subsidence in the area, although where piling is driven in clusters consolidation may be limited to the area immediately surrounding the cluster. Where friction piling is involved, the surface of the piling extending through the silt layer can be counted on for very little support.

The complete removal of organic silts by dredging is often resorted to, but in many cases is not practical either because of the depth and extent or the difficulties of finding suitable disposal areas.

Where depths of organic silt vary between 8 and 20 ft, methods of displacing this material by controlled blasting have been effective. In the case where fill is being advanced across a bed of organic silt, continuous "toe" shooting (see illustration) will displace the silt, permitting suitable fill to replace it. There is, of course, no certainty that substantial quantities of silt will not be mixed with the lower layers of compactable fill, and consequently that later settlement will not occur. For some types of highway work, of course, a limited amount of settlement may be acceptable. Where crushed rock, gravel, or other coarse fill material is being used, this objection does not exist.

The use of sand drains for the consolidation of organic silts has been developed. This method also envisages the possibility of limited settlement and presupposes that the silts are essentially water-soaked.

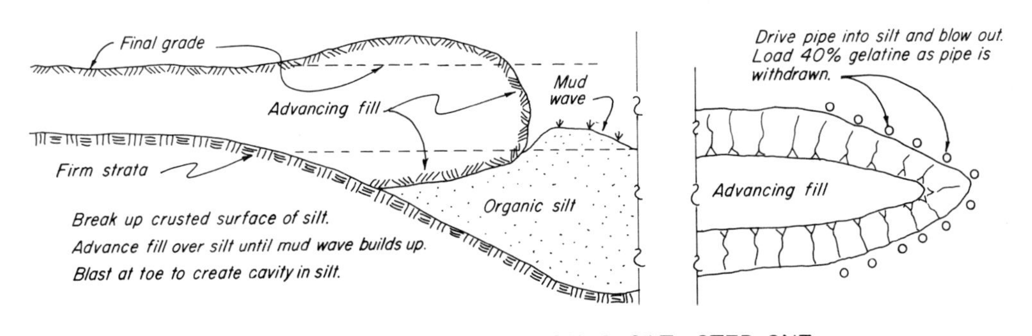

(a) TOE - SHOOTING IN ORGANIC SILT - STEP ONE

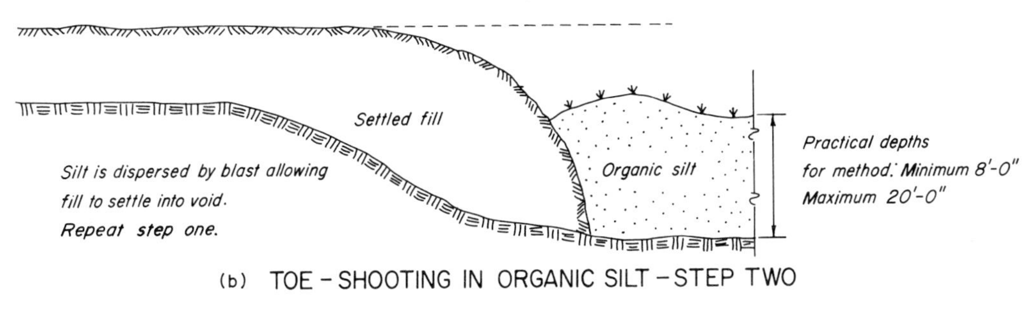

(b) TOE - SHOOTING IN ORGANIC SILT - STEP TWO

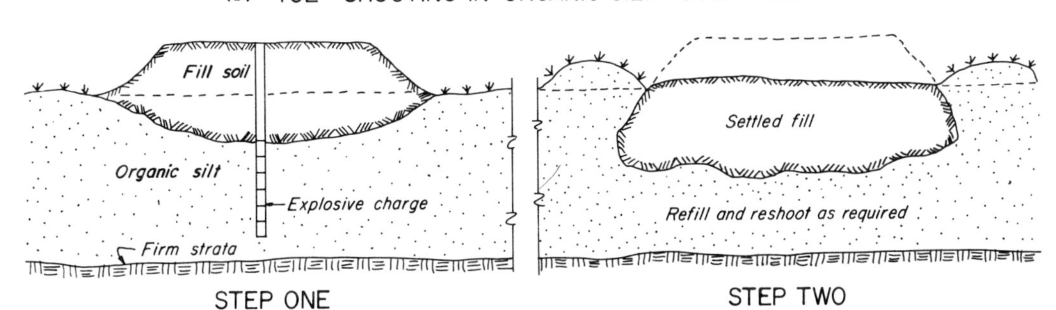

(c) UNDERFILL METHOD IN ORGANIC SILTS

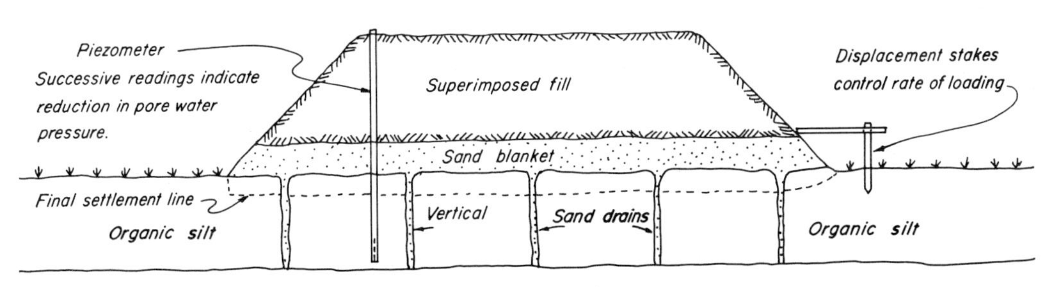

(d) CONSOLIDATION OF ORGANIC SILTS BY VERTICAL SAND DRAINS

3.13 LOESS

Aeolus dwelt on an island massif in the Tyrrhenian Sea within which all the winds were stored. Zeus had confined the winds for fear they would sweep the earth and sea away into the sky, and had given Aeolus their keeping. When a Greek god wanted a wind, Aeolus simply plunged his spear into the proper spot on the face of the massif and let one roar out. For Odysseus, Aeolus had made up a bag of winds which, tied to the ship's bow, prevented that *voyageur* from ever becoming becalmed.

Zeus wasn't just beatin' his gums, for large areas of the North American, European, and Asian continents are covered with aeolian (wind-borne) soils known as loess. Deposits as deep as 200 ft occur in the Nebraska and Kansas portions of the Missouri River basin; in the vicinity of Denver, Colorado; in the state of Washington; and along the Illinois and Mississippi River valleys.

Loess is composed essentially of silt particles ranging in size from 0.01 to 0.05 mm and contains some clay and fine sand in sizes respectively finer and coarser than itself. It is a tan to pale brown in color; is light in weight; crumbles readily; and contains perceptible voids, generally in the form of vertical channels.

Loess owes its major properties to the method by which it was laid down. Picked up by the wind from dry glacial outwashes, its particles were carried forward until the wind died out and they came fluttering down to earth. The dry particles built one above the other, subject to none of the pressures produced by a medium such as water. As a result, in natural masses, loess weighs from 75 to 95 lb per cu ft and occasionally as little as 65 lb per cu ft. Because of its essentially porous nature, rainfall passing through it gradually formed the vertical channels noted.

The natural moisture content of loess is about 10%, and at this moisture content the supporting capacity is high, often up to 4 tons per sq ft with settlements of less than 0.02 ft. This condition prevails up to a moisture content of about 15% with only a slight increment of settlement; but between 15 and 25% moisture content, its supporting capacity falls off rapidly.

Not only the degree of moisture is important, but also the density. Low-density loess (less than 80 lb per cu ft), when saturated, will settle excessively under load. At densities between 80 and 90 lb per cu ft, settlement is much less; and over 90 lb per cu ft, loess will support normal soil loads even with moisture in excess of 20%.

An early failure of a structure on a foundation of loess indicates the problem and offers a solution. A grain elevator in Kansas, completed about July 1, 1950, on a foundation of loess, began to tilt toward the north on July 18 and by August 24 was 26½ in. out of plumb. This was during a period of heavy rainfall and, because of inadequate drainage, considerable ponding occurred along the north side of the elevator where run-off was inhibited by a railroad spur. During the same period, grain was being loaded into the elevator.

To correct the tilt, ponding was artificially induced along the south side of the grain elevator while this side was additionally loaded. This procedure, although dangerous, corrected a substantial portion of the tilt.

Ponding over loessial foundation soils has been used in a number of cases, but the wetting of loess without the addition of superimposed loading is not enough to cause consolidation. In the case of the Medicine Creek Dam in Nebraska, the moisture content of the foundation loess was increased from 12 to 28% by ponding and maintained in that condition while 30 ft of fill was placed. Settlement plates indicated settlement of between 0.8 and 2.0 ft.

Another case is that of the Courtland Canal, where simple ponding 1 ft in depth was tried over a 40-ft-deep bed of loess. Although moisture content was increased to an average of 24%, no appreciable settlement occurred in the test section, indicating that loading as well as wetting is necessary.

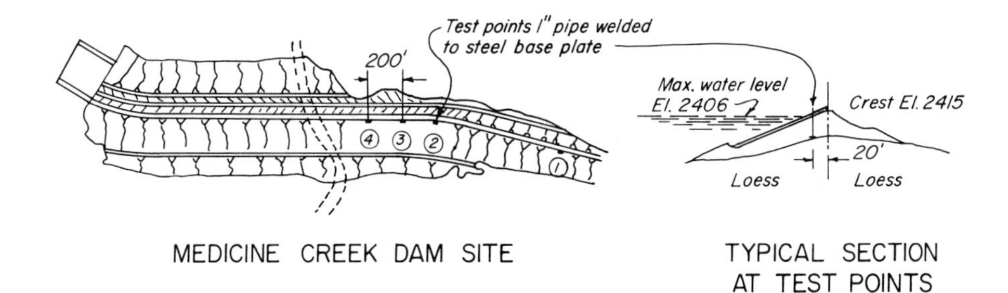

MEDICINE CREEK DAM SITE

TYPICAL SECTION AT TEST POINTS

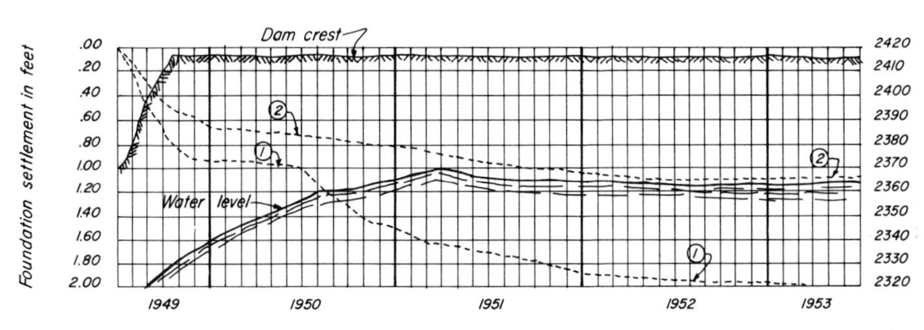

RELATION BETWEEN SETTLEMENT AND RATE OF LOADING AND SOAKING

Depth of loess at (1) = 80'. At other test points depth of loess = 40'. Moisture content of loess raised from 12 % to 28% before filling but no settlement occurred. Load at (1) = 35 p.s.i.; at other points = 45 p.s.i. Note that water level did not get within 40' feet of its maximum.

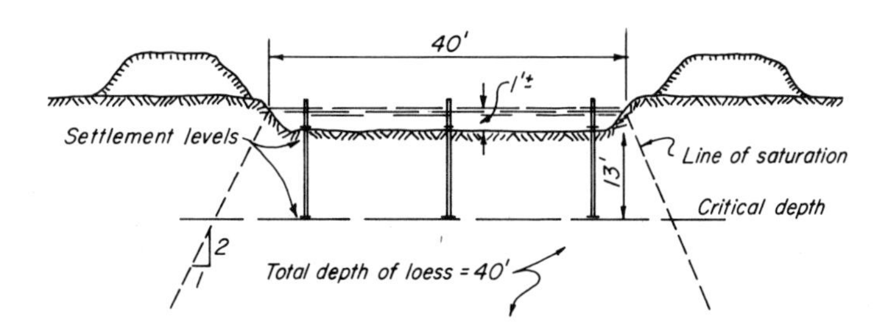

SATURATION TEST AT COURTLAND CANAL SITE

Test area 40' long filled 1' deep with water for 3 weeks permitting 330,000 gallons to permeate loess. Moisture content raised to 24 % but maximum settlement noted only 1/4". Note rather regular extent of saturation of loess.

LOESS AS A FOUNDATION MATERIAL

3.14 PERMAFROST

Permafrost is the name given to that layer of soil below the earth's surface, in arctic regions, in which a temperature below freezing exists and has existed since time immemorial. Its upper limit is known as the permafrost table. That portion of soil lying between the permafrost table and the surface of the ground is known as the active zone. The active zone is responsive to climatological changes, freezing and thawing with the annual variations in air temperature. To some extent, therefore, the depth of the active zone will vary with the length of the summer in a region, although the nature of the soil will also have a bearing on this thickness.

The foundation problems of structures in permafrost areas are complex. Foundations placed in the active zone are subject to the same frost heaving familiar in more temperate climates, an extreme case of which was cited in Sec. 3.5. On the other hand, foundations placed in permafrost are subject to settlement, not from seasonal variations of surface conditions but from the intrinsic characteristics of the structure resting on the foundations. Since permafrost may occur to depths of 200 ft, the founding of structures on rock or non-frozen soils below permafrost is frequently uneconomical.

The combinations of circumstances under which permafrost occurs are numerous. It may occur as a solid, continuous homogeneous mass of frozen ground, or it may consist of ice islands or lens in essentially dry formations which are below freezing in temperature but, not being water-saturated, are not a frozen mass. Permafrost can occur in frozen and non-frozen layers. The combinations in the active zone are similarly variable. The annual freezing and thawing cycle can penetrate the entire active zone down to the permafrost table or it can penetrate only a portion of it, leaving unfrozen ground between the upper frost zone and the permafrost table. This unfrozen ground layer may, and generally will, contain groundwater.

Groundwater is the most important factor relating to permafrost since it is apparent that in perfectly dry ground permafrost could not exist. (Perfectly dry ground will only be encountered, briefly, at and near the surface of the ground in desert areas.) Groundwater may occur above, within, or below the permafrost.

Groundwater above permafrost will originate from rain or melting snow or may seep upward by capillary action from the permafrost table. Since the permafrost represents an impervious stratum, downward drainage of this groundwater will not occur; and where permafrost tables are relatively flat or the active zone is composed of silts or clays, no run-off is possible. This condition produces the vast tundra areas so prevalent in sub-arctic regions. The water, thus retained in the active zone, alternately freezes and thaws without escape, producing bottomless bogs in summer weather. In some instances, because of variations in the nature of the soils, water may accumulate in pockets, freezing in the winter and heaving the ground surface to produce frost mounds, "drunken forests," and crevices.

Groundwater may exist below permafrost occasionally, as a quiescent intrusion, but more frequently it is fed from gravel or creviced rock formations at higher elevations that puncture the continuity of the permafrost layer. Trapped beneath the impervious permafrost and a lower layer of clay or dense rock, it may exist under considerable hydrostatic pressure.

Temperatures within the permafrost layer will generally be 5 to 15° below freezing, and consequently water will be encountered only in the form of ice islands or lens. However, the permafrost belt may be subject to aggradation or degradation, that is, to growth or decline due to either natural forces or the artificial forces incident to construction.

In permafrost areas, nature has established and maintains what is called a "thermal regime." Any disturbance of this regime, as, for instance, by construction operations, disturbs this equilibrium; and, until the original regime can be re-established or a new regime created, foundation construction can be a precarious undertaking.

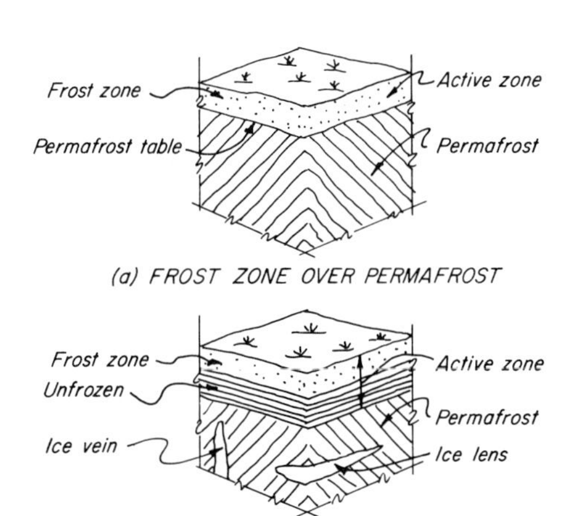

(a) FROST ZONE OVER PERMAFROST

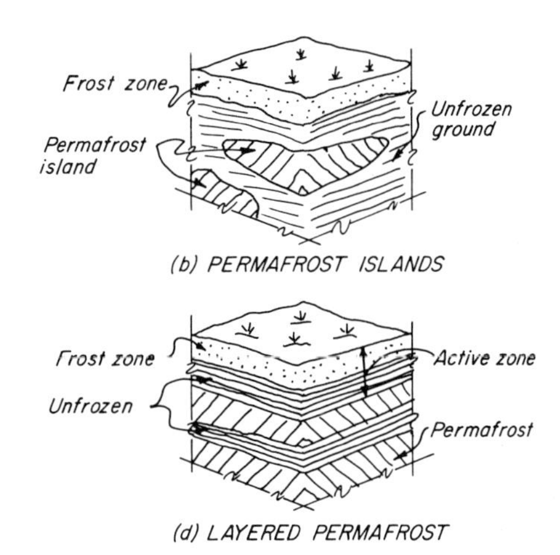

(b) PERMAFROST ISLANDS

(c) ICE INTRUSIONS IN PERMAFROST

(d) LAYERED PERMAFROST

THE OCCURRENCE OF PERMAFROST

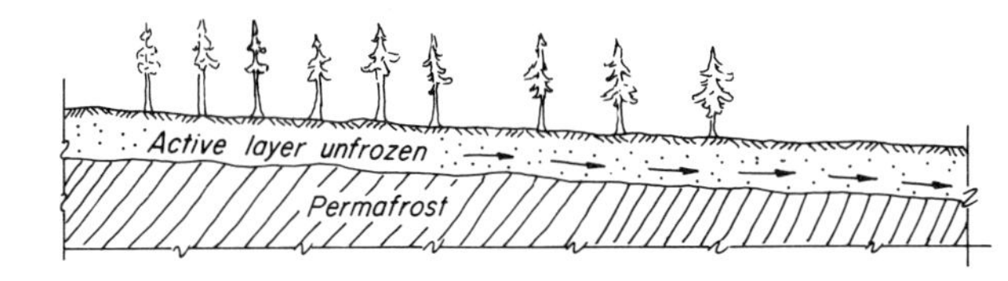

(e) GROUND WATER IN ACTIVE LAYER UNFROZEN PERCOLATES FREELY DOWN SLOPE

(f) ACTIVE LAYER COMPLETELY FREEZES AT LOWER END DAMMING UP WATER. RESULTING HYDROSTATIC PRESSURE PUSHES UP GROUND SURFACE.

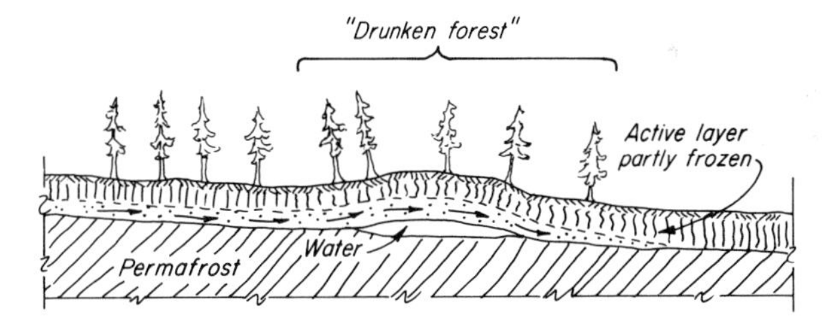

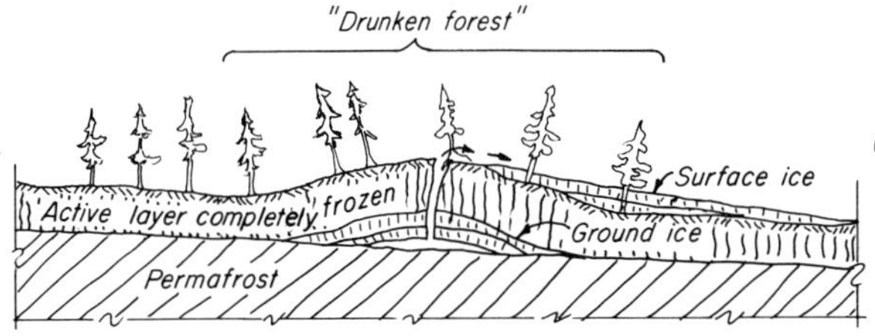

(g) MOUND IS FINALLY RUPTURED BY HYDROSTATIC PRESSURE AND WATER FLOWS OUT TO FORM SURFACE ICE. SOMETIMES A HOLLOW SPACE IS LEFT AT THE CORE OF THE MOUND.

THE TALE OF THE DRUNKEN FOREST

3.15 GROUNDWATER AT FOUNDATION LEVEL

The subject of groundwater and its control in foundation construction is very extensive and is dealt with at length in the author's volume "General Excavation Methods." The discussion here will be limited to placing foundations where limited volumes of water must be handled simultaneously.

In Sec. 3.10 we discussed the effect of water on clay subsoils in the light of its effect on clay; but the combinations of conditions that are possible are quite numerous, and these conditions can affect not only the soils themselves but the foundation which is being placed.

The placing of concrete under water by means of the tremie or related methods is limited to substantial masses of concrete whose primary function is that of providing weight or limited load distribution. Ordinary concrete sections cannot be placed satisfactorily under water or when water is permitted to flow over the fresh concrete. Excess water due to submergence can change the water-cement ratio of the concrete and lower its strength. Variations in water pressure between the two surfaces of a wall or slab can cause channeling of water through the concrete, washing out the fines and producing a porous concrete which is weak. Water flowing over the surface of fresh concrete will produce the same result in addition to contouring it. Once concrete has obtained its initial set, a matter of hours (the number depending on temperature, mix, and placing conditions), the submergence of concrete in water will be definitely advantageous and is used specifically as a curing aid.

Foundations should be poured in the dry unless their mass is substantial. Methods of obtaining dry conditions on which to place foundations will depend on the source and extent of the groundwater. Surface water from streams or run-off can generally be channeled away from the site. Immediate and local rainfalls will require some temporary shelter, not only for the concrete but for its surface finishing as well. Protection of concrete from groundwater will require more extensive provisions.

Certainly the best way of handling groundwater is to lower the groundwater table in the whole area of foundation construction by the use of deep wells or well points. For a number of reasons, this may not be practical or economical, particularly where foundations of limited extent are involved.

Groundwater may enter a pit, excavated for a foundation, through any or all the walls, through the bottom or floor, or through both. Water flowing through the walls, in soil banks, will generally cause sufficient erosion to warrant sheathing the banks. The sheathing will generally not stop the flow of water but will reduce the velocity of it so that it runs down the inside face of the sheathing and can be collected in a ditch and drained to a sump.

In rock formations, seams containing flowing water should be opened sufficiently to prevent the water from spouting into the excavation area, so that all such water flows down the rock face into a gutter cut into the rock at the toe of the slope and is so drained away.

In both the above cases sufficient additional excavation must be provided to accommodate drainage ditches, and in erodible soils ditches should be lined with timber or filled with gravel to prevent their being washed out.

Water flowing into an excavation from the bottom or floor, frequently that on which a slab is to be placed, requires some sort of underdrainage system. The handling of this condition in clay has already been discussed. In general, consideration must be given to the bearing value of underdrainage systems. By carrying the excavation 6 to 8 in. deeper and providing a layer of gravel or crushed stone of that thickness covered with a layer of impervious film or paper (such as Sisalcraft), the drainage problem can be solved. Whether this will also provide the required bearing value for the foundation should be investigated in each case.

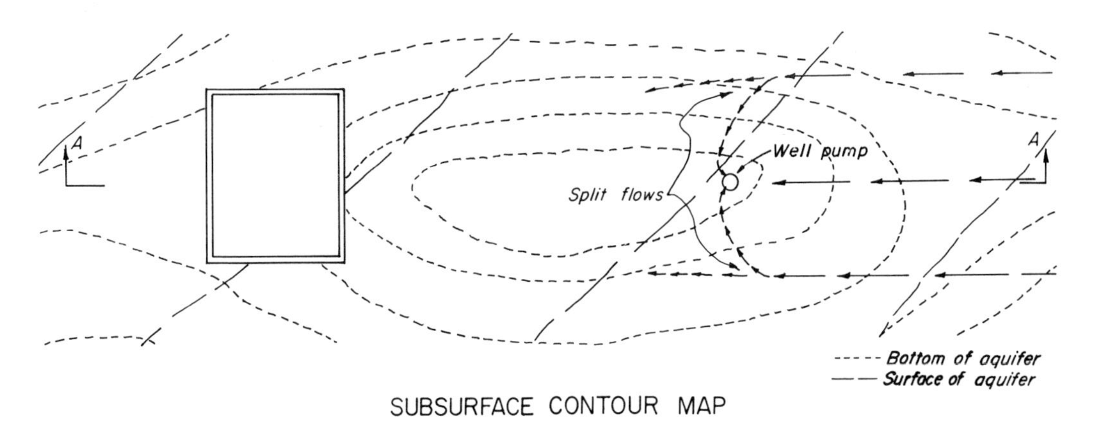

SUBSURFACE CONTOUR MAP

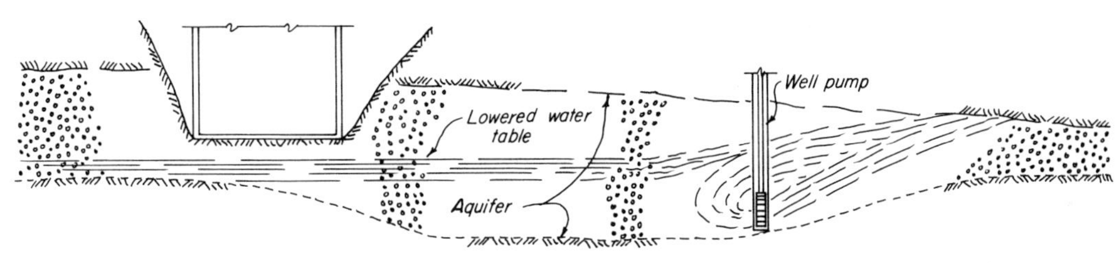

SECTION A-A

(a) A CASE OF DEWATERING BY WELL PUMP

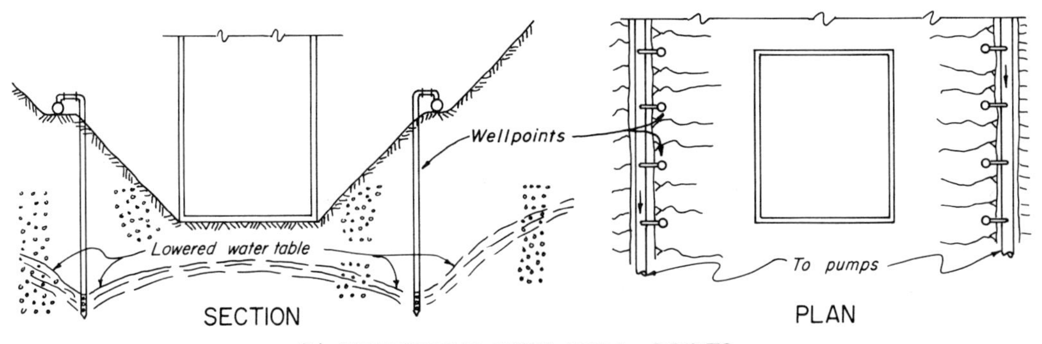

SECTION PLAN

(b) DEWATERING WITH WELL POINTS

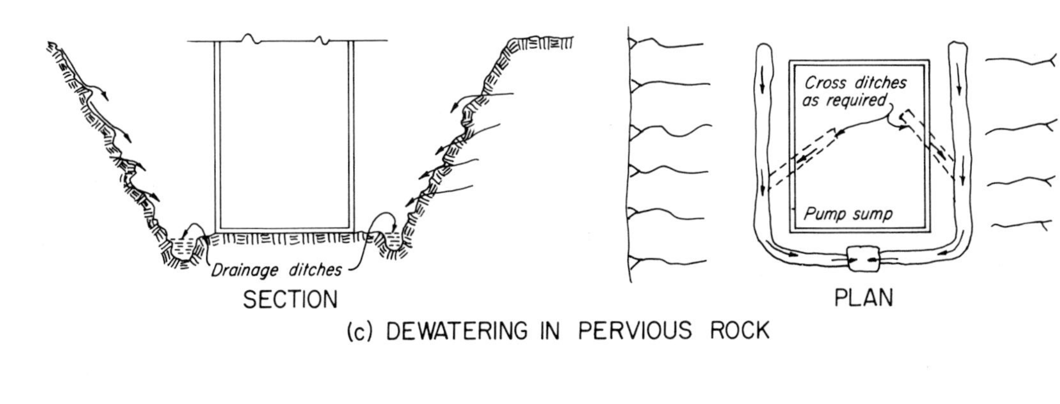

SECTION PLAN

(c) DEWATERING IN PERVIOUS ROCK

3.16 PRE-LOADING FOR SOIL CONSOLIDATION

The techniques involved in constructing fills or embankments to pre-determined densities to provide adequate foundation support are well known. In general, they consist in placing layers of soil mixtures from 6 to 12 in. thick and suitably compacting them by static weight, by kneading action, by vibration, or by combinations of these methods, depending upon the nature of the fill materials involved. In preparing buried soils for supporting foundations, however, these methods—involving excavation, replacement, and compaction—are very often impractical and are always expensive. What, then, can be done about improving the bearing capacity of soils *in situ;* how can they be consolidated in *depth* to improve their density?

The degree of consolidation of clays in their natural state has been found to be related directly to the depth (and hence the weight) of the soil overburden which has accumulated over them since they were first laid down. It is true that present overburdens may be considerably less than those previously existing, because portions have been stripped away by glacial or other eroding action. It is also true that alternate wetting and drying of clays appears to produce greater consolidations than any conceivable overburden would warrant. Another exception to the general rule applies to marine clays laid down in saline waters where chemical solutions have had an influence on the formation of the clay deposit, irrespective of its overburden.

Simulation of natural overburdens to induce consolidation in low-density soils is practical only where considerable depth of fill is required for final grading, but the results obtained have often been very satisfactory.

Several attempts have been made to pre-load soils by using techniques similar to those used for pre-stressing concrete. The results, although successful, have not proved conclusively that the method has general application, and the costs appear disproportionately high. Moreover, their use is dependent on rock or other firm strata at a usable depth below the surface.

In one case the foundation soils supporting piers carrying an arch bridge over the Fraser River in British Columbia were pre-loaded. The original test borings had shown rock at a suitable founding level. Excavation revealed that this was in fact a voided bed of boulders and that solid rock was actually 150 ft and 200 ft, respectively, below foundation level. A 2- to 3-ft layer of clay occurred about 100 ft deep, with sand and gravel formations above and below it. The two-hinged arch abutments would allow no settlement; and although other methods were considered, pre-loading was finally decided on. In this case it was necessary to pre-load the soil not only vertically but horizontally as well. Details are shown opposite.

The concrete for the abutments was poured in place and 12 holes then were drilled vertically downward in the positions shown. These drill holes were from 6 to 8 in. in diameter to provide for casing through the overburden down to rock, and 4½ in. in diameter in the rock. Depth in rock was 80 ft under one abutment and 50 ft under the other. One hundred thirty pre-stressing wires each 0.2 mm in diameter were used in each hole in the shape of an inverted V and secured in place in the rock by pressure grouting. The top of the V was passed over a jacking block as shown.

Two hydraulic jacks were used to stress the cable to a loading of 333 tons, providing a total loading on the soil of 4000 tons. The maximum settlement observed was 0.04 ft, or about ½ in. Recovery, when the load was removed, was about ⅛ in.

The horizontal jacking to pre-load the embankment used eighteen 100-ton jacks applying 1800 tons against a previously cast concrete wall. Here the movement was only 0.007 ft, and the recovery was small.

Since the total permissible settlement was 2 in. and the permissible horizontal movement was 1 in. in the completed structure, it does not appear that this pre-stressing operation, costing $400,000, achieved any necessary results.

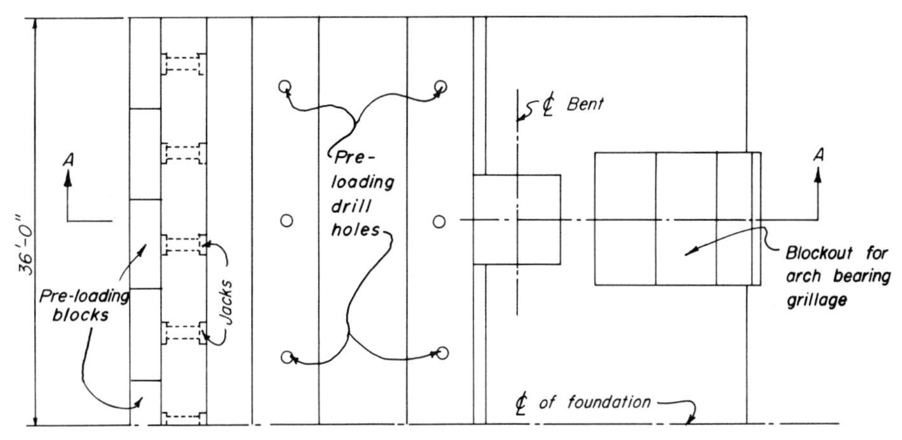

HALF-PLAN

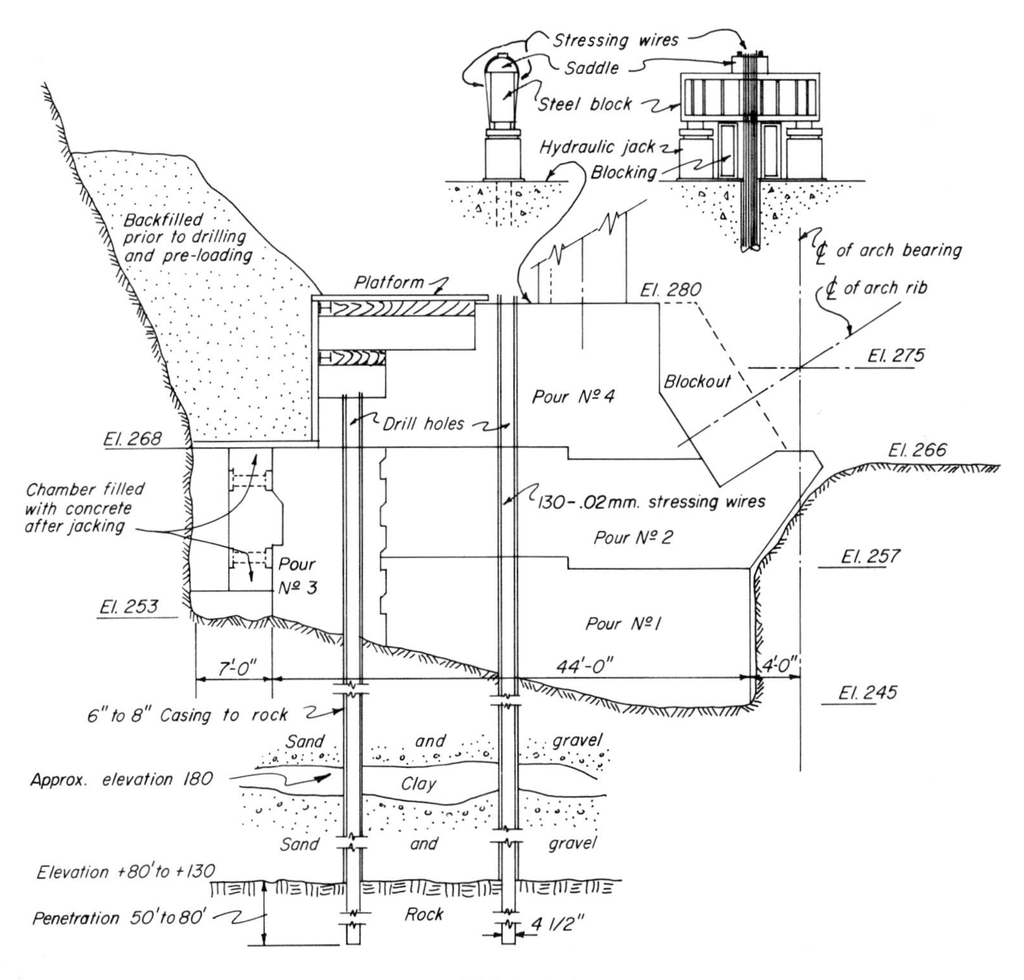

SECTION A-A

3.17 GROUTING

The discussion in Sec. 3.16 dealt with the consolidation of soils in place by pressure and, although it was chiefly concerned with soil sizes ranging from intermediate gradations of sand down through the range of clays, dealt, at least in one case, with the coarser-grained materials. Coarser-grained materials will include coarse sands, gravel, and boulders; or they can be extended to cover fissured or fragmented rock. These materials represent special problems.

If coarse-grained deposits consisted of particles of fairly uniform size and approached the spherical in shape, each particle would be in direct contact with its contiguous fellow and the compressibility of the mass would be that of the particles themselves. Nature seldom produces such a fortuitous concatenation of circumstances. Many coarse-grained aggradations covering a wide range of sizes and shapes have never been subjected to the massive compressive forces of glaciers or deep overburdens and are therefore susceptible to re-arrangement when heavily loaded, when vibrated, or when water flows through their voids. Naturally occurring formations, at other sites, have shown us the solution to stabilizing these mixtures.

One of the more nearly perfect foundation materials occurring naturally is *hardpan,* a mixture of gravel, sand, and clay, well mixed and compressed and containing little moisture. In the case of hardpan we have a discontinuous mass of gravel whose voids are filled with sand and whose smaller voids, in turn, are plugged with clay. This is the essential principle of man-made concrete using cement (which additionally undergoes a chemical change) in lieu of clay. Missing in coarse-grained soils and in fissured, fragmented, or porous (that is, voided) rock formations are the fines necessary to provide a continuous mass. These can be provided by grouting.

In construction parlance, grout is generally thought of as a mixture of sand, cement, and water used to fill voids, perhaps under structural steel base plates or under equipment. More generally applied, the term grout covers a wide range of possible materials and is used for a great variety of purposes.

There are two principal uses of grouting techniques. One of these is that of sealing off the flow of water through pervious soils or rocks, used extensively in the foundations of dams, to control the flow of water into oil wells and to prevent the intrusion of salt water into fresh-water wells. Grouting has been used to plug the bottom of cofferdams and to seal off steel sheet piling where it has terminated in gravel or boulder strata. It has been used in porous formations encountered in tunnel driving to decrease the flow of water into the heading in advance of driving. The second use of grouting is as a consolidation technique to increase the stability of loose-flowing soils and to improve their bearing capacity.

Consolidation grouting is that aspect of grouting which concerns us here, although the two uses are frequently inseparable. This type of grouting not only is used to consolidate foundation soils but also can be used as underpinning beneath foundations, in tunnels to prevent loose soils from flowing into the heading, and to reduce the effects of vibration in soils supporting machinery foundations.

The grouting materials used may consist of cement and water in various combinations, or sand may be added to increase the bulk in very coarse-grained soils. Grouting has been done with clay-and-water mixes, or clay may be added to the cement-water combination. Grouting in fine soils cannot be successfully performed with cement grouts, and several combinations of chemicals have been developed for this purpose.

Where it is necessary to grout formations containing channels of swiftly flowing water, cement grouts have not been successful since the grout is washed away before it can set up. For these purposes hot asphalt is used as a grout which, pumped into the void, forms a progressively expanding bubble upon itself.

Requirements for Grouts

1. *Grouts must be sufficiently fine-grained to pass into very small openings and cracks.*

a. The most common grout is a mixture of cement and water. Its use is limited to conditions where the effective grain size is 1 mm or more.

b. Slurries of clay and water have been used, but have the same size limitation as that for cement-water grouts: 1 mm.

c. Chemical grouts can be used in soils having an effective size down to 0.1 mm. The A.S.T.M. limitation between sand and silt is 0.05 mm, so that even chemical grouting is ineffective in the finer sands and silts. Clays cannot be grouted.

d. Mixtures of cement grout to which sand has been added to economize on cement, in heavily voided soils, are seldom satisfactory. Even where the sand grains are well rounded and many of the voids are large, there is still a percentage of finer voids that the sand grains cannot penetrate.

2. *Grouts must be sufficiently fluid to permit pumping.* (But see Sec. 3.22, "Dry" Grouts.)

a. This qualification is closely allied to item 1 since it is practical to pump mixtures of very considerable densities. One aspect is that of segregation. Water-cement-sand mixes can cause premature plugging of the grout hole, can settle out in supply lines, and will cause excessive wear on the interiors of pumps, piping, and valves.

b. Water-cement grouts will generally be mixed in ratios of from 5:1 to 10:1, depending upon grain size in soils or void size in rock formations. However, in formations with substantial voids, this may drop to 1:1 or, in tighter formations, may rise as high as 20:1.

3. *The grout should form a substance which, when set, will have adequate strength.*

a. A distinction must be made between grouting performed to seal the soil to prevent passage of water and grouting performed to improve its strength or bearing capacity. Cement grouts will set up, with the passage of time, and will increase the strength of the soil or rock mass. However, the amount of increase in strength will depend on the percentage of water. A balance between strength requirements and fluidity requirements, mentioned in item 2, must be drawn.

b. Clay slurries contribute very little additional strength to the soil. Although density is increased in some cases, there is often a tendency for the clay to retain its moisture content. Moreover, there is an additional tendency, when fine clay slurries are used, for a film to form around soil particles, sealing off the voids and preventing complete filling.

c. Chemical grouts vary considerably in the strength of the product produced. They are excellent for sealing where water is involved, or for controlling the flow of loose sand; but the strength of the resulting mass may not be very greatly increased.

4. *Grouts must set with a minimum amount of shrinkage.*

a. Shrinkage control of water-cement grout has been mentioned in item 3*a*, where it was noted that the best control was that of keeping the water-cement ratio low. However, there are a number of admixtures which act to absorb a certain amount of excess water. These increase the cost of the grout and are most generally used where limited areas are to be grouted.

b. There is no effective control of shrinkage with clay slurries.

c. Shrinkage is not a factor with chemical grouts.

5. *Grouting pressures.*

a. Low-pressure grouting is performed at pressures of from 20 to 150 psi for consolidation purposes, generally within depths of from 10 to 50 ft below ground surface.

b. Intermediate- or high-pressure grouting performed at pressures from 150 to 500 psi is generally reserved for grouting at depths of 100 ft or more. This type of grouting is largely limited to providing cut-off curtains under high dams.

6. *Grout admixtures.*

a. Admixtures have been developed for a number of purposes, but their use must be carefully considered since in obtaining one quality, another may be lost.

b. Workability, an aspect of fluidity—see item 2 above—can be improved in water-cement grouts by the addition of pumicite or diatomaceous earth. Bentonite not only will improve workability but will control shrinkage. All these compounds reduce the strength.

c. Control of setting time by retarding or slowing it down is seldom desirable, but speeding it up by accelerators is sometimes necessary. Acceleration can be obtained by a 5% admixture of calcium chloride, by replacing 35% of the portland cement in the grout with a high-alumina-content cement and by a 1:1 mixture of gypsum-base plaster and cement. These can produce flash settings, and their use should be controlled by experienced personnel.

Where consolidation of soils or rocks is the primary purpose, low-pressure or "blanket" grouting will be used, limited in depth to 15 or 20 ft and applied at pressures ranging from 5 to 30 psi. This type of grouting will generally require some containment. Two types of containment must be considered.

In particularly porous formations it may be desirable to provide a curtain of grout around an area, injected at low pressure, to prevent the indiscriminate spreading of grout with attendant loss of density and high cost. Of greater importance is the containment of grout at ground surface.

The shortest escape route for grout is to bubble to the surface in the immediate area around the grout pipe. While this can be controlled by lowering the pressure, this reduced pressure may be insufficient to disseminate the grout properly at lower depths. In rock grouting such as that used by the U.S. Bureau of Reclamation under dams, "packers" which seal the pipe into the rock are often used; but even with these, in blanket grouting, it is possible to raise the whole of a rock layer. A surface coating of pressure grout or concrete 3 to 4 in. thick has been used, but unless careful control of final grouting pressures is utilized, this mat can be raised bodily into the air.

So far as grouting for foundation stabilization is concerned, it is better practice to construct the foundation under which grouting is to be performed before grouting is begun. It is seldom likely that the soils to be grouted will settle under the sole weight of the foundation or, as in the case of caissons, under the lower portion of the foundation. Grout pipes can be cast into the concrete foundation or can be drilled in afterward, with packers in the concrete layer if necessary for sealing.

Spacing of grout holes will depend on the formation to be grouted and the depth of grouting desired. Spacings of grout holes of 5 to 6 ft may be sufficient in soils of low porosity and at shallow depths. For coarser materials, spacings of 10 ft or more for depths exceeding 10 ft may be sufficient. Although grout pipes can be sealed into a pre-drilled hole in rock, they should be driven into soils. The driving will generally produce the same consolidation of soils around the pipe as is encountered in pile driving. The pipe is then cleaned out with air before grouting is commenced.

Single-stage grouting can be used for shallow depths at very low pressures, but successive-stage grouting, in which the upper layer is first grouted and then successive layers below, permits grouting at higher pressures with increased depth and reduces surface losses of grout or distortion of the surface. Essentially this provides a containment mat, as noted above.

In successive-stage grouting the grout pipes are driven to a pre-determined depth and cleaned out, and a water test is put on each pipe to determine the possible grout demand and consistency. After grouting, the grout remaining in the pipe is washed out and the grout pipe is driven to the next depth. Grouting at the next depth is not performed until the upper grout layer has had time to set. Grout holes or pipe will generally be set in a pattern to cover an area with every third or fourth hole grouted first, the balance being grouted to fill in between the initially grouted units. In successive-stage grouting, alternate depth pipes can be pre-set to provide not only the stage feature but the fill-in aspect as well.

One of the more important uses of grout stabilization has been in correcting soil foundation deficiencies *after* construction. Grouting has been used to seal reservoirs and leaking dams, to stabilize large sewers that have begun to settle, to correct settlement of tanks and vibrating equipment foundations, and to halt settlement of footings during the construction period.

The stabilization of soils under existing foundations is performed in the same way as for new foundations. Unless it is desired to actually lift the structure, however, grout pressures must be carefully controlled; and even where jacking-up is involved, minute-by-minute leveling observations must be maintained.

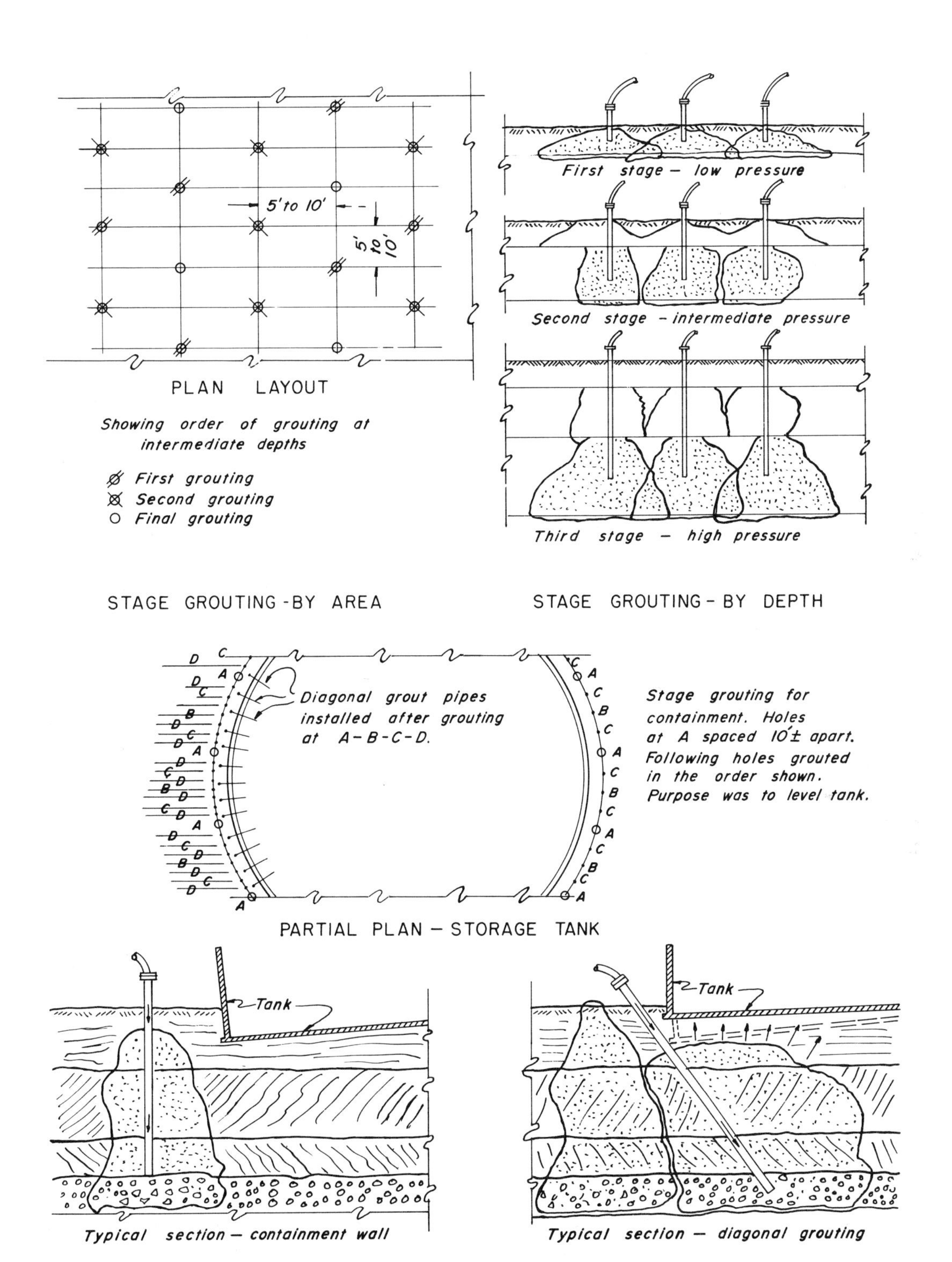

STAGE GROUTING - BY AREA

STAGE GROUTING - BY DEPTH

PARTIAL PLAN – STORAGE TANK

LEVELING A STORAGE TANK BY GROUTING

3.19 GROUTING EQUIPMENT

Because of the wide range of grouts used, grouting equipment is not a standard item but is designed for specific uses and assembled from more or less standard components. Chemical grouts, for instance, dealing with liquid solutions of relatively low viscosity can be handled by any type of pump that will provide the required pressure. At the other extreme are pumps used in the placing of concrete (not a grouting procedure), where segregation can be a major difficulty and abrasiveness is at a maximum. Water-cement grouts have considerable abrasiveness, and the cement tends to settle out unless continuous agitation is provided.

The use of air for placing grout pneumatically is now largely limited to the grouting procedure known as "guniting" (originally a trade name), in which a sand-cement-water grout is sprayed on surfaces to seal them or to provide a new finish. For consolidation grouting, however, the direct use of air is not entirely satisfactory. Pneumatic injectors employing this principle do not provide sufficient flexibility in operation, delivering the grout in intermittent slugs instead of in a continuous stream, and discharges may become segregated before final positioning. Air is occasionally used for very-low-pressure grouting where coarse ingredients such as sand, sawdust, or shavings are contained in the grout.

Grout mixers are barrel-type drums, set either vertically or horizontally, containing a rotating shaft with paddles attached. One end of the drum is open for the loading of dry cement. Since mixing is done by volumes, the water feeding into the mixer is measured by the use of a meter indicating the quantity in cubic feet or fractions thereof. A direct-reading totalizer should also be built into the meter, to provide a check on the number of batches used. On large operations, cement can be fed by mechanical means, but in most cases it is dumped in manually by the sack. For a 5:1 mix, for example, 10 cu ft of water can be fed into the mixer, and two sacks of cement poured into the water.

After mixing, the grout is drawn off into a sump, often a cylindrical tank 36 in. in diameter and 30 in. high, of sufficient size to take the mixer content, equipped with a motorized mixing shaft to prevent segregation. Screening through a No. 4 sieve is desirable between mixer and sump. The grout is drawn from the bottom of the sump by the grouting pump.

The most satisfactory grout pumps appear to be the duplex-type positive-displacement reciprocating pumps similar to the slush pumps used in the oil fields. Although often driven by compressed-air motors, other types of drives, electric or gasoline, are also used. The grout pump consists of a piston operating in a chamber into which a slug of grout is fed as the piston retracts and from which the piston ejects the charge on its forward motion. The duplex feature permits the contents of two chambers to be fed alternately into a single line through an outlet wye, thus providing continuous feed and pressure.

All grouts, save chemical combinations, are highly abrasive; and cylinder liners of a case-hardened steel are therefore generally replaceable. Piston heads are a medium soft rubber molded over a steel core and slightly larger than the cylinder so that the flare sweeps the liner clean at each stroke, preventing grout from building up on the interior face. Control valves within the pump as well as those in the line have hardened steel valve seats against which rubber vanes are compressed. Grout pump capacities range from 20 to 100 gpm, and, in general, greater capacities should be obtained by increasing the number of pumps.

The illustration shows two methods of grouting that have been used, particularly by the United States Bureau of Reclamation. One of these is the single-line system, the other the recirculating system. The tendency to segregation of all grouts has been noted; grouting lines must occasionally be flushed out to prevent this. The single-line system flushes its lines through a blow-off which dumps the grout out on the ground surface. The recirculating system returns this grout to the sump for re-agitating and re-use.

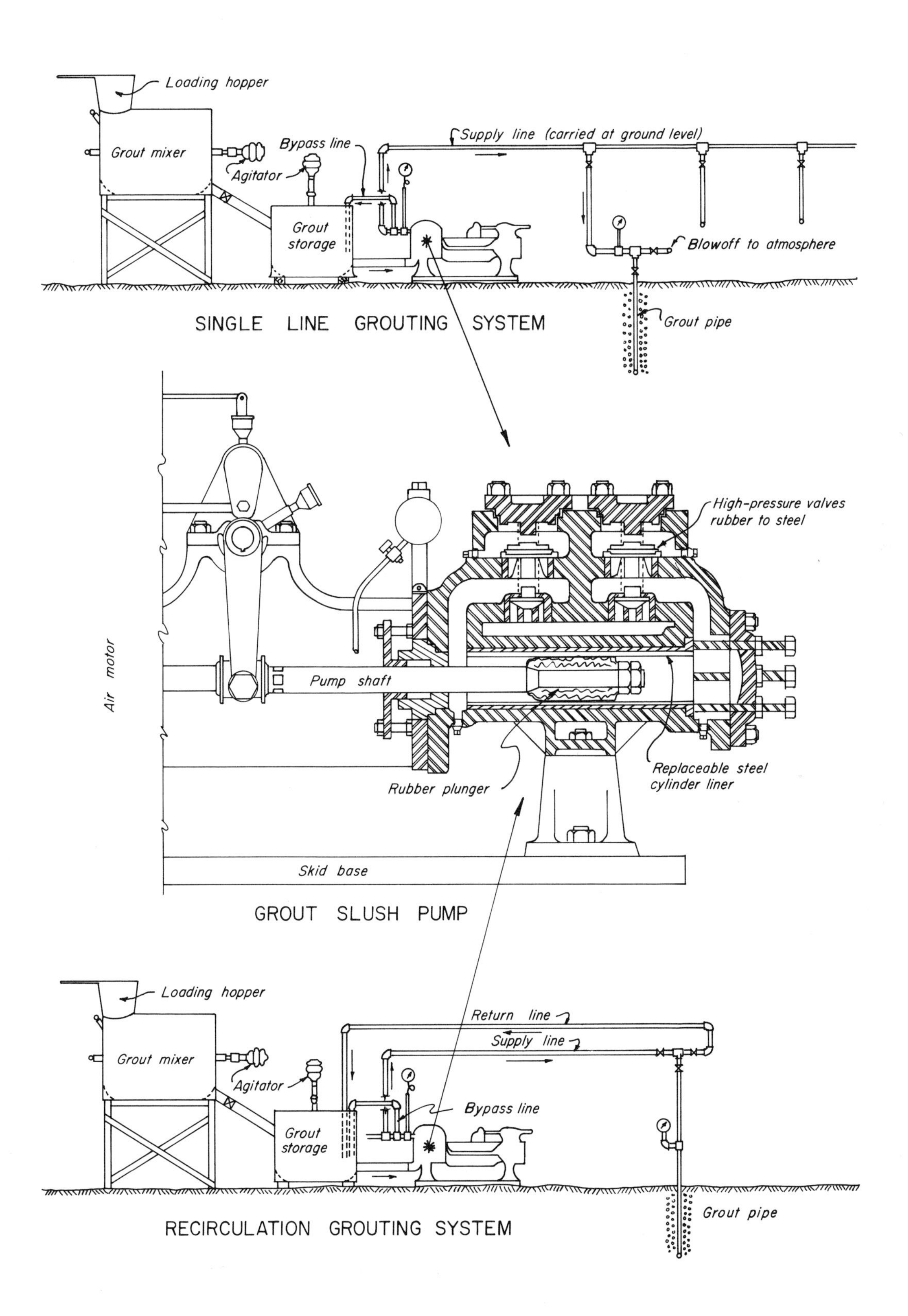
Loading hopper
Grout mixer
Agitator
Bypass line
Supply line (carried at ground level)
Grout storage
Blowoff to atmosphere
Grout pipe
SINGLE LINE GROUTING SYSTEM
High-pressure valves rubber to steel
Air motor
Pump shaft
Rubber plunger
Replaceable steel cylinder liner
Skid base
GROUT SLUSH PUMP
Loading hopper
Return line
Supply line
Grout mixer
Agitator
Bypass line
Grout storage
Grout pipe
RECIRCULATION GROUTING SYSTEM

3.20 CHEMICAL GROUTING

It was mentioned in Sec. 3.17 that chemical grouts can be used in soils whose effective size is greater than 0.1 mm. It was additionally noted that since, by definition, the silt grain size starts at 0.05 mm, very fine sands, silts, and clays could not suitably be stabilized by chemical grouting. Bear in mind, however, that the effective size of a sample containing disparate particle sizes is defined as that grain size than which 90% of the grains are larger and 10% are smaller. Many silts will contain substantial quantities of larger particles with corresponding voids, into which chemical grouts can be forced. (The permeability of a soil is a better criterion than particle size but is difficult to evaluate in underground deposits.)

The upper limit of particle size for chemical grouts is an economic one since chemical grouts are far more expensive than cement grouts. Another factor is that in substantial masses, the gel formed by chemical grouts will have a lower compressive strength than cement grouts.

The earliest chemical grouting was by the Joosten Process (now patented in the United States), consisting of a combination of sodium silicate and calcium chloride. The sodium silicate (a viscous liquid once widely used for the preservation of whole eggs and known as water glass) is diluted with water in a ratio of 3 parts to 2 parts of water. The gelling agent, a 5 to 15% solution of calcium chloride, is added just before injection. A small quantity of a synthetic ammonia derivative is used to inhibit the reaction during the pumping interval.

A number of variations have been developed using the basic sodium silicate with different gelling agents, some of these being the François, Gayard, Rodio, Langer, and KLM methods. Some of these additionally involve adjustments to compensate for the acidity of the soil and to permit wide, but controllable, variations in the setting time.

AM-9, a trademarked product of the American Cyanamid Company, uses a somewhat different combination of materials. The AM-9 is a mixture of two organic compounds, acrylamide and *N-N′*-methylene bisacrylamide, available as a water-soluble powder. Made up in solutions varying from 7 to 15%, the AM-9 is gelled by the use of two catalysts, ammonium persulfate and dimethylaminopropionitrite. By varying the relative quantities of these catalysts, the setting time can be controlled from a few minutes to 2 hr.

Another approach to chemical grouting involves the use of lignin, a waste product resulting from paper manufacture. The lignin, combined with a dichromate, forms a soluble chrome-lignin powder which gels to a dark brown substance of rubbery consistency. Terra Firma, a chemical grouting compound controlled by Intrusion-Prepakt Inc., is a chrome-lignin compound in current use. It is mixed with 4 or 5 parts of water, the amount of water controlling the setting time.

Chemical grouting procedures differ from those for cement grouting in a number of ways. The equipment to be used can be standard types of pumps, pipes, and tanks, for, although there may be some chemical action on linings, this will not be so severe as that caused by the abrasiveness of cement grouts.

Injection is performed through pipe ½ to ¾ in. in diameter which is driven into the ground to full depth and then retracted slightly. Occasionally the lower portion is perforated. Pipes are spaced on 18- to 30-in. centers, generally staggered, since the area of dispersion is considered to be within a radius of about 2 ft.

Grouting is begun at the bottom of the layer to be grouted, the grout pipe being gradually withdrawn as an increase in the pressure required indicates that the maximum quantity of grout has been absorbed. Grouting pressures vary with the nature of the soil, ranging from 100 to 200 psi.

Chemicals can be injected into soils which contain water under a hydrostatic head, providing sufficient pressure is used and providing there are no channels which would tend to wash the grout mixture away.

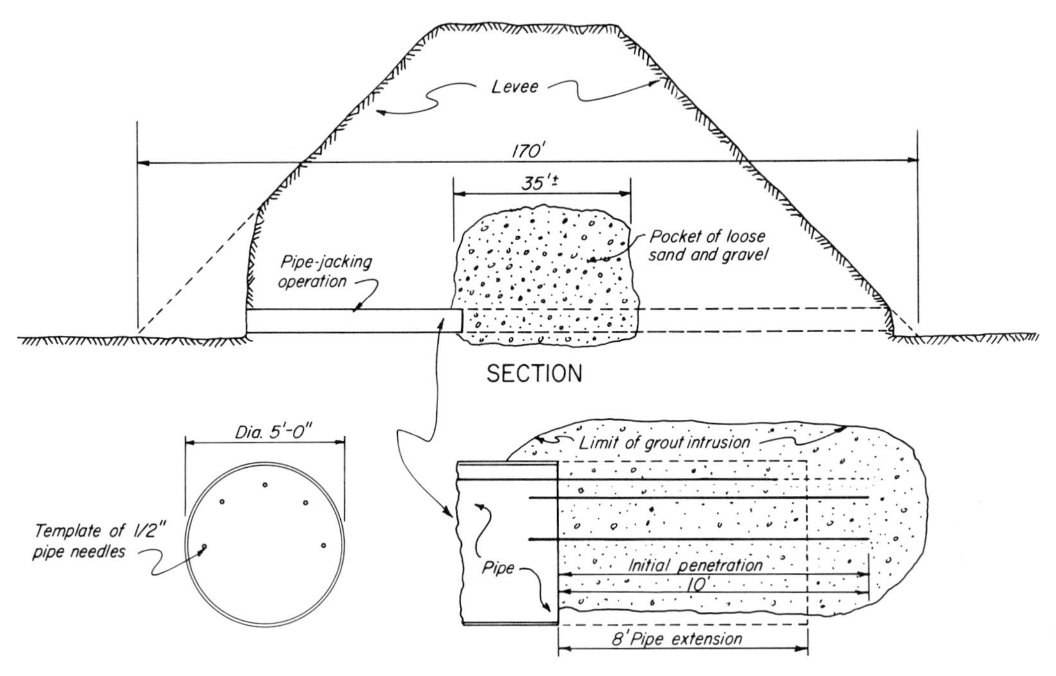

CHEMICAL GROUTING AHEAD OF PIPE-JACKING

Sodium silicate base · Needles withdrawn 1' at a time · Grout used = 10 gal./foot
Injection at night with 2-hour setting time · Pipe-jacked during day

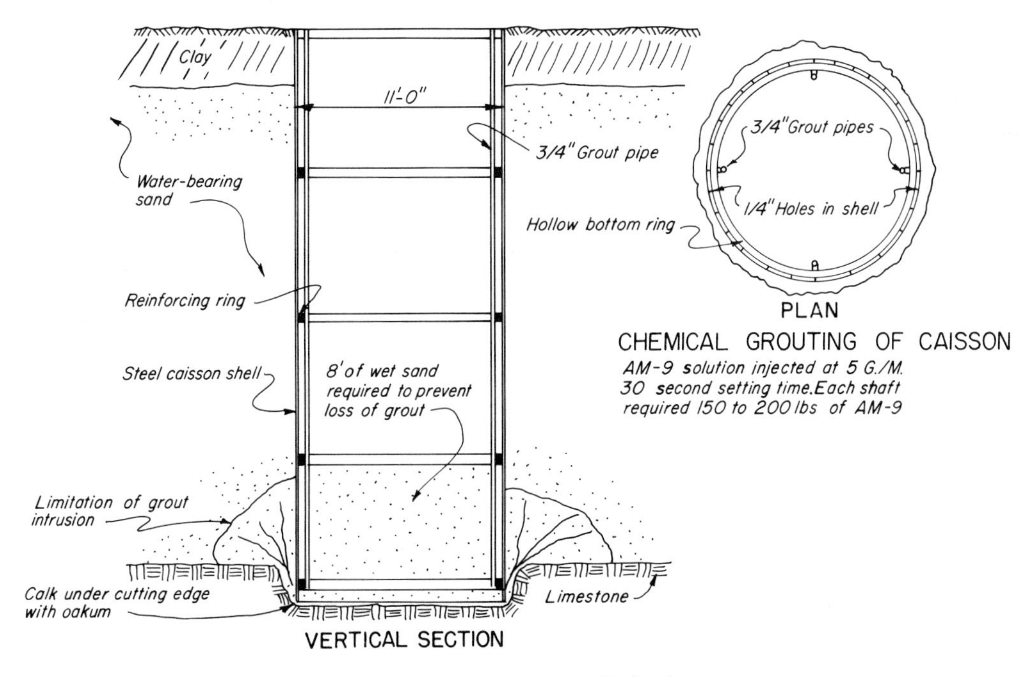

CHEMICAL GROUTING OF CAISSON

AM-9 solution injected at 5 G./M.
30 second setting time. Each shaft required 150 to 200 lbs of AM-9

CHEMICAL GROUTING APPLICATIONS

3.21 INTRUSION GROUTING

The discussion of grouting to this point has been concerned with filling the voids in natural soils and rocks to consolidate them into a continuous mass. In all these cases it can be said that grout is "intruded" into the natural voids. However, "intrusion grouting," a technique developed by Intrusion-Prepakt Inc., has a somewhat different meaning.

Intrusion grouting is, in reality, a method of placing concrete in which the gravel alone is placed in position and is then converted to concrete by intruding the mortar component into its voids. One of the chief advantages of the method is that it permits the placing of concrete under water. In this case, the grout is forced in at pressures higher than the prevailing water pressure, displacing the water in the voids; but the method has many other uses.

Effective intrusion grouting depends first on the careful control of void sizes in the gravel. This is accomplished by reducing the permissible smaller sizes. A.S.T.M. specifications for coarse aggregate permit varying percentages of ¼-in. sizes. For intrusion grouting the limiting size is raised to ⅜ in., thereby maintaining a larger percentage of voids (from 40 to 50%) and preventing pockets of smaller sizes through which the grout cannot readily penetrate.

The mortar used for intrusion is also especially designed. The A.S.T.M. specifications for fine aggregate include a limiting percentage of ¼-in. or No. 4 screen sizes, but for intrusion grouting the larger sizes are reduced to those passing a No. 8 sieve in order to avoid segregation and the clogging of voids, and a more uniform and somewhat finer sand results.

Shrinkage must be limited, the density of the mortar must be maintained, and the rate of setting must be controlled. Intrusion-Prepakt Inc. has developed a number of patented admixtures to control the performance of the mortar. Alfesil is a finely divided amorphous silica (actually a puzzolan, derived from volcanic ash) designed to react with the lime produced by the hydration of the portland cement, strengthening the hardened grout and improving its density.

Intrusion Aid, another patented admixture, is a colloidal suspension designed to improve the workability by inhibiting an early set in the grout. It also controls the rate of shrinkage.

For intrusion grouting it is desirable that the gravel be contained within a form, within a steel shell, or against solid banks not subject to grout penetration in order to prevent the dissipation of grout and grouting pressure. Intrusion grout is placed through 1¼-in. pipe driven vertically into the gravel (or set in place before the gravel) on about 10-ft centers. As the intruded grout level rises, the jetting pipe is raised. Slotted pipe wells staggered between jetting pipes indicate the level of the rising grout. Pressures of 5 to 6 psi are adequate under atmospheric pressure, but for grouting under water, pressure will increase with depth. An additional increment of pressure amounting to 10 or 20% may be required for grouting under water.

The actual procedure of intrusion grouting is not dissimilar to that already described for consolidation grouting except that once grouting is begun the entire operation must be continuous so that the forward extension of the grout does not have time to obtain a set. The air-driven, positive-displacement slush pump is used for intrusion grouting since it provides better control of application than other types. Careful records of the quantity of grout pushed in are maintained and compared against the estimated voids to be filled.

In general, intrusion grouting resembles the placing of concrete under water by tremie except that control of operations with mortar is much closer than where the full concrete aggregate must be handled. Its use is limited to sections of substantial size (where placing is to be done under water), and the method is not fully accepted for use where the structural value of the member is critical, because of the limitations on aggregate sizes.

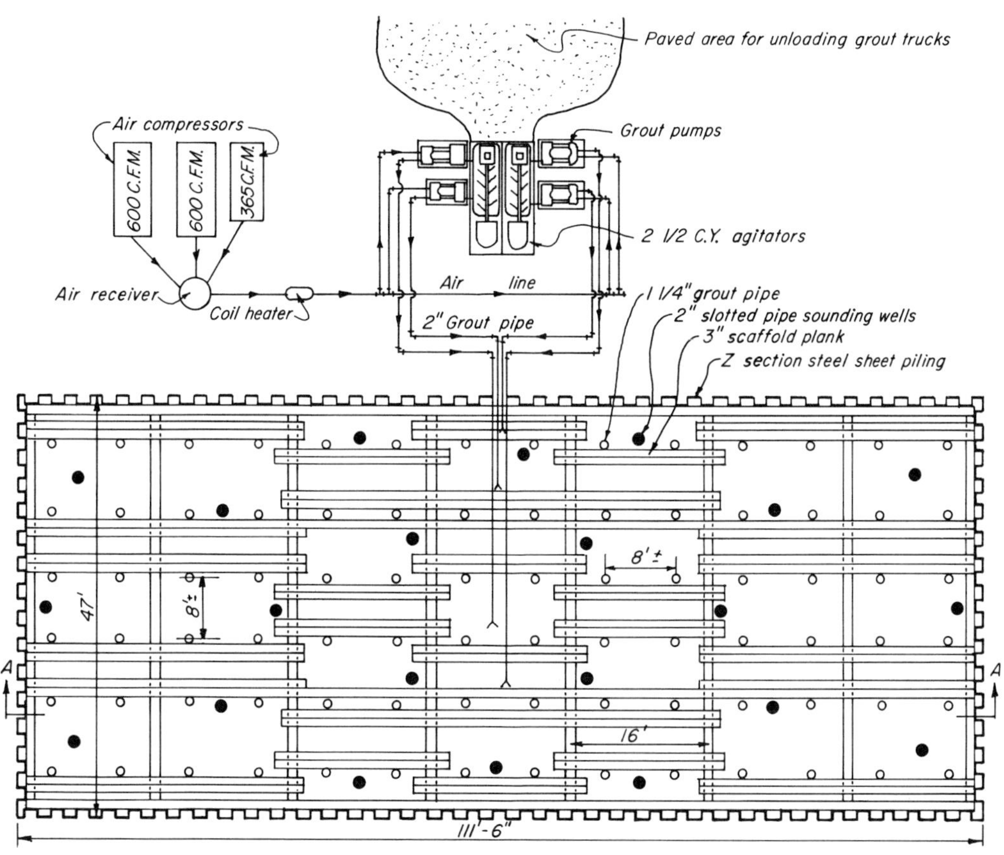

INTRUSION GROUTING METHOD UNDERWATER

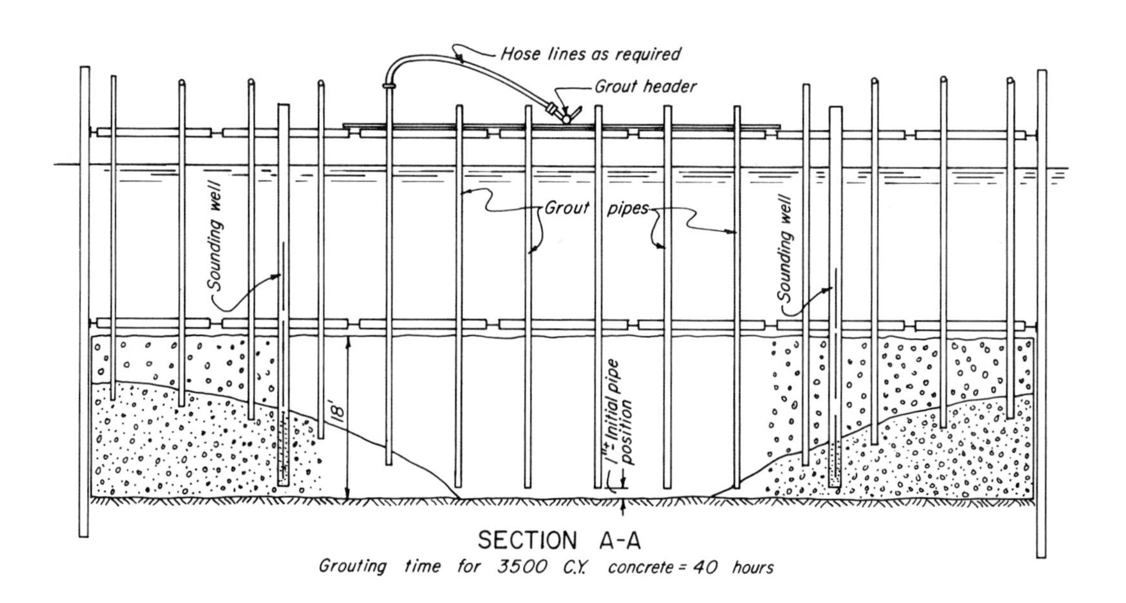

SECTION A-A

Grouting time for 3500 C.Y. concrete = 40 hours

3.22 "DRY" GROUTS

The term "dry grout" has been coined for use here to distinguish those grouts which are deployed in an essentially solid mass from those grouts, otherwise discussed here, which are essentially liquid. Although the basic purpose of the treatment of the subject of grouting has been that of the preparation of sub-grades, their consolidation and densification to receive and support structures, it has been necessary to refer to a number of allied types. Underpinning, for instance, consists primarily of poured mass concrete, but the final closure between it and the existing foundation is a grouting operation. The grouting of drill holes for rock anchorage or tie rods requires, certainly, a more liquid grout than an underpinning closure, but it is, nevertheless, far from being the 10:1 water-cement mixture used in consolidation grouting.

The prime necessity for dry grouting springs from the fact that standard concrete mixes shrink during setting. It is the cement, in a well-designed concrete mix, that produces the shrinkage; consequently, the stronger the concrete (that is, the higher the percentage of cement) the greater the shrinkage. (Poorly designed or mixed concretes, having an excessive water content, will also have a high shrinkage.)

"Dry-pack" was the earliest type of grout used for closures such as those for underpinning, and it is still used. Dry-pack consists of a mixture of 1 part of portland cement to 2 or 3 parts of fine sharp sand (usually obtained by screening the larger sizes from concrete sand) with only so much water added as will moisten it. A fistful of dry-pack, squeezed, should exude no water and should retain the impression of the fingers when the fist is opened. Theoretically the dry-pack should contain just enough water to interact with the cement used—without surplus.

Dry-pack is most effectively used where the void to be filled is not much more or less than an inch wide and where it is practical to ram it into place with a stick or a blunt-end tool such as a caulking iron. The harder and tighter it can be rammed into position, the more substantial is its bearing capacity.

When properly mixed and placed, dry-pack is a very satisfactory and economical void filler, but there are a number of situations in which its use is not altogether desirable and often even impractical. Where the void to be filled is irregular or angular or obstructed so that direct ramming is impossible, there is no assurance that voids will be adequately filled. This condition occurs particularly under machinery, where it is necessary to close the gap between the foundation and the base plate. A liquid mix of cement and sand that will flow into place and fill such voids has a high shrinkage factor and is not suitable for heavy, vibrating machinery. For such purposes a non-shrinking or expandable grout has been developed.

There are several products available which, added to a water-cement-sand mix, produce a non-shrinking grout. The oldest of these, Embeco, a trademarked product of the Master Builders Company, was first developed and is still used extensively for the grouting of machinery. A similar material, Ferrolith G DS, a trademarked product, is manufactured by Sonneborn Chemical and Refining Corporation. Both of these, and others, are composed basically of finely divided iron particles which, on exposure to water and oxygen (air), convert to iron oxide, expanding in volume in the process. Additional elements in these compounds accelerate and control the action so that expansion balances shrinkage and excess water is mopped up.

For most purposes the grout is prepared by mixing 100 lb of compound to 1 bag of cement to 100 lb of sand. To this, 5.5 gal of water is added to produce a "flowable" mix. For stiffer mixes, less water is used.

The 28-day strength of the resulting grout will depend on the consistency of the mortar. A stiff mix will average about 12,000 psi; a flowable mix will run about 10,000 psi; a fluid mix, about 8000 psi.

The term "expandable grout" should be considered as synonymous with "non-shrinking grout."

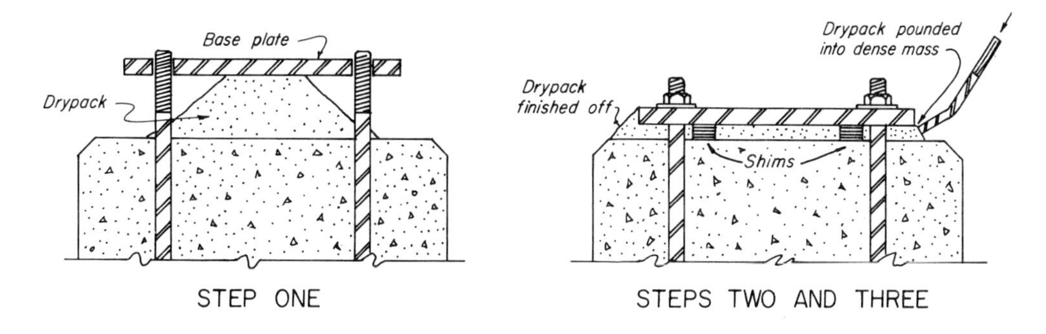

DRYPACK FOR GROUTING BASE PLATES

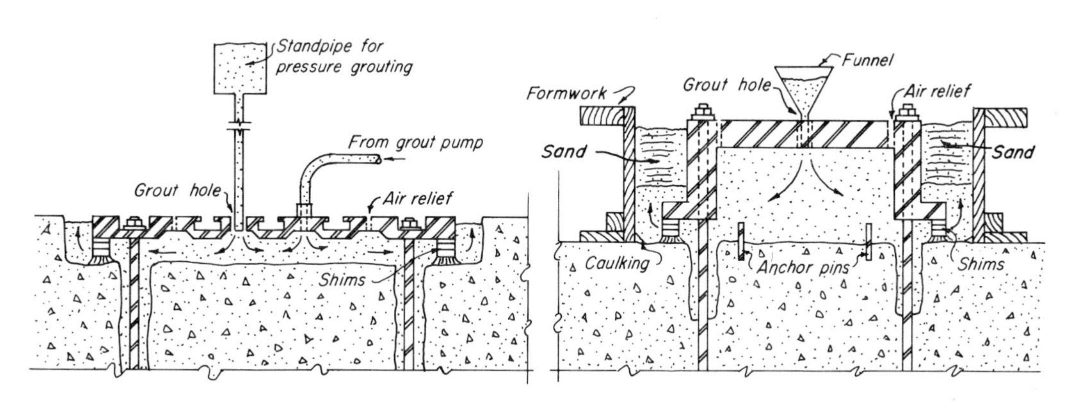

METHODS OF PRESSURE GROUTING WITH NON-SHRINK GROUTS

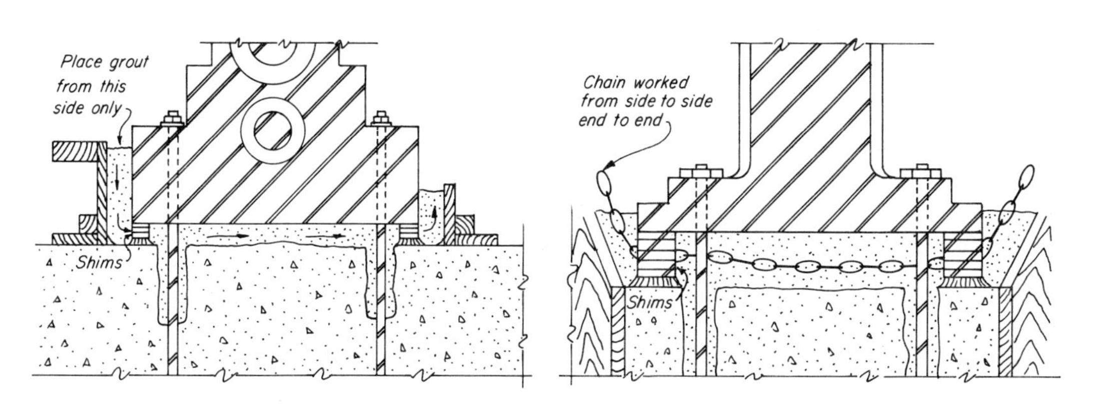

METHODS OF PLACING FLOWABLE NON-SHRINK GROUT

4 Underpinning

The facades of the cities of the past have owed much to the party-wall principle, a concept old in England at the time of the founding of the 13 American colonies and known to the surveyor Thomas Holme when he laid out the city of Philadelphia for William Penn in 1668. Its legal background lies in the principles of private property, although its actual operation seems to contravene that principle.

In essence and as applied over the centuries, the first person to build on a single lot in a row of lots (even in a wilderness of free land, the original lots laid out in Philadelphia were quite small, matching custom in the crowded city of London) may build his structure so that the walls enclosing it on either side (but not on the street face) straddle the respective property lines. For instance, if a 12-in.-thick wall is required to support and enclose the building, 6 in. of this wall can be built on the adjoining property, thus adding a foot to the space available within the dwelling. However, in doing this, the original owner loses some of his rights to this half thickness of wall. The adjoining owners acquire the right to use the walls previously constructed for the support and enclosure of *their* properties, thus saving the cost of building separate walls and yet gaining their own additional foot of space.

However, when the basic incumbent comes to remove or replace his structure, he is bound to leave the *entire* wall standing since, although only half the wall lies on the adjoining property, the neighboring property owner has acquired the right to the use of the whole wall.

Down to the advent of the twentieth century the construction of private dwellings within cities had not varied much. They were built in rows to house workers, or other dependents of an estate owner, with dividing walls separating them into sections. The party-wall problem did not arise until either the exigencies of the owner's finances or the importunities of an entrepreneur (even the blossoming of a workers' finances or the sad decay of the ancient structures) resulted in the sale of two or three units in a row. Where these old units were merely

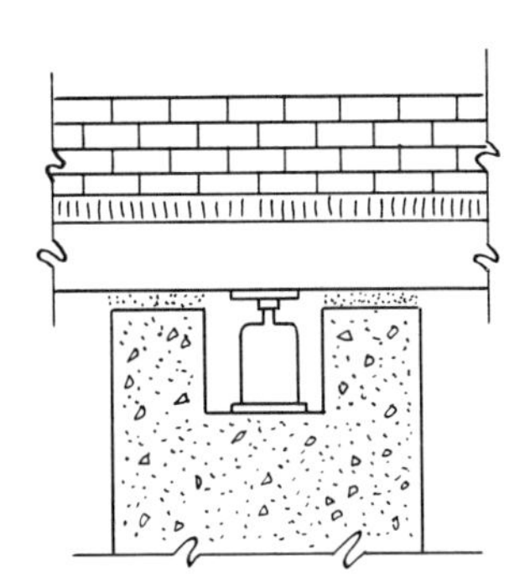

demolished and rebuilt in a more sprightly style (perhaps to obtain more revenue from the same acreage) no special problems were involved; but as story heights increased or the area changed to commercial structures of greater weight and pretensions, the condition of the remaining party walls, and more particularly the nature of their foundations, cast the existing party walls into oblivion. Since they could not be torn down yet could no longer support the increased loadings, they were first isolated and used solely as curtain walls and finally ignored completely, the new owner resigning himself to the loss of a foot of land by throwing an extra story on the top of the new building.

While buildings crept higher, they also went deeper. The basement, once a hideout for a furnace and its debris, was found to be finishable and usable, and a sub-basement followed by a sub-sub-basement was developed. All this burrowing had a profound effect not only upon the foundations for the new structure but upon the old foundations of the irremovable party wall. The new builder was obligated not only to retain that old wall but to prevent its damage. To do this it was often necessary to extend the footings of the old wall deeper into the ground, to more substantial soil bearing or even to the full depth of the new structure. This had to be done without disturbing or endangering the party wall (no small feat with an old masonry wall slapped up with lime mortar), and the techniques of underpinning were born out of this necessity.

Then railroads came along and went underground, followed by subways, and a miscellaneous range of subsurface structures now endangered not only ancient party walls but structures of all kinds, even those which, left standing by themselves, had more than substantial foundations. Then finally came highways diving beneath underground obstructions—even subways—or rising above the old ground surfaces to be wedged tightly between old walls with puny foundations.

In all these cases and many more the only alternative to damage suits and distress was underpinning and yet more underpinning.

4.1 STRAIGHT-WALL UNDERPINNING

A generation ago the design of footings was given much less attention than it is today; and as we go progressively backward in time, this neglect is more apparent. A century ago concrete, first coming into use, was tossed carelessly into a rough trench dug in the ground without forms or reinforcing of any kind. Before that there was masonry, often adequate, but just as often little more than a pile of rock flushed in with mortar. Some of these foundations, adequate when built, deteriorated with time, cracking erratically at the slightest adjacent disturbance to their delicately balanced *status quo.*

The first problem, therefore, in planning underpinning of a straight wall is an estimate of its intrinsic strength and characteristics at foundation level. Is it strong enough to act as a short beam, or is it so poor that it must be replaced completely? How short is a short beam, 2, 3, 4, 5 ft? To a great extent the capacity of a section of foundation to act as a beam will depend upon the nature of the wall above it. If this wall is a tight, well-bonded wall (we are speaking in terms of masonry here, for concrete will generally represent fewer strength problems than masonry), we can depend on a certain arching action in the brick or stonework that will, at least temporarily, help to carry the loads of the upper stories.

Next to be considered (although there is no order in which these things ought to be considered) is the total weight resting on the foundation. The weight of the wall above is not difficult to figure rather exactly; but since many walls are bearing walls supporting floors, with their additional loadings, as well as the roof, considerable judgment is required in determining, not what load *ought* to be transmitted via the wall to the footing, but what load does, in fact, rest on the footing.

General excavation of a site adjoining a wall to be underpinned will depend on the appraisals of the structure noted above. Where investigation discloses that the wall is sound, general excavation can be carried to a depth a few feet above the bottom of the footing and underpinning can be commenced from that level. In cases of weak walls or walls in poor condition, removing the earth from their face may cause damage, and underpinning should begin at the existing ground level.

The actual underpinning consists in digging a series of pits, at spaced intervals, against the face of the wall. When the bottom of the foundation has been reached, excavation is carried under the foundation wall down to the final level to which underpinning is required. From this level a new support is constructed up to the underside of the old foundation.

The spacing of the initial pits may be arbitrarily set at 10 or 12 ft, but should be sufficient so that the ground left between pits will still be adequate to carry the wall. The size of each pit must provide sufficient working space, usually not less than 3 ft, since the excavation is generally a manual operation. Where a narrower width is mandatory along the wall, the pit can be increased in dimension at right angles to the wall to provide working space.

Once the first set of pits has been dug and concrete or occasionally masonry columns have been extended down under the footing to the required depth (see illustration), adjoining or intervening sections are then carried down successively until the individual columns become a continuous wall.

It has been assumed that ground conditions are satisfactory and that, with minimum shoring, the pit walls can be retained; but where ground conditions are not satisfactory, where the foundation wall is weak, or where the final depth of underpinning desired is considerable, it may be necessary to proceed in two steps. The first of these consists in constructing what in effect will be a beam under the existing wall, 2 or 3 ft deep with each section doweled to the adjoining section so as to form a continuous member or with stepped sections interlocking to produce the same effect.

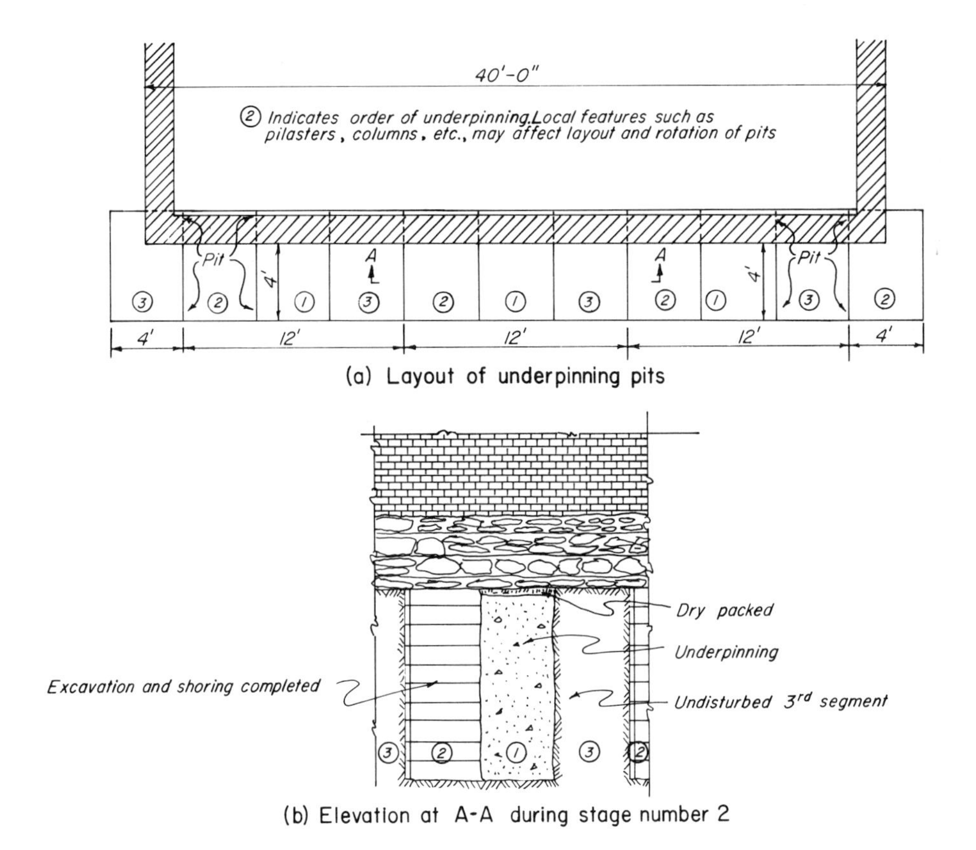

(a) Layout of underpinning pits

(b) Elevation at A-A during stage number 2

Step one

In general; place 2 side planks and wedge two end planks between them.

Use trench jacks where temporary additional bracing is required.

Step two

Step three

Step four

At W-W

At X-X

At Y-Y

At 2nd pit

STRAIGHT WALL UNDERPINNING

4.2 COMBINED PIER AND WALL UNDERPINNING

Although underpinning has for its primary function the preservation of existing walls, it does not necessarily follow that these walls belong to someone else—are party walls. Expanding industrial construction will frequently require the complete operating maintenance of an existing building while a new building or an extension to the old is being constructed.

Underpinning an existing wall will generally not make it any more satisfactory as a bearing wall. To construct a new bearing wall, adjacent to the old, consumes valuable space. The solution is the use of columns, sometimes steel, built on the face of the underpinned curtain wall, but requiring footings that it may be desirable to place *under* this old wall. (There are substantial design problems in placing a column on footings which are non-concentric, and it is generally avoided. Alternative schemes in which the column is wholly within the new building and floors are cantilevered out to the old walls need not concern us here.) A typical case requiring column footings underneath a wall is illustrated.

The first step was similar to that described in Sec. 4.1 for straight-wall underpinning. The space between columns was divided into three parts of approximately equal length. In this case each section is 5 or 6 ft long, it having been estimated that the masonry wall above it, of no great age, would satisfactorily arch over an opening of this width.

The 6-ft sections on either side of the middle section are next underpinned, with a slight variation in procedure. These sections extend over the area of the spread footing for the column by half their length. To provide support for this 3-ft section, a steel column with base plate resting on the ground at the level of the bottom of the spread footing is cast into the section.

The wall, with the exception of the area to be occupied by the column pedestals, has now been underpinned by using a series of pits as described in Sec. 4.1, and bulk excavation of the area in front of the underpinned wall can now be performed. Coincidentally, the volume of soil in the pedestal area is excavated by hand and tossed outward to be handled by shovel or front-end loader. The back face of the excavation at the pedestal is supported by planking wedged behind the two masses of underpinning on either side.

It has already been noted that soil conditions will determine when and to what extent bulk excavation can be performed prior to completion of underpinning. Since pit excavation, chiefly by hand, is expensive, it would be desirable to perform that bulk excavation before any underpinning is attempted, but this is practical only in rock formations. In very compact soils or shales it may be possible to bulk-excavate, leaving a berm 2 to 3 ft wide at the top with a 1:1 sloped face. This permits excavation for underpinning to be cut in horizontally with the excavated soil tossed out into the bulk pit, but the method is hazardous in formations of doubtful stability.

In this case, where a spread footing is to be placed under the column pedestal, and where only the earth within the width of that pedestal remains unsupported by final concrete, there can be little question regarding the sequence of bulk excavation in the overall operation.

Although these discussions of straight-wall or column-inserted underpinning have stressed the use of care in excavation and in the placing of timber forms on the *inside* of the underpinning wall, the possibility of movement in the soil under the existing structure cannot be ignored. Voids can occur under the ground floor slab of the old building which could lead to settlement or to at least the cracking of the slab. As a precaution against this contingency it will be advisable to insert grout pipes horizontally through the upper portion of the underpinning and extending slightly into the earth under the existing floor slab. Where signs of distress in the slab are later discerned, standard grouting procedures can correct the difficulties (see 3.18).

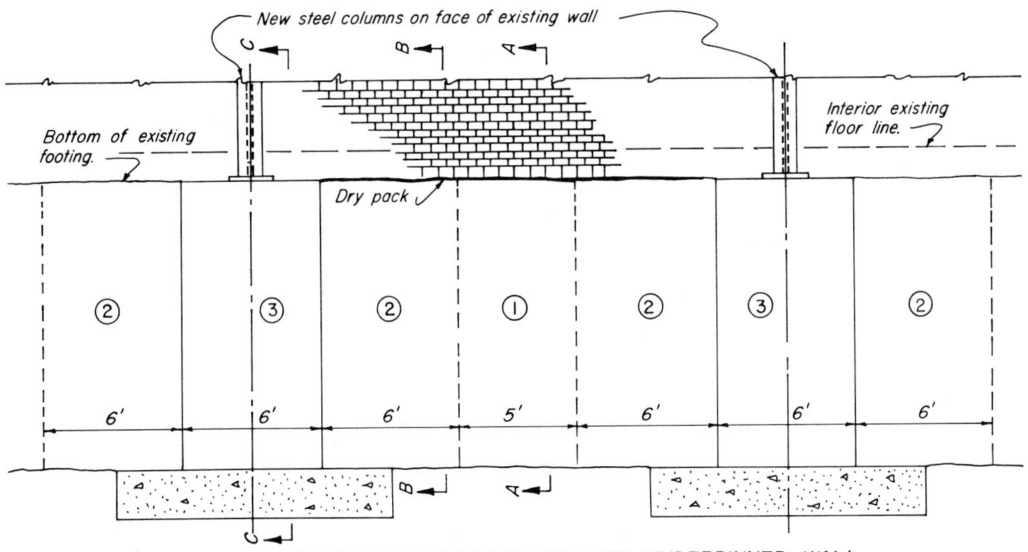

ELEVATION OF COLUMN-INCLUDED UNDERPINNED WALL

① Indicates underpinning order

Pre-excavated

Dry pack

D

Grout pipe

Underpinning pit

A - A

Column ℄

E

B - B

Bulk excavation and pier excavation performed together.

Column ℄

F

C - C

D - D

Set-up for pouring

E - E

Foundation slab forms

F - F

COLUMN - INCLUDED UNDERPINNING

4.3 VARIATIONS IN WALL UNDERPINNING

The predominance of clay as a subsurface soil in the areas where many large cities have been founded has made possible the ordinary methods of underpinning walls. The cohesiveness of clay has permitted excavation to limited depths prior to the placing of sheathing without undue danger of the loss of soils by flowing from under the structure to be supported. The time that sands or silts will stand alone is variable, generally depending upon their moisture content and degree of compaction. In some instances these soils will act the same as clays, but if they are very dry they may tend to flow, and if they are saturated the same thing can occur.

Where underground water exists, the only satisfactory recourse is that of drying up the area where underpinning will occur, by using either deep wells or well points, but this procedure will not necessarily work in formations of silt or rock flour (such as "bulls liver") since water will not drain from them in all cases.

Although considerable stress has been placed on underpinning performed for the purpose of carrying existing foundations to a deeper level where adjacent structures are to be built, one important purpose of underpinning is that of re-supporting structures where settlement is taking place due to underground conditions. Such settlements are apt to occur precisely in the types of soil which are not well-compacted clays and are very often attributable to changes in underground conditions which have developed since the original date of foundation construction.

Although all the possibilities are numerous, two special conditions suggest themselves. In one case a rise in the water table has changed the consistency of the soil so that the weight of the structure, acting as a unit, is slowly squeezing the newly liquefied soil out from under it. This action may be uniform throughout the underlying area, or it may be taking place under only a portion of the foundation. The other case is where the underground flow in an aquifer at some depth below the foundations is slowly leaching away a portion of the soils on which the structure is resting.

Conditions such as these may often be solved by consolidation grouting where the characteristics of the soil will permit this. In cases where the soils are partially or wholly too fine for effective grouting or where other factors are involved, a continuous row of cast-in-place or drilled-in-place piling may act as a bulkhead to permanently stem the flow of soils.

One variation on the drilled-in support as underpinning is shown in the illustration. Its application is limited to those cases where the cause of the trouble lies well below foundation level while the overlying soil permits open excavation, with minimum sheathing. Because of the expense of the method, it is practical only where the existing foundation wall is of such design as to permit substantial spanning of the distance between the supports indicated.

The method consists of a pile or caisson drilled down to solid bearing in a casing and concreted in place as the casing is withdrawn. Drilling is performed flush with the exterior face of the footing, and concreting of the pier is carried up to a point below the underside of the footing in the first stage. Excavation is then carried away from the pier and under the footing in the form of a bracket. The bracket is poured as an extension of the pier, to which it is tied by reinforcing steel. The remainder of the pier is then completed to the desired height on the outside of the footing.

The principal problem with this method of underpinning is that of preventing the pier from being displaced laterally at the bottom because of the eccentric loading. A 30-in. pier might present sufficient surface so that the surrounding soil would prevent displacement. The pier could be drilled several feet into the supporting strata, in which case it might function as a beam spanning the space between footing and bearing stratum. Loading of the pier outside the footing, using it as the foundation for another structure, could also prove beneficial in preventing tilting.

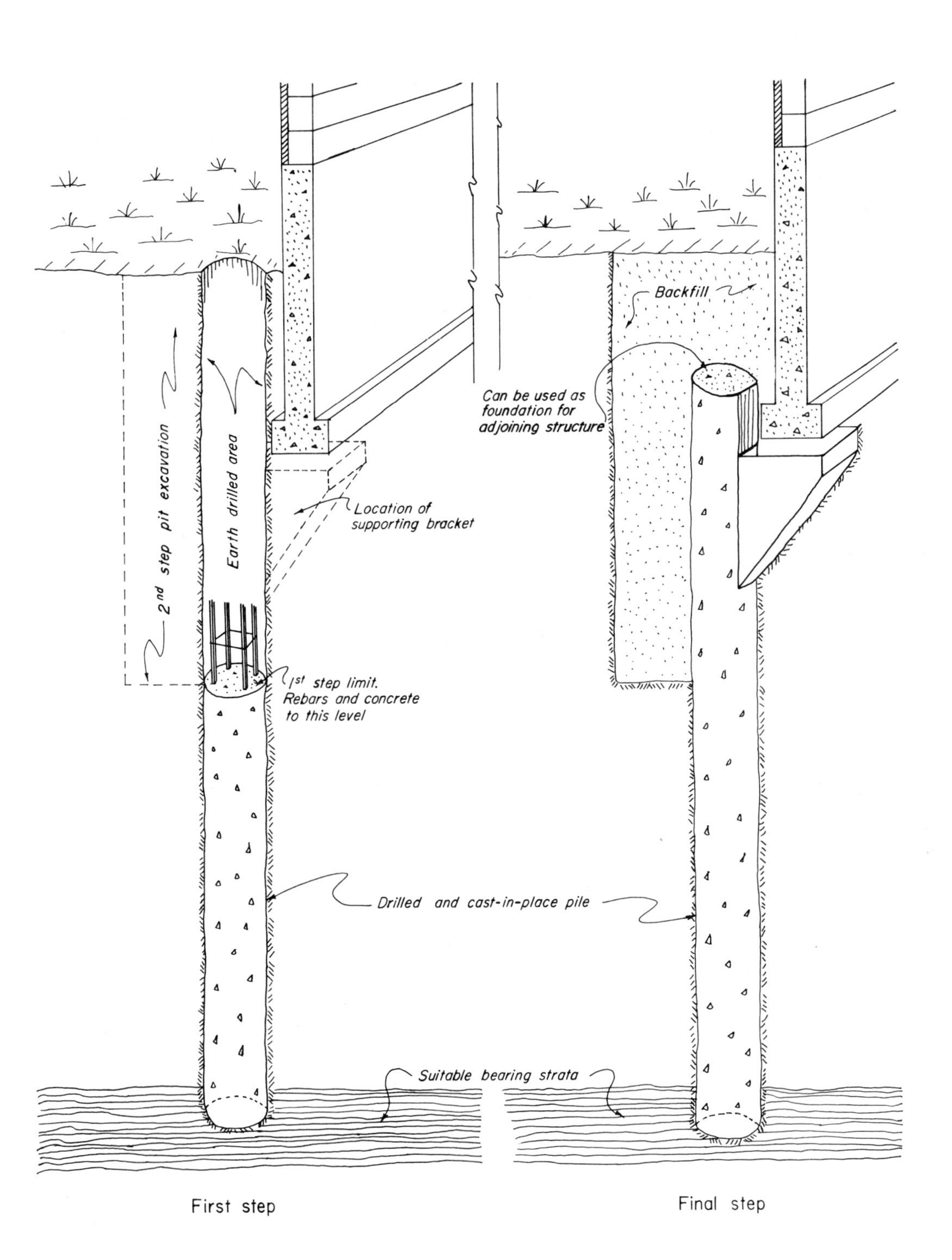

UNDERPINNING WITH DRILLED-IN PILES

4.4 NEEDLE BEAMS IN UNDERPINNING

The term "needle beam" is frequently used in connection with underpinning. The term has no precise significance and may be either a wooden timber of almost any size or a steel beam or column section suitable to the job at hand. The term derives from the fact that earliest underpinning operations often involved pushing a beam through a hole cut in a wall to transfer support, and this operation is still required in many cases. From this origin, any long thin member used in transferring structural loads during the underpinning operation came to be dubbed a needle beam.

The original use of the needle beam involved supporting a masonry wall (often brick) while an opening for a door or window was "let into" the wall. Figure *a* shows such a use. The spacing of needle beams may be 3 to 4 ft and occasionally more, depending on the characteristics of the wall, its thickness, bond strength, and type of brick and coursing. (Stone walls present greater problems, especially if of ashlar or rubble construction.) This type of underpinning depends, almost entirely, on the arching effect of the brickwork to carry a portion of the loading and is, of course, useless in poorly bonded masonry.

The location of openings, such as windows above the needles which interfere with the arching transmission of loads, may make the method impractical or limit the location and spacing of the needles. The method requires firm footing on which to rest the mudsills; and once the needle has been wedged in place, substantial bracing from the A-frame to the needle should be installed to hold the support steady.

Needle beams have been used in underpinning wall footings. This method generally has no advantage over that described in Secs. 4.1 and 4.2 but may have the disadvantage of putting the needle-beam supports in the way of operations incident to excavation. Practical in the very hardest ground, other soils will generally require sheathing to prevent the concentrated loading under the supports from breaking down the bank.

Figure *b* shows a special use of the needle beam. In this case the exterior columns along one side of a building had settled disastrously because of saturation of the soil by the adjoining lake. The next row of interior columns still rested on firm undisturbed soil and had, moreover, sufficient bearing so that additional loading could be applied without failure.

Outside the building and 10 to 12 ft away a drilled and cased concrete pile was placed, carried down to a solid gravel stratum 32 ft deep. The top was cut off level with the interior foundation and the needle beam laid over the two footings. To accomplish this the old exterior columns, which had settled considerably, had to be burned off at the bottom, and so much of the concrete footing as interfered had to be broken out. A jacking angle was welded to the face of the exterior column.

With the needle beams in place, the columns were now jacked up to proper roofline elevation, and a continuous beam, carried longitudinally under them, acted as a pedestal as well as serving to support the wall between columns, replacing the damaged grade beam.

Figures *c* and *d* show another application of a needle beam, this time carried longitudinally with the wall to be supported. Here the requirement was that of extending the depth of the existing column footings down an additional 15 ft. This was a needle beam in the true sense since it pierced the column, a somewhat unusual procedure. This is practical only where the position of the vertical reinforcing steel in the columns is known.

Here, temporary footings were provided in sheathed pits midway between columns and carried to solid bearing depth. The two 18-in. WF 60# beams bolted together were passed through the column to rest on the temporary supports. Steel wedges drove the beams tight against the underface of the plate grouted into the column hole. When the load had been transferred, an underpinning pit was dug on the outside of the building down to requisite level and the column footing was underpinned.

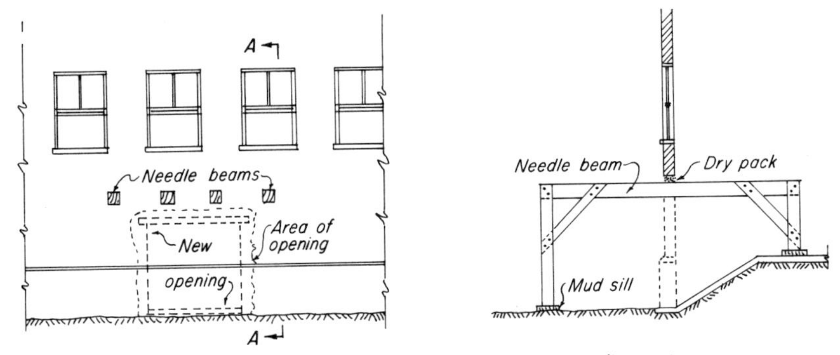

(a) Needle beam used in cutting wall opening

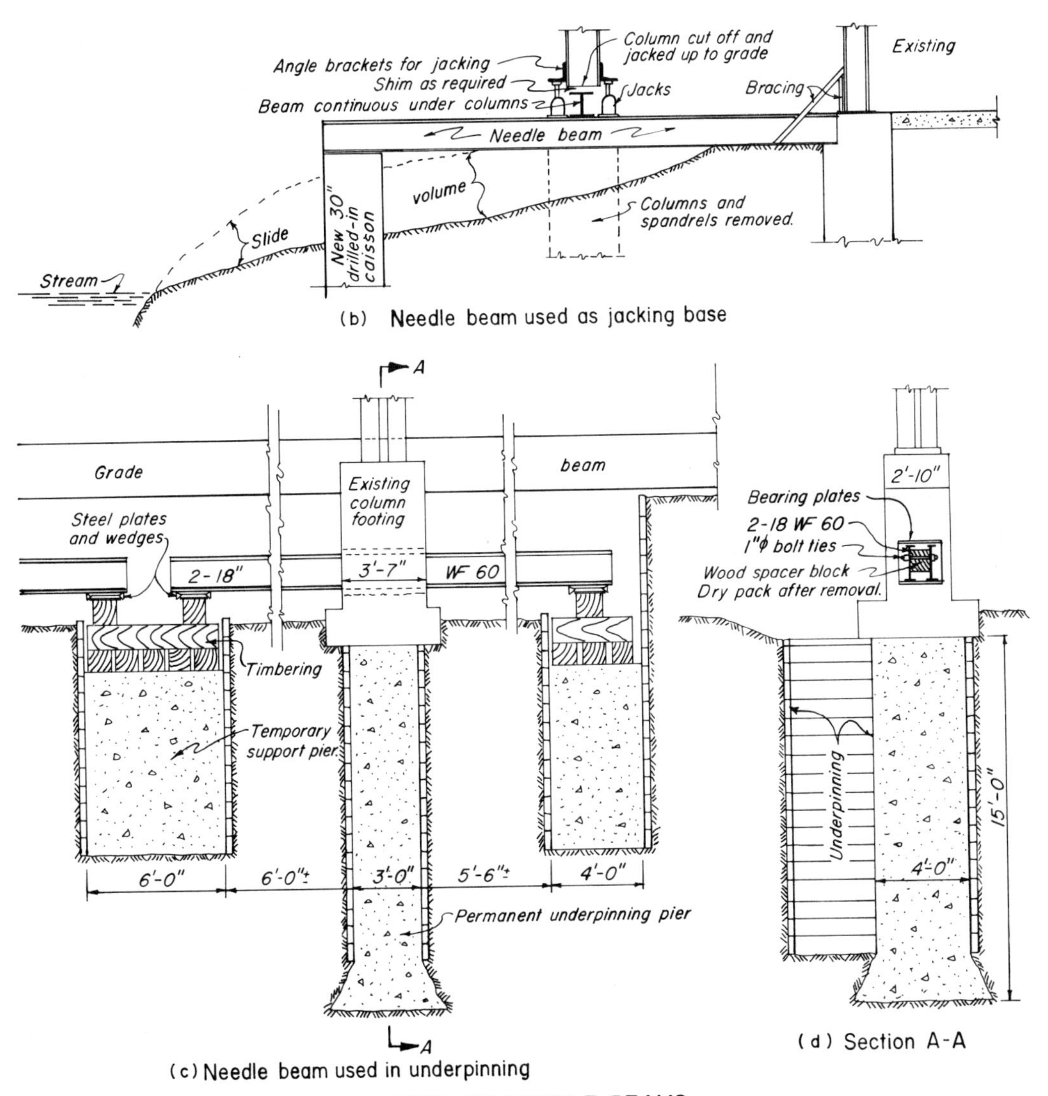

(b) Needle beam used as jacking base

(c) Needle beam used in underpinning

(d) Section A-A

USES OF NEEDLE BEAMS

4.5 AN OPENING IN A WALL

Some years ago, in connection with the revamping of an old water-supply pumping station, it was necessary to run a new 84-in. concrete pipe through an existing wall foundation. The opening required was too great for the existing simple footing to span safely, and it was desired to relieve the proposed pipe of the additional load of any portion of the structure. Underpinning of the wall was therefore necessary and was specified, not merely to be done as a part of the contract but with each step in the process listed.

The illustration shows the problem and the method, step by step, that the contractor chose in executing it. The engineer had very carefully listed his own version of the method of execution. True, it was proposed only as a "suggested" procedure, but it illustrates how widely the approach to underpinning can vary.

The existing foundation was presumed to rest on rock, and it was noted that additional underpinning might be required if the bearing conditions turned out to be otherwise, but they did not.

Engineer's Suggested Steps in Underpinning

1. Cut opening for 10-in. needle beams (openings for needles shall not be less *than 2 ft 6 in. center to center).*

This is curious since one would assume they should not be *more* than 2 ft 6 in. on center. The intent may have been to leave just so much brickwork for bearing between them, but the limiting size of opening is not specified.

2. Place 10-in. needles at elevation shown and grout in place.

In this instruction the leveling of the needles end to end is presumed if not stated. Type and method of grouting are not indicated.

3. Place 30-in. beams as shown.

Again the leveling of the beams is not shown, nor their method of support.

4. Concrete and grout a bearing for 30-in. beams at location of underpinning piers.

This also is somewhat vague.

5. Insert wedges between needles and top flange of 30-in. beams and drive wedges to transfer wall load to 30-in. beams.

See later discussion.

6. Concrete and grout under bottom flange of 30-in. beams full length.

Again ambiguous as to concreting, but see later discussion.

7. Excavate for underpinning piers.

This operation now removes the support from beneath the 30-in. beam at either end or both ends.

8. Concrete underpinning piers to construction joint shown on plans, placing 6-in. H-beams in top of pier.

9. Wedge between bottom flange of 30-in. beams and 6-in. H-beams.

10. Excavate for 84-in. conduit outside building and suction flume inside building and proceed with construction as shown.

It is difficult to understand the reasoning behind this procedure. Since the present footings are presumed to be on rock, the brick wall above is substantial, and no additional loading is now being placed on it, there would seem to be no reason why a 4-ft-wide underpinning pier could not be constructed *first,* both being done simultaneously in fact. (The spacing between them is 15 ft.) Since the interior floor is to be torn up and replaced at a slight change of elevation, the piers can be attacked from both sides of the wall, providing space for forming the resulting pier. If rock, in fact, exists, no shoring may be needed.

Once the piers are complete and their 6-in.-beam bearings have been placed, the 30-in. beams could be set, leveled, and wedged in place to provide complete bearing. This involves cutting away a portion of the footing over the pier but again would not appear to endanger the wall.

Once the 30-in. beams are in place, the openings for the needle beams can be cut and set. By wedging, the wall load can be transferred to the 30-in. beams, and the concreting of the footings outside the piers can be done.

With the 30-in. beams carrying the wall, the entire space between piers can be excavated, the concrete wall constructed, and the 84-in. conduit connection into it completed.

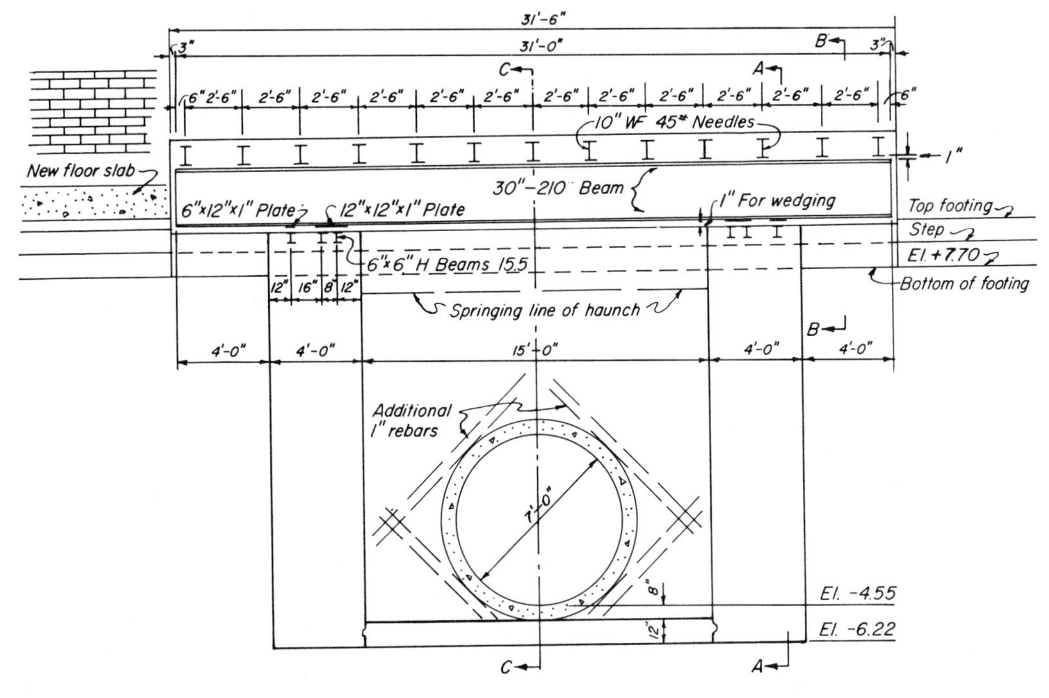

LONGITUDINAL SECTION ALONG UNDERPINNING

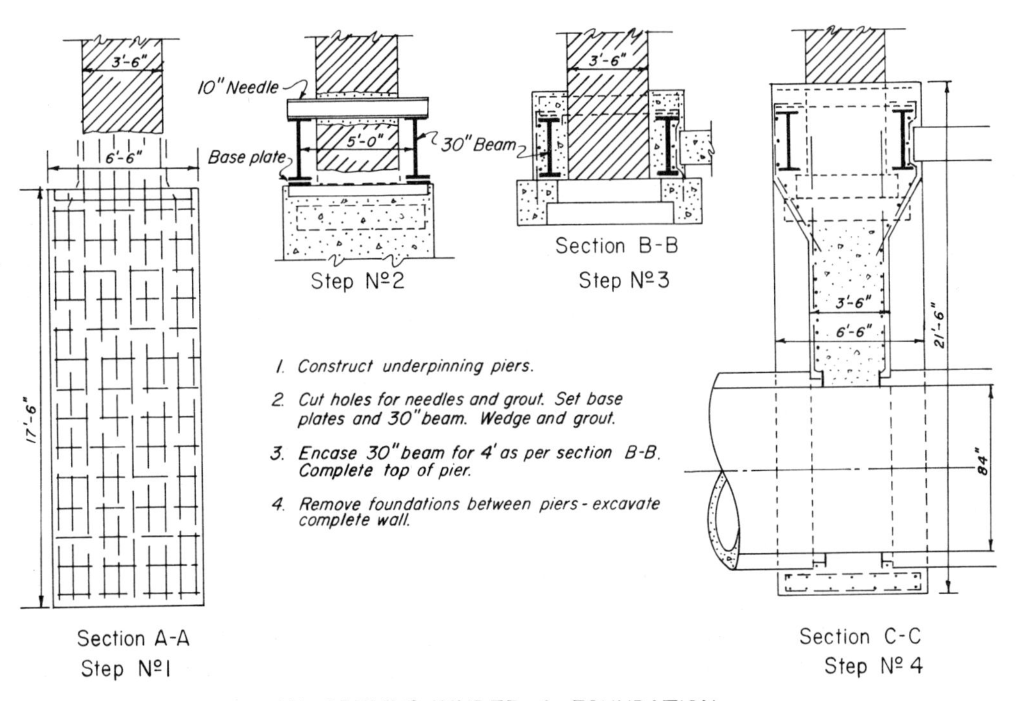

Section A-A
Step Nº1

Section C-C
Step Nº 4

AN OPENING UNDER A FOUNDATION

4.6 COLUMN UNDERPINNING

Some instances of column underpinning have been discussed previously, principally in cases where the column has been a part of an exterior wall, but a very important aspect of column underpinning is that involving the independent or isolated column. The underpinning approach in these cases depends on the type of column, the soil conditions, and the reason for underpinning, whether it be for the installation of structures below foundation level adjacent to the column or whether it is an operation to jack up a column which is settling.

The simplest case is that involving a steel column, whether exposed or requiring the stripping away of the fireproofing or other encasement. Loading is transferred to a point outside the foundation by needle beams, in this case suitably sized channel sections whose back face is bolted against the flange of the column on either side with high-tensile-strength bolts. The loading carried by an interior column, where settlement has begun, may be difficult to properly assess, and a substantial factor of safety should be used in determining the size of the needles and the number of bolts.

The length of the column needles will depend on ground conditions and the underpinning operation contemplated under the column; for if substantial excavation is required, the ends of the needles must be extended to a point where the ground will have sufficient bearing capacity.

In the past, the load carried by the end of the needles was transmitted down to bearing through wooden posts set on blocking and mudsills. The transference of load was then accomplished by driving wood (or sometimes steel) wedges under the posts. This method, although still feasible for light loading, has largely been done away with in favor of jacks and steel plates in lieu of the wood posts and mudsills. Screw jacks are often used, although hydraulic jacks are more desirable. In any case, care must be taken to avoid pushing the column *up* unless a leveling operation is involved, and even here excess upward thrust may damage the surrounding structure.

It is not good practice to use diagonal shores carried up to the underside of beams where they tie into columns (Fig. *e*). The length of such shores, even if of steel, makes them additionally subject to failure, and the difficulties of getting diagonal seating of the mudsills on the ground are considerable. There is also the possibility of failure by shearing of the soil mass.

Changes in foundation conditions are frequently due not to new construction but to changes in soil conditions. Although changes in soils are often thought of as resulting from increases in the quantity of moisture in a soil due to rising water tables or changes in surface or subsurface drainage, a number of cases occur because of the *loss* of water or the lowering of groundwater.

In one case a hot-air heating unit set at floor level supplied a hot-air duct running under the floor. The initial concentration of heat had dried out the clay in the area, causing it to crack and fissure, and, when the duct ruptured, fed hot air additionally into the fissured clay. Four columns in the vicinity settled in excess of 3 in. to a point where they hung from the superstructure.

In addition to other remedial measures (the use of controlled recharge wells to maintain moisture), the footings of the four affected columns were carried 18 ft deeper into the clay, where the effect of drying would not be noticeable. Three of the smaller columns were needled and jacked up to their original positions. The fourth footing, was handled without needles.

An approach pit was dug along half of one side to a depth 4 ft below the footing. Excavation was then carried over horizontally and continued down, directly under one corner of the footing, to the full 18-ft depth. This pit was then poured full of concrete to a level beneath the foundation which would permit the insertion of a jack. The jack was installed and placed under tension. The other four quarters of the footing were treated similarly.

With the four hydraulic jacks in place, the column was jacked up into position and concreted.

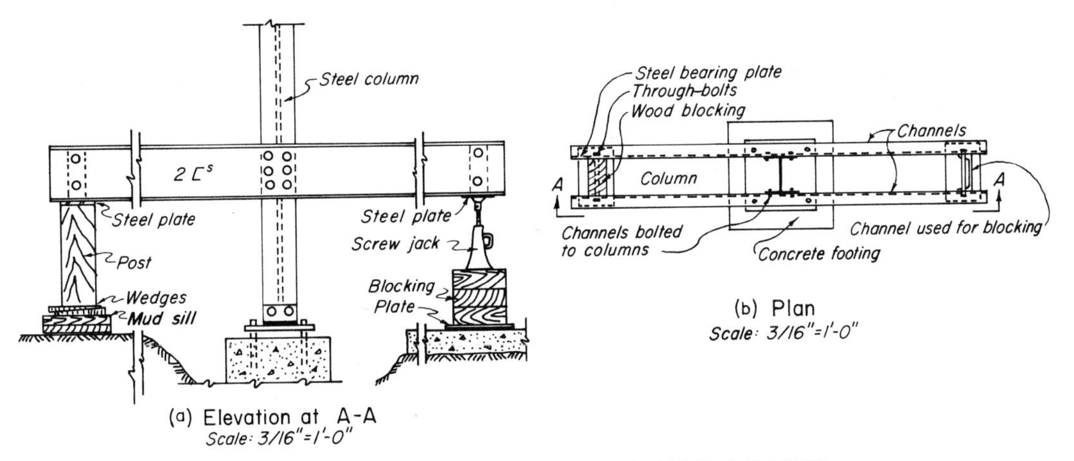

(a) Elevation at A-A
Scale: 3/16"=1'-0"

(b) Plan
Scale: 3/16"=1'-0"

GRABBING A STEEL COLUMN FOR UNDERPINNING

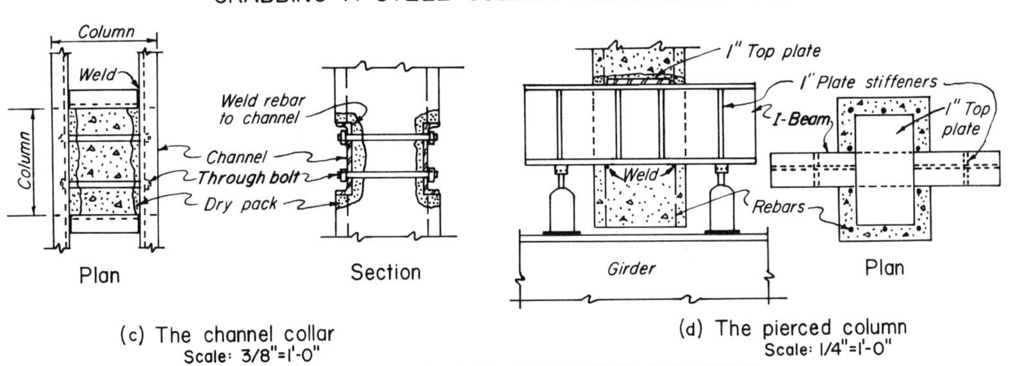

(c) The channel collar
Scale: 3/8"=1'-0"

(d) The pierced column
Scale: 1/4"=1'-0"

PICK-UP METHODS FOR CONCRETE COLUMNS

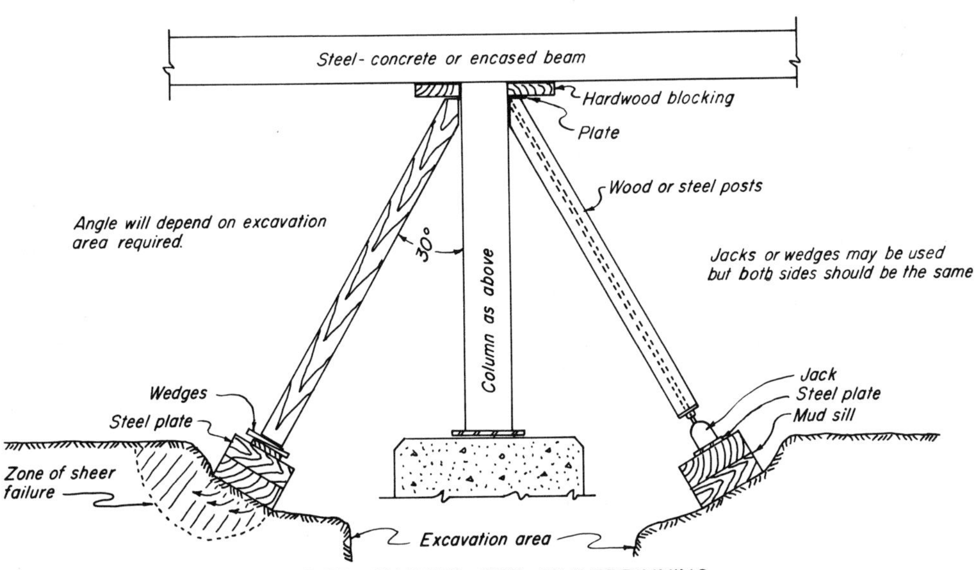

(e) INCLINED SHORES FOR UNDERPINNING

4.7 UNDERPINNING THE WHITE HOUSE

When, in the spring of 1962, Mrs. John F. Kennedy took the world on a tour of the White House and its treasures, via television, notably absent from the sound track was any indication of creaking floors such as might be expected in an old house. This was largely due to the reconstruction of a major portion of the building, completed just 10 years earlier by Mr. Kennedy's Democratic predecessor in the office of President who, 4 years before that, had set the reconstruction in motion and who had, during its progress, risked assassination by his temporary sojourn in nearby Blair House.

The whole account of this very extensive reconstruction does not belong here, but the earliest stages of the operation involved underpinning of the exterior walls and temporary underpinning of portions of the interior which cover facets of the subject not otherwise considered.

After considerable initial study it was decided to retain the exterior walls of the White House, originally constructed in 1792 of limestone and lime mortar masonry. These walls, 4 ft thick and spreading to 8 ft at the bottom, rested on a clay stratum 4 ft below ground surface. They showed little indication of settlement, although interior walls 3 ft thick, not spread, showed appreciable movement.

The necessity of underpinning these exterior walls arose not only from the increased loadings that would be developed by the weight of the reconstructed interior but from the need for providing basement areas for mechanical installations, storage space, public lavatories, and other facilities. Previous basement space had been a miniscule area in one corner, a single floor in depth. The new basement would involve the whole area within the exterior walls and would be two stories deep.

The depth required could have been obtained by going about 17 ft deep with the footings but, at this depth, test borings showed a layer of silt 6 to 7 ft thick. Below the silt was a layer of sand and gravel bound together with silt extending down to rock; and on this layer, 24 ft below ground surface, the underpinning was founded. Two test pits dug to this depth and load-tested indicated a satisfactory safety factor for a loading in excess of 3 tons per sq ft.

The wall underpinning, of 4-ft-thick concrete, was performed as discussed in Sec. 4.1 by taking alternate sections of wall 4 ft long and carrying these down to final depth in a single operation. Because of the delicacy of the operation, previous laboratory studies had been made to determine the possible settlement which might occur during the construction period. It was estimated that this settlement would be about ⅓ in., and precise leveling during the operation substantiated this figure.

The second step in the reconstruction process was the underpinning of the interior portions of the structure which were to be retained. The first floor consisted of steel framing with concrete slabs; the second floor was framed in timber. The first floor required removal for basement construction, and the removal of the second floor was indicated on structural grounds. The third floor and roof, of terra-cotta blocks supported by steel beams, were to be retained.

A temporary steel framework, constructed on temporary footings carried to the full depth of the new basement, was built up to the underside of the third floor. In addition to supporting the third floor and roof, this framing replaced the lateral support of the exterior walls formerly provided by the old floors and walls.

With the upper portion of the structure thus supported, excavation of the 10,000 cu yd in the basement area, using front-end loaders and trucks, could be carried out. Permanent footings on new centers were constructed and the permanent steel framing, threading through the temporary members, was then erected.

In Sec. 3.11 reference was made to earth rebound in connection with excavation, and as a precaution rebound was precalculated in this instance. It was predicted to be about ¹⁄₁₆ in., which precise leveling later indicated did, in fact, occur.

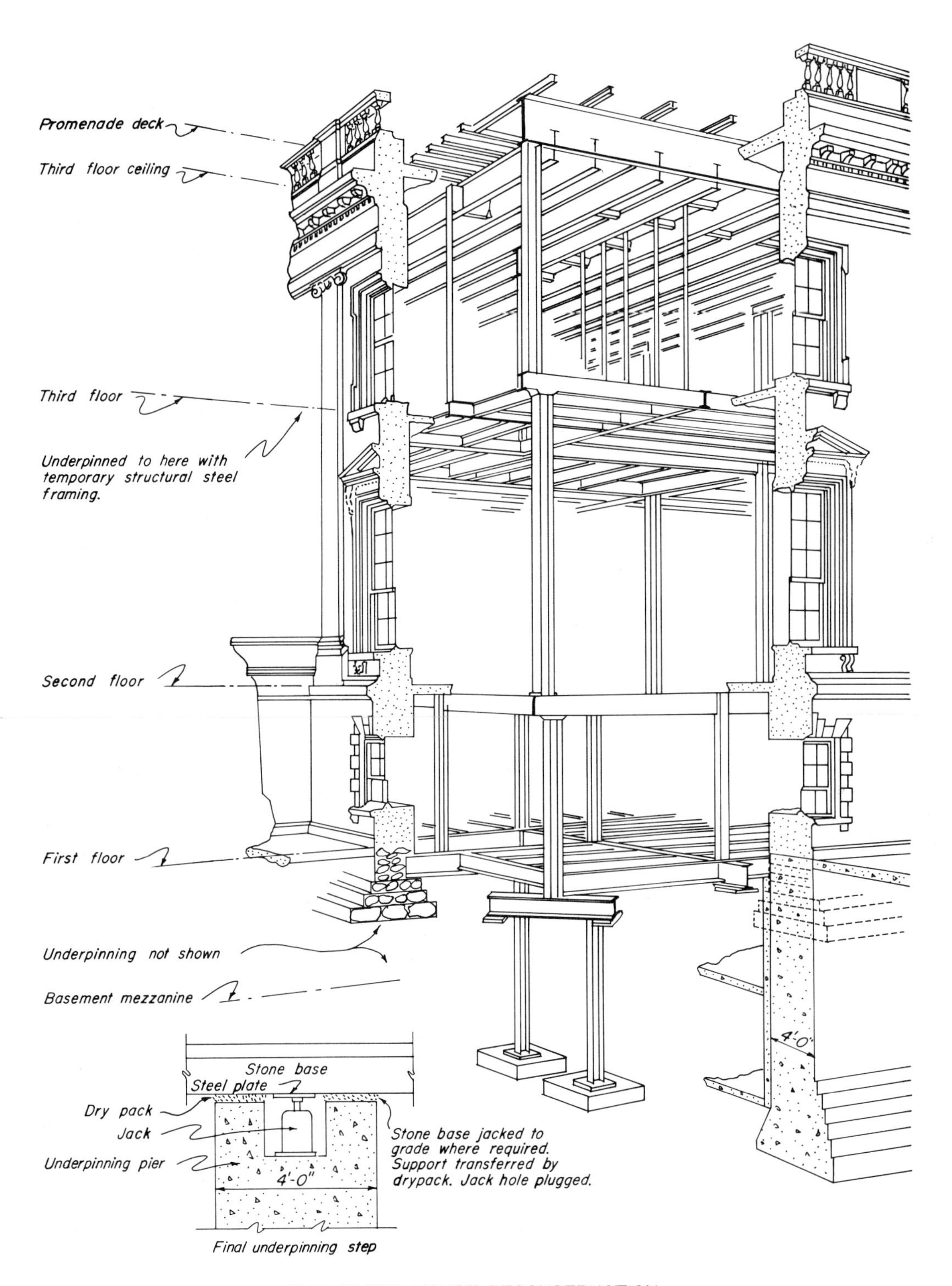

THE WHITE HOUSE RECONSTRUCTION

4.8 SHEET PILING IN UNDERPINNING

The use and design of steel sheet piling will be discussed at length in Chap. 6, Cofferdams, but the use of steel sheet piling involving the transference of loads during the construction of a deeper foundation seems suitably to belong to the subject of underpinning.

The function of steel sheet piling in underpinning is, basically, that of supporting the soil mass on which the foundations rest. Any disturbance of this soil mass can lead to partial or even complete foundation failure. Disturbance of the foundation soil mass can be caused by two factors. The act of driving steel sheet piling produces vibration and, under certain conditions, can lead to soil consolidation. It has been seen that in the presence of water, sand can be consolidated by vibration, and the same thing can be true for other soils with or without the existence of water. Although care in driving may limit the vibratory oscillations, it cannot eliminate them, especially since the piling must be driven to specified depth. The linear transmission of the vibration through the soil is, however, limited, so that if the line of piling is kept away from the foundation edge, the vibratory effect can be negligible.

The second factor is movement of the piling after it has been installed, permitting the foundation soil to sag or settle. The prevention of this requires absolute control, by bracing, of the alignment of the sheathing. Ordinarily this can be controlled by driving a second line of sheathing on the opposite side of the proposed excavation and bracing horizontally to this. However, if the intervening distance is considerable or the opposing ground level is lower, close control of deflections is difficult and uncertain. The use of "rakers" or diagonal bracing can be satisfactory, but it is generally necessary to transfer their thrust from member to member during the course of construction. In the illustration shown, this method has been adopted.

The addition of a fourth unit to the Philips Power Station on the Ohio River required foundations 22 ft deeper than those under the existing unit. The foundation for the existing unit was a mat 3 ft 9 in. thick resting on a fine silty sand of medium density, supporting a loading of 3000 psf. The area of proposed excavation to be spanned was 120 ft, and the density of soil on the opposite side was considered insufficient to take the thrust, so diagonal bracing was decided upon.

General excavation was carried down to the top of the base slab and a row of Z-section steel sheet piling was driven tightly along and against the slab. A 12-in. angle was placed along the outside of the piling as a waler; and by means of 1-in. bars welded horizontally to the top leg and punched through the piling to be welded into the exposed reinforcing bars in the top of the mat, the sheeting was tied back to the concrete. This tie-back prevented deflection of the top of the sheet piling; but, were the piling to be exposed for the required depth, excessive deflection could occur in its length, possibly near the center of the vertical span.

Excavation was then carried 6 ft deep along the outside face of the piling and sloped down on a 1:2 slope to a point 34 ft away from the piling where another row of steel sheet piles was driven. Between the rows of piling, the earth was undisturbed. On the opposite side of the excavated area, no second row of sheeting was driven.

The central slab section was then poured with projections as indicated; diagonal rakers were installed; and excavation between these, down to slab level, was completed. The loading was then transferred to horizontal shores set above the surface of the final slab. These were jacked into position in order to eliminate any deflection that might have developed.

Finally a row of horizontal shores, braced against the central slab, were placed, jacked into full compression, and concreted into the base slab.

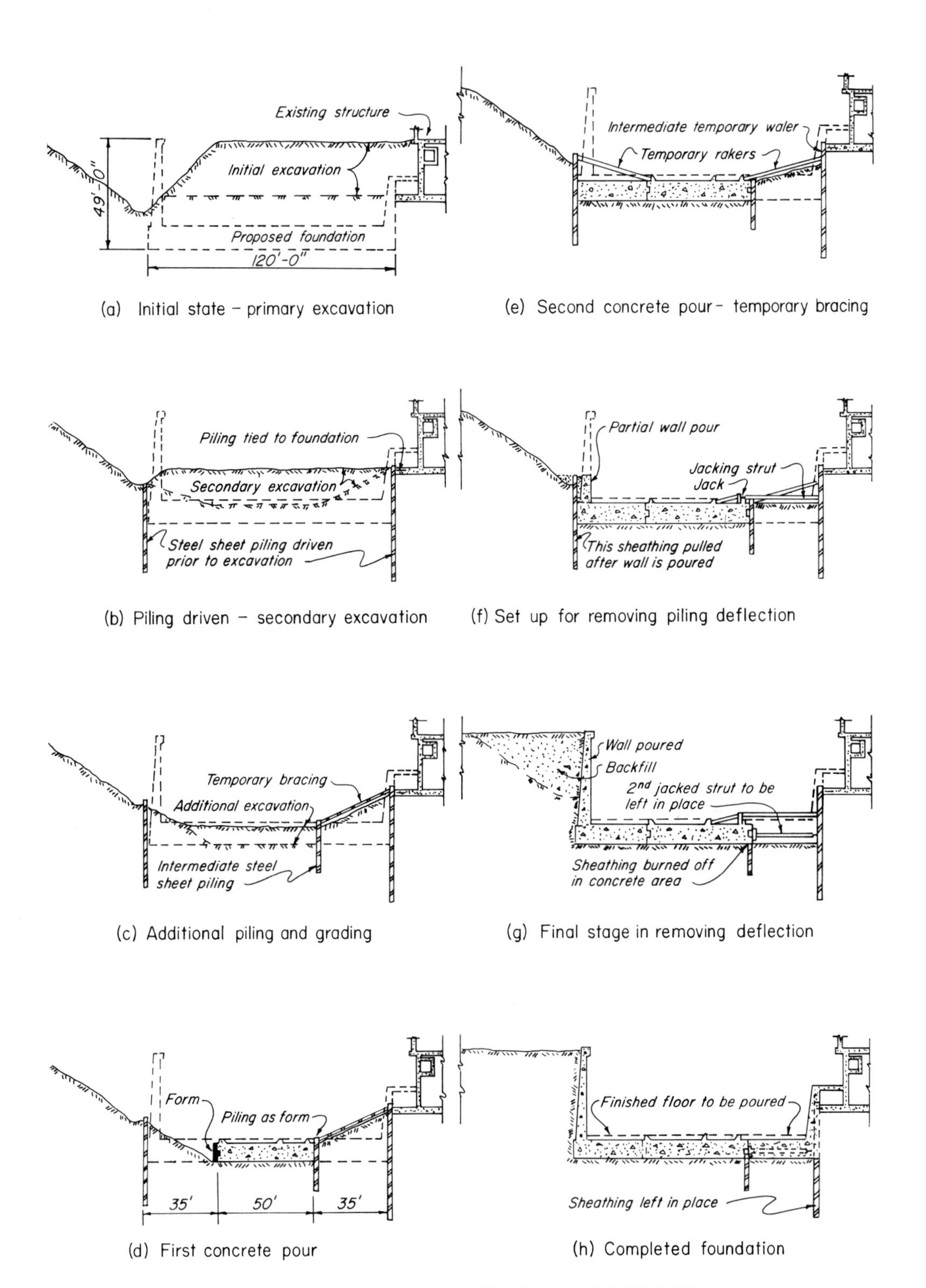

STEEL SHEET PILING USED FOR UNDERPINNING

4.9 PREPARATIONS FOR A SUBWAY UNDERPINNING

Every tenet of logic and language tells us that we should underpin a *high*way to run a *sub*way under it. But as the playwright Andreyev noted: "How marvelously everything is turned about in this world; the robbed proves to be a robber, and the robber is complaining of theft and cursing!" In Philadelphia, when Roosevelt Boulevard was extended through the city, it was found to be necessary to run it under the existing Broad Street Subway, without interrupting its operations.

The Broad Street Subway consists, at this point, of a four-lane concrete box carried on a mica schist formation below the line of weathering. It is about 57 ft wide and 16 ft in height with 7 ft of fill to the surface of the street above it. The two outer lanes at that time carried the traffic while the two inner lanes, provided for future express service, carried no rails and could be blocked off if the need arose. The basic structure was a steel frame with the outer wall and roof steel encased in concrete while the three interior rows of uncased steel columns were carried on independent, continuous concrete footings.

A review of the proposed operation indicated two separate steps to the underpinning operation. The first of these was the supporting of the three interior rows of independently founded columns and the tying of them into the whole structure. The second step involved the entire structure after it was tied together, and the transference of its loading to foundation supports on either side of the proposed highway.

To individually support each interior column would be costly and uncertain. It was therefore proposed to use the columns as vertical members in trusses carried longitudinally to independent foundations behind the proposed opening for the highway. The concrete in the footings and center bottom slabs was accordingly removed at the foundation locations and new footings poured 7 to 8 ft into the rock below. The six columns that would be the vertical end members of the three trusses were individually shored. After these foundations were in place, a row of holes was drilled into the rock down to the final highway excavation depth along the inner faces of the foundation—on 3½-in. centers. This would control rock breakback when the final excavation was made.

The truss consisted of top and bottom chords of 18-in. channel placed back to back against the column faces with diagonal beams cut in between the existing vertical columns. Load transference at contact points was accomplished by the use of gusset plates. These members were prefabricated to fit existing column conditions from full-size templates and were pre-drilled for assembly by using high-tensile-strength bolts instead of welding.

All these operations were performed by means of access from the top of the subway. Holes 12 ft square were cut in the paving and sheeted down through the 7 ft of fill to the top of the subway roof slab. The concrete of the roof slab was then broken out between the 20-in. steel beams leaving two openings about 4 ft wide and from 10 to 12 ft long. The beams were not disturbed and all material, equipment, and supplies had therefore to be passed down through the limited opening available, with the exception of the channel members, which could not be maneuvered through the limited space (in any satisfactory lengths) and were brought in on flatcars over the subway tracks.

Once the trusses were completed and the load transferred to the new foundations, excavation under the center truss and through the floor of the two center lanes was begun, leaving the rock support under the outer trussed columns temporarily intact.

Highway excavation had progressed up to the face of the subway wall on either side, although not to full depth. The 7 ft of fill over the subway had been retained by soldier columns bolted to the face of the exterior walls carrying horizontal wood sheathing, so that traffic on Broad Street could be maintained. Utilities (sewer, water, gas) parallel to the subway and outside its faces were either temporarily supported on cables or rerouted. The second stage of underpinning could begin.

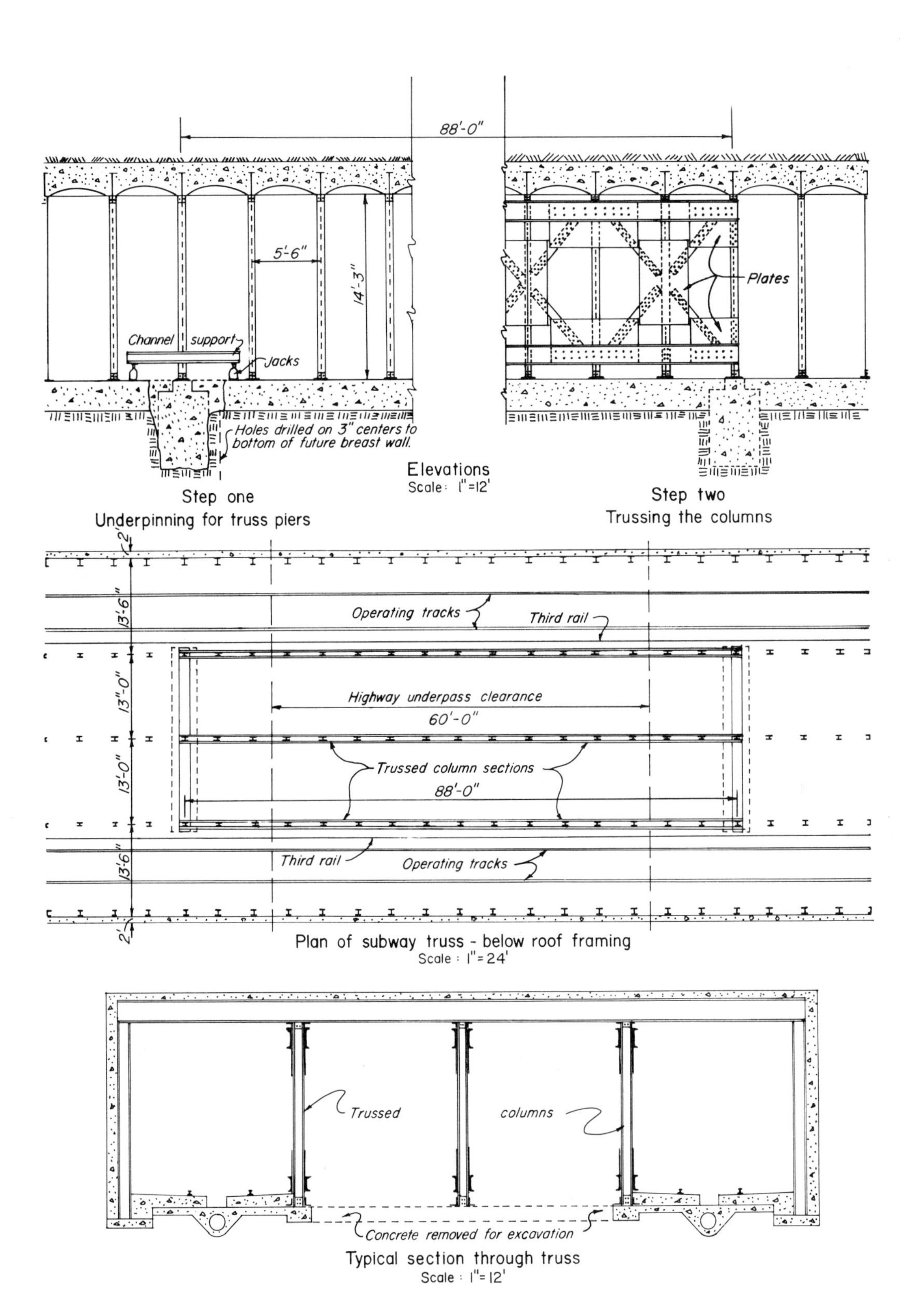
88'-0"
5'-6"
14'-3"
Channel support
Jacks
Holes drilled on 3" centers to bottom of future breast wall.
Plates
Elevations
Scale: 1"=12'
Step one
Underpinning for truss piers
Step two
Trussing the columns
2'
13'-6"
Operating tracks
Third rail
13'-0"
Highway underpass clearance
60'-0"
13'-0"
Trussed column sections
88'-0"
13'-6"
Third rail
Operating tracks
2'
Plan of subway truss - below roof framing
Scale: 1"=24'
Trussed
columns
Concrete removed for excavation
Typical section through truss
Scale: 1"=12'

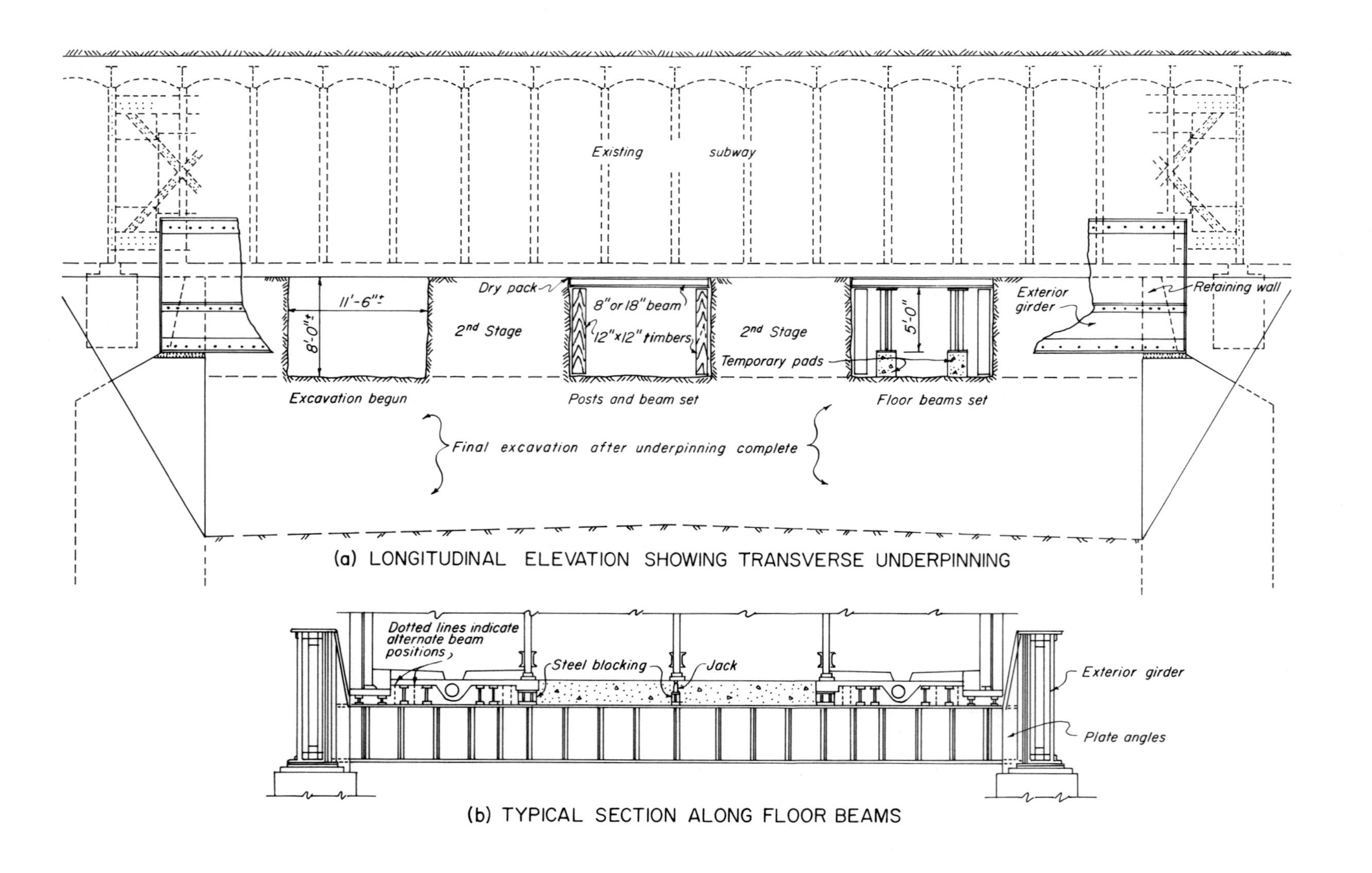

(a) LONGITUDINAL ELEVATION SHOWING TRANSVERSE UNDERPINNING

(b) TYPICAL SECTION ALONG FLOOR BEAMS

4.10 UNDERPINNING THE SUBWAY

The starting condition which had been reached, as discussed in Sec. 4.9, consisted of two outer sections of subway still supported on rock with excavation proceeding under the two inner lanes and under the center row of columns. Excavation up to the exterior faces of the subway walls had progressed, on both sides, to a depth of about 8 ft below track level.

The next step was the careful excavation of equally spaced tunnels from side to side under the subway, proceeding not only from the exterior but from the interior as well. Tunnels were about 7 to 8 ft high and about 10 ft wide. The spacing provided was that resulting from the interior column spacing, the center of each tunnel being the center between every third row of subway columns. The general scheme was to be that of supporting the subway on grillage beams, in turn supported by floor beams centered under the subway columns, which transferred the load to box girders spanning the highway on the outside faces of the subway.

The tunneling through rock was largely a hand operation using controlled blasting sparingly, since the face had to be maintained essentially vertical. As tunneling progressed, the grillage beams were set successively against the underside of the subway concrete, as poling boards are set. The bottom of these beams had to be set to a precise elevation suitable to the top of the floor beam on which they would later rest. They were supported at their outer ends on 12 × 12 in. timbers wedged vertically against the tunnel walls and cross braced. Once the beams were in position, the gap between the top of the grillage beam and the underside of the subway slab or foundation was closed with dry-pack.

When the two sections of a single tunnel had been holed through and all the grillage beams had been set on their temporary timber supports, the floor beams, 5 ft deep, were threaded through the tunnel, set in position, and jacked up against the underside of the grillage beams. As will be seen, each tunnel carried two floor beams and thus could take the loading previously carried by the 12 × 12 in. temporary timbers. Temporary concrete pads, resting on the rock floor of the tunnel, were placed under the floor beams, not at their ends but inside the outer face line of the tunnel.

With the first three tunnels completed and the six floor beams set, the remaining four openings were excavated similarly. Here, larger grillage beams were used to span between the floor beams already positioned, although the increased span was temporarily reduced by timber posts. As indicated, the positioning of the original set of grillage beams had been staggered to leave space for the second set to slip through. After completion of floor-beam setting, the loose ends of the overlapping grillage beams would be tied together by welding.

While the floor beams were being set, abutments for the box girders were being constructed in pits on both sides of the subway. Since these were outside the face of the subway wall, no underpinning was involved, only care in excavation to prevent disturbance to the rock on which the adjacent section of subway was supported.

The box girders were pre-fabricated off the site and brought in on low beds for which access ramps had to be specially constructed to reach the lower level of the excavation. Each 60-ton unit was set in position on its abutment by using a separate crane lift near each end. Since lifting and swinging was impractical with this loading, a bulldozer was employed, thrusting at the center, to swing the girder forward into position.

Each box girder was set to pick up the ends of the floor beams, which were then connected to it by bolting with two angled plates per beam. The girders were jacked up to take the full load of the floor beams and grouted and bolted down.

The initial interior trusses were left in, but their loading was transferred to the floor beams by placing a jack under each column and loading it to provide ¼ in. of deflection in each floor beam. These jacks were left in position and buried in the concrete of the replaced subway floor.

4.11 PILE JACKING

Pile jacking is a method of underpinning first introduced as long ago as 1916 by members of the firm of Spencer, White and Prentis Inc., the foundation contractors, and still successfully used by them. Although the possible applications of the method are numerous, these can be roughly grouped into three main categories: (1) the placing of piling under existing foundations to extend their bearing to a deeper stratum; (2) the extension of existing piling under a structure to a deeper level without disturbing the structure it supports; (3) the reconstituting of damaged piling under a structure.

The use of pile jacking depends upon the existence of a structure under which the work is to be performed; and, although in some respects it is responsive to the general aspects of a driven pile, it is an underpinning operation rather than a piling operation *per se.* For general underpinning purposes it has certain advantages over other methods. It is not necessary to excavate to the full depth of bearing, and the method can therefore be used in running sands or silts or where water-bearing strata would intervene between the foundation and the bearing strata. The dewatering of an area can often change the bearing value of a soil or produce settlement under loads, and pile jacking avoids this possibility of disturbance.

In pile jacking, an approach pit is excavated down to the bottom of the existing footing and then carried about 6 ft below it. As in the excavation for straight-wall underpinning, excavation is then carried under the foundation to the full 6-ft depth, and sheathing, as required, is installed. The location of the pile to be placed, and hence the pit size, will depend on the factors discussed in Chap. 5, Piling.

A 4-ft section of 12-in-diameter steel pipe is placed in the desired position and capped, and a hydraulic jack is wedged between the top of the cap and the underside of the foundation. (Because in many cases the foundation was poured against earth and may be quite rough, a steel plate may be required, grouted level under the concrete.) This section of pipe is then jacked down into the earth, carefully, so that absolute plumbness is maintained.

When the top of the first section has been jacked down to within a few inches of the bottom of the pit, the jacks and cap are removed, a special coupling is placed (usually provided with an internal collar or spacer), another section of pipe is placed, and the coupling is welded to top and bottom sections of pipe. Here again a truly plumb condition must be maintained.

Additional sections of pipe are added as the jacking proceeds. Since the bottom section has an open end and the whole pipe will fill with soil as it goes down, it may be necessary to clean out the pipe from time to time. This can be done with a hand auger or a small grab bucket lowered and operated by cable, or the earth can be blown out by air. When final depth has been reached, the pipe is cleaned out completely and filled with concrete.

One advantage of this method is that the pile can be tested for bearing in its final position. After filling with concrete, a short section of steel beam is laid across the top of the pile of sufficient length to permit the placing of two jacks side by side with a space provided between them. These are calibrated hydraulic jacks, operated jointly to provide uniform lifting. Jacking is now resumed to provide 150% of the design loading on the pile. (A 40-ton pile would be loaded to 60 tons.) When proper test loading is reached, a section of steel beam is placed between the top of the cap beam and the underside of the foundation and wedged in place with steel wedges to pick up the load. The jacks are now removed, and the sections of beam and upper portion of the pile are encased in a concrete block.

Great care must be taken in estimating the existing loading on a foundation to prevent jacking the foundation *up* instead of the pile down. Where only a single pile is to be placed, the 150% testing procedure cannot be performed without uplift danger. Precise check leveling should be maintained at all times to detect any possible upward movement of the footing.

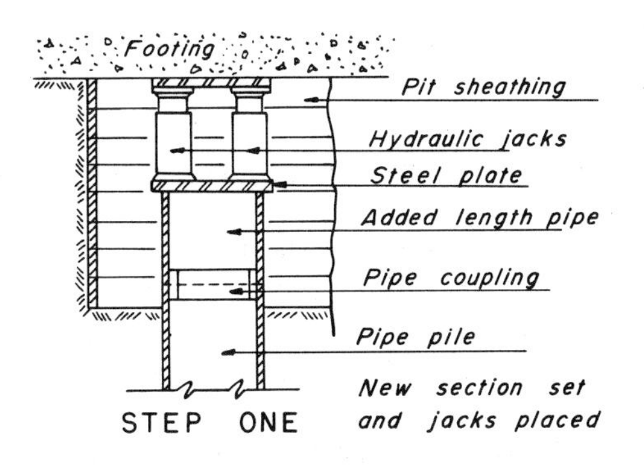

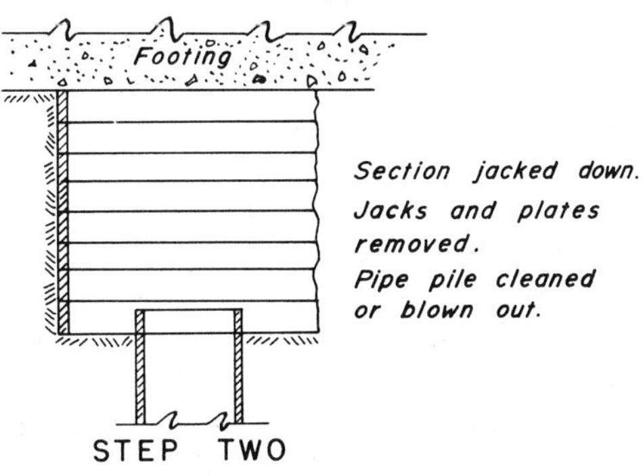

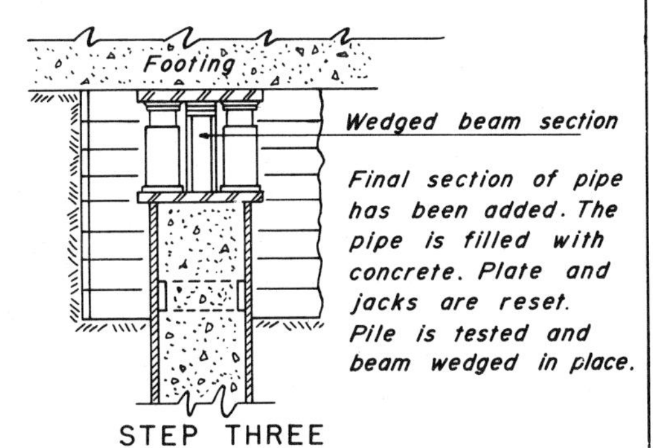

Footing
Wedged beam section
Final section of pipe has been added. The pipe is filled with concrete. Plate and jacks are reset.
Pile is tested and beam wedged in place.
STEP THREE

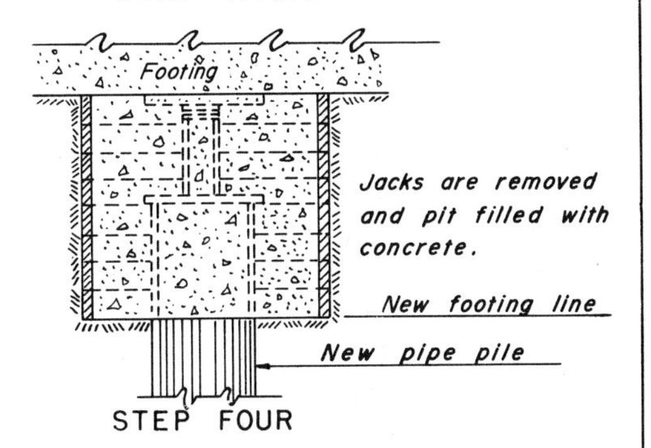

THE METHOD

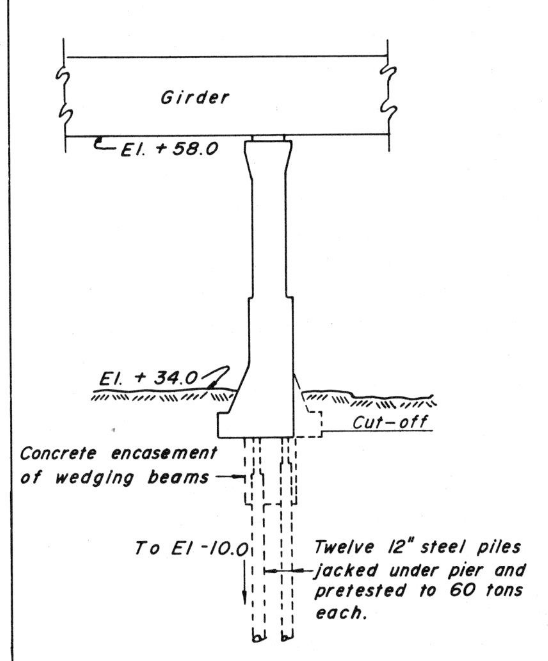

BEFORE EXCAVATION WAS PERFORMED

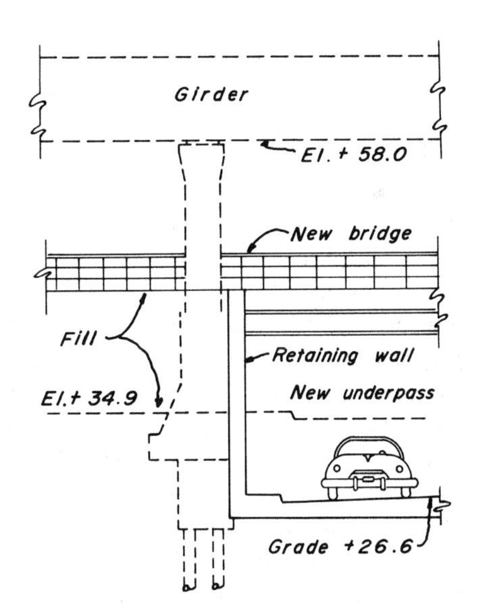

COMPLETED PROJECT

THE RESULTS

4.12 A SUBWAY UNDER A SUBWAY

New York City is famous for the tallest buildings and the deepest subways; but whereas the height of a building is chiefly a matter of free choice, the depth of a subway is more often a matter of grim necessity. An old Kansas law once said, in essence, that when two trains approach a track intersection both shall stop and *neither* shall start up again until the other has cleared the intersection. This sort of thing can be pretty frustrating and just won't do for travel in underground tunnels. When one subway meets another, the last one there must simply go deeper, get under the earlier one, and keep moving.

In 1958 the new Chrystie Street Subway found it necessary to dive under the existing four-track BMT Subway—on a curve yet. The method devised for this involved no interference with the existing structure, nor work within it, but did combine several of the underpinning techniques which have been discussed.

The existing structure consisted of rows of steel columns bearing on a non-reinforced pad. Along the outer walls, these were subject to possible damage or disturbance arising from the proposed underpinning operation, and the first step was to protect the support of these individual columns. Those columns which lay across the line of the operation were supported on individual pipe piers, pile jacked into position. These supports would be exposed in the lower subway and would be cut away when that structure received the loading. Adjoining sections of the outer wall of the BMT subway were also carried on pile-jacked columns, but these would be left in place for permanent stability. All this pile jacking was performed in individual sheathed pits carried down from street level prior to beginning the main underpinning operation.

The second step was to run tunnel headings or drifts directly under the BMT foundations along the line of the new subway. These were about 6 ft wide and 6 ft high and were supported by a post-and-beam sequence, the beams jammed tightly under the existing concrete slab and the posts serving to retain horizontal sheathing. For this operation it was necessary to dewater the area by using wells, strategically located, when well points had proved ineffectual in the fine silt.

From the floor of the drift, at alternate 4-ft intervals, 4-ft-square pits were excavated vertically to a depth below the bottom of the new subway. The sheathing required was used as formwork to pour concrete piers which would be portions of the proposed subway walls. This work involved transference of post loading in the drift support from pier to pier as work progressed.

With a section of the 4-ft-thick wall in place, a 15-in. wall was carried from the row of piers up to the underside of the existing slab in line with the outside of the 4-ft-thick wall, using intrusion grouting methods. On the inside, on top of the 4-ft wall, a steel beam seat consisting of a 10-in. BP 42# was laid down, leveled, and concreted into place.

The next step was to connect the 6 × 6 ft heading on each side of the proposed subway by a continuous heading, 6 ft deep and the full subway width. The underside of the existing structure was supported by temporary timber cribbing. At 4-ft intervals the steel-beam structural roof members were placed successively on the beam seats and wedged under the BMT slab by temporary timbering.

With these in place the remainder of the depth of the new subway could be excavated. Wall columns for facing the 4-ft walls were hung from the roof beams and, as excavation progressed, were connected at their bottom by permanent steel floor beams to which any thrust at the bottom of the 4-ft walls was transferred.

With the floor beams in place the bottom slab could be concreted, the 4-ft walls waterproofed, and the interior wall columns then encased in an interior concrete wall.

The final step was that of replacing the blocking between the steel roof beams and the underside of the BMT slab with a reinforced concrete mat which additionally encased the roof beams.

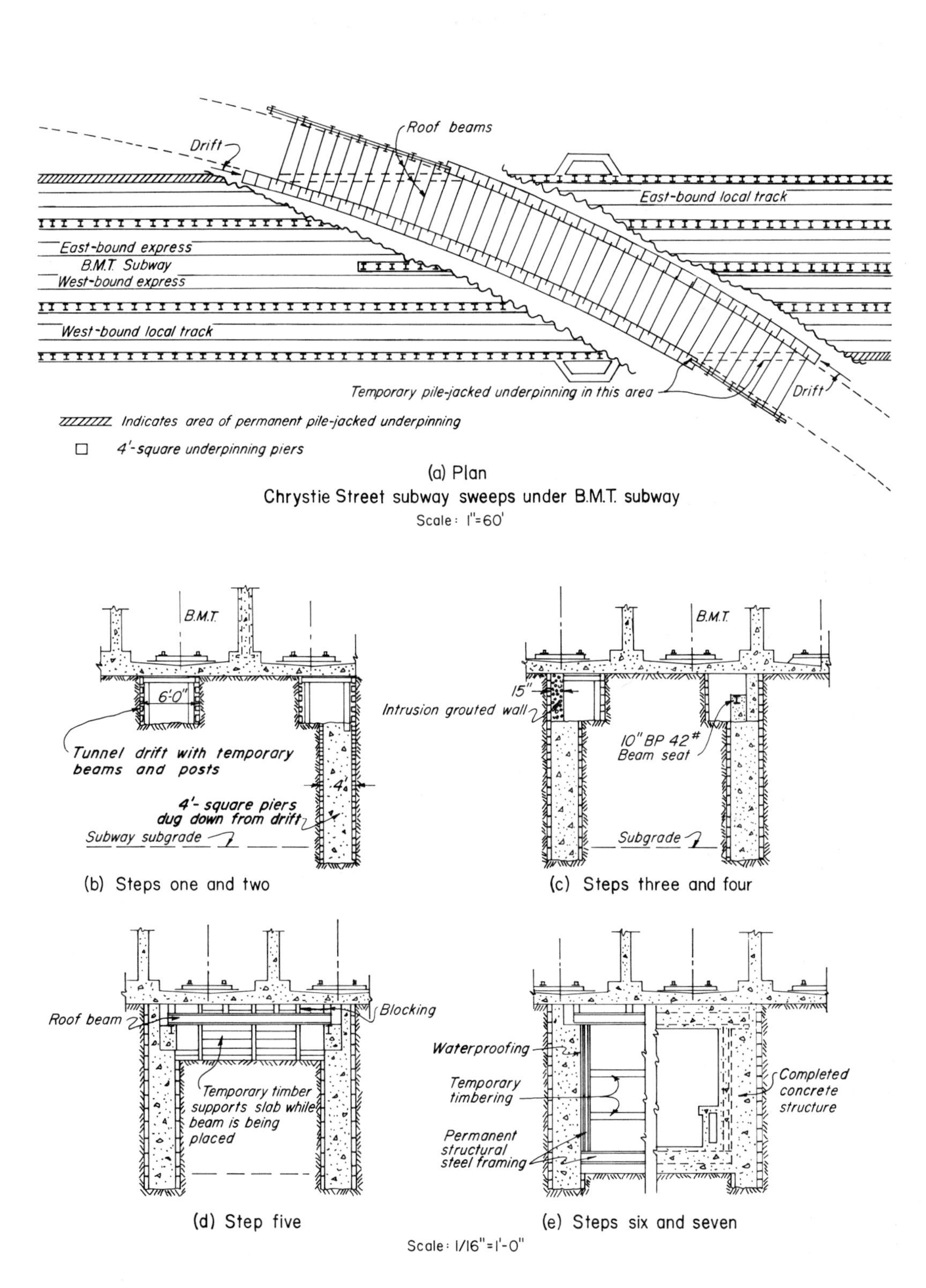

(a) Plan

Chrystie Street subway sweeps under B.M.T. subway

Scale: 1"=60'

(b) Steps one and two

(c) Steps three and four

(d) Step five

(e) Steps six and seven

Scale: 1/16"=1'-0"

A SUBWAY UNDER A SUBWAY

4.13 UNDERPINNING A CAISSON

In 1955 the Texas-Illinois Natural Gas Pipeline Company desired to throw two 30-in. gas pipelines across the Mississippi River at a point some 80 miles below St. Louis, from Grand Tower, Illinois, to Wittenberg, Missouri. The method called for a narrow suspension bridge with a tower on either shore, providing a main span of 2150 ft at a minimum height of 75 ft above maximum flood level in the Mississippi River.

On the Illinois side, rock lay not far below the surface, and the tower foundation offered no special problems. On the Missouri side, a caisson was required. The sinking of this caisson by open well and pneumatic means is discussed in Sec. 8.24. When the last phase of the sinking under air was within 5.5 ft of the planned seating level, the cutting edge came in contact with rock ridges at several points near its center. A crack had developed horizontally along one wall of the working chamber. Blasting to remove the rock ridge appeared dangerous, and it was decided to underpin the remainder of the cutting edge.

The means of underpinning here were not dissimilar to the wall underpinning already described except that working under compressed air at considerable depth introduced new problems. As has been discussed, air pressure, when working under water, must be kept proportional to the depth of water. Therefore it is not practical to dig to the full depth required for an underpinning pier, since where substantial depth is involved the air pressure is too variable, and blow-ins or blow-outs may develop. In this instance, therefore, to control air pressure, underpinning was employed working from the underside of the cutting edge down, in 2-ft increments, instead of working from the bottom up.

A curtain wall closing the space between cutting edge and rock surface was constructed using blocks of poured-in-place concrete, 2 ft high, 5 ft long, and 1 ft thick. The soil lying over the rock was a medium fine sand, which even under the pressure of over 40 psi would not stand vertically. Consequently, small areas of a square foot or less were exposed and immediately plastered over by hand with a wet clay. As the whole 2 × 5 ft area was developed, it was necessary for one man to keep smoothing the clay surface with a wetted hand.

Once a 2 × 5 ft area had been opened, a form was set, light reinforcing placed, and concrete, mixed by hand on the floor of the chamber, was shoveled into the forms. As for all underpinning, this procedure was followed for alternate 5-ft sections, working away from the exposed rock area. The intervening 5-ft sections were then filled on the next round.

After a 2-ft-high ring had been completed, excavation over the whole floor of the chamber was continued down to the level of the bottom of the strip and another 2-ft strip was begun.

Where there is any chance that underpinning to any extent may become necessary, provisions should be made in the steel cutting edge for fastening reinforcing steel to it. Welding under compressed air is always dangerous; and in this instance, time did not permit extensive welding operations. Yet hanging reinforcing from the cutting edge can effectively prevent these blocks from falling away as a second layer is placed, even though the joints of the second layer are staggered.

As work progressed, a fault in the rock surface was uncovered crossing the caisson near the south end. Underpinning was carried down into it for a depth of 16 ft, but no rock surface was reached. At 16 ft the legal limit for pneumatic work was reached and it was necessary to place a continuous 12-in. slab over the sand surface on which to terminate the underpinning.

Rock surfaces on the floor of the chamber were now cleaned off and concrete was placed up to the level of the cutting edge, lowering it with relative care in muck buckets to prevent damage to the underpinning. The concrete seal was finally carried 24 ft above the cutting edge.

This case is noted because there are relatively few instances where caisson underpinning has been attempted.

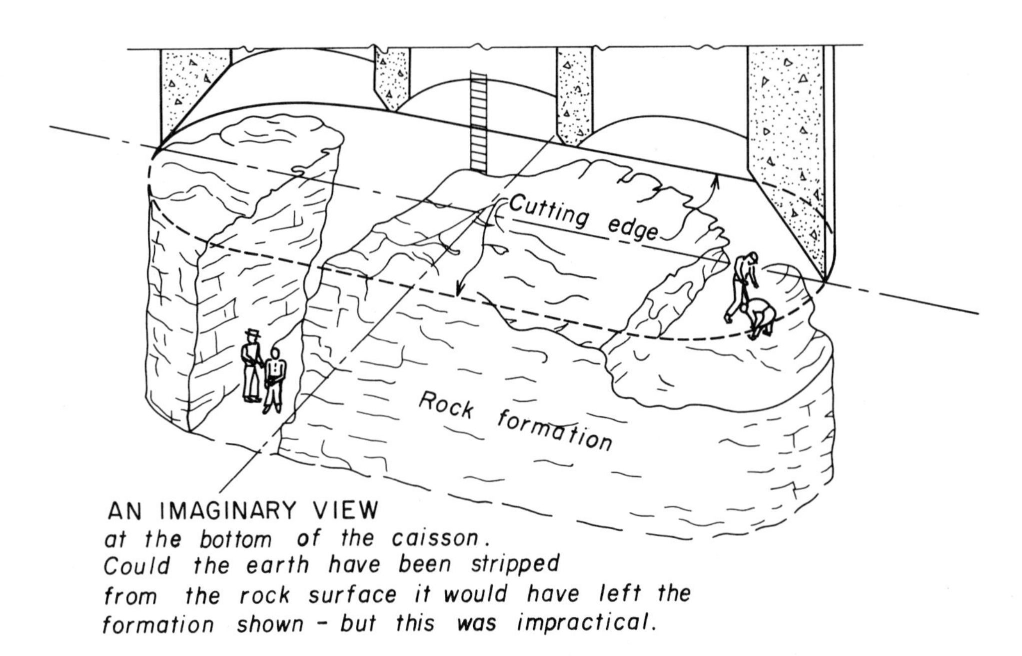

AN IMAGINARY VIEW
at the bottom of the caisson.
Could the earth have been stripped from the rock surface it would have left the formation shown – but this was impractical.

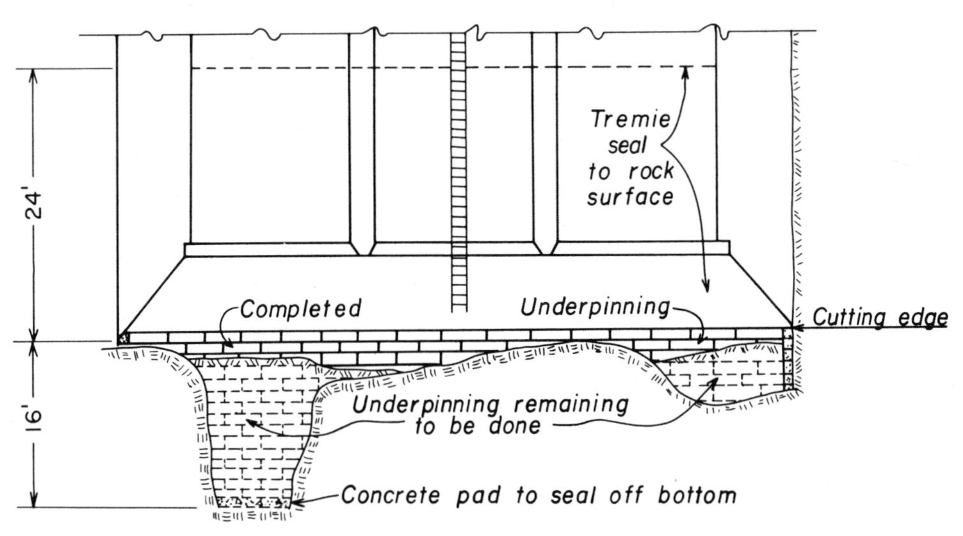

THE UNDERPINNING OPERATION
AT AN INTERMEDIATE STAGE

4.14 UNDERPINNING THE GODS

Some 40-odd years ago in a slim small volume of verse, John Cowper Powys published a poem with the title "The Twilight of the Gods." Some randomly remembered lines now seem presciently appropriate.

> In a long sad row the old gods come;
> They come and bow to me.
> Like candle-flames in a raftered room,
> Like trees in an avenue of doom,
> They bend in unity.
> . . . .
> They nod and mutter; they sway and bend
> Like monoliths of stone,
> . . . .

They are indeed "monoliths of stone," cut from the solid pink sandstone cliff on the bank of the Nile in the now desert wasteland at Abu Simbel, in the Egyptian Sudan. "The old gods," Re-Harmachis of the morning sun, Amon of Thebes, Ptah of the underworld, and a deified Ramses II (who had them all carved out 3200 years ago) are to be found in a rock-hewn sanctuary within the cliff, approached down a 200-ft-long corridor lined with a 30-ft-high "long sad row" of colossi.

The corridor is so oriented that only twice each year the sun strikes brightly down its length to light up three of the figures, leaving Ptah, appropriately, perpetually in the shadows. But always in the sun and flanking the entrance to the corridor, in duplicate, are the seated figures of Ramses II and his Hittite wife Nefertari, also cut, 67 ft tall, from the same pink sandstone.

The "avenue of doom" is the Nile River, always wild in flood, but now to be tamed by the construction of the High Aswan Dam (a lower, older, Aswan Dam already exists) 175 miles down river; for the lake which will be formed behind the dam will be 200 ft deep at Abu Simbel and will completely inundate these ancient figures.

To avoid this "doom," seemingly in store for the old gods when the new dam is completed in 1968, several methods had been presented to that committee of UNESCO which was assigned to the salvage task. The method under consideration in 1962 had been proposed by a firm of Swedish engineers. They proposed to cut the complete ensemble from the solid rock, raise it 200 ft, and re-create the general aspects of its present setting at the higher elevation. The first step was an underpinning operation.

A series of tunnels about 13 ft wide and 18 ft high will be cut into the rock from the front end, below the level of the corridor. A base slab will be laid down and continuous underpinning piers constructed on this as excavation proceeds. A slab will be carried under the rock over the top of the piers and the load transferred to the piers. When this step has been completed, a trench will be cut around the outside and the inner rock wall will likewise be sheathed in reinforced concrete.

It is apparent that no blasting can be permitted for either the tunneling or the trenching operation. Experiments have shown that this Nubian sandstone can be excavated by the use of air-driven paving breakers to within a distance of 24 ft of the figures. Closer than this, power saws and hand chisels will be required.

Within the cavern there are evidences of substantial cleavage planes which must have existed when the original monuments were carved out 3200 years ago; and part of a figure of Ramses II, guarding the entrance, has slid off along such a fault. It is expected that considerable rock bolting will be required to tie portions of the basic rock together as excavation proceeds.

Once the entire monolith has been cut loose from the mother rock in its casing of concrete, the cavern filled with sand and the entrance figures likewise buried in it, after a roof and front wall have been constructed over and across it to stiffen the mass and contain the sand padding, the rigging stage begins. This involves jacking up a load estimated at 250,000 tons, almost three times greater than any before attempted, to a height of 200 ft, also greater than man has tried previously, and doing this rapidly enough to keep ahead of the slowly rising lake water level, which will only be denied access by a temporary dam for the first 55 ft of this rise.

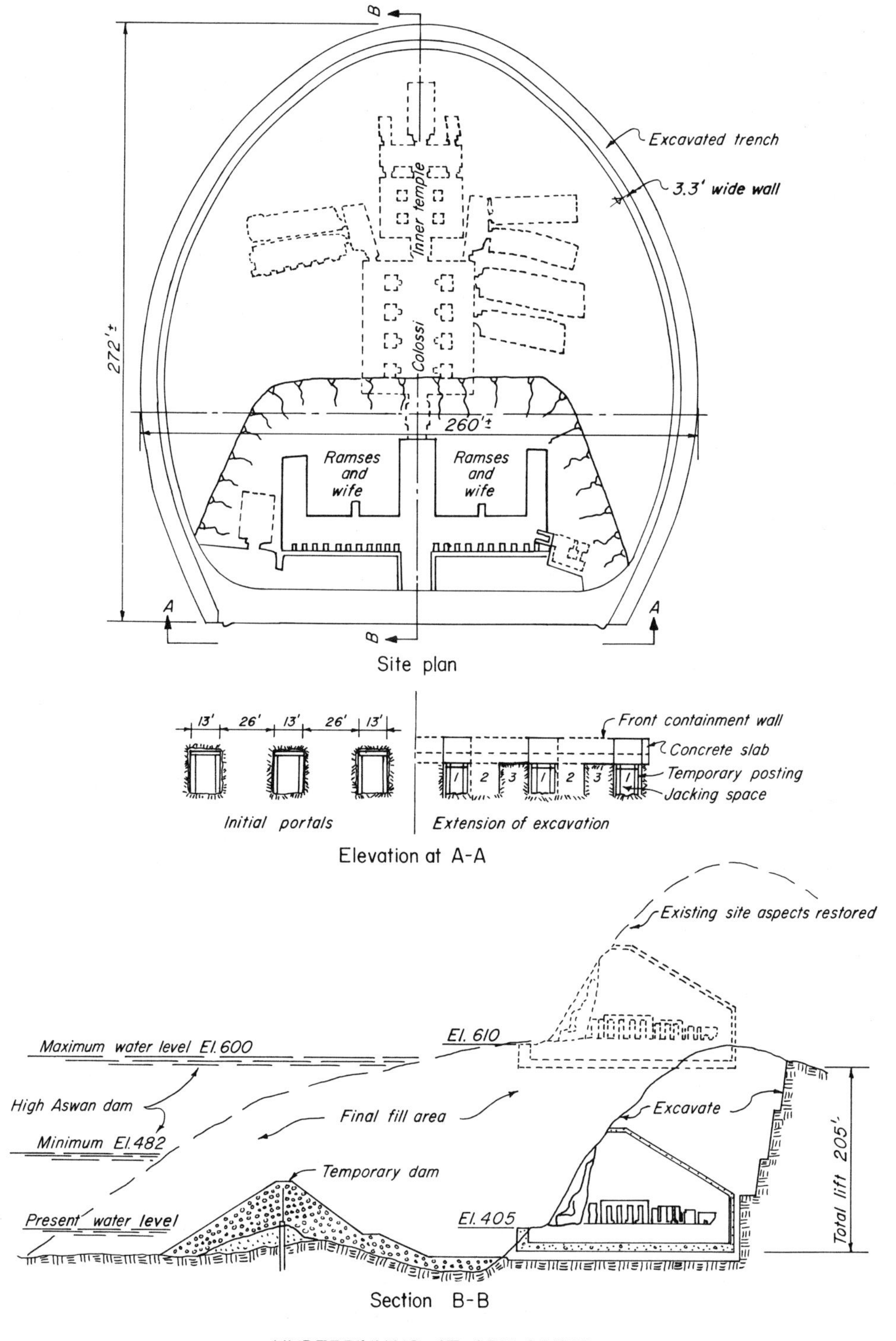

UNDERPINNING AT ABU SIMBEL

4.15 THE JACKS OF SPADES

The basic, underlying operations involved in underpinning, the spadework so to speak, consist in raising, holding, and lowering of loads, functions once performed largely by wedging action but now more satisfactorily performed by a variety of jacking devices.

There are three general types of jacks used in underpinning operations: the ratchet jack, the screw jack, and the hydraulic jack, utilizing respectively the lever-and-fulcrum principle, the screw-thread principle, and the principle of incompressible fluid pressure. For underpinning construction or rigging purposes these principles must be incorporated into a compact, portable, rugged shell. This eliminates several of the variations which might find industrial uses.

The ratchet jack (Fig. *a*) is the simplest type of jack and the oldest, operating on a principle well known to the Greeks several milleniums ago. Limited to load capacities of less than 20 tons because of the physical human effort involved, it is the fastest of the several jacks discussed. The downward stroke of the lever raises the rack bar one notch at a time and pawls hold the load, releasing the lever for the next lifting stroke. This limits the lifting or lowering ability to a fixed increment which is often not fine enough for underpinning operations in their final stage. Generally equipped with a foot lift, a ratchet jack can be operated in a relatively narrow opening.

The track jack is a ratchet jack used for leveling railroad rails but, unlike other ratchet jacks, has no provision for lowering the load. The track jack releases the load completely when tripped and is therefore not suitable for other than trackwork.

The capacities of screw jacks, employing the screw-and-nut principle, are dependent, like the wedge, on sliding friction along the pitch of the thread. For lighter loads a lever bar is employed, similar to the lever of the ratchet jack, and capacity is limited by manpower strength. Larger screw jacks are actuated by power, frequently air motors operating through gear-reduction mechanisms and ratchet devices. When these devices are employed, however, they increase the size of the jack and reduce its portability.

Hydraulic jacks exert a unit pressure on a small area of incompressible fluid which is transmitted undiminished to a larger area in the same closed circuit. Hydraulic jacks are somewhat slower in operation than either screw or ratchet jacks but can lift heavier loads and can be controlled within very small tolerances. They have an additional advantage in that most models can be equipped with a calibrated gage to show the weight of the load being raised. (This is a pressure gage which converts the fluid pressure in the cylinder into pounds or tons of weight.) The lighter load models (Fig. *c*) have a threaded ram extension which can be turned out by hand to fill up a large gap quickly. In heavier models, dual pumps are used: a speed pump which rapidly extends the jack and a power pump which automatically takes over when the loading point is reached.

The model shown is hand-operated, but hydraulic jacks are frequently operated in multiple by the use of small hydraulic pumps controlled through a console. Hydraulic fluid is pumped to or withdrawn from each of a number of jacks in order to control the level of a lifting operation. This method is used for concrete-lift slab construction as well as jacking tunnel shields and in large underpinning operations such as those proposed at Abu Simbel.

A new principle in jacking termed a Roll-Ramp, was introduced in 1962 by the Philadelphia Gear Corporation. Curiously enough, this returns to the principle of the wedge except that in this case the wedge moves, intermittently, between two ball-bearing races. The mechanical advantage is very great, once sliding friction has been eliminated. It is motor-operated through a gear-reduction mechanism, thus converting rotary to linear motion. Considerable speed of raise is claimed for it, up to 5 fpm, but in most underpinning operations speed is often not particularly desirable. Because of the Roll-Ramp's cost, it is contemplated that capacities will start at 50 tons and go upward.

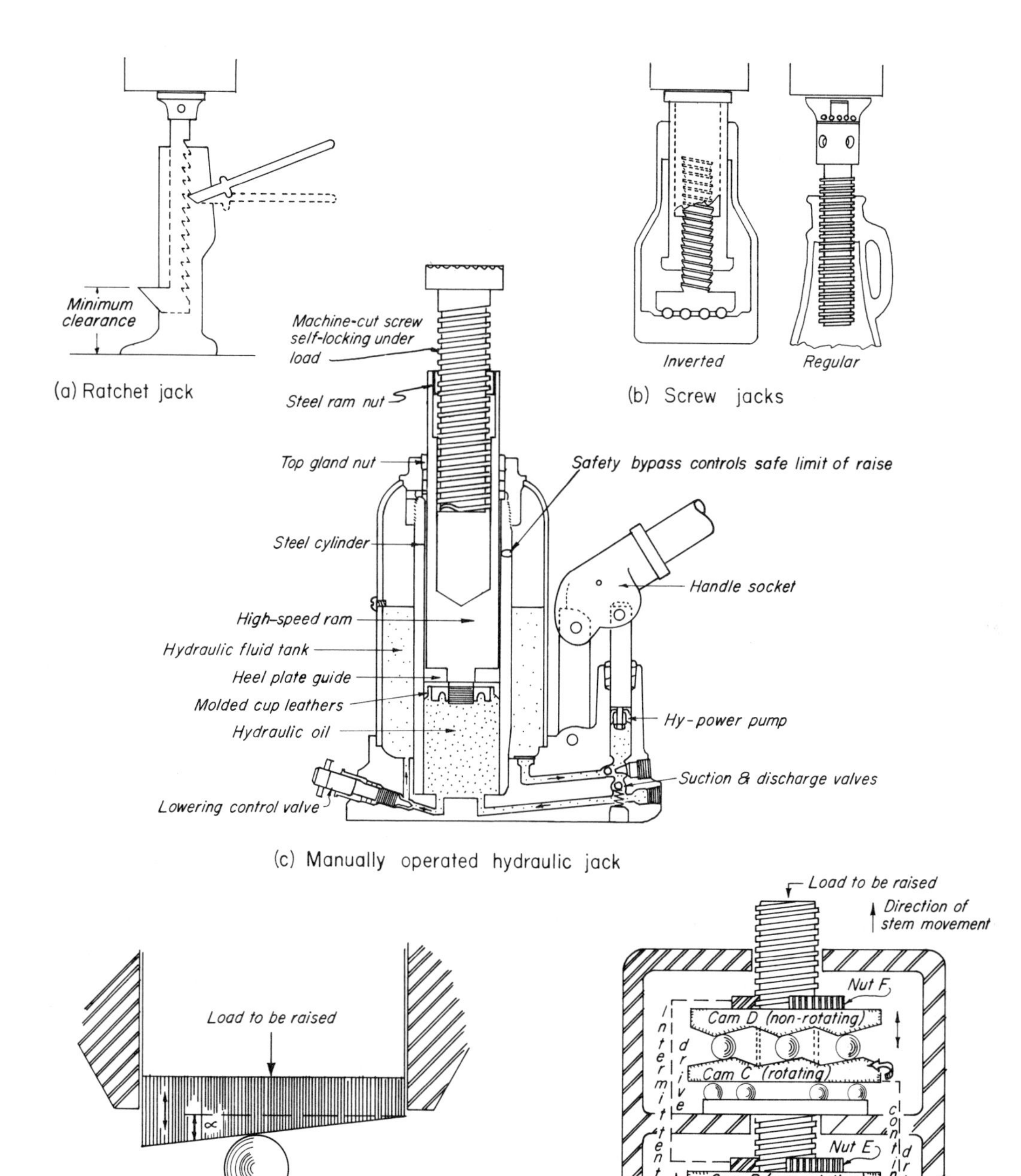

(a) Ratchet jack

(b) Screw jacks

(c) Manually operated hydraulic jack

The principle

The mechanism

(d) The roll-ramp

5 Piling

Piling was once a very simple matter. Those ancient Swiss pile experts simply cut down a tree, knocked off its branches, floated it out into the lake, stood it in position, and either beat it or wiggled it down into the soft muck until it would stand by itself. These piles were then tied together with other poles by lashings, and a platform laid down over the poles served as a surface for the construction of huts.

Those early pile dwellers had, perhaps, just gotten out of the habit of sleeping in trees; and their pile-supported dwellings, swaying gently in the breeze, were more comfortable, surely, than the crotch of a Druidic oak and more sleep-inducing than even a modern motel vibratory mattress.

Built for safety from more predatory tribes, these pile-dwellings-in-the-lake were ideally situated to resist a long siege. These tribesmen had below them plenty of fresh water and, so long as the fish kept biting, a supply of food: they "had it made" indeed, and in splendid isolation. But this idyllic scene faded, and these early peoples disappeared. Whether the fish stopped biting or whether the pile dwellers simply got tired of a diet of fish every day and twice on Friday; whether in this milieu of peace and tranquillity they bred so fast that their overloaded pilings collapsed; whether they consumed their structures, log by log, in cooking the fish; or whether some sneaky cave dweller, resenting their security, invested their pilings, at waterline, with termites or wood-boring beetles who, quietly crunching away, ate the pilings right out from under them is not known.

Piling is a very ancient device that has permitted mankind to develop harbor sites, cities, and, indeed, whole countries on soft swampy sites that even the birds would founder on. The city of Venice owes its existence to wood piling some of which, put down 1000 years ago, is still perfectly serviceable. We associate the existence of Holland with the dykes that hold *back* the North Sea; but vast areas, including the entire city of Amsterdam, are held *up* on wood piling.

Until the nineteenth century all piling was wood; and how many ancient port cities of the Greeks and Romans were swept

away because of infestation of their piling supports by marine borers has never been noted. With the dawn of the twentieth century, steel and concrete piling began tentatively to replace wood piling, but wood remains a highly considered piling material.

In 1962 a shortage of wood piles in the 80- to 120-ft length range seriously hampered pile-driving operations in Metropolitan New York City, and in that period the 2-year demand for these lengths alone was 80,000 pieces. This is a substantial forest.

A 100-ft-long pile will require a straight tree as much as 170 ft in length with a natural taper of roughly 1 in. in 10 ft. First the ground swell at the base of the tree must be removed to the extent of 5 to 10 ft of the length. The point diameter of the finished pile may account for another 60 ft of the upper tree. The mere shipment of this forest of trees, trimmed and cut to length, will require some 30,000 flatcars.

The trouble is, of course, that wood piling is cheaper than steel or concrete and, so long as the forests hold out, will continue to hold a preferential position.

Substituting for the commonly considered *driven* pile are a number of alternatives which have developed in the past half century. These bear a variety of names but still perform the same function as piling and can therefore be considered in the same general category. One of these is the caisson having a limited diameter, sometimes not much greater than that of a large pile, consisting of shells forced into the ground as excavation is performed within them. (These of course are not in the same class as the open caissons discussed in Chap. 8.)

The development of drilling equipment has led to what is known simply as a drilled-in pile, where a hole is drilled down through varying soils to firm bearing and filled with concrete. Then there is the caisson which is drilled in instead of being excavated within a shell. There is no clear distinction between a drilled-in pile and a drilled-in caisson except the whim of the contractor applying the nomenclature.

The discussion of piling in this chapter will be limited to *driven* piling, leaving the various alternative types for later discussion.

5.1 TYPES OF PILING SUPPORT

There are two types of piling: those which support the load placed on them by *end* or *point* bearing on a firm stratum existing at some depth below the ground surface; and *friction* piles, which support their load by the friction developed between the surface of the pile and the soil through which it is driven. The combined effect of end bearing and friction to support a load might be considered a third type save for the fact that there is seldom the one type of support without some indication of the other, so that the designation of either type merely indicates the *principal* source of support which the pile furnishes.

The type of piling used will depend almost entirely upon ground conditions as revealed by the logs of test borings. End-bearing piles will be driven to rock, hardpan, or other firm stratum, but the important consideration here is the depth at which such a stratum occurs and the effect of placing a concentrated load on that stratum. The nature of the earth's surface being what it is, it would always be possible to drive piling to a solid rock bearing except for three contravening factors: (1) The depth at which rock could be found might be so great as to make piling of that length economically infeasible. (2) Overlying strata might be so dense that piling could be battered out of functional shape before reaching rock. (3) Even in relatively soft intervening layers, sufficient friction could develop between the soil and the pile surface to prevent the further driving of the pile.

All end-bearing piles function as columns. Where extremely long piles are required to reach a deep stratum through intervening soft soils, water, or both, not only the strength of the pile acting as a column must be considered, but its lateral stability as well. Is there enough solid material surrounding it to hold the pile upright without depending upon tying the tops of piles together, and is there enough soil support to resist bending in the unsupported length of pile? This question has limited the lengths of piling in some materials, and batter piles have been devised to assist in resisting lateral (i.e., horizontal) forces.

Rock may lie beneath successive layers of hardpan, sand, or gravel, any one of which is not considered quite adequate for end bearing. Two possibilities present themselves here. These dense layers may be drilled through, the pile dropped into place and then merely driven through the last thin stratum to the final point-support layer. This avoids heavy and damaging driving. The other alternative, where the depth is sufficient, is to employ the pile as a friction pile, driving it only to a sufficient depth to support the load by friction *and* partial end bearing, so that a minimum of those intervening dense layers have to be pierced.

In the third case no effort is made to reach a "rock" stratum even though the intervening soils are not dense, and this pile is limited in function to providing support as a friction pile. This is the more usual case, wholly justifying the use of piling. The soils materials will be soft wet clays, silts, or fine to coarse sands. Lengths of piling will vary according to the kind and stratification of the various soils.

Test borings indicate the existence of various subsurface conditions; and the selection of a type of piling, whether end-bearing or friction, will generally be determined by a review of this information. (Note that the term "type" is used here; "kinds" of piles—wood, steel, concrete, etc.—may be dictated by other considerations. See Sec. 5.9.)

The length of end-bearing piling is usually determinable from test-boring logs, often within fractions of a foot. With friction piling the precise depth at which sufficient friction will be developed between pile and soil to support a given load is not so readily ascertained. Theoretically a friction pile might be driven down indefinitely so long as the required energy was applied and the pile remained undamaged; but, since cost increases with length, the limiting factor is the depth at which only sufficient friction to support the designed load will develop. This is the function of pile-driving formulas to be discussed in Sec. 5.10.

FRICTION PILES

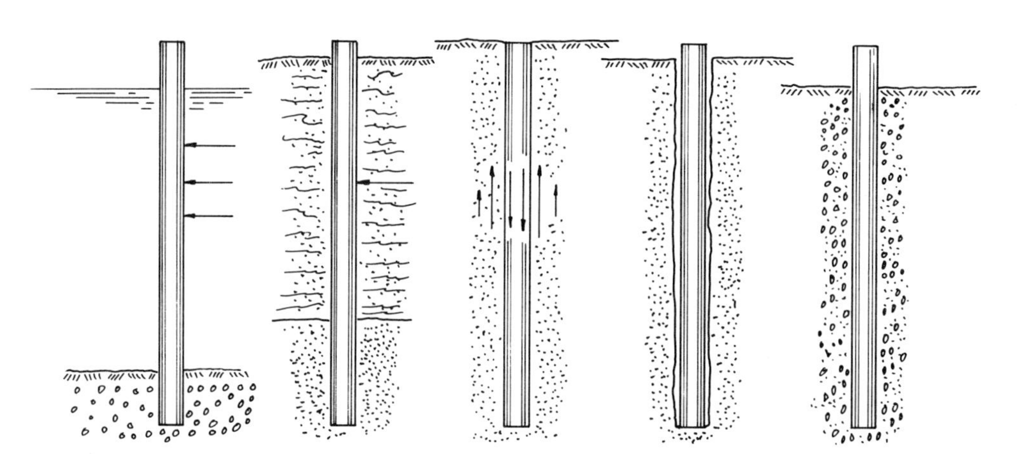

UNDER WATER	LIQUID SOIL	MOIST CLAY	DRY CLAY	SAND - GRAVEL
Not a true friction pile unless water is shallow in relation to depth of penetration.	*May offer some support but depth of penetration below liquid is important.*	*Ideal friction pile. Clay bonds to pile. Shear value of clay determines bearing.*	*Pile may simply punch hole in clay providing inadequate bond and pile damage.*	*Hard to drive. Pile may be damaged. Can be jetted in sand but not in gravel. Is this pile necessary?*

END BEARING PILES

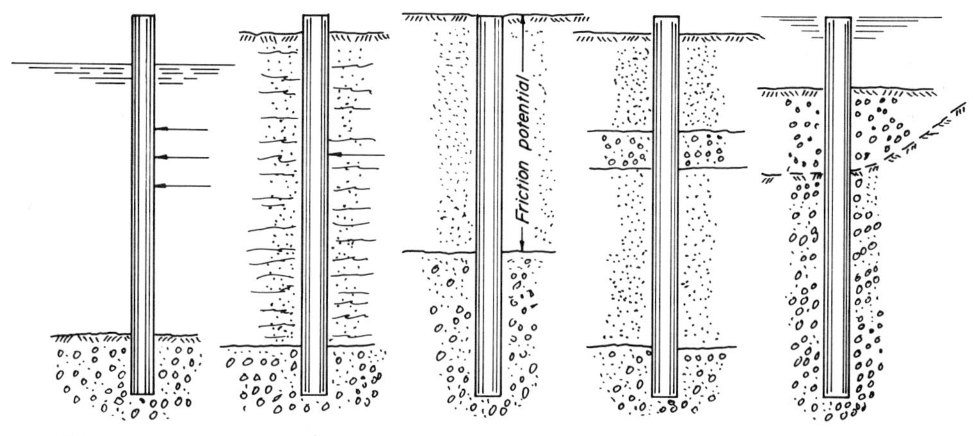

UNDER WATER	LIQUID SOIL	MOIST CLAY	STRATIFIED SOIL	GRAVEL-HARDPAN
Satisfactory use if size of pile or cap provides adequate lateral stability.	*Check bending as free-standing column. Dynamic formula should not be used.*	*Carried through supportable layers to increase bearing potential. Some friction allowable.*	*Pile punched through thin hard stratum to reach better support below.*	*Piling needed here only as protection from scour. Bearing carried well below scour line.*

5.2 TIMBER PILES

Wood pilings are considered as friction piles rather than as end-bearing piles for two reasons: (1) the wood pile, being relatively soft, cannot be driven through hard layers without the danger of brooming (that is, the shattering of its cellular structure) either at the top or bottom; (2) wood piling is driven with a pointed tip (or shaped metal shoe) to reduce the impact necessary to get it to a satisfactory depth so that even should it reach rock, the pointed tip is seldom depended upon for substantial bearing.

The splicing of wood piling is not a recommended procedure although, of course, it is possible. As a consequence, wood piling is driven in a single length. To determine the required length it is necessary to drive test piles in order to determine the length required under the specific driving conditions.

Since wood piling is used in a single length, the dimensions of a pile can be specified for the tip end and for the butt end. The diameter of the tip (that is, above any pointing) varies from 5 to 9 in., depending upon the class of pile—A, B, or C. The diameter 3 ft from the butt end is specified as a minimum diameter of from 12 to 14 in. Since we are primarily concerned with the *volume* of wood available at any cross section rather than its conformity to a true circle, growth variations are bypassed by specifying a circumference which will produce the required average minimum diameter. The 3-ft limitation is intended to reach past the ground swell of the smaller tree.

Other controlling factors in the selection of timber for piling are:

1. The growth characteristics which determine the straightness of the tree.
2. The degree of taper throughout its length.
3. The number and kind of knots.
4. The condition of the bark.

Decay of wood piling is caused by plant life, by fungi, which break down the cellular structure of the wood. To grow and flourish the fungi must have air, the moisture must be limited, and the temperature must be optimum. Wood piling *submerged* in fresh water will not decay; but at the other extreme, decay will not advance where the moisture in the soil is less than 20%. In the first case there is insufficient air; in the second, there is insufficient moisture. Fungi will not develop at freezing temperatures, but thrive in relation to the rise in temperature above freezing.

To counteract the effect of decay fungi, as well as chemical contamination which may exist in soils or waters and the action of numerous marine borers in salt water or of other insects which attack non-buried timber, wood piling is frequently pressure-treated with creosote. One of the chief reasons for creosoting is that piling driven in wet areas and presumed to be submerged will frequently become exposed to the effects of air by the inadvertent lowering of the water table in the area.

Pressure-creosoted wood piling must have all its outer bark removed, generally in the forest, and at least 80% of its inner bark as well, to permit proper penetration of the creosote. Full debarking is generally accomplished by machine peeling after the outer bark is removed. Piles are seasoned or conditioned before being fed into the treating retort.

There are two methods of creosoting lumber, in each of which controlled temperatures and pressures will vary depending upon the kind of lumber being treated. The "empty cell" method is applied to wood which has been under atmospheric pressure. The "full cell" process first subjects the timber to a substantial vacuum after which the creosote is applied under pressure.

The degree of creosoting is measured in terms of pounds of creosote per cubic foot of wood and, for piling, a retention of not less than 12 lb per cu ft is generally specified. The degree of penetration into the creosoted member, measured in inches, or in the degree of sapwood penetrated, is also used as a criterion. Penetration may vary from ¾ in. for Douglas fir to 3.5 in. for southern pine; from 65% of annual rings in red oak to 85% in red pine.

Summary of Specifications For Round Timber Piles

(A.S.T.M. Designation D25–52T)

USE

Class A: Suitable for heavy railway bridges or heavy frame construction.
Class B: Suitable for docks, wharves and building foundations.
Class C: Suitable for foundations which will be perpetually submerged, for light construction or for falsework.

QUALITY

Classes A and B: Shall be free from red heart, decay, splits (in treated piles) splits exceeding butt diameter (in untreated piles), unsound or clustered knots, shakes exceeding one-third the smallest diameter.
Class C: Shall be sufficiently sound to stand driving, free of decay, bad knots and imperfections that will reduce strength.

REQUIREMENTS

Classes A and B: Shall exclude wind-felled, blighted, fire-killed timber or timber attacked by decay or insects. Cut above ground swell. Tip and butt shall be sound. Gradual taper from tip to 3 ft below butt. Trim and finish knots, lumps or humps. Square cut butt and tip unless tip is pointed.

PEELING

Classes A and B: For treated piles remove all bark and inner bark. When untreated, remove outer bark and up to 80 per cent of inner bark.
Class C: Not peeled when driven untreated.

STRAIGHTNESS

All classes: A line drawn from center of tip to center of butt shall lie within the taper. Piles shall be free from short or reverse bends; crook shall not exceed one-half pile diameter at the middle of the bend.

Sizes of Timber Piles

(From A.S.T.M. Book of Standards, 1954 Supplement)

	Class A						Class B					
	3 ft from butt				At tip, minimum		3 ft from butt				At tip, minimum	
	Minimum		Maximum				Minimum		Maximum			
Length, ft	Circ., in.	Dia., in.	Circ., in.	Dia., in.	Circ., in.	Dia., in.	Circ., in.	Dia., in.	Circ., in.	Dia., in.	Circ., in.	Dia., in.
Douglas fir, hemlock, larch, pine, spruce, or tamarack												
Under 40	44	14	57	18	28	9	38	12	63	20	25	8
40–50 incl.	44	14	57	18	28	9	38	12	63	20	22	7
51–70 incl.	44	14	57	18	25	8	41	13	63	20	22	7
71–90 incl.	44	14	63	20	22	7	41	13	63	20	19	6
Over 90	44	14	63	20	19	6	41	13	63	20	16	5
Oak and other hardwoods, cypress												
Under 30	44	14	57	18	28	9	38	12	57	18	25	8
30–40 incl.	44	14	57	18	28	9	41	13	63	20	22	7
Over 40	44	14	57	18	25	8	41	13	63	20	19	6
Cedar												
Under 30	44	14	69	22	28	9	38	12	63	22	25	8
30–40 incl.	44	14	69	22	28	9	41	13	69	22	25	8
Over 40	44	14	69	22	25	8	41	13	69	22	22	7

All diameters approximate.

5.3 STEEL H-PILES

The earliest use of I-beam sections for piling is not recorded, although Bethlehem Steel Company suggests the year 1899. Their earliest usage seems to have been to perform functions for which wood piling had failed. Wood piling cannot be successfully driven through considerable layers of dense sand and gravel or hardpan. In pure sands, wood piling can be jetted into place; but where gravel abounds, this is impractical. Under these hard driving conditions, wood piling quickly acquires a formulary resistance to driving before sufficient depth of penetration has been attained.

Bridge foundations can rest on short piling with adequate supporting ability; but if scour is heavy, substantial depths of pile can be exposed, relieving their lateral stability. Ice jams, accumulating against these foundations, can exert sufficient pressure to tilt the bridge. Furthermore, where piling is thus exposed the ice can grind away at the soft wood.

The first use of steel sections was in supporting bridge pier foundations in western rivers where conditions of scour alternated with ice jams. Steel could be driven through those dense layers of gravelly sand to sufficient depths to offer lateral stability regardless of scour and pressure.

When, in 1908, the H-column section was first rolled, it naturally superseded the earlier sections since piling driven under the conditions noted is essentially end-bearing. Steel H-sections are seldom used solely as friction piles, since because of their smooth, limited-area surfaces, little friction can be counted upon unless special protuberances such as lagging, bolted or welded to the faces of the web and flanges, are used.

Corrosion of steel sections under conditions of moisture, air, and chemical action has always been of concern to those employing these members as piling. Tests of steel piling sections which have been in place over a period of years indicate that to reduce the size of the section by 1/16 in. on all faces would require an *in situ* time period of from 80 to 150 years. This is far in excess of the 20- to 30-year life span of structures in the twentieth century, so that obsolescence will generally occur before substantial corrosion. However, H-sections are available with 0.2% of copper added, which increases the resistance to corrosion without greatly increasing the cost per ton.

Steel H-piles, point-bearing on rock, act as columns and are so designed. The intervening strata above rock surfaces are depended upon for lateral stability. However, where the rock is overlain by soft materials such as mud, lateral stabilty can be obtained only by driving into the rock or hardpan layer itself. In these cases it may be necessary to reinforce the bottom section of pile to prevent curling of the leading ends of the flanges. The H-section itself, however, should not be tapered or pointed since this weakens it precisely where the greatest strength is needed.

Where the underlying rock is too hard for any substantial penetration and little support can be expected from the soil above it, the H-pile should not be used; and in certain similar marginal cases its use is not recommended. Under hard driving conditions where no intervening support is available, there is a tendency of the pile to bend throughout its vertical length under driving impact, thus depriving the point of the full value of the blow.

Where the bottom supporting stratum is highly irregular, H-piles have considerable advantage since they are readily field-spliced to extend their lengths; and where they are driven in too long a section, the cut-off can be salvaged for re-use. Where boulders exist above seating level, however, the pile is often forced out of plumb. Here again the stiffness of the intervening layers is necessary to hold it in position.

Although various devices have been suggested or offered for increasing the H-pile area in its bottom section and thus providing greater support in strata of limited density, it will generally be desirable to increase the length of pile to carry it to a full support stratum or to turn to an alternative type of pile.

Characteristics of Steel H-Bearing Piles

Section number	Weight per ft, lb	Area of section, in.^{2}	Depth of section, in.	Flange		Web thickness, in.	Axis X-X			Axis Y-Y		
				Width, in.	Thickness, in.		I, in.^{4}	S, in.^{3}	r, in.	I′, in.^{4}	S′, in.^{3}	r′, in.
BP 14	117	34.44	$14\frac{1}{4}$	$14\frac{7}{8}$	$\frac{13}{16}$	$\frac{13}{16}$	1228.5	172.6	5.97	443.1	59.5	3.59
BP 14	102	30.01	14	$14\frac{3}{4}$	$\frac{11}{16}$	$\frac{11}{16}$	1055.1	153.4	5.93	379.6	51.3	3.56
BP 14	89	26.19	$13\frac{7}{8}$	$14\frac{3}{4}$	$\frac{5}{8}$	$\frac{5}{8}$	909.1	131.2	5.89	326.2	44.4	3.53
BP 14	73	21.46	$13\frac{5}{8}$	$14\frac{5}{8}$	$\frac{1}{2}$	$\frac{1}{2}$	733.1	107.5	5.85	261.9	35.9	3.49
BP 12	74	21.76	$12\frac{1}{8}$	$12\frac{1}{4}$	$\frac{5}{8}$	$\frac{5}{8}$	566.5	93.5	5.10	184.7	30.2	2.91
BP 12	53	15.58	$11\frac{3}{4}$	12	$\frac{7}{16}$	$\frac{7}{16}$	394.8	67.0	5.03	127.3	21.2	2.86
BP 10	57	16.76	10	$10\frac{1}{4}$	$\frac{9}{16}$	$\frac{9}{16}$	294.7	58.9	4.19	100.6	19.7	2.45
BP 10	42	12.35	$9\frac{3}{4}$	$10\frac{1}{8}$	$\frac{7}{16}$	$\frac{7}{16}$	210.8	43.4	4.13	71.4	14.2	2.40
BP 8	36	10.60	8	$8\frac{1}{8}$	$\frac{7}{16}$	$\frac{7}{16}$	119.8	29.9	3.36	40.4	9.9	1.95

Maximum Loads and Lengths Recommended For H-Piles

Section number	Weight per foot, lb	Maximum load-tons end-bearing	Maximum load-tons friction	Recommended maximum length, ft
BP 8	36	48	48	60
BP 10	42	55	55	70
BP 10	57	75	75	80
BP 12	53	70	70	90
BP 12	74	98	70	100
BP 14	73	96	70	125
BP 14	89	118	70	Over 125
BP 14	102	120	70	Over 125
BP 14	117	120	70	Over 125

Slenderness Ratio Versus Loads For Unsupported H-Piles

Unbraced length, ft	BP 14								BP 12				BP 10				BP 8	
	117 lb		102 lb		89 lb		73 lb		74 lb		53 lb		57 lb		42 lb		36 lb	
	l/r	Load-tons	l/r	Load-tons	l/r	Load-tons	l/r	Load-tons	l/r	Load-tons	l/r	Load-tons	l/r	Load-tons	l/r	Load-tons	l/r	Load-tons
30	In excess of recommended maximum								123.8	104	125.9	72	147.0	59	150.0	42	184.6	22
32									132.1	94	134.2	65	156.7	52	160.0	37	196.9	19
34									140.4	84	142.6	58	166.5	46	170.0	32	209.2	15
36					122.3	127	123.8	102	148.6	75	151.0	52	176.3	40	180.0	28		
38					129.1	117	130.6	94	156.9	68	159.4	47	186.1	35	190.0	24		
40			134.8	124	135.9	107	137.5	86	165.1	60	167.8	42	195.9	30	200.0	21		
42			141.6	114	142.7	98	144.4	79	173.4	54	176.2	37	205.7	26				
44	147.1	122	148.3	104	149.5	90	151.3	72	181.6	48	184.6	33						
46	153.8	111	155.1	95	156.3	82	158.1	65	189.9	43	193.0	29						
48	160.4	102	161.8	87	163.1	75	165.0	59	198.1	38	201.4	26						
50	167.1	93	168.5	79	169.9	68	171.9	54	206.4	33								
52	173.8	85	175.3	72	176.7	62	178.8	49										
54	180.5	77	182.0	66	183.5	56	185.6	45					l/r exceeds 200					
56	187.2	70	188.8	59	190.3	51	192.5	40										
58	193.9	63	195.5	54	197.1	46	199.4	36										
60	200.6	57	202.2	49	203.9	41	206.3	33										

SEAMLESS STEEL PIPE PILES

Nominal size, in.	Diameter, in. O.D	Diameter, in. I.D.	Wall thickness, in. Decimal	Wall thickness, in. Fraction	Weight per ft, lb	Moment of inertia, I	Section modulus, I/Y	Area of metal, A, sq in.	Radius of gyration, R	Inside area, sq in.	Internal volume per lin ft	Allowable load in tons: Joint Committee Formula: Concrete only	Joint Committee Formula: Grade 2 steel: Steel only	Grade 2 steel: Steel & concrete: Open	Grade 2 steel: Steel & concrete: Closed	Grade 3 steel: Steel only	Grade 3 steel: Steel & concrete: Open	Grade 3 steel: Steel & concrete: Closed	Proportionate Method: Open-end piles	Proportionate Method: Closed-end piles
8	8.625	7.981	0.322	5/16	28.55	72.489	16.809	8.399	2.938	50.03	0.347									
10	10.750	10.136	0.307	5/16	34.24	137.42	25.567	10.072	3.694	80.69	0.560	27.2	55.8	83.0	58.1	71.8	99.0	69.3	65	46
10	10.750	10.020	0.365	3/8	40.48	160.73	29.903	11.908	3.674	78.85	0.548	26.6	68.7	95.3	67.7	88.3	114.9	80.4	72	50
10	10.750	9.874	0.438	7/16	48.19	188.95	35.153	14.190	3.649	76.57	0.532	25.8	84.6	110.4	77.3	108.8	134.6	93.2	80	56
10	10.750	9.750	0.500	1/2	54.74	211.95	39.433	16.101	3.628	74.66	0.518	25.2	98.0	123.2	86.2	126.0	151.2	105.8	87	61
12	12.750	12.126	0.312	5/16	41.51	235.91	37.005	12.191	4.399	115.49	0.802	39.0	67.9	106.9	74.8	87.3	126.3	88.4	86	60
12	12.750	12.090	0.330	21/64	43.77	248.38	38.962	12.876	4.393	114.80	0.797	38.7	72.7	111.4	77.9	93.5	132.2	92.5	89	62
12	12.750	12.000	0.375	3/8	49.56	279.34	43.818	14.579	4.377	113.10	0.785	38.2	84.6	122.8	86.0	108.8	147.0	102.9	95	66
12	12.750	11.874	0.438	7/16	57.53	321.42	50.419	16.942	4.356	110.74	0.769	37.4	101.2	138.6	97.0	130.0	167.4	117.2	103	72
12	12.750	11.750	0.500	1/2	65.42	361.54	56.712	19.242	4.335	108.43	0.753	36.6	117.3	153.9	107.7	150.8	187.4	131.2	112	78
14	14.000	13.250	0.375	3/8	54.57	372.76	53.251	16.052	4.819	137.89	0.958	46.5	93.2	139.7	97.8	119.8	166.3	116.4	109	76
14	14.000	13.124	0.438	7/16	63.37	429.50	61.357	18.662	4.797	135.28	0.939	45.7	111.5	157.2	110.0	143.3	189.0	132.3	119	83
14	14.000	13.000	0.500	1/2	72.09	483.76	69.109	21.206	4.776	132.73	0.922	44.8	129.3	174.1	121.9	166.2	211.0	147.7	129	90
16	16.000	15.250	0.375	3/8	62.58	562.09	70.261	18.408	5.526	182.65	1.268	61.6	107.0	168.6	118.0	137.5	199.1	139.4	136	95
16	16.000	15.124	0.438	7/16	72.72	648.75	81.094	21.414	5.504	179.65	1.248	60.6	128.0	188.6	132.0	164.6	225.2	157.6	147	103
16	16.000	15.000	0.500	1/2	82.77	731.94	91.492	24.347	5.483	176.72	1.227	59.6	148.5	208.1	145.7	191.0	250.6	175.4	158	110
18	18.000	17.250	0.375	3/8	70.59	806.63	89.626	20.764	6.233	233.71	1.623	78.9	120.7	199.6	139.7	155.2	234.1	163.9	164	115
18	18.000	17.124	0.438	7/16	82.06	932.24	103.58	24.166	6.211	230.30	1.599	77.7	144.5	222.2	155.4	185.8	263.5	184.4	177	124
18	18.000	17.000	0.500	1/2	93.45	1053.2	117.02	27.489	6.190	226.98	1.576	76.6	167.8	244.4	170.1	215.7	292.3	205.3	189	132
20	20.000	19.250	0.375	3/8	78.60	1113.5	111.35	23.120	6.940	291.04	2.021	98.2	134.4	232.6	162.8	172.8	271.0	189.7	196	137
20	20.000	19.124	0.438	7/16	91.41	1288.2	128.82	26.918	6.918	287.24	1.995	96.9	161.0	257.9	179.8	207.0	303.9	212.7	209	146
20	20.000	19.000	0.500	1/2	104.13	1456.9	145.69	30.631	6.897	283.53	1.969	95.7	187.0	282.7	197.9	240.4	336.1	235.3	226	158
22	22.000	21.250	0.375	3/8	86.61	1489.7	135.43	25.476	7.647	354.66	2.463	119.7	148.2	267.9	187.5	190.5	310.2	217.1	229	160
22	22.000	21.124	0.438	7/16	100.75	1725.0	156.82	29.670	7.625	350.46	2.434	118.3	177.5	295.8	207.1	228.3	346.6	242.6	245	171
22	22.000	21.000	0.500	1/2	114.81	1952.5	177.50	33.772	7.603	346.36	2.405	116.9	206.3	323.2	226.2	265.2	382.1	267.5	260	182
24	24.000	23.000	0.500	1/2	125.49	2549.4	212.45	36.914	8.310	415.48	2.885	140.2	225.5	365.7	256.0	289.9	430.1	301.1	299	209

{Outer 1/16-in. allowance for corrosion
Closed-end piles computed at 70% of open-end piles}

By the Joint Committee Formula

$$P = 0.225f_c' \times A_c + 0.4f_{syp}A_s$$
$$= 0.225 \times 3000 \times 132.73 \times 0.4 \times 45{,}000 \times 18.47$$
$$= 89{,}592 + 332{,}460$$
$$= 422{,}052 \text{ lb} = 211.0 \text{ tons}$$

The problem

Using 14-in. pipe pile, 1/2-in. wall thickness
Concrete = 3000 psi, $n = 10$
$A_c = 132.73$ sq in. = area of concrete
$A_s = 21.21$ sq in. − 1/16 in. = 18.47 sq in. = area of steel

By the Proportionate Method

$$P = 1.2 \times 0.225f_c'(A_c \times nA_s)$$
$$= 1.2 \times 0.225 \times 3000(132.73 + 10 \times 21.21)$$
$$= 279{,}312 \text{ lb} = 139.6 \text{ tons}$$

5.4 PIPE PILES

Although spiral or lap-welded pipe is occasionally used for pipe piles, seamless steel pipe is generally preferred because of its greater strength. Filled with concrete, the seamless steel pipe pile combines the strength and rigidity of both the steel and the concrete. Seamless steel pipe is available in two yield strengths, 35,000 and 45,000 psi, and in diameters from 8 to 24 in. with a range of wall thicknesses in most sizes. (Twelve-inch pipe is available in five wall thicknesses, for instance, whereas 8-in. and 24-in. pipe are available in only one thickness.)

Pipe piles can be driven with either open or closed bottom, depending upon ground and driving conditions. When driven open-end, the pipe is cleaned out with an air or water jet (or with a miniature grab bucket) before filling with concrete. Closed-end pipe piles are driven with a cast or forged steel driving shoe and can be inspected before concreting in the same way as tubular piles.

Steel pipe piles have considerable structural strength, and open-end piles can be driven through obstructions, using heavy hammers, that would broom or damage other types of piles. Because of their high strength in flexure, they are easier to handle in long lengths without the potential bending incident to some other types of piling and can be used effectively as batter piles.

Steel pipe piles are customarily driven with a standard type of pile hammer rather than with a mandrel.

Open-end steel pipe for piling is driven to rock or other firm bearing, where such bearing lies within 40 to 50 ft of the surface, and in these cases acts as a column. Buckling action may be controlled by the lateral resistance of the soil through which the pile is driven, or it may be necessary to consider the length-to-slenderness ratio where intervening soils are soft. The pipe pile can also be designed as a braced column for axial loads.

The load-bearing capacity of the open pipe pile is figured on its supporting capacity at the point of bearing in firm strata, disregarding any friction induced by the soil between the ground surface and the level of bearing. Because of this, open-end pipe piles are driven to refusal, and the use of driving formulas is not required. Theoretically the critical load necessary to produce buckling in the unbraced length of a filled pipe pile exceeds the load necessary to crush the pile, but in very soft intervening soils this should be checked.

Closed-end piles are used where no firm stratum exists within reasonable depth, and they are consequently driven to a definite penetration based on driving formulas. Closed-end piles are occasionally driven to rock to avoid the necessity of blowing out the confined material or in cases where dewatering may be necessary before the concrete fill can be placed. Since their bearing capacities are not governed solely by their structural column strength but by the length of pile required to develop the assigned load through the resistance of the penetrated soil, the allowable design loads are reduced to 70% or to two-thirds that of an open-end pile.

The load-bearing capacity of pipe piles can be rationally determined on the basis of the known material strength of the steel tube and the enclosed concrete core. When the structural load-bearing capacity of a steel pipe pile is computed, the area of the steel in the pipe is generally reduced by a corrosion allowance amounting to 1/16 in. of shell thickness.

For open-end piles under normal driving conditions the accepted practice is to use a standard wall thickness (indicated opposite) and then to determine the diameter which, taken with the thickness, will provide the desired load-bearing capacity. Roughly, the pile diameter is frequently taken at about one-sixteenth of the total length. For difficult driving conditions pipe thickness is increased by from 1/16 to 1/8 in., as seems appropriate.

For closed-end piles the diameter is first selected and then the thickness, reversing the order above. Economy dictates the selection of larger diameters to increase the friction surface and decrease the length.

5.5 TUBES FOR CAST-IN-PLACE CONCRETE PILES

Midway between the pipe pile and the concrete pile are a large variety of pilings consisting of a tube of metal driven into the ground and filled with concrete. These shells or tubes range from the heavy Monotube pile through the thinner tubes of the Raymond standard or step-tapered pile to the Armco corrugated pipe pile. At the heavier end, the shell is designed to provide part of the support; whereas at the lighter end, the shell provides only a casing for the concrete. Although suitable for point bearing, they are chiefly used as friction piles.

The Raymond standard concrete pile was the first of the tube piles to be developed, and the tube was designed to resemble a wood pile in several particulars. The tube, of light sheet metal, is wall-reinforced with a spiral wire on a 3-in. pitch, forced into the sheet so as to leave a spiral extrusion on the exterior of the shell. The tube is tapered outward from a point diameter of 8 in. at the rate of 1 in. in each 2½ ft of length. The point of the tube is closed with a steel boot, and the tube is driven by using a collapsible mandrel, or core, resting on the boot and fitting tightly within the tube when expanded. Tubes are made in fixed lengths ranging from 15 to a maximum of 37 ft.

To increase the length yet to provide a pile similar to their standard pile, Raymond Concrete Pile Company in 1931 introduced the step-tapered pile, driven not with a collapsible mandrel but with a solid steel core, stepped down in increments of diameter to suit the stepped sections of piling tube. The tapered shells used are spirally corrugated in 8-ft sections with a specially formed steel ring (a plow ring) welded to the shell at the bottom of each section. The stepped offset in the pile core rests on this ring during driving so that no more than 8 ft of shell is dragged into the ground by each point of impact. Sections are screwed together below each plow ring by a short section of spirally corrugated shell left hanging for that purpose. Instead of using a pointed boot, this pile shell is closed at the 8 ¼-in.-diameter tip with a steel plate welded to the shell.

The maximum length of the Raymond step-tapered pile is 80 ft, conditioned once again by mandrel limitations, but Raymond has extended the possible length of pile up to 150 ft by using a 10-in. pipe pile first driven below the step-tapered section. A specially designed joint between the pipe and the tube permits the mandrel core to directly impact the pipe section.

About 1928 the Union Metal Manufacturing Company, who had been making tapered, fluted columns for street-lighting standards, introduced the Monotube pile; like the light standards, this is a vertically fluted shell ranging in metal thickness from 11-gage to 3-gage. These also start at a point diameter of 8 in., the end protected by a solid, round, pointed nose welded to the shell. Individual lengths of shell vary from 10 to 75 ft in length and taper at the rate of either 0.14, 0.25, or 0.40 in. per ft, up to a maximum diameter of 18 in. For greater depths, non-tapered sections are added at the top.

The fluting of the Monotube pile provides sufficient vertical strength to permit driving with standard hammers without the use of a mandrel, and these tubes are therefore frequently used by contractors to provide cast-in-place concrete piles when mandrels and the necessary rigs to handle pile cores are not available. Because of their appearance they are frequently used where the pile is to be carried exposed, above the ground surface.

The Armco Steel Corporation has adopted its corrugated metal pipe for use as a pile shell, which is sold under the trade name of Hel-Cor Pile Shell. This is a thin-gage (from 14 to 18) shell consisting of a continuous lock seam, corrugated metal pipe with spiral corrugations ¼ in. in depth on a 1½-in. pitch for 10 in. pipe diameters and ½-in. corrugations on a 2-in. pitch for pipe diameters of from 12 to 22 in.

The Armco Hel-Cor Pile Shell is not tapered but of uniform diameter throughout. It is driven with a pneumatically expandable mandrel which, when open, hugs the corrugations of the shell and provides continuous contact with the tube.

STANDARD MONOTUBE WEIGHTS AND VOLUMES

Extension type N

Tapered section type F, J, or Y

8"

Type	Length, ft	Nominal diameter, in.	Theoretical weights of steel					Estimated vol. of concrete, cu yd
			11 gage, lb	9 gage, lb	7 gage, lb	5 gage, lb	3 gage, lb	
F Taper 0.14 in./ft	30	12	404	500	589	685	778	0.55
	40	14	597	749	895	1055	1215	0.95
	60	16	955	1190	1412	1645	1872	1.68
	75	18	1289	1624	1957	2307	2653	2.59
J Taper 0.25 in./ft	17	12	235	289	338	392	444	0.32
	25	14	372	460	541	629	714	0.58
	33	16	529	656	775	902	1025	0.95
	40	18	690	857	1023	1191	1355	1.37
Y Taper 0.40 in./ft	10	12	142	173	203	235	266	0.18
	15	14	227	278	330	383	434	0.34
	20	16	323	399	476	553	628	0.56
	25	18	432	535	641	746	848	0.86
N12	10	12	172	214	253	295	338	0.26
	15		252	314	372	436	499	0.38
	20		334	416	494	480	644	0.51
	25		413	515	616	719	822	0.64
	30		495	618	735	863	986	0.77
	35		578	722	860	1009	1152	0.89
	40		658	821	986	1150	1314	1.02
N14	10	14	201	249	295	345	395	0.35
	15		293	365	434	506	579	0.52
	20		388	484	574	673	769	0.70
	25		480	598	715	839	958	0.87
	30		574	716	858	1001	1143	1.05
	35		670	836	996	1169	1336	1.22
	40		762	952	1141	1339	1530	1.40
N16	10	16	228	283	334	390	446	0.45
	15		333	415	490	575	657	0.68
	20		439	548	651	763	872	0.90
	25		544	678	810	950	1085	1.13
	30		650	812	966	1133	1295	1.35
	35		755	941	1128	1322	1511	1.58
	40		863	1077	1285	1506	1722	1.80
N18	10	18	258	320	378	443	505	0.58
	15		375	468	556	648	740	0.87
	20		496	618	736	859	982	1.16
	25		615	767	912	1069	1222	1.45
	30		733	914	1093	1281	1464	1.75
	35		854	1066	1276	1488	1700	2.04
	40		976	1218	1454	1703	1945	2.33

Theoretical weights should not be used for final determination of shipping costs.
The gage of the various sections in a pile may vary to suit requirements.

5.6 PRE-CAST AND PRE-STRESSED CONCRETE PILES

The use of pre-cast concrete piling was well advanced when pre-stressing was introduced to the concrete industry in the United States. Since pre-stressing reduces or eliminates tensile stresses to which piles are subjected during transportation, driving, and in service, the pre-stressed pile has largely replaced the simply reinforced pre-cast concrete pile.

As with all pre-stressed concrete, piling may be either pre-tensioned or post-tensioned. Pre-tensioning, in which stress is applied to the wires prior to the casting of concrete around them, is used for piling up to 24 in. in diameter. Post-tensioning, in which the wires are stressed against a shell of concrete previously poured, is used for pile diameters of 30 in. and over.

The illustration shows the various types of pre-stressed concrete piling currently in use; and a joint committee of representatives of the American Association of State Highway Officials and of the Prestressed Concrete Institute, set up in 1961, has established standards for the square and octagonal pre-stressed piles.

The larger, circular, post-tensioned piles have chiefly been pioneered by Raymond Concrete Pile Company using the Cen-Vi-Ro process for centrifugally spinning what is essentially a pipe. This process, similar to that employed in the centrifugal manufacturing process for cast-iron pipe, uses a dense, dry concrete spun outward against a form shell by the centrifugal force developed within a spinning cylinder. This force, plus vibration, produces a concrete with a compressive strength of 7500 psi at 28 days.

Cylinder piles are manufactured in 16-ft lengths with 5-in-thick walls in which 12 longitudinal holes have been cast on steel rods with rubber tubing stretched over them. After casting, the cylinders are steam-cured and the rods and tubes removed. The sections of pile are then lined up on roller assembly racks to provide the total pile length required, and pre-stressing wires 0.192 in. in diameter are threaded through the holes left by the steel rods. The sections are first pulled together and the joints are sealed with a plastic compound. The wires are then tensioned to 150,000 psi and secured with steel locking cones. The core holes are filled with grout pumped in under a pressure of 100 psi.

Pre-stressed concrete piles generally provide a working compressive stress of 700 psi in the concrete, although this is occasionally varied with the size and shape of pile or the driving conditions.

Splicing of pre-stressed piles is a difficult procedure and is generally avoided if possible. This means that careful logs of soil strata must be developed and that test piles must be driven and correlated with these data. Additional length of pile should be provided (within controlled limits) so that the upper section can be chipped away to provide anchorage in the cap.

Solid pre-stressed piles have points tapering down to a 6-in.-square tip in a length equal to twice the diameter of the pile. Hollow piles, particularly the cylindrical piles, are driven with a 6-ft-long plug of concrete in the end. After driving and cut-off, the upper section of hollow piles is filled with concrete to varying depths depending upon the type and thickness of cap to be used.

The earlier, plain, pre-cast concrete piling was seldom driven in lengths exceeding 55 ft since the problems of handling and driving longer sections were considerable. The introduction of pre-stressing, however, has more than tripled potential lengths. In some cases, 20-in. octagonal hollow-core piles have been driven in lengths up to 132 ft, and 36-in. cylinder piles have been driven in lengths up to 192 ft.

Regardless of length, it is generally desirable to limit the number of pickup points to three to facilitate handling the concrete pile from the horizontal to the vertical position. As the length of pile increases the weight also increases, and this requires a heavier hammer for effective driving. Piles of 36-in. diameter, 192 ft long, driven in Lake Maracaibo, Venezuela, for oil-well drilling platforms, required a special hammer delivering 75,000 ft-lb of energy (see Sec. 5.20).

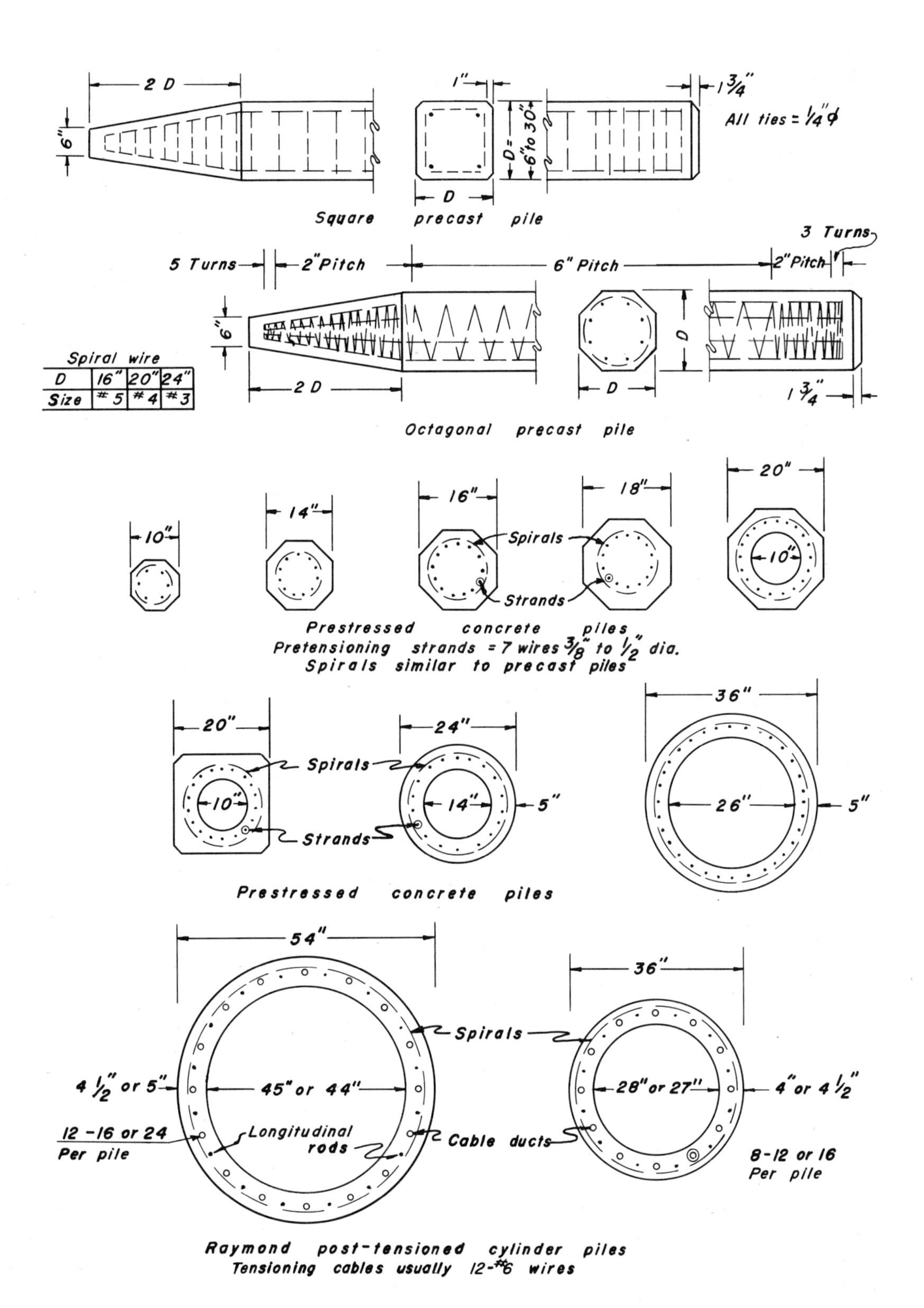

2 D
6"
1"
1 3/4"
All ties = 1/4"φ
D = 6" to 30"
D
Square precast pile
3 Turns
5 Turns
2" Pitch
6" Pitch
2" Pitch
6"
D
2 D
D
1 3/4"
Spiral wire
D | 16" | 20" | 24"
Size | # 5 | # 4 | # 3
Octagonal precast pile
10"
14"
16"
18"
20"
Spirals
Strands
10"
Prestressed concrete piles
Pretensioning strands = 7 wires 3/8" to 1/2" dia.
Spirals similar to precast piles
20"
24"
36"
Spirals
10"
14"
5"
26"
5"
Strands
Prestressed concrete piles
54"
36"
Spirals
4 1/2" or 5"
45" or 44"
28" or 27"
4" or 4 1/2"
12 -16 or 24
Per pile
Longitudinal rods
Cable ducts
8-12 or 16
Per pile
Raymond post-tensioned cylinder piles
Tensioning cables usually 12-#6 wires

5.7 UNCASED CONCRETE PILING

The introduction and development of rotary drill rigs for earth boring in the middle decades of the twentieth century, sparked by the restless drilling of the petroleum industry, have led to a type of piling which is essentially a concrete pile but which is neither cast in a driven pipe, or tube, nor pre-cast and then driven into place. As so often happens in the construction industry, developed techniques outpace available words that are lucid, pat, and comprehensive for describing them. The term "uncased" pile could apply to a pre-cast pile, and the use of the term "unformed" pile to define our subject here is to leave an unfortunate connotation. Nor is this the end of the difficulty, since a caisson is defined as a box or casing and when some of these are filled with concrete the industry refers to them as drilled-in or dug-in caissons and these resemble nothing so much as a big fat pile.

The cast-in-place concrete pile is usually defined in specifications as a steel pipe, Monotube, or corrugated shell driven to bearing and filled with concrete; but if we drive a steel pipe with an open end, clean it out with air or water, and then fill the *void* remaining as the pipe is slowly withdrawn, this could also be called a cast-in-place concrete pile, although in construction parlance it is termed an uncased pile. Such a pile, where the concrete surface conforms exactly to the surface inside the hole should provide maximum friction.

The basic problem with the type of uncased pile described above is that pulling the pipe may be difficult and that, in the operation, the hole may collapse or may be partially caved in. To overcome this, a pipe just smaller than the original pipe is dropped into the void, the outer shell is withdrawn, and the inner shell is then pulled as concreting proceeds. Even here, to protect the wall of the hole, the concrete level in the pipe must be maintained above the bottom rim of the shell. Adequate vibration to densify the concrete is difficult under these conditions, and several pile-driving firms (MacArthur Concrete Pile Corporation and the Franklin Compressed Pile Company) use a ram operating within the shell to compact or "compress" the concrete as it is being placed.

The drilled-in concrete pile, excellent for firm soils, is produced by simply drilling a hole, either with a continuous-flight auger or with a bucket-type drill, and then filling the hole with concrete through a pipe whose bottom height is maintained just below the level of the rising concrete. (In the larger dimensions, carried to firm bearing, these are often dubbed "caissons.") The method cannot be used where soils are soft, crumbly, or sandy. For these conditions Intrusion-Prepakt has devised a hollow-shaft earth auger which first drills the hole to the proper depth and then fills the hole with conditioned grout pumped down through the hollow shaft. Other contractors (notably Peter Kiewit & Sons) have developed similar rigs which provide a standard concrete fill placed under a pressure of 30 psi.

A variation of this general type of pile, also developed by Intrusion-Prepakt, is essentially the same as that described for soil consolidation in Sec. 3.21. Here the soil is not removed from the hole, as in other augering methods, but a combined mixing and drilling head injects a conditioned mortar into the disturbed soil as the auger travels slowly upward. Obviously this type of pile will vary in bearing capacity with the kind of soil being intruded, having the greatest value in sand-gravel mixes.

Uncased concrete piles have not been fully accepted by the engineering profession as being wholly desirable. Their bearing capacity cannot be fully appraised without testing, a costly process when applied to *all* piles. There is also the uncertainty involved in any process where visual inspection of the supporting member cannot be conducted at any stage of the operation. Any pile which is driven is considered to be tested by the application of a formula and although, as will be noted in Sec. 5.10, formulas have been proved shady if not, indeed, bad, still the engineering profession would rather trust to them than to the allegations of some bruiser in a safety helmet.

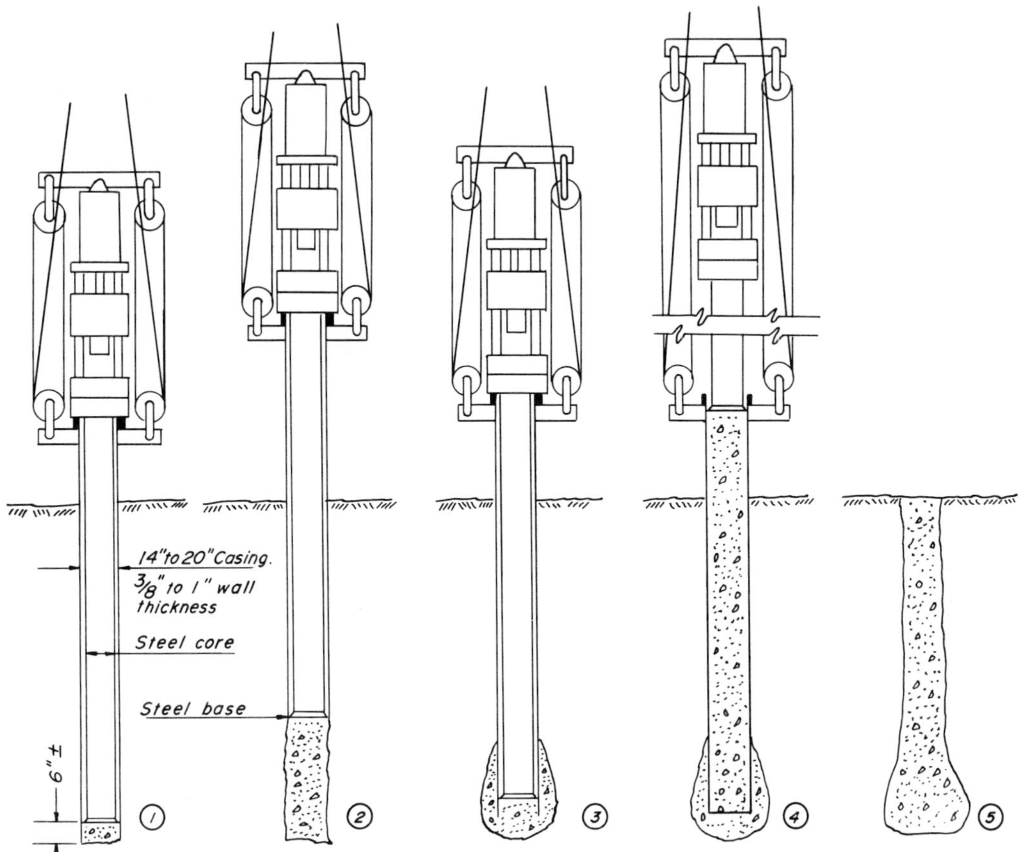

① Steel core, casing and concrete plug driven to pre-determined resistance.
② Core removed. Charge of dry concrete inserted. Core forced down. Casing drawn up.
③ Core and casing driven down until resistance meets test load requirements.
④ Core removed and casing filled with concrete compacted in increments.
⑤ Casing withdrawn while core continues to apply pressure to concrete.

CONCRETE PEDESTAL PILE INSTALLATION
WESTERN FOUNDATION CORPORATION

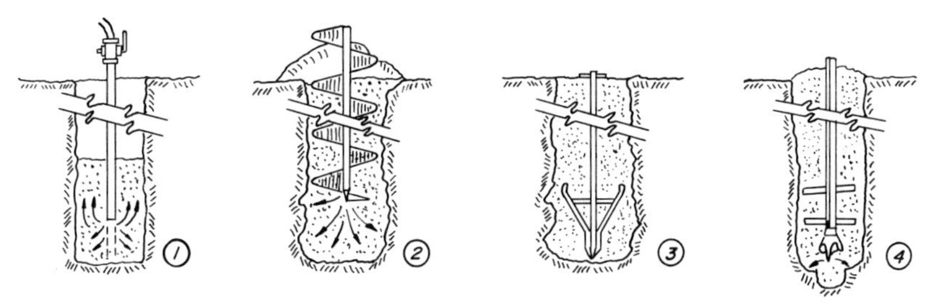

① Cast-in-place pile for solid soils. Augered hole filled with intrusion mortar.
② Pakt-in-place (T.M.) pile for unstable soils. Intrusion mortar injected through auger shaft.
③ Lock-in-place (T.M.) pile for uplift. Rod and sleeve assembly inserted in ② method.
④ Mixed-in-place (T.M.) pile for sands and gravels. Intrusion mortar injected into native soil.

CAST-IN-PLACE CONCRETE PILES
INTRUSION-PREPAKT INCORPORATED

5.8 PILING ADAPTATIONS

The standard kinds of piling have been discussed on the previous pages, but it has been found to be desirable to adapt these single standards either by combining two kinds of pile to improve the economics of long piles or to meet special conditions, and to modify other kinds of piling, generally concrete, to improve their supporting value.

It has been noted that wood piles up to lengths of 80 ft certainly, and in longer lengths, possibly, are cheaper than other types of piles. It has also been noted that wood piling kept submerged in water has an indefinite life, and that many failures of piling have been due to the lowering of natural water tables. To meet these requirements of economy and submergence the composite pile was early developed, combining the wood pile with a concrete extension.

There are two approaches to a composite wood-and-concrete pile. In the first of these, dewatering and excavation of the site is involved, and in the second the composite is driven essentially as a single unit. Pre-excavation (and dewatering) of a site may be required for other purposes than that of constructing a concrete extension to a wood pile; and where this is the case, a form placed around the wood pile is the simplest expedient. More frequently, it becomes necessary to place the concrete section under water or to drive the combination through water-saturated soils. Of the various means of accomplishing this, the following seems the most practical.

A casing of sufficient size to encompass the maximum diameter of the wood pile is driven to the depth, or slightly below the depth, required for the concrete section. The wood piling is then driven through the casing, using a mandrel, to the desired penetration. The top of the wood pile has been reduced from its full section to form a tenon and reinforced by a winding of iron wire. After the pile has obtained the desired supporting value, a corrugated metal shell is dropped through the casing to rest on the shoulder formed by the tenon. The bottom of the shell contains a rope seal which is forced down on the shoulder to make the shell watertight; and a cage of reinforcing, fastened to the shell, slides over the tenon to provide proper positioning. The shell is then concreted in the usual way, and the original casing is withdrawn.

When the concrete section is short it is possible to affix the shell to the wood pile prior to driving, so that the shell follows the wood pile down. Driving is done on the butt of the wood pile using a solid-core mandrel as in the previous case.

Where it is desirable to have some penetration of a concrete pile into an underlying coral or other soft rock formation, a section of steel H-pile can be set into the tip of a concrete pile to a distance of 5 ft, the surrounding concrete being then specially reinforced. Projection of the steel H-pile will depend on the depth of penetration required.

Another type of adaptation of standard piles, particularly of concrete piles, is that involving an increase in the lower diameter of the pile to provide improved bearing. From a practical standpoint, most piling is driven with a pointed tip or shoe to facilitate penetration. With friction piles it is presumed that the outer diameter of the pile will provide all the supporting value. It has been reasoned, however, that in some borderline cases were the tip to be enlarged, additional support would be obtained from the tip, and the length of the pile correspondingly shortened.

Methods of spreading the tips of concrete piles are several, are ingenious, and are generally patented. They involve driving a casing to firm bearing, lifting it a few feet, and then driving fresh concrete, with a core compactor, until it forces itself out into the subsoil. In some cases, a portion of the soil is removed by an expandable bit in order to provide concrete space. In other cases an oversize pre-cast concrete button is set on the bottom of the casing to provide the spread, with a light shell placed inside the casing after it has been driven. The shell is filled with concrete and the casing withdrawn.

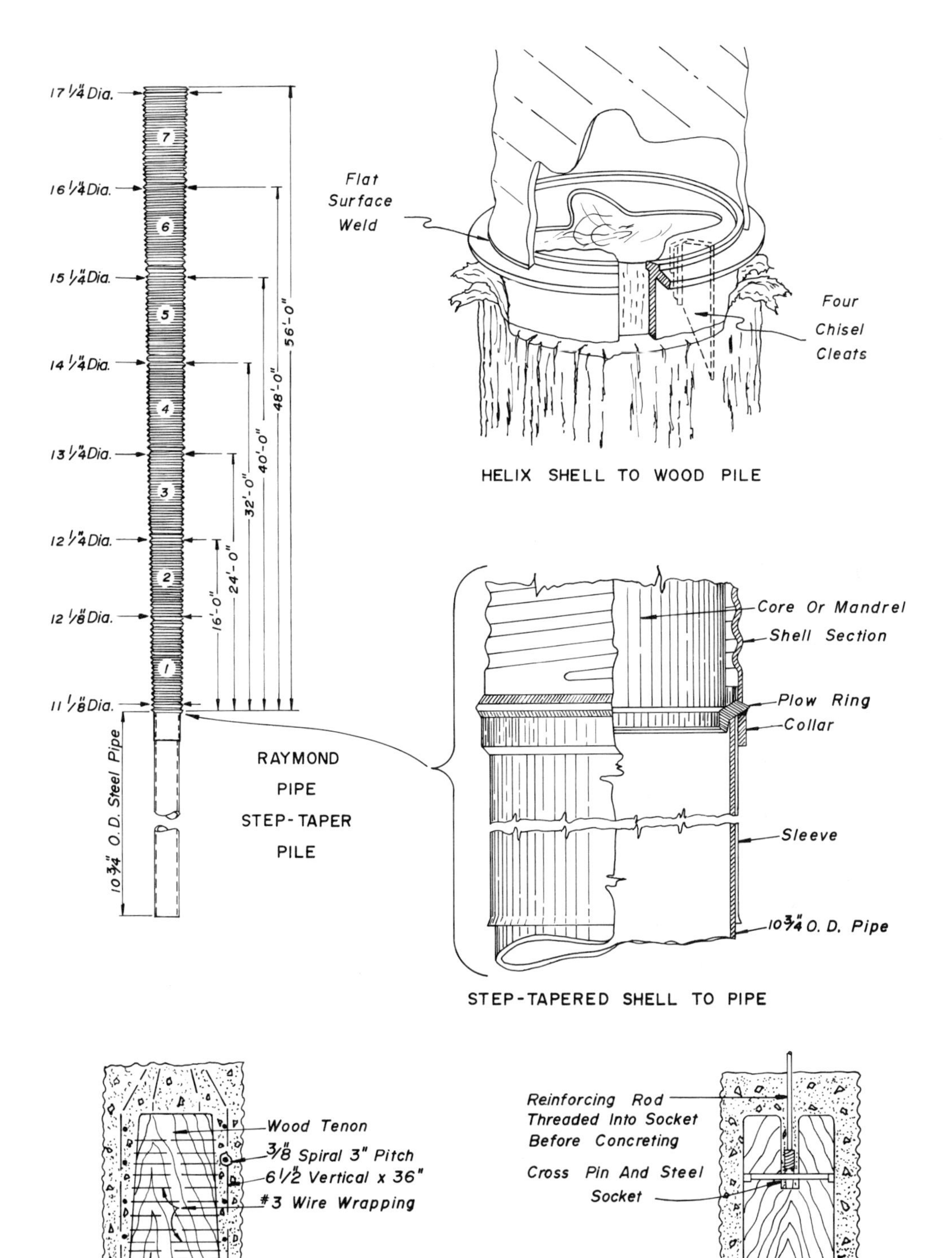
17 1/4" Dia.
16 1/4" Dia.
15 1/4" Dia.
14 1/4" Dia.
13 1/4" Dia.
12 1/4" Dia.
12 1/8" Dia.
11 1/8" Dia.
7
6
5
4
3
2
1
56'-0"
48'-0"
40'-0"
32'-0"
24'-0"
16'-0"
10 3/4" O. D. Steel Pipe
RAYMOND
PIPE
STEP-TAPER
PILE
Flat
Surface
Weld
Four
Chisel
Cleats
HELIX SHELL TO WOOD PILE
Core Or Mandrel
Shell Section
Plow Ring
Collar
Sleeve
10 3/4" O. D. Pipe
STEP-TAPERED SHELL TO PIPE
Wood Tenon
3/8" Spiral 3" Pitch
6 1/2" Vertical x 36"
#3 Wire Wrapping
Rope Grommet
Reinforcing Rod
Threaded Into Socket
Before Concreting
Cross Pin And Steel
Socket
Metal Sealing Ring
CONCRETE TO WOOD COMPOSITE

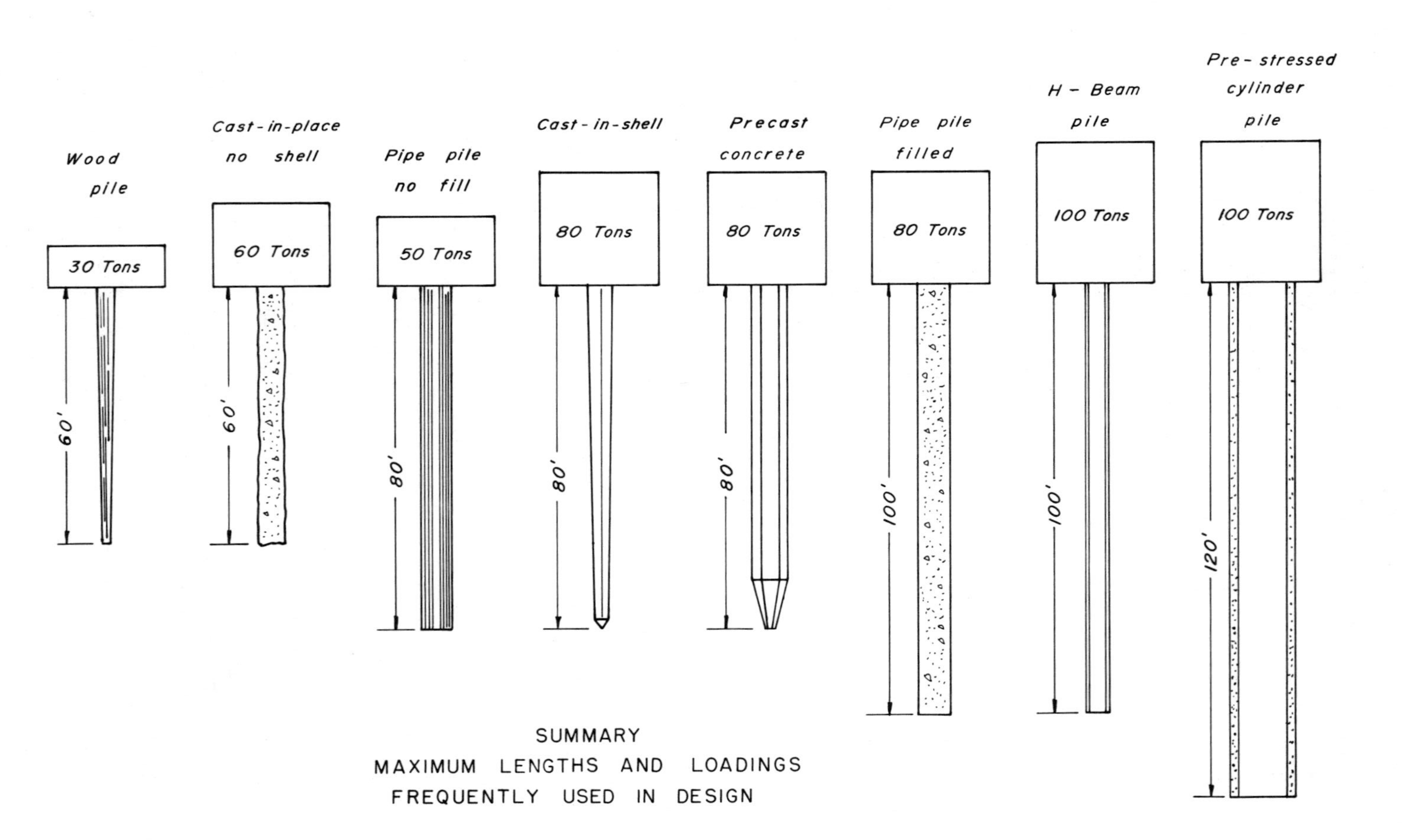

SUMMARY
MAXIMUM LENGTHS AND LOADINGS
FREQUENTLY USED IN DESIGN

BUT NOTE THAT THESE ARE OFTEN EXCEEDED
WHERE CONDITIONS WARRANT

5.9 SELECTION OF KIND OF PILE

Although the contractor is not often the deciding authority on the kind of pile to be used, he should be conversant with the relative advantages and limitations of any piles that he may be called upon to drive. Where delays in delivery of materials loom, where ground conditions prove to be other than those indicated by test borings, where the economics of the operation require improvement, or where special equipment (a mandrel, for instance) may be involved that is not readily available to the contractor, it may become necessary to switch from one kind of piling to another.

One basic question that frequently receives insufficient attention is whether piling should be used at all. Although much is reported concerning the maximum lengths of pile that should or can be used, no attention is paid to minimum lengths that are practical or desirable. Piling should not be used in dense stable soils where there is any possibility of obtaining adequate bearing with a different type of footing. Piles specified to be 8, 10, or 12 ft long, particularly end-bearing piles driven through dense overburden, have frequently had to be abandoned and the footings redesigned, when the driving of such piling became impossible. This is particularly true where an attempt is made to drive pile shells or tubes, under these conditions, without a mandrel.

The first condition in selecting a pile for a particular use is derived from an examination of the test borings. It is always necessary to select a pile to suit the driving conditions so that the completed pile will not have been damaged during the driving operation.

The second criterion in pile selection is the ability of the pile material to sustain the designed load. Although of particular importance with end-bearing piles, the compressive strength of the pile member cannot be ignored in any case. Nor can the necessary support be obtained merely by adding piles, unless proper spacing can also be provided (see Sec. 5.24).

Timber piles are usually designed for 25-ton loadings in Federal and state specifications: private consultants frequently use 30-ton loadings. But tests sponsored by the American Wood Preservers Institute in Chicago, in 1960, would indicate that 40-ton loadings would not be unrealistic. The piles tested were Class A (with an 8-in. tip and a 14-in. butt) and Class B (with a 7-in. tip and a 12-in. butt) creosoted wood piles driven 36 ft down to a layer of hardpan. All were tested to a 100-ton loading without failure, providing a factor of safety of 2.5. When load testing was resumed, two failed at 122 tons, one at 142 tons, one at 151 tons, and one at 235 tons. Neither the class of pile nor the kind of lumber (oak, southern pine, Douglas fir) seemed to have much influence on the result.

Cast-in-place concrete piles are generally designed for loadings of from 60 to 80 tons per pile when the steel is left in place or 50 to 60 tons when the shell is withdrawn. This distinction as to *casing* seems questionable. Some of the cast-in-shell piles for the Terminal Building at La Guardia Airport in New York City were found to be defective because of substantial voids and extensive segregation of the concrete fill. On the other hand, improvements in the placing and compaction of concrete for uncased piles would indicate that they might be equal to cased piles.

Steel pipe piles, as specified by public bodies, are generally limited to 50-ton loadings when only pipe is used or 80 tons when filled with concrete. However, private designs have used loads up to 150 tons for a 20-in. pipe when filled, and this size has at least once been tested to 300 tons without signs of distress.

Steel H-beams are seldom used as friction piles. This is due less to their having an insufficient surface area than to the fact that they are used chiefly to punch through hard layers which may not close back tightly around the pile. Steel H-piles seem to be chiefly limited to 100-ton loadings, although any specific case will depend on the size of section and the nature of the subterranean strata.

5.10 THE GENERAL PILE-DRIVING FORMULA

A pile may be driven either to absolute refusal or to what might be termed "substantial" refusal, where the final *rate* of penetration is recorded and related to the bearing capacity of the pile. Point-bearing piles are driven to refusal into, or onto, a stratum of such firm bearing that any real penetration into it is impossible. A light pile driven with a heavy hammer may broom or telescope at the tip, indicating continued penetration when none exists in fact. The best protection against overdriving of end-bearing piles is a continuous check against the test-boring log.

Piles driven to *substantial* refusal are essentially friction piles even where their tip pulls up in a dense stratum. The supporting value of such piles is measured by the number of blows required to produce penetration of a specified distance, usually 1 in. The conversion of the rate of penetration in blows per inch to tons of supporting capacity is accomplished by means of a dynamic pile-driving formula.

In theory, dynamic pile-driving formulas are very simple. The hammer or weight used to deliver the blow imparts so many foot-pounds of energy to the pile. Part of this energy is lost on impact when the moving hammer strikes the stolidly standing pile. A cap is generally used between the hammer and the pile, and additional energy is lost in transmittal through the cap. A further loss of energy occurs in transmission from the butt to the point of the pile, while the final loss is that produced by the soil through which the pile is being driven. Three of these losses can be said to be relatively fixed, whereas the fourth, the loss of energy due to the resistance of the soil, will increase as the pile is driven. When these losses of energy to the soil (termed the dynamic resistance of the soil) approach the total energy put in, less the other three losses, we have a measure of the supporting capacity of the pile.

More than a score of dynamic pile-driving formulas have been suggested and used. Some have been abandoned as unreliable, others have persisted. The credit for persistency perhaps should go to that known as the Engineering News formula introduced in 1893. A national survey conducted in 1962 indicated that virtually all Federal and state agencies specified the use of this formula and that many privately prepared specifications also contained the Engineering News formula. This, despite the fact that every investigator or authority on the subject has proved that it is unreliable, that it consistently indicates higher potential loadings than tests have confirmed, and that its factor of safety, consequently, is considerably less than 1.

The charm of the Engineering News formula is its simplicity, its dearth of multiple factors, and the non-complexity of its mathematics, well within the comprehension range of an average sixth-grader. Several other formulas have been found to give more accurate results on the basis of test loads, and the Hiley formula seems to be the most desirable of these. One advantage of the Hiley, Eytelwein, or so-called SO formulas is that they are applicable, without change, to all types of piling under all conditions, whereas the Engineering News formula was designed originally for wood piles driven solely as friction piles, in clay, and was only later modified for other conditions.

The general practice is to drive, initially, a certain number of piles in positions where they can be used for the final structure. These test piles are driven by using assumed values in a dynamic driving formula under carefully observed conditions. The piles are then tested to 100, 150, or 200% of the design load, and the values originally assumed are then adjusted to the test results obtained. Using this method, of course, almost any formula would do and, in cases where piles are tested, would justify the use of the Engineering News formula.

Static pile formulas which would determine the bearing capacity of a pile from static load tests have been offered, but from a practical standpoint they are no better than the dynamic formulas since they do not convert the result into terms that can be used to drive subsequent piles.

The Hiley Formula:

$$eWH = SH + \frac{CR}{2} \qquad \text{or} \qquad R = \frac{eWH}{S + C/2}$$

where R = resistance of pile to penetration in tons. In this formula the resulting value of R is the ultimate load and, where a safety factor is required, should be divided by the required safety factor of 2 or 3. Thus, for a value $R = 150$, the design load for a safety factor of 2 would be 75 tons; for 3 it would be 50 tons.

W = the weight of the ram or striking parts of the hammer, in pounds. These are listed for various hammers in Sec. 5.16.

H = the distance that the hammer falls, or the stroke of the hammer, in feet. W and H are usually combined and termed the rated energy of the hammer. These are also listed in Sec. 5.16 and for diesel hammers in Sec. 5.17.

WH = the effective hammer energy in foot-pounds. For double-acting hammers, the rated energy is used. For single-acting hammers use 0.9 of the rated energy.

S = average penetration per blow in inches for the last 5 blows. This is also known as the set. It is generally taken as 0.10 in. per blow or ½ in. for the last five blows.

C = allowance for temporary compression and is composed of the following. Values of these coefficients will be found in the tables on the following page.

C_1 = temporary compression in the pile in inches.
C_2 = temporary compression in the soil in inches.
C_3 = temporary compression in the pile cap in inches.

e = efficiency of the blow of the hammer upon impact and depends on the relative weight of the pile and the striking parts of the hammer (W above); also upon the nature of the intervening cushion.

$$e = \frac{W + PN^2}{W + P}$$

P = weight of the pile in pounds.

N = coefficient of restitution, that is, the relative recovery of the cushion after impact (see table on following page).

The Engineering News Formula:

$$R = \frac{2WH}{S + C}$$

Originally designed for wood piles driven with a drop hammer, the value of C was assumed as 1.0. This was later revised to 0.10.

The Modified Engineering News Formula:

$$R = \frac{2WH}{S + C \times (P/W)}$$

This takes into account the pile to hammer weight ratio.

Both the *Engineering News* formulas are generally considered to give a value of R with a safety factor of between 2 and 3.

5.11 THE USE OF PILE FORMULAS

Dynamic pile formulas can be solved for any one of their major factors, but they are usually solved for the value of R = resistance, the other factors being either assumed or obtained from standard sources.

The first determination to be made is that of the ratio of pile weight to hammer weight. The contractor may often have one or more hammers available or at least have access to several; however, if they are not of suitable size, being either too light or too heavy, it may be desirable to rent or buy one more in accord with the economics of the operation.

Determine the weight of the pile to be driven.
Determine the value of P/W where W is the weight of the striking parts of the hammer.
The greatest driving efficiency exists when this ratio is less than 1.0.

With the hammer selected and the pile characteristics known, a cushioning device is selected.
The value of e can now be determined (see bottom table opposite).
S is assumed to be 0.10 in.
The value of C (in the Hiley formula) is obtained by adding together the values for C_1, C_2 and C_3 as listed in the tables opposite.
The value of C (in the *Engineering News* formula) is arbitrarily chosen as 0.10.
The Hiley formula requires conversion of rated energy, in foot-pounds, to inch-pounds.
The *Engineering News* formula, being empirical, uses foot-pounds without conversion.

EXAMPLE

Determine value of R for Pre-cast Concrete Pile

Use 20-in. square pre-cast concrete pile where A = area = 400 sq in., L = length = 40 ft., P = weight of pile @ 412 lb/lineal ft = 16,480 lb = 8.25 tons; S-8 McKiernan-Terry hammer available (see Table 5.16).
W = weight of striking parts = 8000 lb = 4 tons
Rated energy = WH = 26,000 ft-lb = 13 ft-tons = 156 in.-tons
$P/W = 8.25/4.0 = 2.06$; driving conditions medium; consolidated wood cushion.
From table opposite where $N = 0.50$, $e = 0.50$.
$C = C_1 + C_2 + C_3 = 0.15 + (0.004 \times 40 \text{ ft}) + 0.10 = 0.41$

Hiley formula:

$$R = \frac{eWH}{S + C/2} = \frac{0.50 \times 156 \text{ in.-tons}}{0.10 \text{ in.} + 0.41/2} = \frac{88}{0.305} = 288 \text{ tons ultimate load}$$

Modified *Engineering News* formula:

$$R = \frac{2WH}{S + (C \times P/W)} = \frac{2 \times 13 \text{ ft-tons}}{0.1 + (0.10 \times 2.06)} = \frac{26}{0.306} = 85 \text{ tons}$$

EXAMPLE

Determine value of R for Pipe Pile

Use 12-in. pipe pile where outside diameter = 12.750, wall thickness = 0.375, L = length = 60 ft, P = weight of pile @ 49.56 lb/lineal ft = 3000 lb = 1.5 tons; S-5 McKiernan–Terry hammer available (see Table 5.16).
W = weight of striking parts = 5000 lb = 2.5 tons
Rated energy = WH = 16,250 ft-lb = 8.125 ft-tons = 97.5 in.-tons
$P/W = 1.5/2.5 = 0.6$; driving conditions hard; well-compacted wood cushion.
From table opposite $e = 0.72$ (by interpolation).
$C = C_1 + C_2 + C_3 = 0.12 + (0.009 \times 60 \text{ ft}) + 0.10 = 0.76$

Hiley formula:

$$R = \frac{eWH}{S + C/2} = \frac{0.72 \times 97.5 \text{ in.-tons}}{0.1 \text{ in.} + 0.76/2} = \frac{70.2}{0.48} = 146 \text{ tons ultimate load}$$

Modified *Engineering News* formula:

$$R = \frac{2WH}{S + (C \times P/W)} = \frac{2 \times 8.125 \text{ ft-tons}}{0.1 + (0.1 \times 0.6)} = \frac{16.25}{0.16} = 101 \text{ tons}$$

VALUES OF C_1, COMPRESSION IN PILE HEAD AND CAP

Material in pile	Easy driving 500 psi on cushion or on pile butt if no cushion, in.	Medium driving 1000 psi on head or cap, in.	Hard driving 1500 psi on head or cap, in.	Very hard 2000 psi on head or cap, in.
Head of timber pile	0.05	0.10	0.15	0.20
3 to 4 in. packing inside cap on head of pre-cast concrete pile	0.05 to 0.07	0.10 to 0.15	0.15 to 0.22	0.20 to 0.30
½ to 1 in. mat pad only on head of pre-cast concrete pile	0.025	0.05	0.075	0.10
Steel-covered cap containing wood packing; for H-beam or steel pipe	0.04	0.08	0.12	0.16
3/16-in. red fiber disk between two 3/8-in. steel plates; for Monotube piling	0.02	0.04	0.06	0.08
Head of H-beam or steel pipe	0.00	0.00	0.00	0.00

VALUES OF C_2, COMPRESSION IN PILE

Kind of pile	Easy driving 500 psi for wood or concrete 7500 psi for steel, in.	Medium driving 1000 psi for wood or concrete 15,000 psi for steel, in.	Hard driving 1500 psi for wood or concrete 22,500 psi for steel, in.	Very hard 2000 psi for wood or concrete 30,000 psi for steel, in.
Timber piles	0.004 × length	0.008 × length	0.012 × length	0.016 × length
Pre-cast concrete piles	0.002 × length	0.004 × length	0.006 × length	0.008 × length
Steel H-beam or pipe piles monotube or mandrel pile	0.003 × length	0.006 × length	0.009 × length	0.012 × length

VALUES OF C_3, COMPRESSION OF GROUND

For piles of constant cross section the value of C_3 can be taken as 0.10 for all conditions of driving. For easy driving it may vary from 0.00 to 0.10.

VALUES OF e FOR VARIOUS TYPES OF CUSHIONING

Ratio of pile weight to hammer weight = P/W	Fresh wood cushion, $N = 0.25$	Medium-compacted cushion, $N = 0.40$	Well-compacted cushion, $N = 0.50$	No cushion steel to steel, $N = 0.55$
½	0.69	0.72	0.75	0.77
1	0.53	0.58	0.63	0.65
1½	0.44	0.50	0.55	0.58
2	0.37	0.44	0.50	0.54
2½	0.33	0.40	0.45	0.51
3	0.30	0.36	0.42	0.48
4	0.25	0.32	0.36	0.44
5	0.21	0.27	0.31	0.42
6	0.19	0.24	0.27	0.40

5.12 TEST PILES AND PILE TESTS

There are two distinct aspects of pile testing, although both will apply to the same pile or group of piles. The first aspect of pile testing is that conducted by the contractor to determine the driving conditions and probable lengths of pile involved. These test piles will be driven in locations where piles are to be placed under the contract and, if otherwise suitable, will be accepted as part of the contract. In addition to the length of pile, these preliminary piles will determine, for the contractor, the length of time required per pile for driving, what may be required to punch through intermediate dense layers, what happens physically to the pile under prolonged driving, whether the steam or air supply is adequate, and whether the hammer is operating at proper efficiency.

The second aspect of testing involves the pile's capacity to perform its function after driving. These tests are the pile tests frequently required by contract terms for which separate payment is made. They will generally be performed on the first piles driven in selected locations throughout the entire site in which the piling is to be driven.

Piling is generally paid for on a unit price, per foot of pile in place after cut-off, so that excessive lengths of cut-off are expensive for the contractor. This is particularly true for wood or pre-cast concrete piles where splicing is difficult and generally undesirable and where, consequently, too short a pile becomes equally expensive.

Test borings may show a firm layer of soil at a specific depth into which it is presumed that the pile tip will penetrate and in which the pile will "pull up," but with friction piles the intervening strata may provide sufficient resistance to driving to prevent the pile's ever reaching the stratum depth indicated. Again, with friction piles, the actual driving may so alter the condition of the soil through which the pile is being pushed as to change the anticipated driving condition. The density of the soil immediately surrounding the pile may be greatly increased by driving, so that a shorter pile is adequate; or the liquidity of the soil may be increased, requiring a much longer pile.

Changes in the density of soils will frequently be due to changes in pore-water pressure, where the soils are saturated, or may be due to the compaction of sands resulting from the vibratory effect of driving. Where piles are driven in closely spaced clusters, the changes in soil density are sufficient to cause rebound upward of piles already driven (see Sec. 5.24).

The second aspect of testing, to compare the actual load-bearing capacity with that indicated by the dynamic pile-driving formula, involves fewer factors but is more costly and time-consuming. Indeed, the element of time is extremely important in all load testing.

There are two views regarding the support of a friction pile. The first is that the maximum support offered by the pile derives from the friction between the surface of the pile and surrounding media. The second view is that consolidation of the soil around the pile is such that this area of consolidation and the pile act as a unit and that actual support depends on the shearing resistance of the soil (particularly in soft clays) at some radial distance from the pile.

Although the choice of viewpoint is largely academic so far as load tests are concerned, the second view presents a rational explanation of the importance of time in testing. Increasing soil consolidation apparently continues over a considerable period of time so that, with age, the load-bearing capacity of the pile will be increased. In one case, a pile tested at 6-month intervals showed an increase in load-bearing capacity of 4.5 tons after 6 months and 7.5 tons at the end of a year, in both instances having settlements of only ¼ in.

The same effect has been noted so far as the *rate* at which the pile is loaded during the initial test period, a slower rate of loading often providing a higher maximum load.

Since time improves the test results (although often only fractionally), the piling contractor is well advised not to hurry his pile testing despite the well-recognized fact that in construction "time is of the essence".

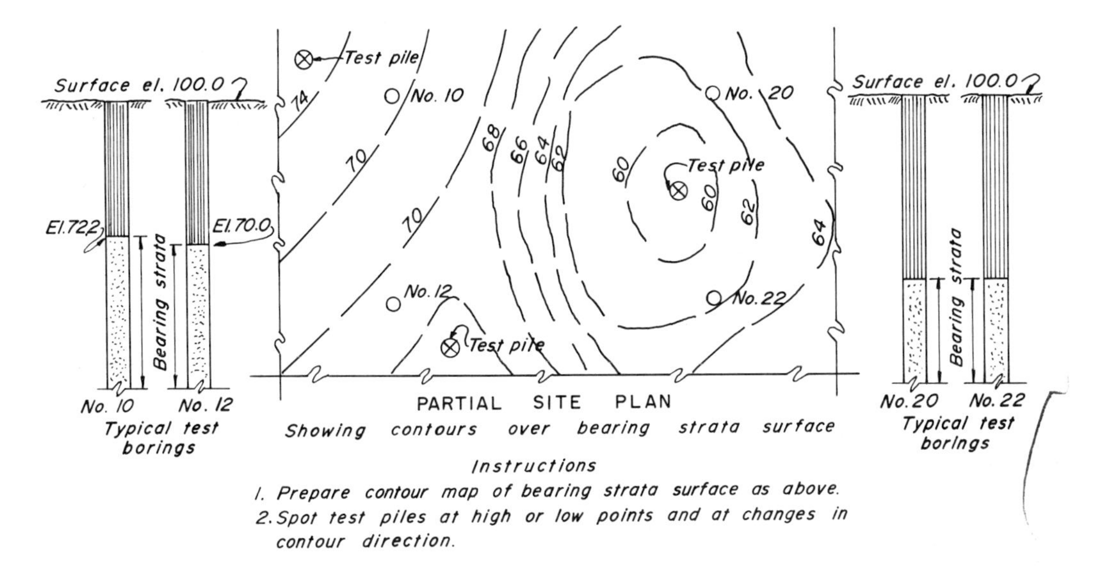

LOCATING TEST PILES

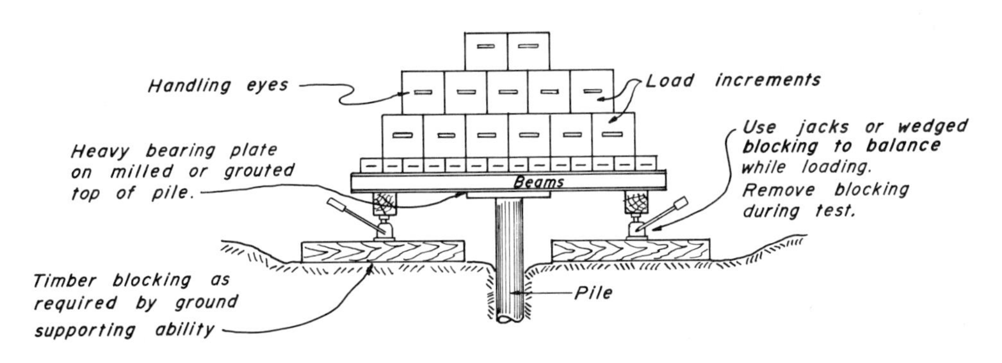

PILE TESTING BY BALANCED LOADING

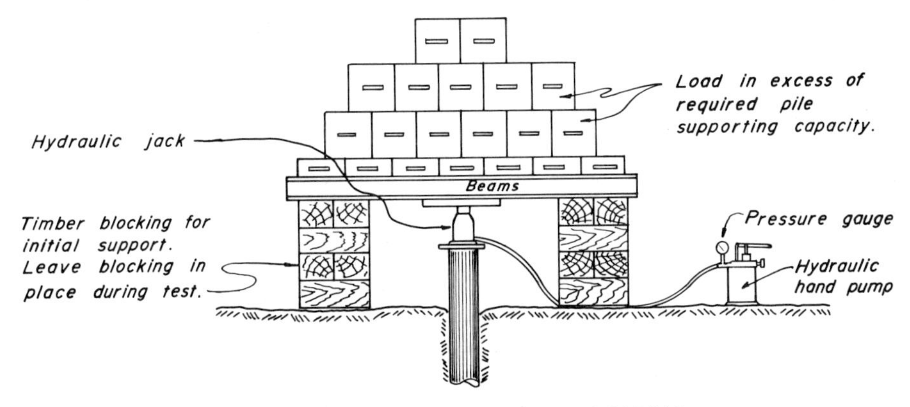

PILE TESTING BY RECORDED LOADING

5.13 PILE LOAD TESTS AND TESTS FOR UPLIFT

There are two sorts of tests conducted on driven piles, the first and most common being that required to verify the dynamic pile-driving formula and prove the load-bearing capacity. These tests involve loading the pile to 150% (occasionally 200%) of its design capacity and noting the resultant settlement, if any.

The second test is that required to determine the pile's resistance to uplift, that is, its resistance to being extracted. This factor is of considerable importance in a number of instances, particularly where hydrostatic pressure might tend to lift a hollow structure or tank constructed partly below water table. Another major use of piles in uplift is to resist overturning forces at the base of structures due to wind loads or other lateral forces or due to frost or ice action. The test procedures for uplift are similar to those for load capacity and are sometimes referred to as tension and compression tests, although this terminology is only strictly applicable to the pile member itself.

Balancing a 60-ton load on the 12-in. butt of a pile partakes of the elements of inscribing the Lord's Prayer on the head of a pin. Although some load tests are still performed by balancing a platform over a pile and loading the platform progressively with weights, the more usual procedure in the United States is to jack the pile downward (or upward for uplift) against a fixed overload, using a calibrated hydraulic jack. Occasionally, because of site limitations, a beam is cantilevered over the pile from a fixed fulcrum, but this application is rare.

Where ground-support conditions warrant, a jacking resistance platform can be built, centered on the pile but supported, at its corners, on the ground. Loading is concentrated on a single beam of suitable strength placed directly over the pile, and the jack is placed between the beam and the pile with suitable bearing plates, top and bottom, to distribute the load. The platform is then loaded with weights of iron ingots, concrete blocks, structural steel sections, or other materials of determinable weight. This loading must be well in excess of the test load required of the pile.

Where insufficient ground support is available to initially support the platform load, or over water, other piling is used to provide the jacking resistance. This piling may be especially driven for testing purposes or may be piling to be used as part of the structure, the decision, most often, depending upon the relative position of the contract piling.

If we assume that 25% of a friction pile load is taken by the tip, then only 75% of the design load is available to resist uplift. This would mean the use of at least two piles to match the loading on a single pile. To provide a suitable margin and prevent any movement in the holding piles, at least four piles should be used, and six is more frequently the number selected.

The procedure in testing is to add increments of load, maintaining the resulting load until any evidence of settlement has stopped, after which additional increments are added. In general, 50 to 75% of the design load can be added before evidence of settlement will occur. From this point, 1-, 2-, or 5-ton increments can be added and the resulting settlement observed.

The amount of settlement must be carefully noted. The pile is marked in units of 1 in., and these are compared with a fixed point of reference. The reference point should be placed several feet from the pile and may be an iron pin or stake driven into the ground. Measurements are made using a line level carried from the top of the pin across the markings on the face of the pile. The pin should be carefully checked from time to time, by transit or tripod level, to be certain that ground movements have not altered its elevation. In some cases direct readings by transit on the pile from a remote point may be necessary.

Specifications will frequently require that the maximum load remain in place for one or more days, with further settlement readings taken at regularly spaced intervals.

Pile load tests to failure are seldom required as part of a pile-driving contract.

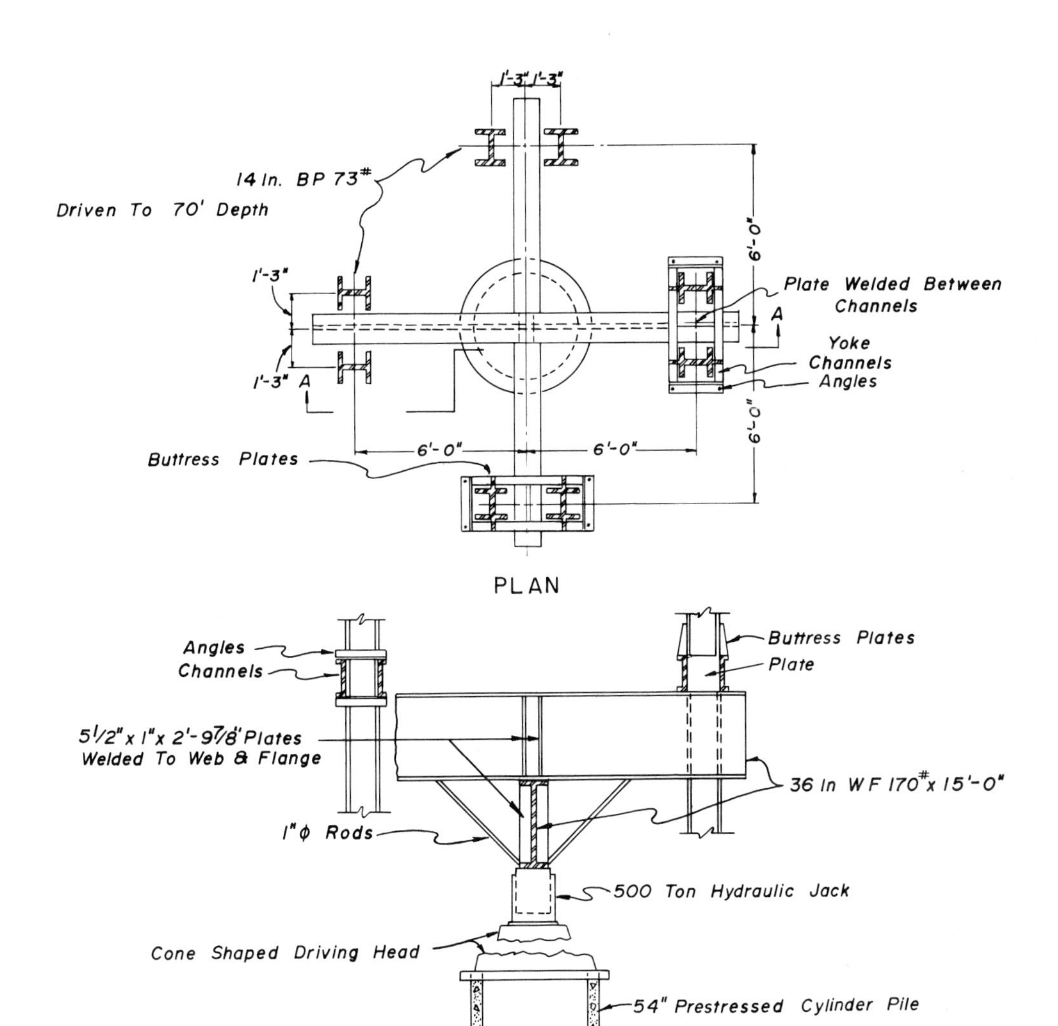

A 425 TON PILE TEST OVER WATER

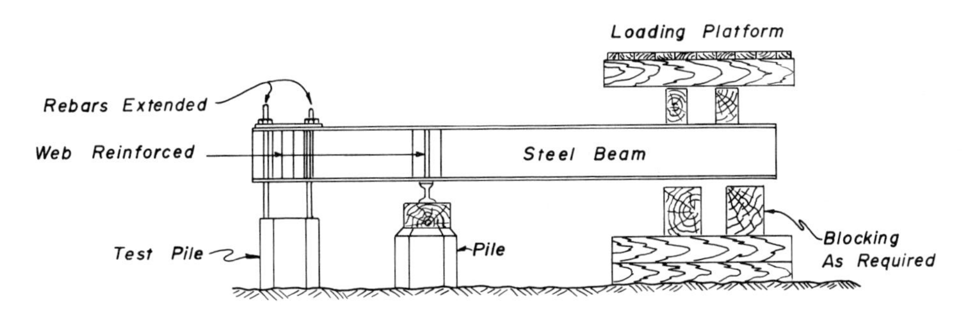

AN UPLIFT TEST SET-UP ON A PRECAST CONCRETE PILE

Pile-Driving Procedure

In order to visualize the various pieces of equipment used in pile driving it will be advisable to list the basic steps in the driving of a pile. The following steps constitute the procedures in placing a pile.

1. The pile, lying horizontally on the ground, is picked up by means of a 1-, 2-, or 3-point suspension, raised and stood vertically upright resting on its point. The suspension system is changed so that the pile is supported by a hanging rope with a hitch near the butt end of the pile.

2. The location of the pile has been spotted by a short stake driven lightly into the ground. This is removed, the pile is lifted slightly, and the point of the pile is centered over the hole left by the stake and lowered to rest on the ground, but with the weight carried by the rope hitch.

3. The pile is carefully plumbed and the weight of the pile is transferred from the hitch to the point, permitting it to sink under its own weight. With large piles, suspended over soft ground, this sinking may be considerable. On harder surfaces no appreciable sinking may occur.

4. With the pile resting on its tip, either at or below ground surface, the plumbness of the pile is again checked and adjusted, if need be, the upper portion of the pile being restrained and controlled by the rope hitch.

5. The hammer is now lowered onto the head of the pile with a cap placed between the top of the pile and the hammer. Frequently the cap is hung from the hammer, and in these cases all that is needed is the centering of the cap over the pile. In other cases some portion of the cap must be placed and adjusted before the hammer is lowered onto the pile. In these cases the cap must be manually controlled for positioning, requiring a man to reach this elevation in order to perform the operation.

6. The hammer must be suspended by a cable independent of the line supporting the pile and must be susceptible to being raised and lowered in a truly vertical line; it must be guided.

7. The full weight of the hammer is allowed to rest on the pile, in some cases pushing the pile further into soft ground, before a few tentative taps are given the pile to gage driving resistance. In this operation, considerable slack must be permitted in the hammer cable to prevent the pile from falling away from it and losing contact between pile butt, cap, and hammer.

8. Before full driving is begun, the pile must again be checked for plumbness.

9. Under driving, as the resistance to penetration increases, the rate of penetration is noted, generally by 1-ft markings on the pile, and when this rate becomes sufficiently limited, driving is momentarily halted while the pile is marked off in inches. From this point on, the foreman, or more usually the inspector, notes the number of blows required to force the pile deeper into the ground per inch of pile length.

10. The solution of all dynamic pile-driving formulas for friction piles is for the "set," the number of blows per inch which should produce the load-bearing capacity required. When observation and/or measurements indicate that the required penetration has been reached, driving is terminated.

11. With end-bearing piles the above procedure is somewhat altered. The use of a formula has no meaning, and a point is reached where no further penetration is possible. This point will often be determinable, not only by lack of penetration but by a change in the sound produced when the hammer strikes the pile. However, there is an area of considerable danger here since, although the battering of the top of the pile is visible, what occurs at the point is not.

12. The expected pull-up point of an end-bearing pile will be checked and established from test borings; and when this level is being approached, more careful driving should be employed. All kinds of piles can be damaged by overdriving, and the damage occurs, generally, precisely where point-bearing support is required. On the other hand, insufficient driving at this point can also fail to provide the support desired.

13. Once the pile has reached its final position, the hammer is lifted off the pile; the cap, if separate, is removed; and the hammer and the rig supporting it are moved to a new driving location.

5.14 PILE-DRIVING EQUIPMENT

On the opposing page are listed the generalized steps in driving a pile. These impose certain minimum requirements on the equipment used, and although the basic rigs employed to lift the pile and to support the hammer are numerous in kind and various in incidental attributes, these minimum requirements must first be met.

There must be, first of all, a mast whose upper end is high enough above ground to accommodate the length of the longest pile plus the length of the hammer, when two-blocked, with some slight additional margin of length. The upper end of the mast must be provided with sheaves to handle running cable. The cable over one sheave will support the hammer, while a second sheave will support the pile-holding cable. Occasionally a third sheave will be used where a multiple suspension is required for lifting a long pile, so that the suspension line and the pile-holding line are separate.

The two or three cables used must be activated by running them onto or off of drums, and a source of power must be supplied to the drums. The power source and drums must be mounted on a platform to which the mast will be rigidly attached; and the entire assembly must be, if not self-mobile, at least movable. There are three general types of rigs employed to meet these requirements, their selection depending chiefly upon site conditions.

The most usual rig employed is a standard crane, generally crawler-mounted (although for short piling, truck mountings have been used), fitted with a boom of sufficient length. The size of crane used will, of course, depend on the length and weight of pile to be handled. So long as the ground conditions at the site are firm enough to support crawlers or crawlers riding on mats, the crane is the most suitable rig.

On softer terrain, the vibrations of driving may tend to tilt the crane rig during driving or may be too soft to provide any substantial support. For these cases a large platform is often used, mounted on timbers which spread the loading over a wider area, somewhat removed from the pile being driven. This is a skid-mounted platform whose position from pile to pile can be changed by sliding the platform, on pipe rollers, over its timber skid frame.

The third condition is where piling is to be driven over water. For this condition the pile-driving rig will be barge-mounted with spuds driven into the underwater soil to prevent its movement during driving. The barge may be self-powered, moved by tugboat, or may be re-positioned by swinging-winches from fixed anchorage points (see Sec. 2.14).

The next requirement of pile-driving equipment is a method of maintaining the hammer in position during the driving, and this is generally accomplished by the use of pile-driving "leads."

Essentially, pile-driving leads consist of two rails or guides in which the pile-driving hammer moves vertically. The kind of rails and their spacing will depend on the specific hammer being used, although several sizes of hammer may be suitable for a single set of leads. The rails must be held rigidly in position but in such a way as to permit the pile to be placed between them. This requirement results in a U-shaped frame, constructed variously of steel pipe, steel flats, or channels in which the rails are set on the inner faces of the prongs of the U. With skid rigs, timber frames are occasionally used.

Simple swinging leads, used chiefly with cranes, are hung from an extended boom-point sheave pin at the top of the boom. The bottom of the leads terminate in long steel points. The points are positioned astride the pile location and the crane boom maneuvered so that the hammer center is vertically above the center of the pile. The boom is then lowered slightly to permit the relieved weight of the leads to force the points into the ground, holding it in position.

Fixed leads are also frequently used by crane rigs, by running horizontal struts from the leads back to extended boom foot pins, thus providing a fixed triangular frame consisting of the boom, the leads, and the struts.

5.15 ASPECTS OF PILE-DRIVING EQUIPMENT

The discussion in Sec. 5.14 was limited to the general aspects of pile driving and to what might be termed the traditional items of pile-driving equipment, derived from earlier experiences with wood piling. Variations in kinds of piling and in the conditions of driving have led to some important modifications.

In discussing the driving of pile tubes in Sec. 5.5, reference was made to the *mandrel,* sometimes *mandril* (but never *mandrill,* which is a "large, gregarious, ferocious West African baboon," seldom encountered in pile driving), originally the term applied to a spindle forced into a metal section to prevent distortion during machining operations or a core on which metal was cast, molded, or shaped. The use of a mandrel is necessary with all thin shell piling not only to prevent telescoping of the shell under vertical impact but to prevent collapse of the shell from horizontal earth pressures.

The mandrel exterior must have the same shape as the interior of the shell during driving yet must be removable on completion. Mandrels currently in use are proprietary items controlled by their developers and are of three kinds. The tapered core used to drive the shell for the Raymond standard concrete pile is expandable (or retractable) in size, by means of mechanical wedging. The mandrel used in driving the shell for the Raymond step-tapered pile is a stepped solid steel core whose spaced offsets rest on plow rings on equivalent spacings within the shell. The Cobi mandrel, used in driving corrugated Hel-Cor pipe piles, is expanded by the application of compressed air at 125 psi, and its inner surface carries projections matching the pipe corrugations.

Shells used with mandrels are assembled from short lengths since, otherwise, the point of hammer suspension would have to be twice the pile length above the ground surface. The mandrels must also be assembled from limited length units to permit their transportation to the site.

The use of leads for pile and hammer guides can frequently be eliminated, although some sort of alternative guide method must be employed. The position of the hammer on top of the pile, and in line with it, must be maintained, and access to the head of the pile to seat the hammer may be required.

In some cases fixed timber frames can be used, where facilities exist for holding them sufficiently above ground level. This method can be used where piling is driven inside a cofferdam, for instance. Timbers are secured in both directions over the top of the cofferdam, providing a square just slightly larger than the diameter of the pile. The pile is dropped through, and the tip is then placed in position at the bottom of the cofferdam to provide the second point in the required line.

Timber frames constructed on ground surface would be insecure and costly and would interfere with maneuvering the pile-driving rig; but if a frame were constructed on the front of a crawler mounting, in place of a boom, no interference would ensue. Such a rig has been employed to hold large concrete piles vertically in position on firm ground surfaces.

When driving over water, pile leads have been replaced by a holding frame constructed on the forward end of a separate barge. The barge is anchored, or tied up, to hold position until the first few piles have been placed, after which a collar, bolted to the driven piles, supports guides for driving the remaining piles in the group.

Pile hammers driving piles without leads will generally require a frame which can be slid a short distance down over the pile to provide alignment of pile and hammer. Such a frame is fastened to the hammer and called a "skirt."

One factor that will determine the feasibility of driving piling without leads will be the allowable deviation from the true position, permitted by the specifications. This in turn will frequently depend on the type of structure being supported by the piling and the design of pile cap to be employed.

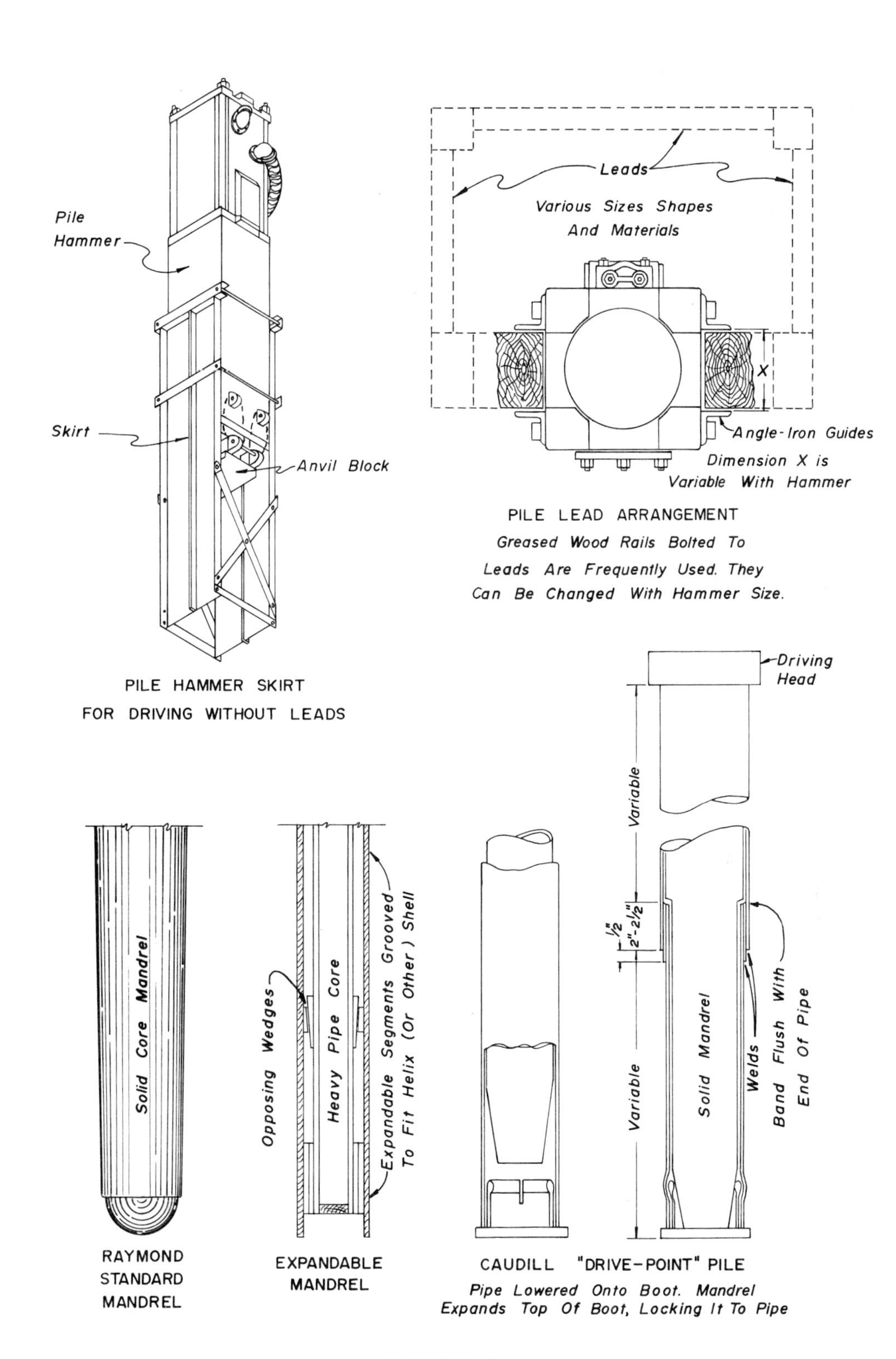
Pile Hammer
Skirt
Anvil Block
PILE HAMMER SKIRT
FOR DRIVING WITHOUT LEADS
Leads
Various Sizes Shapes
And Materials
X
Angle-Iron Guides
Dimension X is
Variable With Hammer
PILE LEAD ARRANGEMENT
Greased Wood Rails Bolted To
Leads Are Frequently Used. They
Can Be Changed With Hammer Size.
Solid Core Mandrel
RAYMOND
STANDARD
MANDREL
Opposing Wedges
Heavy Pipe Core
Expandable Segments Grooved
To Fit Helix (Or Other) Shell
EXPANDABLE
MANDREL
Driving
Head
Variable
1/2"
2"-2 1/2"
Variable
Solid Mandrel
Welds
Band Flush With
End Of Pipe
CAUDILL "DRIVE-POINT" PILE
Pipe Lowered Onto Boot. Mandrel
Expands Top Of Boot, Locking It To Pipe

MANDRELS

5.16 AIR- OR STEAM-OPERATED PILE HAMMERS

The oldest and the simplest of pile-driving hammers is the drop hammer, now rarely used but still available in weights ranging from 1000 to 8000 lb. In effect, it is a freely falling weight dropping between guides and raised after each fall by winding its supporting rope on a powered winch. Sometimes this is done manually over a cathead, but various devices have made the operation almost automatic, paying out rope as the pile goes down. Here, the weight of the ram, less frictional and drag resistances, times the height of fall, provide a quick estimate of the foot-pounds of energy produced.

With the drop hammer, a long fall would produce higher impact energy but would also increase the time between blows, much of which is consumed in raising the ram. Pile-driving experience indicates that rapid, lower-energy blows are more effective since they tend to keep the pile moving downward continuously instead of in surges, and the recognition of this fact led to the development of the first single-acting steam hammer.

The single-acting steam hammer uses steam acting against a piston to raise the ram, now moving within containing guides, to a limited height, at which the steam pressure is exhausted and the ram falls of its own weight. This increases the speed at which blows can be delivered as well as controlling the length of drop so that it is always the same. However, the short fall decreases the striking velocity and the resulting foot-pounds of energy.

The double-acting steam hammer was developed to improve the striking energy of the single-acting steam hammer. This hammer not only uses steam to raise the ram but, by steam pressure on a separate piston, increases the downward acceleration of the ram by adding steam pressure to the force of gravity. Not only are the foot-pounds of impact energy increased, but the increased acceleration provides a swifter fall; and greater speeds, stated as number of blows per minute, are available.

Double-acting hammers alternately exhaust the steam from the raising and driving chambers so that, at each stroke, they must be re-filled; and, although adequate pressure will accomplish this rapidly, there is still some loss of time as well as steam. The differential-acting steam hammer was devised to overcome this.

The differential-acting steam hammer contains two piston heads operating in a single chamber on the same piston rod. The lower piston head is of small diameter while the upper head is of considerably larger diameter. On the up stroke, steam enters between the piston heads, the same pressure acting against the larger upper head producing a greater force to move the piston upward. On the down stroke, steam is added *above* the upper piston, forcing it downward. The steam remaining from the up stroke is forced through a channel into the upper chamber without being exhausted. On the up stroke, steam is exhausted from the *upper* chamber but the chamber between heads remains full at all times. Steam flow is controlled by valving contained within the hammer.

In the differential-acting hammer, there is no drop from the entering pressure to the mean effective pressure moving the piston on the down stroke, and as much as a 20% increase in striking power and greater speeds can be obtained.

Pile-driving hammers indicated as "steam" hammers were originally designed for the use of steam but are now generally driven by compressed air. This changeover followed upon the development of portable air compressors of adequate size to provide the quantity of air required. It is doubtful whether compressed air provides any saving in cost; and on large jobs steam is still frequently used, but on smaller jobs the certainties of a plant-engineered compressor, and the reduction of potential hazards, make air a more attractive source of power.

Opposite is listed information relating to hammers currently available, and data required for solving the formulas discussed in Sec. 5.10 are taken from these tables.

Pile Hammers and Their Characteristics

Rated energy, ft-lb	Make of hammer	Type	Size	Blows per min	Weight striking parts, lb	Total weight, lb	Length of hammer	Cfm air per min	A.S.M.E. boiler H.P.	Steam or air, psi	Size of hose, in.	$\sqrt{E \times W}$ rating
Energy over 100,000 ft-lb												
113,478	Super–Vulcan	Differential	400C	100	40,000	83,000	16′9″	4659	700	150	5	67,378
Energy 50,000 to 100,000 ft-lb												
60,000	Vulcan	Single-act.	020	60	20,000	39,000	15′0″	1756	278	120	3	34,640
60,000	McKiernan–Terry	Single-act.	S20	60	20,000	38,650	18′5″	1720	280	150	3	34,640
50,200	Super–Vulcan	Differential	200C	98	20,000	39,050	13′2″	1746	260	142	3	31,685
Energy 30,000 to 50,000 ft-lb												
42,000	Vulcan	Single-act.	014	60	14,000	27,500	14′6″	1282	200	110	3	24,248
37,500	McKiernan–Terry	Single-act.	S14	60	14,000	31,600	14′10″	1260	190	100	3	23,000
36,000	Super–Vulcan	Differential	140C	103	14,000	27,984	12′3″	1425	211	140	3	22,449
32,500	McKiernan–Terry	Single-act.	S10	55	10,000	22,200	14′1″	1000	140	80	2½	18,027
32,500	Vulcan	Single-act.	010	50	10,000	18,750	15′0″	1002	157	105	2½	18,027
Energy 20,000 to 30,000 ft-lb												
26,000	Vulcan	Single-act.	08	50	8000	16,750	15′0″	880	127	83	2½	14,422
26,000	McKiernan–Terry	Single-act.	S8	55	8000	18,100	14′4″	850	119	80	2½	14,422
24,450	Super–Vulcan	Differential	80C	111	8000	17,885	11′4″	1245	180	120	2½	13,985
24,450	Vulcan	Differential	8M	111	8000	18,400	10′6″	1245	180	120	2½	13,985
Energy 10,000 to 20,000 ft-lb												
19,875	Union	Double-act.	0	110	3000	14,500	10′1″	800	. . .	125	2	6360
19,850	McKiernan–Terry	Double-act.	11B3	95	5000	14,500	11′1″	900	126	100	2½	9785
19,500	Vulcan	Single-act.	06	60	6500	11,200	13′0″	625	94	100	2	11,258
19,200	Super–Vulcan	Differential	65C	117	6500	14,886	12′1″	991	152	150	2	11,201
16,250	McKiernan–Terry	Single-act.	S5	60	5000	12,375	13′3″	600	84	80	2	9000
16,000	McKiernan–Terry	Compound	C5	110	5000	11,880	8′9″	585	56	100	2½	8944
15,100	Super–Vulcan	Differential	50C	120	5000	11,782	10′2″	880	125	120	2	8689
15,100	Vulcan	Differential	5M	120	5000	12,900	9′4″	880	125	120	2	8689
15,000	Vulcan	Single-act.	1	60	5000	10,100	13′0″	565	81	80	2	8660
13,100	McKiernan–Terry	Double-act.	10B3	105	3000	10,850	9′4″	750	104	100	2½	6269
12,725	Union	Double-act.	1	125	1600	10,000	8′2″	600	. . .	100	1½	4530
Energy 5000 to 10,000 ft-lb												
9000	McKiernan–Terry	Single-act.	S3	65	5000	8800	12′4″	400	57	80	1½	5200
8750	McKiernan–Terry	Double-act.	9B3	145	1600	7000	8′2″	600	85	100	2	3742
8280	Union	Double-act.	1½A	135	1500	9200	8′4″	450	. . .	100	1½	3524
7260	Vulcan	Single-act.	2	70	3000	7100	12′0″	336	49	80	1½	4666
7260	Super–Vulcan	Differential	30C	133	3000	7036	8′11″	488	70	120	1½	4666
7260	Vulcan	Differential	3M	133	3000	8490	7′11″	488	70	120	1½	4666
Energy under 5000 ft-lb												
4900	Vulcan	Differential	DGH900	238	900	5000	6′9″	580	75	78	1½	1897
3660	Union	Double-act.	3	160	700	4700	6′4″	300	. . .	100	1¼	1600
3600	McKiernan–Terry	Double-act.	7	225	800	5000	6′1″	450	63	100	1½	1697
445	Union	Double-act.	6	340	100	910	3′10″	75	. . .	100	¾	210
386	Vulcan	Differential	DGH100A	303	100	786	4′2″	74	8	60	1	196
356	McKiernan–Terry	Double-act.	3	400	68	675	4′10″	110	. . .	100	1	155
320	Union	Double-act.	7A	400	80	540	3′7″	70	. . .	100	¾	160

E = rated striking energy in foot-pounds; W = weight of striking parts in pounds.

5.17 DIESEL PILE-DRIVING HAMMERS

Early in World War II, the Bureau of Yards and Docks, the construction arm of the United States Navy, realizing that they would be called upon to provide docking facilities throughout the world, many of them built on piling, saw the advantages of an integrally powered pile hammer in reducing shipping tonnage and cubage, as well as savings in equipment procurement and manpower. Internal-combustion engines had been tried, by others, but had not proved successful; the impact-induced vibrations threw them out of adjustment. The Bureau, therefore, pioneered research in a diesel-fuel-powered hammer. Actually, the basic principles had been developed in the late 1930s, in Germany, but had not been applied to an actual hammer on the North American continent. The Bureau's program did not produce a working hammer until after the war, and Yards and Docks first displayed a diesel hammer at the Centennial Convention of the American Society of Civil Engineers, in Chicago, in 1952.

Since 1952 the use of diesel hammers for pile driving in the United States has increased rapidly. By the early 1960s three makes of hammer were available, each having some variations on the basic principle. The McKiernan-Terry Corporation manufactures two models, the Link-Belt Speeder Corporation three models in the United States. The Delmag hammer, a German product, is available through Special Construction Machines Ltd. of Toronto, Canada.

Diesel fuel oil, when vaporized ("atomized" is the word frequently used here), can be ignited simply by higher temperatures without need for a spark or flame. In Sec. 2.1 it was noted that the act of compressing air raised its temperature. In the diesel hammer the rise in temperature resulting from compressing a small volume of air is used to explode a charge of atomized diesel oil.

A ram is raised vertically in a long cylinder, released, and permitted to fall freely onto an impact block or anvil. The anvil itself rests on the pile and may be secured to the bottom of the cylinder or may be free to move within it. As the ram falls, it activates a fuel pump which injects a measured quantity of diesel fuel into a ball-shaped pan in the top of the anvil. As the ram continues its fall it closes exhaust ports in the cylinder wall, trapping a small volume of air, which the continued ram fall compresses, with a resulting rise in temperature.

The shaped head of the ram fits into the ball pan in the top of the anvil and, when it strikes the pool of oil, atomizes it. The resulting diesel vapor is exploded by the higher temperatures of the compressed air. The explosion serves first to provide additional impact to the anvil and so to the pile and also provides the force to raise the ram, once again, to the top of the cylinder. In rising, the ram clears the exhaust ports so that the products of combustion can be scavenged.

The ram of the diesel-powered hammer projects through the cylinder head in two makes of hammer and is lifted, to start, by an outside hoisting line. The Link-Belt models have an enclosed ram surmounted by a bounce chamber, in which the rising ram compresses air for additional push on the down stroke. Link-Belt also uses a high-pressure fuel-injection system and battery-operated glow plug instead of depending upon the heat from the compression of air to explode the vaporized fuel.

There are two difficulties with diesel hammers, the major one being that they are impractical in soft driving. As can be seen, the action of the hammer depends on the ram's striking the impact block with sufficient force to vaporize the fuel. But in soft driving, the impact block can move away from under the ram without providing sufficient resistance, and the explosion can be expended downward without providing sufficient force to raise the ram in the cylinder. Where pile penetration per blow is high, other types of hammers are more satisfactory.

The second objection to diesel pile hammers is the difficulty of rating the energy delivered per blow. This will be discussed in Sec. 5.20.

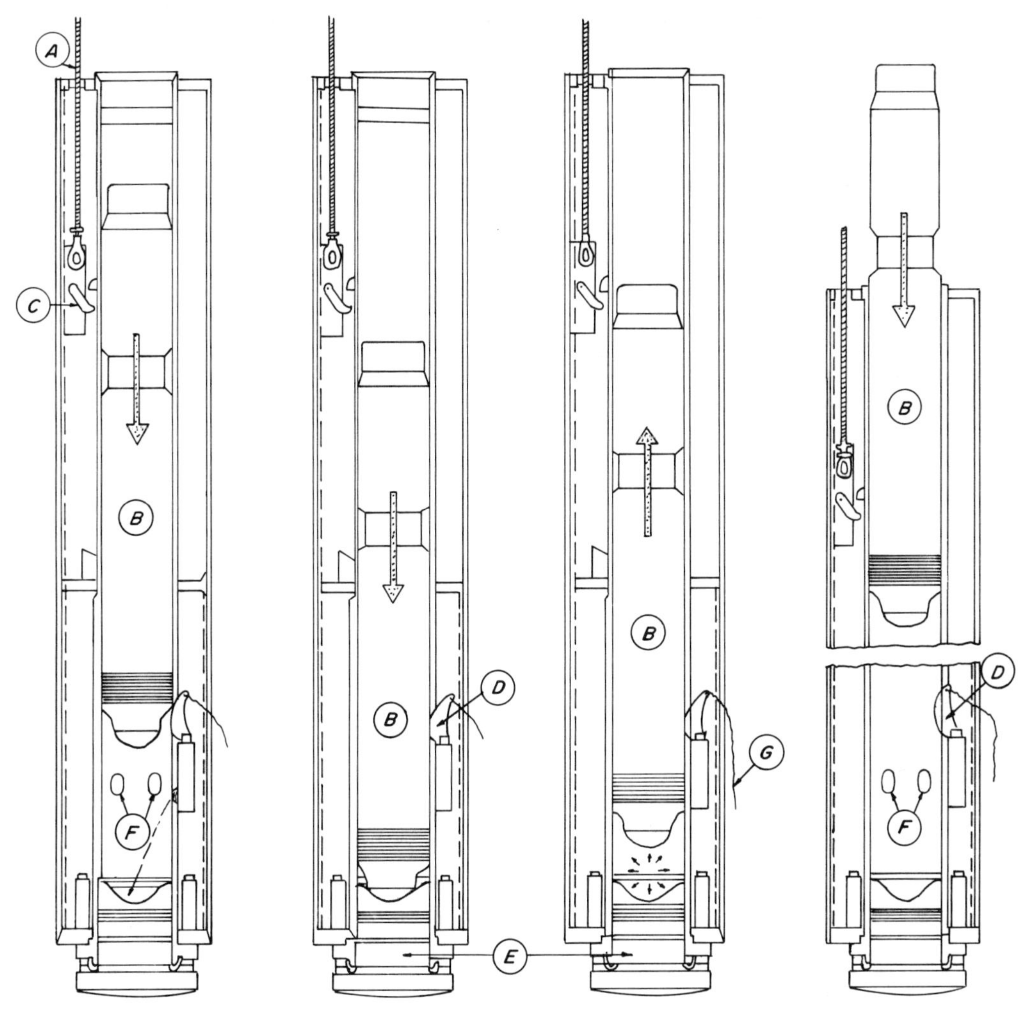

DIESEL PILE HAMMER OPERATION

1. Crane hoisting line lifts starting line *A*. Trip mechanism automatically releases ram-piston *B*.
2. Falling ram-piston actuates fuel pump cam *D*. Liquid fuel is fed into cup on top of anvil block *E*.
3. The falling ram-piston blocks exhaust ports *F* and starts compression of air.
4. Impact of ram-piston in anvil cup splatters fuel which ignites in the high-pressure, hot air.
5. Explosion drives ram-piston upward and forces anvil downward. Ports are opened and exhaust is released.
6. Rise of ram-piston is counteracted by gravity, resistance of pile determining the length of stroke.
7. Driving is stopped by pull on rope *G* which disengages fuel pump cam *D*.

MAXIMUM ENERGY RATINGS—DIESEL PILE HAMMERS

McKiernan–Terry Corp.	Delmag	Link–Belt Speeder Corp.
Model DE-20 = 16,000 ft-lb Model DE-30 = 22,400 ft-lb	Model No. D-5 = 9100 ft-lb Model No. D-12 = 22,500 ft-lb Model No. D-22 = 39,700 ft-lb	Model No. 105 = 7500 ft-lb Model No. 312 = 18,000 ft-lb Model No. 520 = 30,000 ft-lb

5.18 SONIC AND HYDRAULIC PILE HAMMERS

The Bodine Sonic Pile Driver was the invention of Albert G. Bodine, Jr., of California and was patented as an "acoustic method and apparatus for driving piles." The driver was first demonstrated in 1961 after a rig had been put together, based on Bodine's patent, by the C. L. Guild Construction Company, Inc., of Providence, Rhode Island.

Vibratory hammers with somewhat similar principles of operation had been tried out in Russia, Germany, and France, but the Bodine driver was the first to be built in the United States.

The principle employed by the sonic pile driver is that of delivering vertical vibrations to the head of a pile in alternating up and down cycles at a rate of 100 cps. (The hammer is rigidly secured to the pile butt.) These vibrations set up high-amplitude waves of tension and compression in the pile, producing alternate expansion and contraction in minute amounts. The elongation of the pile in expansion displaces the soil at the pile tip; and the weight of the pile, hammer, and added loads shoves the pile into the miniscule void. Since this action is occurring at the rate of 100 times a second, the individual movements need not be of great magnitude to produce rapid penetration of the pile.

As a structural member (here, the pile) increases in length, it contracts in cross-sectional area. The rate at which this occurs (Poisson's ratio) is a fixed value for a particular kind of material. As the pile contracts, longitudinally, a corresponding increase in area occurs. This alternating increase and decrease in size serves to relieve and engage the frictional resistance along the face of the pile, further aiding in its sinking.

The result of these end and side movements is to provide a continuous downward motion to the pile which, as has been previously noted, is the aim in all pile driving. The ideal soil for this type of driving is sand, with possibly silt or gravel intrusions (see Secs. 3.6 and 3.7), but the method has been used successfully in clays as well. The speed of driving with sonic hammers is measured in seconds, so that a pile which might require an hour and a half to sink with a steam hammer can be sunk in a minute and a half with the Bodine hammer.

Bodine's original patent papers called for a hydraulic turbine, built into a cylindrically shaped driver, operated by fluid from a separately hung reservoir, but the earliest test model employed a 500-hp engine rotating eccentrically mounted weights on a cam shaft. The hydraulic principle, however, had simultaneously been receiving attention from another source.

The hydraulic principle, applied to pile drivers, had been the subject of experiment by many manufacturers over the years but always at limited pressures of up to 1500 psi, inadequate for the purpose. As has happened so frequently in the research-poor construction industry, application of the hydraulic principle to pile drivers lagged until the aviation and missile industry had developed hoses and accessories that could handle hydraulic pressures of 5000 psi.

In 1962, the Raymond Concrete Pile Company developed and field-tested the first successful hydraulically operated pile hammer. In developing this hammer, Raymond first adapted to this use a standard differential-acting type of steam hammer whose action has already been described in Sec. 5.16.

The specific hammer adapted was also of Raymond's design but was similar to a Vulcan 50C hammer and, used with steam, provided a 19,500 ft-lb blow at the rate of 117 blows per minute. The original 16-in. piston operating the ram was replaced by a 2½-in. piston for hydraulic use, and various other dimensions were adjusted accordingly.

The hydraulic hammer developed by Raymond is more closely allied to the steam hammer, from which it was derived, than to the sonic hammer. It delivers the same foot-pounds of energy as the original hammer but is faster, is less expensive to operate, and largely eliminates the noise and dirt of a steam hammer. It can be used effectively in soft driving.

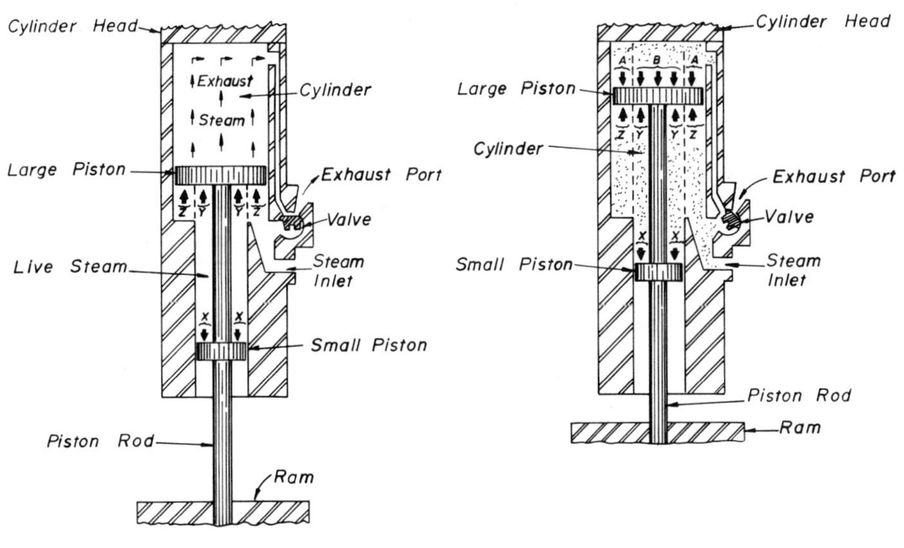

UP STROKE DOWN STROKE

SCHEMATIC OF DIFFERENTIAL HAMMER OPERATION

On Upstroke - Force X X on small piston is balanced by force Y Y on large piston. Remaining force Z Z lifts piston.

On Downstroke-Force X X on small piston is still balanced by force Y Y. Force Z Z is balanced by force A A

Force B, equal to area of small piston pushes ram down

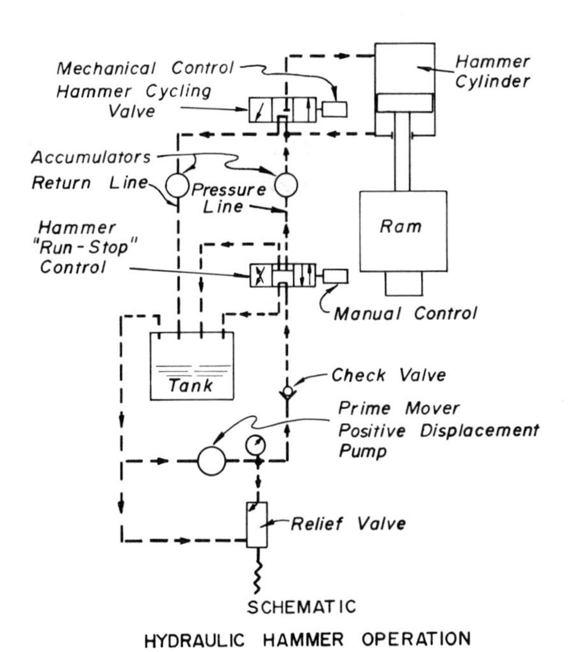

SCHEMATIC

HYDRAULIC HAMMER OPERATION

Hydraulic Hammer

In 1963, Raymond International Inc., placed on the market the first pile driving hammer employing hydraulic power at 5000 psi.

This hammer was produced by converting a differential type hammer, as shown above, into a hydraulic hammer operating as shown to the left.

Two models were initally available:

Model 80-CH-24,000 Ft-lb energy

Model 65-CH-19,500 Ft-lb energy

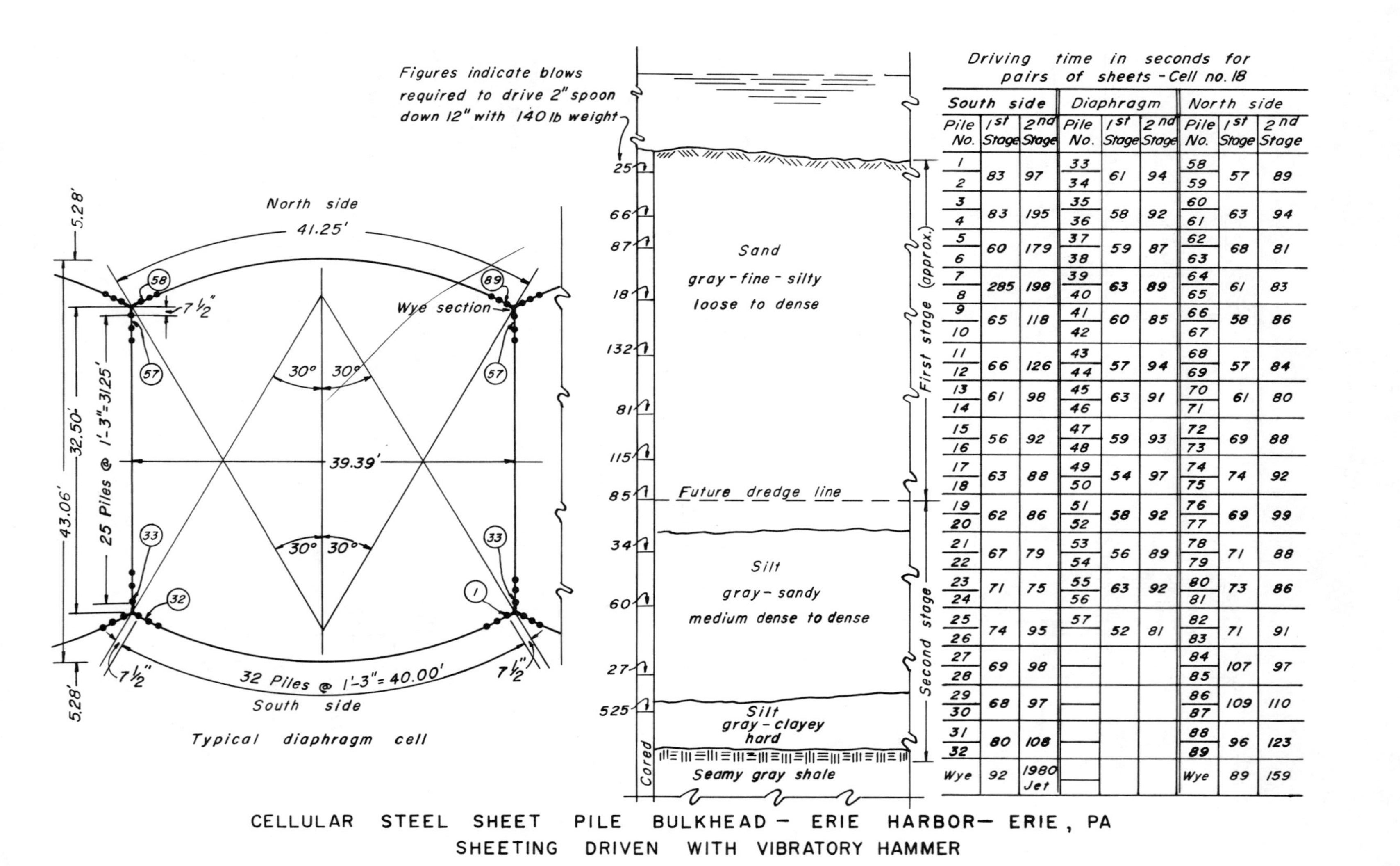

Driving time in seconds for pairs of sheets - Cell no. 18

South side			Diaphragm			North side		
Pile No.	1st Stage	2nd Stage	Pile No.	1st Stage	2nd Stage	Pile No.	1st Stage	2nd Stage
1, 2	83	97	33, 34	61	94	58, 59	57	89
3, 4	83	195	35, 36	58	92	60, 61	63	94
5, 6	60	179	37, 38	59	87	62, 63	68	81
7, 8	285	198	39, 40	63	89	64, 65	61	83
9, 10	65	118	41, 42	60	85	66, 67	58	86
11, 12	66	126	43, 44	57	94	68, 69	57	84
13, 14	61	98	45, 46	63	91	70, 71	61	80
15, 16	56	92	47, 48	59	93	72, 73	69	88
17, 18	63	88	49, 50	54	97	74, 75	74	92
19, 20	62	86	51, 52	58	92	76, 77	69	99
21, 22	67	79	53, 54	56	89	78, 79	71	88
23, 24	71	75	55, 56	63	92	80, 81	73	86
25, 26	74	95	57	52	81	82, 83	71	91
27, 28	69	98				84, 85	107	97
29, 30	68	97				86, 87	109	110
31, 32	80	108				88, 89	96	123
Wye	92	1980 Jet				Wye	89	159

CELLULAR STEEL SHEET PILE BULKHEAD – ERIE HARBOR – ERIE, PA
SHEETING DRIVEN WITH VIBRATORY HAMMER

5.19 A VIBRATORY HAMMER

In Sec. 5.18 we discussed the sonic hammer and noted in this connection that vibratory hammers had been in use in Russia, Germany, and France. One such vibratory hammer, designed in France, was introduced into the United States in 1963 and used, in that year, for driving steel sheet piling on a bulkhead project at Erie, Pennsylvania, and for driving steel pipe piles for the foundations of the Saturn V lunar rocket building at Cape Kennedy.

The vibratory hammer used in both instances was manufactured by the Procédés Techniques de Construction of Paris, France, and was distributed in the United States by the L. B. Foster Company under the trade name Vibro-Driver Extractor. The hammers are available in three sizes ranging from 34 to 100 hp, each adaptable to any standard head.

The hammer used at Erie was actuated by two 35-hp 440-volt 60-cycle electric motors providing a variable operating frequency of between 840 and 1100 cycles per min with amplitudes of from ⅜ to ½ in. The unit weighed 4.5 tons including the driving head.

The hammer is secured to the pile by means of a hydraulically controlled clamp that grips the top of the pile for both driving and extracting. Current is generally supplied from a 200-kw generator through a central ground control cabinet which also provides control of the clamp. A hand-held remote-control panel permits the operator to observe pile driving from any ground position. Extraction requires only a pushbutton touch plus the application of tension to the hammer holding line.

The project at Erie consisted of a cellular sheet pile cofferdam wharf constructed along the south shore of the inner entrance channel to Presque Isle Bay, formed by the enveloping arm of Presque Isle thrusting into Lake Erie. The cofferdam cells, of the diaphragm type, were 43 ft in length and 39.39 ft wide, each end of the 1600-ft length terminating in a circular cell. MP-101 steel sheet piles about 52 ft long were used throughout, driven through 40 ft of silt to rock, a shale of varying hardness.

Driving of the 40 cells was begun in June of 1963, starting with cell No. 40. Piling was driven around a spudded template in the first stage, the template was removed, and driving was completed in a second stage, using a Super-Vulcan 50C steam hammer for each stage. Driving was difficult, the first eight cells requiring a month of double-shift driving and considerable auxiliary jetting. On cell No. 32 the vibratory hammer was introduced.

Pile lengths had been largely pre-determined from core borings, and limits for the vibratory hammer called for the total required penetration plus zero penetration for 10 sec. It was stipulated that lacking total penetration the Super-Vulcan 50C hammer should be used, but this was not necessary.

The illustration shows the test-boring log with the times required to drive a number of sheets, in pairs, in that same area on the Erie job. Very little can be deduced from this evidence. It appears that the time required for driving was largely independent of the depth of pile, but that any sharp change in soil characteristics could change the driving time.

At Cape Kennedy, 16-in. o. d. pipe piles were driven with a cast steel shoe. Here the formation consisted of sand to a depth of 118 ft where a 3-ft-thick stratum of limestone was encountered. Vibratory driving through the 118 ft of sand required an average driving time of 17 min or about 8 sec per ft. Here, where two stages were used, with a splicing interval between, the second half of the pile required roughly twice the time of the first stage. At Erie the driving time ran between 2 and 3 sec per ft, with little difference between the time required for the two stages.

At Cape Kennedy, diesel or steam hammers replaced the vibratories at the 118-ft depth to drive the piles through the limestone shelf and another 45 to 55 ft to rock. At Erie, seating of the steel sheet piling was also on rock, so that end-bearing piles were involved in both cases. There is no record of the use of vibratory hammers in driving friction piling.

5.20 THE TROUBLE WITH PILE HAMMERS

For centuries man drove piles with a drop hammer. Even before he had learned to translate the weight of the hammer, times the length of the drop, into foot-pounds of energy, he understood that dropping a heavier weight a greater distance would root a pile more firmly into the ground.

A greater number of factors entered the picture with first the double-acting and then the differential-acting steam hammer, but there was still the visible ram moving below the cylinder, plus the known action of steam pressure under controlled conditions.

With the advent of the diesel hammer a new element was introduced. The height to which the ram was kicked upward depended upon the resistance of the pile. If the ram was pushed up to its full height then the length of fall was determinable, but in soft driving this did not occur. In addition, the action of the explosion was variable. A fixed amount of fuel, under specific impact conditions, would produce a uniformly explosive vapor; but if the ram's fall varied, might not this also vary the energy of the explosion?

With the Delmag and McKiernan-Terry diesel hammers there was some clue as to what was happening, since the ram was visible, thrusting out of the upper end of the cylinder; but with the Link-Belt hammer, terminating in an enclosed bounce chamber, there was no visible clue. To correct this, Link-Belt introduced a gage which recorded on a dial the stroke of the ram and the pressure within the bounce chamber. Connected to the bounce chamber through a hose of appropriate length, the gage tells us what has happened but does not control the energy being applied.

When we come to something like a sonic hammer, which does not drive piling down by impact energy, none of the old criteria remain for determining the load-bearing capacity of the pile. In Sec. 5.10 it was noted that although some pile-driving formulas give more predictable results than others, the surest criterion is that of driving and loading test piles. May it become desirable to test-load *all* piles driven at a site?

The time and cost involved in test-loading every pile driven is obvious from the discussion of the procedure on previous pages. Most of the time involved, and resultant cost, by any method of testing, is in the set-up necessary to do the testing, whether straight loading is involved or reaction loading from previously driven piles.

It has been pointed out that the bearing value of friction piles tends to increase with age, at least in certain soils. It would therefore appear that pile tests run in the conventional manner, with load increments applied over a period of some hours and with resultant settlements recorded, are more accurate than what might be termed "instantaneous" tests. However, is the cost of such tests any less than the cost of a few extra feet of piling?

Experiments with load tests applied using the maximum desired loading (by hydraulic jack) at the start, in which the *rate* of penetration of the driven pile was noted rather than the *amount* of penetration (or settlement), have been under investigation at the Building Research Station in Watford, England. By this method a test can be conducted in from 10 min to 1 hr. Since this is essentially a static load test similar to the loading to which the pile will be subjected in service, it would appear to provide a satisfactory test. It still does not overcome the costs incident to setting up the test.

The California Division of Highways has approached the matter, in at least one case, from a different point of view. Faced with providing a 200-ton test on a 24-in. pile they found that they lacked the necessary equipment for such loading. They therefore applied a 150-ton test to a 16-in.-diameter pile and equated the one to the other. (See opposite for procedure.)

The one problem that still remains incommensurable is how much driving to perform on a pile before testing and what happens in re-driving a pile which does not meet the load-bearing capacity required. Is a pile pushed into the ground at a uniform rate able to sustain the same loading as a pile banged down with a hammer?

Load Testing By Constant Rate of Penetration (CRP)

Test applied to 12-in. square reinforced concrete piles driven 39.5 ft into soft compressible clay. The rate of penetration varied from 0.009 to 0.015 in. per min and load envelopes were obtained, independent of the rate of penetration. Maximum resistance was obtained on the first thrust.

Six tests each lasting from 30 to 45 min produced essentially the same result as a static load test lasting for 40.5 hr. Initial maximum loading in these cases was 16 tons per pile which fell to a final figure of 14 tons.

Tests in clays indicate that satisfactory results can be obtained in 10 min, whereas tests in sand or granular soils require CRP testing for a period of an hour or longer.

Values, when plotted, show a "break" in the load settlement curve indicating failure of the pile-soil system, but to produce this, much higher leadings may be required than with the static test system.

CRP tests, so far conducted, have been to failure and no co-relation between failure loading and safe loading has been established. As a result, the CRP method can only be used for test loading and not as a means of establishing the quality of a pile which it is intended to use.

Load Testing By Comparative Values

The problem was to obtain a 200-ton loading on a 24-in. concrete pile cast in a drilled hole, by comparing it with a test loading on a similar 16-in. pile.

If B = ultimate loading determined by load test, E = end-bearing, and P = skin friction, then

$$B = E + P$$

If the perimeter area of

16-in. pile = 4.2 sq ft per foot
24-in. pile = 6.29 sq ft per foot

and the end area of

16-in. pile = 1.4 sq ft
24-in. pile = 3.14 sq ft

then the value of B for the 24-in. pile would be

$$B_{24} = \frac{3.14}{1.4} E_{16} + \frac{6.29}{4.2} P_{16}$$

and

$$B_{24} = 2.24\, E_{16} + 1.5\, P_{16}$$

Since a B_{24} of 200 tons is desired, the equation can be reduced to

$$133 = 1.5\, E_{16} + P_{16}$$

This is the equation of a straight line and can be plotted graphically. The 16-in. pile is load tested to a safe bearing of 150 tons. Then $B_{24} = 1.5 \times 150 = 225$ tons, assuming that only skin friction is considered.

The worst condition would be a failure of the 16-in. pile at 133 tons which would represent the 200-ton minimum loading assumed.

Obviously this method of comparison can only be applied where soil conditions are uniform and piles are of constant diameter.

Most hammers (drop hammers excepted) terminate at the bottom in an anvil block which is struck by the falling ram. In order to provide continuous alignment, the anvil block is built into and secured to the hammer so that, since the hammer is fixed in the leads, the anvil is similarly fixed. Even where leads are not used, controlling the position of the hammer controls the position of the anvil, although the anvil is frequently permitted limited movement in line with the hammer.

Under the anvil, and above the pile-driving cap, is a block, sometimes termed a pillow block but more generally indicated as a cushion block. Although in some cases the cushion block is omitted, where used it may be a 2-in.-thick oak disk whose function it is to take up the irregularities that exist between the bottom of the anvil and the top of the pile-driving cap. Considered an expendable item, one pillow, or cushion, will be used for each pile driven; and in hard driving two or three may be required for a long pile. Since the cushion block is often removed smoking hot, it is quickly apparent where some of the energy of the hammer was dissipated.

The pile-driving cap (as distinct from the pile cap constructed over the pile on completion of driving) is generally a cast or forged steel casting designed to fit snugly over, and around, the top or butt end of the pile. Because of the variety of, and the special requirements for, each kind of pile, special pile-driving caps are used for each type and, in many cases, each size of pile.

Pile-driving caps for wood pile are generally shaped as a helmet to fit down over the head of the pile, and the butt of the pile is then shaped by adz or ax to fit the shape of the helmet before it is stood in position for driving. The purpose is not only to confine the head of the pile but to spread the impact over a greater area.

The bottom of a pile-driving cap for driving H-piles or pipe piles is often grooved to match the shape of the pile being driven or is fitted with cast-in lugs which approximate the shape of the pile. The fit, in these cases, is not designed to be tight since any slight mushrooming at the top could wedge the pile irremovably in the cap.

With concrete piles it is desirable to have a cap fitting around the outer perimeter of the pile head closely enough to contain the head of the pile. A steel plate is set over the top of the pile, inside the cap, to distribute the impact uniformly over the contact area. In some cases a cushion block is added, squeezed between two steel plates.

Pile-driving caps are heavy and, to facilitate their handling, are usually hung from the underside of the hammer by rope or cable slings carried up to hooks on the sides of the hammer or slung over its top. They must be hung loosely enough to provide access for replacing cushion blocks, if used, but securely enough to prevent their falling away when the hammer is raised at the completion of driving.

Pile shoes were originally developed for wood piles, the purpose being less to provide a point for ease of driving than to prevent mushrooming of the tip when hitting obstructions or hard strata. Made of cast iron or steel plate, they take a variety of shapes, of which the "Arrow Head" is perhaps the most common.

All tubes driven for cast-in-place concrete piles are provided with a tapered cast or forged steel shoe of sufficient strength to take mandrel impact if one is used. With Monotube piles, a shouldered shoe is used, welded in place.

Pipe piles (closed) are also driven with a cast or forged steel shoe with a shoulder on which the pipe rests and which is welded to the pipe. These may be pointed, in the case of friction piles, or slightly concave where end bearing is required. The use of flat steel plates welded to the ends of pipe should be avoided. The amount of weld that can be used is often insufficient to prevent their being torn loose.

Concrete piles are frequently tipped when solid, and shoes can be used with hollow pre-stressed piles even when the bottom is plugged with concrete.

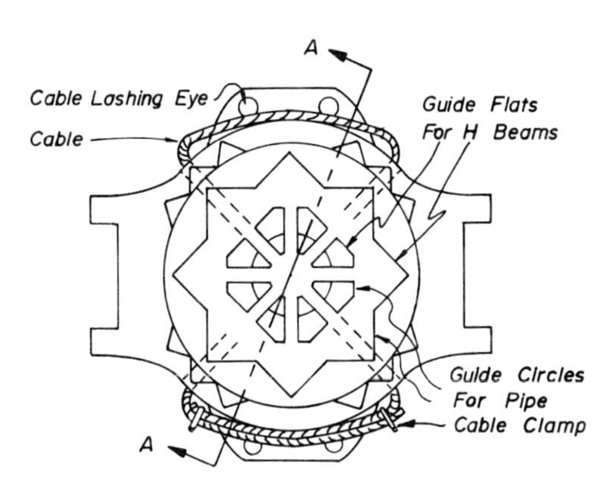

UNIVERSAL DRIVE CAP

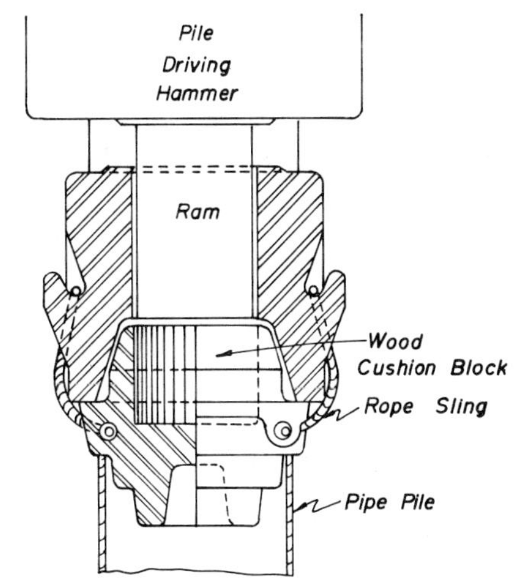

PIPE PILE DRIVING CAP

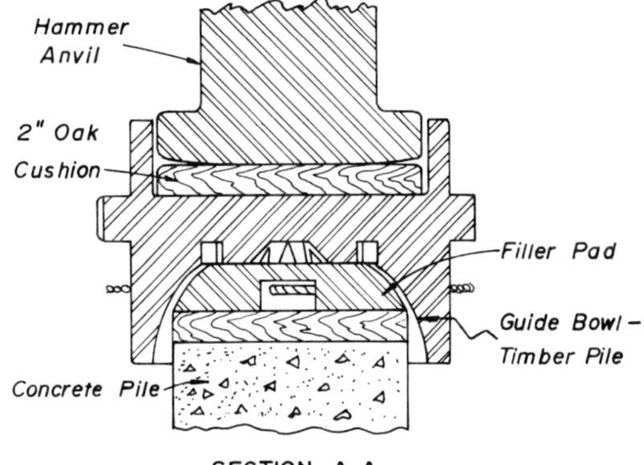

SECTION A-A

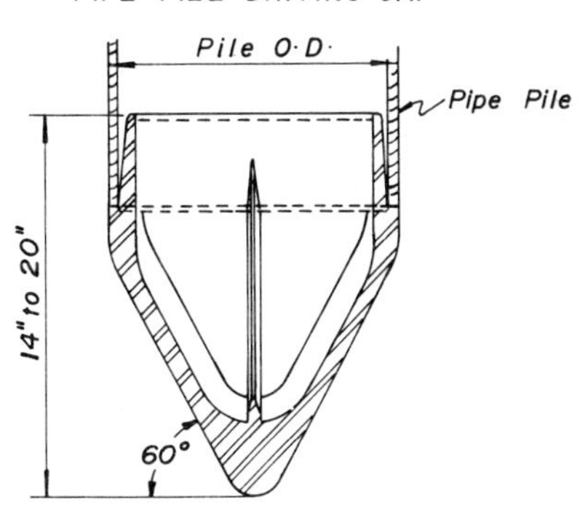

CAST STEEL DRIVING POINT

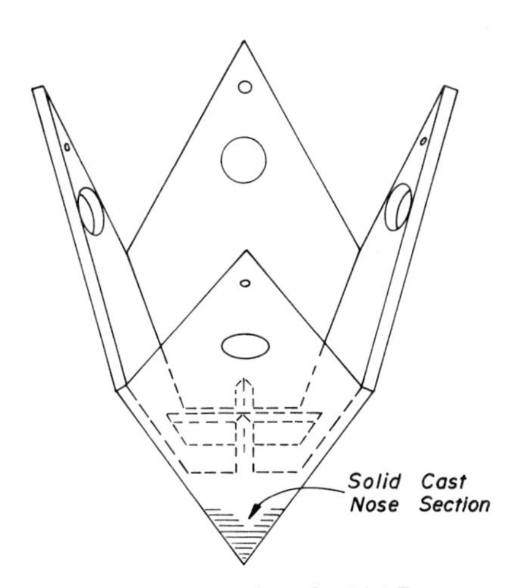

WOOD PILE DRIVING POINT

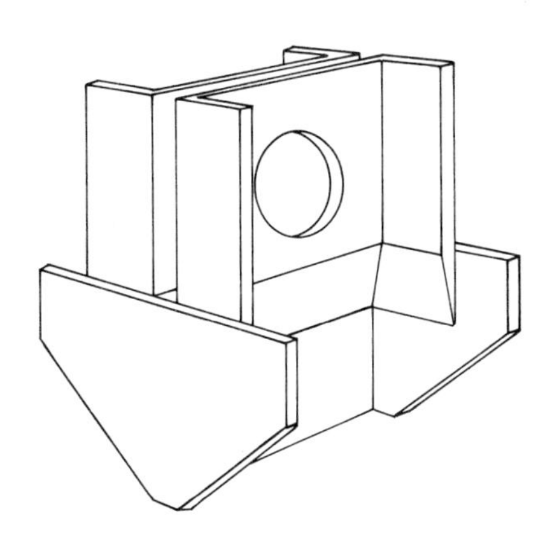

STEEL H BEAM PILE DRIVING POINT

5.22 SPLICING PILES

The field splicing of piling is generally undesirable, but, the pile-driving operation being what it is, it occasionally becomes necessary to increase the length of a pile after it has been partially driven. In some cases the driving operation itself requires a temporary extension.

Whenever the pile-driving hammer reaches the bottom of the leads, it cannot be used farther unless some extension piece is added; the "follower" is such a piece. Several followers are manufactured, for use where piling is to be driven to a top level below water, but followers can be improvised at the site by using a steel band fastened to a separate section of piling that will fit down over the head of the pile already driven, like a pile-driving cap. This method of approach, suitable where the additional depth required is minimal, avoids the necessity of changing the pile-driving cap. Such devices are generally provided with an additional cushion block between the two banded sections of piling and with outside hooks and slings for lifting the follower.

In driving through solid ground surfaces it is advisable to excavate down to the probable pull-up level of the top of the pile, but under water or on soft surface soils the additional drag of the follower is not material.

In discussing pile tests it has been previously pointed out that the first tests are those required to determine the probable length of piling under actual driving conditions. At many sites subsurface conditions will be sufficiently uniform so that actual lengths of piling driven will vary only a few feet. Under these conditions the cheapest solution is that of providing sufficient additional length of piling to provide for these relatively minor variations. In some instances, as with wood or concrete piles, the resulting cut-offs may be entirely wasted, but substantial cut-offs of H-piling and pipe piling may be re-usable.

The splicing of wood piles is almost completely avoided where additional driving, after splicing, is required. If the pile diameter is not to be increased, a portion of the two sections must be cut away to effect the splice, weakening the section; and where splicing techniques increase the diameter, the effect of friction on the portion above the splice may be lost.

Steel pipe piles offer the best and easiest method of splicing, providing a connection that may actually be stronger than the pipe itself. One end of the cast-steel pipe sleeve used (referred to in Sec. 4.11) fits down into the bottom section of pipe, resting on a midpoint shoulder. The upper section of pipe drops down over the tapered upper portion of the sleeve to also rest on the sleeve shoulder. A welding bead is then run around both points of sleeve and pipe contact. In addition to use where headroom is limited, this facility of splicing is of value where leads of limited length are used.

The process of splicing steel H-piles is somewhat more time-consuming but is, on the whole, just as satisfactory in the result as the splicing of pipe piles. Positioning plates are welded to the interior flanges and web of the bottom section; the upper section is then dropped into the grooves thus formed, is plumbed, and the plates are fillet-welded to the upper section. A full butt weld is then run around the complete connection (see details).

The splicing of pre-stressed concrete piling is, as has already been noted, a chancy operation, to be avoided if possible. Solid concrete piles, either pre-cast or pre-stressed, have been spliced by doweling. The lower section is driven as a solid mass. Dowel holes are now drilled in the top of the driven pile, within the area of the spiral cage—to a template. Dowel bars are dropped into the holes, and the upper section of pile, with cast-in holes, is dropped down on the dowels. A plasticized cement or an epoxy compound, in liquid form, fills the oversized dowel holes and is squeezed out to form a tight section as the top portion is lowered. Splices also can be made in concrete piles by casting steel plates into the concrete ends of the pile sections, which are then welded together to form a joint. After welding, the interval between concrete faces is sealed with an expandable grout.

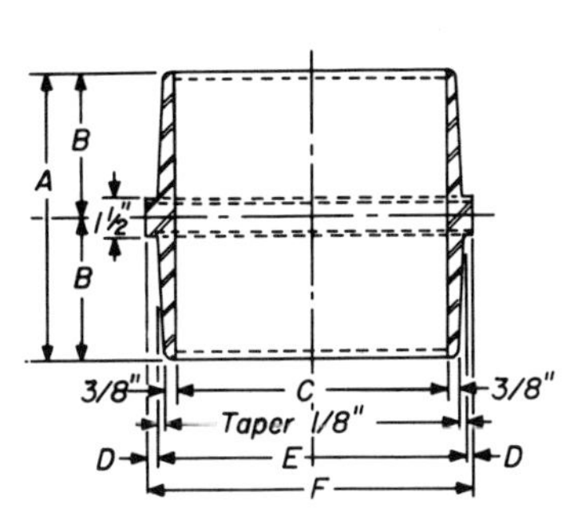

① Cast Steel Pipe Sleeve

Pipe pile		Dimensions (inches)						Est. wt. lbs.
Pipe O. D.	Thickness	A	B	C	D	E	F	
10 3/4	.365	11	5 1/2	9 1/8	3/8	10 1/8	10 7/8	47
	1/2	11	5 1/2	8 7/8	1/2	9 7/8	10 7/8	47
12 3/4	3/8	11	5 1/2	11 1/8	3/8	12 1/8	12 7/8	56
	1/2	11	5 1/2	10 7/8	1/2	11 7/8	12 7/8	57
14	3/8	12 1/2	6 1/4	12 3/8	3/8	13 3/8	14 1/8	70
	1/2	12 1/2	6 1/4	12 1/8	1/2	13 1/8	14 1/8	71
16	1/2	15	7 1/2	14 1/8	1/2	15 1/8	16 1/8	96
18	1/2	18	9	16 1/8	1/2	17 1/8	18 1/8	129

H-Pile		Dimensions (inches)					Est. wt. lbs.
Nominal size	Weight #/Ft.	A	B	C	T	L	
8"	36	1/2	1 3/4	7 1/16	5/16	10	20
10"	42	15/32	2	8 13/16	5/16	12	27
10"	49	3/8	2	8 13/16	3/8	12	28
10"	57	5/8	2	8 13/16	3/8	12	28
12"	53	1/2	2 1/2	10 27/32	5/16	14	41
12"	65	7/16	2 1/2	10 27/32	3/8	14	42
12"	74	21/32	2 1/2	10 27/32	3/8	14	42
14"	73	9/16	2 3/4	12 9/16	3/8	16	65
14"	89	21/32	2 3/4	12 9/16	3/8	16	65
14"	102	3/4	2 3/4	12 9/16	3/8	16	65
14"	117	27/32	2 3/4	12 9/16	3/8	16	65

② Steel H-Beam Bearing Pile Splicer

(A patented product sold by Associated Pipe and Fitting Co.)

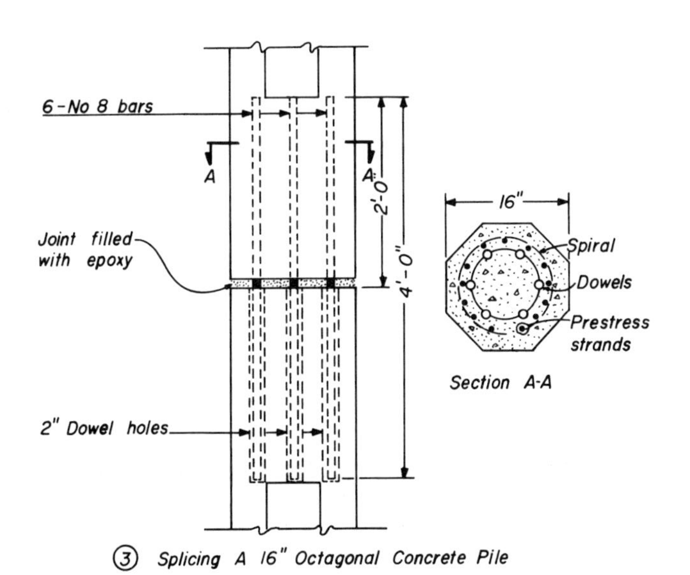

③ Splicing A 16" Octagonal Concrete Pile

Notes:

① Welding of pipe pile sleeves is not necessary unless extraction may be required.

② Notch upper beam web 7/8"X2" for key. Kerf outside edges of top and bottom beam flanges. Top section dropped into place with splicer welded to it.

③ Make template of top dowels as guide for drilling bottom section of pile.

5.23 BATTER PILES

The general function of batter piles is to resist lateral forces beyond the capacity of vertically driven piles to absorb. Long narrow structures (piers, bridges), set on piling which extends considerably above embedment level, are subject to sway which can gradually loosen the soil grip on a vertical pile at ground level. This increases the bending stresses throughout the length of pile. Structures such as retaining walls and dams, which have an overturning tendency, may be held down by the resistance of vertical piles to uplift; but a more satisfactory method is to provide the horizontal component of a batter pile to resist this overturning. The horizontal increment of motion of vibrating machinery can be successfully dampened by batter piles driven in pairs in opposite directions.

A batter pile is one driven at an angle to the vertical, the degree of angle being specified as the ratio between the horizontal and the vertical leg lengths of a right-angle triangle, the hypotenuse of which lies along the center line of the pile (4 in. on 12 in., for instance). The batter of a pile is rarely specified in angular degrees and minutes.

A method of batter-pile analysis proposed by Prof. A. Hrennikoff in 1949 (*Transactions of the American Society of Civil Engineers,* Volume 115, 1950) was applied to transmission tower foundations in 1954 and is shown in detail opposite.

Generally speaking, piling batters are limited to deflections from the vertical not much in excess of the 4 in. on 12 in. shown; but in one case, at least, piling was used, driven closer to the horizontal than the vertical. In this case 8-in. I-beams were driven on angles ranging from 43 to 62 deg from the vertical under existing foundations to serve as tie-backs for a retaining wall (see illustration) along a section of the West Berlin Expressway in West Germany. This case required special adaptations of the equipment generally used for driving batter piles.

Leads are usually employed for driving batter piles and have been fixed in place on the supporting rig at the angle required. Swinging leads can be swung forward and fixed by horizontal struts in the proper position, providing what is known as an "in and out" setting. They can also be tilted from side to side of the vertical line of the supporting boom; and, in some cases, where batter is specified in two directions, both in-and-out and side-setting combinations may be required.

In 1962 McKiernan-Terry Corporation introduced a set of leads to hang from a boom point by means of a swivel frame, permitting side swing. The strut or bottom brace consisted of a parallelogram frame with hinged ends which could be angled by means of diagonally set hydraulic cylinders to follow the side set of the leads. Furthermore, the strut was extendable along its axis by additional hydraulic cylinders so that in-and-out settings could simultaneously be obtained.

When hammers are used, in tilted leads, to drive batter piles, the formulas given in Sec. 5.10 must be adapted to compensate for a reduction in the vertical drop of the ram and for the increase in friction in the guides. For drop, single-acting, or diesel hammers, h^1 would replace h, where

$$h^1 = h(\cos\theta - 0.1 \sin\theta)$$

Here θ is the angle of batter and 0.1 is the coefficient of friction. For double-acting or differential-acting hammers where the available steam pressure does not change, $E_n{}^1$ would replace E_n, where

$$E_n{}^1 = E_n - W_r h \frac{1 - \cos\theta}{12}$$

In Sec. 5.15 there is a discussion of driving vertical piles without leads, and the alternative methods mentioned there are perfectly adaptable for driving batter piles without leads and have, in fact, frequently been used for that purpose. Perhaps the final criterion of method is the deviation from theoretical position permitted by the specifications at the point of cut-off. It will seldom be practical to pull a batter pile and re-drive without the danger of its following the previous path and being deflected a second time.

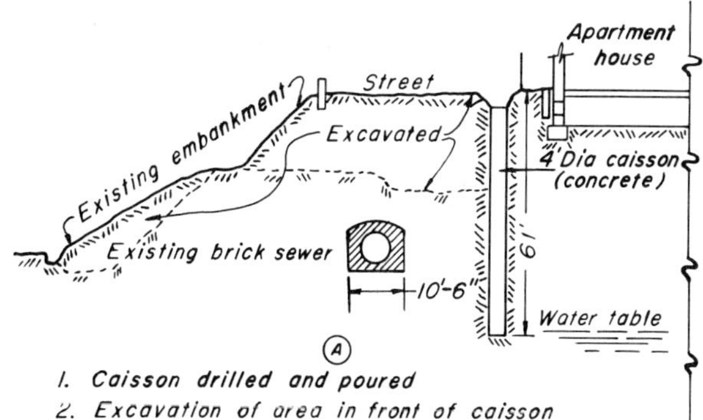

1. Caisson drilled and poured
2. Excavation of area in front of caisson

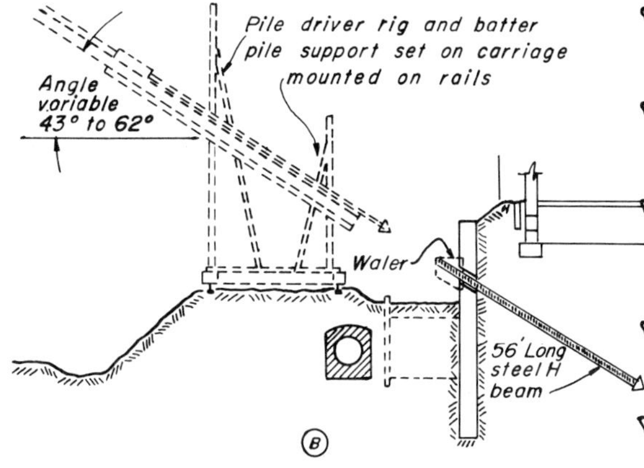

3. Batter piles driven through openings in caissons.
4. 3'X 3' Concrete waler poured over pile heads.
5. Sheeted trench excavated in front of caisson line for new sewer.

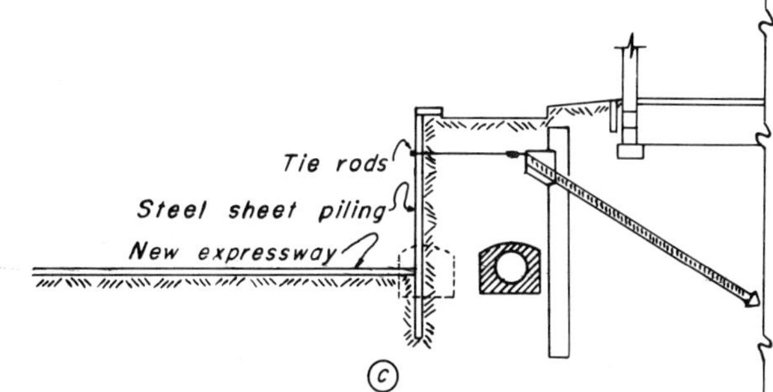

6. New sewer built. Old sewer demolished.
7. Steel sheet piling driven and tie rods installed.
8. Excavation for expressway made.
9. Backfill placed behind steel piling.

BATTER PILES FOR ANCHORAGE WEST BERLIN EXPRESSWAY

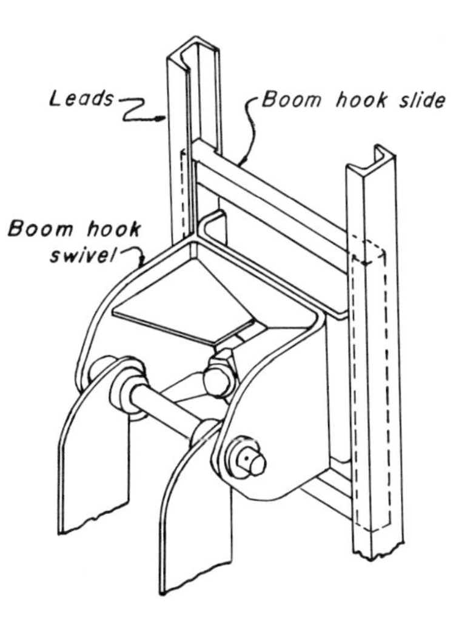

BOOM POINT ADAPTER

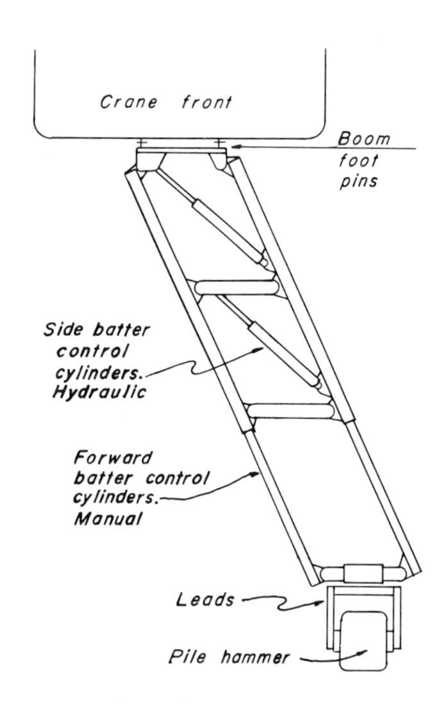

BOTTOM BRACE FOR PILE LEADS

PARALLELOGRAM BOTTOM BRACE AND BOOM POINT ADAPTER

(McKiernan – Terry Corporation)

5.24 PILE CLUSTERS

To this point, the driving of piling has been considered only as though each pile were remote from its fellows. It has been noted that, at least in certain soils, the actual frictional support of a pile acts in the soil mass at a distance from the pile and is gaged by the unconfined compressive strength of the soil rather than by the friction between soil and pile surface. This leads to a consideration of what happens when we drive a second pile within the zone of soil influence of the first pile, and what happens to the soil displaced by each pile. What is the general effect of concentrating a number of piles within a limited area?

It has been long recognized that when a number of piles are driven at close spacings the load-bearing value of each pile is correspondingly reduced. The amount of reduction in the bearing value of individual piles, thus clustered, has been the subject of many theories which agree only in observing that there *is* a substantial reduction. Many building codes have established a minimum-spacing criterion of 2½ times the diameter of the pile butt, at cut-off, so that for 12-in.-diameter piles their centers would be spaced not closer than 2 ft 6 in. More recent studies indicate that this is too close and that from 3 ft 6 in. to 5 ft, as minimum spacings, are more desirable.

Ignoring the design aspects, the pile-driving contractor is very much concerned about the practical considerations involved in putting piles down in groups. A pile driven close to an adjoining pile may cause the earlier pile to heave; and where this occurs, re-driving may be necessary. (It should be noted that only friction piles are referred to here. Heaving is less likely with end-bearing piles.) When piling is driven in groups, at close spacings, the earliest piles driven should be marked with a fixed point that can be checked, after the driving of subsequent piles, by means of a remotely positioned level. The proper sequence of driving will help to avoid heaving. Start driving at the interior or center of a pile group, driving successive piles toward the outside of the group.

It will generally be found that, as piles are driven in a group, later piles drive harder and pull up sooner than the first pile driven. This is due to increased compaction of the soil, not only near the tip but throughout the length of the pile. This compaction results from the displacement of the soil mass by the pile mass. In some cases the pile is not heaved upward but the soil in the surrounding area is.

Although wood piles are particularly prone to heaving, the phenomenon can and does occur with piles of any type; and with thin shells or tubes, used for cast-in-place concrete piles, there is the additional hazard of damage to the shells due to the noted horizontal stresses induced in the soil. The thin walls of tubular piles can be crushed in either to the extent of creating a "neck" in the pile or being otherwise distorted in shape, often to the point of collapse. This can occur as the mandrel support is withdrawn or, more frequently, may occur at a later time as subsequent adjacent piles are driven.

Inspection of the shell, as discussed in Sec. 5.26, may indicate that the pile shell has become distorted, but damage can as readily occur after the shell has been filled with concrete and prior to the concrete's having obtained its initial set. In these cases, necking of the concrete can occur, or sufficient separation of the concrete to leave a void, sadly lowering its value in compression. Unfortunately, this type of damage cannot be detected unless a load test is put on the completed pile. The same thing can occur with cast-in-place concrete piles where the shell or pipe is withdrawn as the concrete is placed.

To combat distortion of shell or concrete it is standard practice to drive all shells, in a small group, before any is filled with concrete and to make an inspection before actual filling begins. With a larger group, driving all shells within some radius such as 5 ft offers some protection. Extraction of damaged pile shells, for re-driving, is generally the only recourse where substantial crushing has occurred.

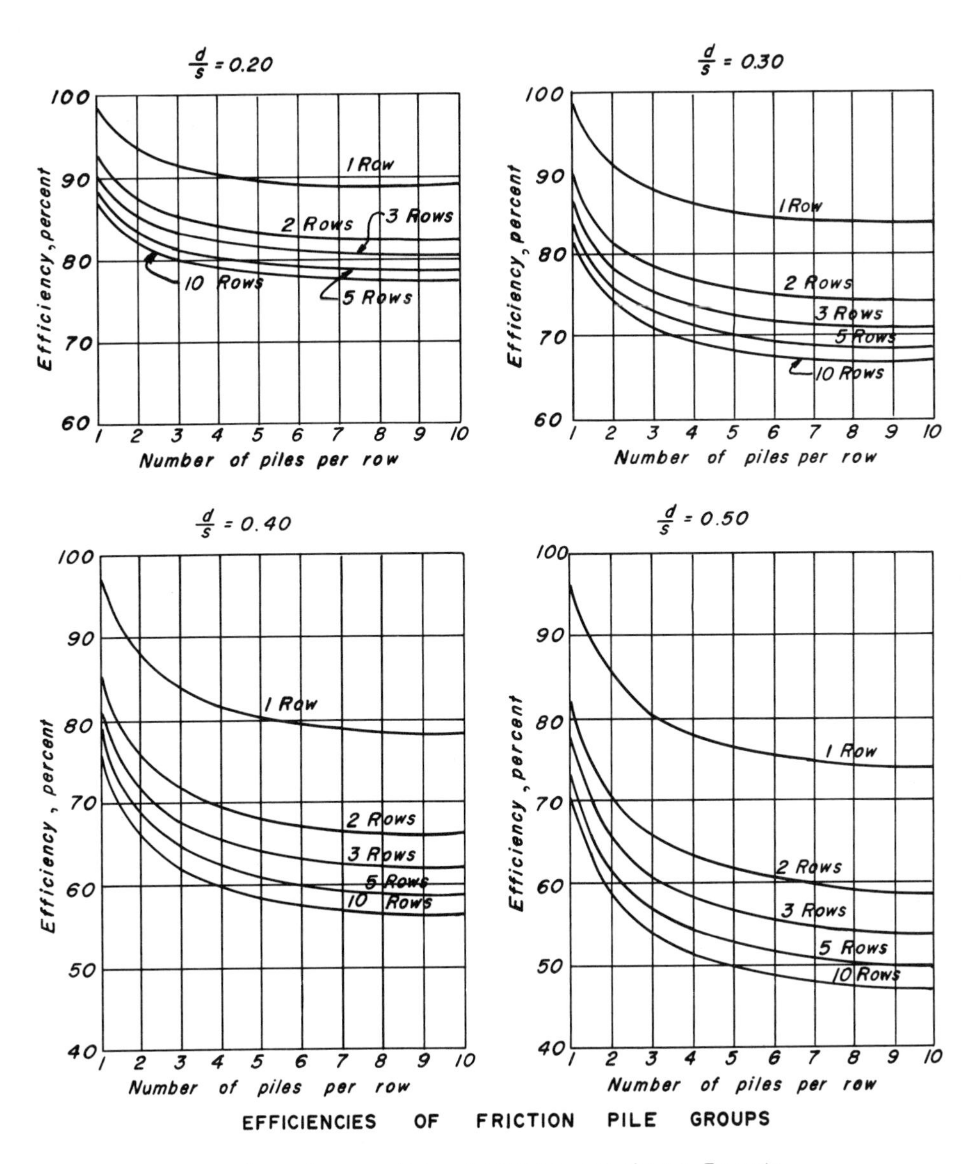

EFFICIENCIES OF FRICTION PILE GROUPS

Based on the Converse–Labarre Formula

$$E = 1 - \phi \frac{(N-1)M + (M-1)N}{90MN}$$

E = efficiency of a pile in a group
N = number of piles in each row
M = number of rows in the group
ϕ = tangent d/s in degrees
d = pile diameter
s = pile spacing

Efficiency of single isolated pile is taken at 100 %.
These values are chiefly applicable to piles in an elastic soil such as clay.
The efficiencies shown should not be accepted too literally.

5.25 PILE JETTING

Discussions of pile driving, heretofore, have dealt chiefly with clays as a medium through which piling is driven; but many thousands of feet of piling are driven along ocean fronts where sand is the predominating soil material, and this offers special difficulties in driving piling.

In Sec. 3.7 it was noted that the compaction of sands involves two steps: sinking the Vibroflot to depth, using a bottom jet plus vibration, and then using a top jet, with vibration, to compact the sand as the Vibroflot is withdrawn. Where piling is driven into sand, the driving itself furnishes vibration; and since in many cases the sand is saturated by water, either lying over it or percolating through it, the harder the piling is driven the more resistant becomes the sand stratum. The hard driving needed to sink a pile under these conditions can severely damage wood piling and, since concrete piling offers a larger area, can sometimes make the sinking of concrete piles impossible without jetting.

The jetting of piles into place is not restricted to locations where sand exists but can also be used to punch holes in other formations. It is not recommended in cohesive soils where the pile is to act as a friction pile, since the soil around the pile, on which friction depends, is changed to a consistency of soft mud and the time required to re-establish its cohesiveness is indefinite. Where, in mixed soil formations, there is a tendency to wash fines from the soil, this may reduce the bearing capacity of the pile, requiring a longer pile to provide the support needed.

Jetting is frequently done with a 1½-in. pipe with an open end; but larger sizes have been used, frequently with a nozzle on the end, reducing the opening to from ¾ to 1½ in. The jet pipe is not fastened to the pile but is independently suspended alongside it. Where an adequate volume of water is used, at sufficient pressure, the jet pipe can generally be punched down by hand and is never driven, since driving could plug the nozzle.

Jet pipes will frequently be used to prepunch the soil at the center of the pile before the pile is placed, loosening the soil sufficiently to start the pile. As the pile is driven, the jet pipe is punched down alongside it, its nozzle generally kept above tip level. Since there is a tendency for piling to drift toward the jet, it is better practice to use two jets, one on either side of the pile. One of these follows the tip down, while the second is kept 5 or 6 ft above the tip to aid in the continuous upward motion, or "boiling," of the sand particles.

In sand beds of indefinite thickness it is possible to place piling, by jetting, to great depths if sufficient water and pressure are used. To limit pile length, the theoretical depth required to provide the necessary friction between sand and pile is estimated and the pile is jetted to a point higher than this. From this point driving is continued without jetting. When jetting is stopped, the sand settles back around the pile, giving it support throughout its length. This reduces damage to the tip to a minimum.

For substantial depths, jet pipes may need to be sectionalized. In adding sections of jet pipe, however, not only will jetting be stopped but driving of the pile should also stop. It is generally preferable to provide jets in a single length.

Pressures used will vary from 100 to 300 psi, and these pressures are provided by high-pressure pumps since no public water supplies operate within this range of pressure. Because of the quantity of water required, some natural source of supply is desirable, and along seashores the use of salt water is indicated. Bronze pump linings and fittings should be used where the sea serves as a source of water.

The connection between the pump and the top of the jet pipe will be hose, designed for high pressures, and a swivel fitting at the top of the jet pipe makes for ease in handling.

Although high-pressure air is useful in limited cases for jetting in sand, its use should be combined with water so that both are discharged together at the tip of the jet. Compressed air can be used to increase the action of low-pressure water supplies.

Selection of Jetting Equipment

There is no practical way to predetermine jetting requirements either as to the equipment required or as to the pressure that may be needed. Although contractors working in limited areas will often have accumulated guides based on their previous experiences in those areas, these may not remain valid if piling is carried to greater depths or if new conditions are encountered.

For example, wood piling is frequently used in seashore areas where sand predominates, especially along the oceanfront. But many of these sand spits contain clay cores where jetting is either inappropriate entirely, or requires somewhat lower pressures to prevent the creation of a mud-filled cavity.

It is suggested that the contractor start with a 2½-in. jet pipe and hose using a 1-in. nozzle, usually a short length of 1-in. pipe connected to the jet pipe by a reducing fitting. The starting pressure should be about 50 psi. Where much gravel is contained in the soil higher pressures, up to 100 psi, may be required, but these may also require an increase in pipe size to 4 in. with a 1½-in. or 2 in. nozzle, to provide needed volume.

The tables given below should be used with caution and judgment. The delivery of water through hose or pipe can vary depending on the interior condition of the conduit and such factors as whether the hose is kinked or flattened. However, Table A will give some indication of the probable quantity of water which must be available and which must be disposed of unless the soil is very porous.

Table B, giving the theoretical discharge of nozzles, should also be used with caution. The pressure available from a nozzle will vary with the flow conditions at the transition from pipe to nozzle, the length of nozzle and the condition of the interior of the nozzle.

The values in both tables should be used only on a comparative, not an absolute, basis.

Table A. Pounds Pressure Required to Deliver Indicated Gpm

Gpm	100 ft of hose with inside diameter of						
	1¼ in.	1½ in.	2 in.	2½ in.	3 in.	4 in.	5 in.
100	88	37	12.5	3.5	1.7	0.4	0.1
150	183	78	27	7.5	3.5	0.7	0.3
200		133	46	13	5.9	1.4	0.5
250			70	19	9.1	2.1	0.7
300			95	27	12	2.9	1.0
350			126	36	17	4	1.3
400				46	21	5.1	1.7
450				57	26	6.3	2.1
500				70	32	7.4	2.6

Table B. Theoretical Discharge Nozzles, Gpm

Psi	Nozzle opening					
	1 in.	1⅛ in.	1¼ in.	1½ in.	2 in.	3 in.
50	211	267	330	475	845	1320
75	259	327	404	582	1036	1618
100	299	378	467	672	1196	1870
125	334	423	522	751	1338	2090
150	366	463	572	824	1466	2290
175	395	500	618	890	1582	2473
200	423	535	660	950	1691	2645

5.26 INSPECTION AND CONCRETING OF TUBULAR PILES

A preference for tubular, shell, or pipe piles has occasionally been expressed by engineers responsible for specifying piling, on the ground that, once driven, they can be visually inspected to determine their condition and alignment. Unfortunately, such inspection does not assure the inspector of the quality of the finished pile nor preclude the desirability of load tests. The situation that arose in 1961 in connection with the piling support for the passenger terminal at New York City's La Guardia Airport is an excellent case in point.

The design called for the support of the structure on some 1500 piles, 100 to 130 ft long, of the combined pipe and shell type described in Sec. 5.5, with a 12-in. shell carried above a bottom 10-in. steel pipe, a type of composite pile developed, and in this case driven, by Raymond Concrete Pile Company. The pile shells (and pipe) were checked for alignment before concreting, and about 7% or 110 piles were rejected because of misalignment or, in a few cases, because of visible vertical damage in the casings. The allowable misalignment in a casing pile is usually specified as 2%, or 2 ft in 100 ft.

The question of the potential support value of tilted piles was examined at some length by James D. Parsons and Stanley D. Wilson, both consulting engineers, in a paper to be found in the *Transactions of the American Society of Civil Engineers,* volume 121, 1956, under the title "Safe Loads on Dog-leg Piles." The dog-leg piles referred to were identical with those used on the airport job: a 10-in. pipe pile surmounted by a 12-in. shell. The dog-leg resulted when the pipe tilted out of line with the shell section. In a second case cited in the same paper, 10-in. pipe piles had been driven to replace wood piles, battered beyond salvage in driving through a boulder formation. Here the pipe was not kinked at a single point but was bent as much as 6 ft from the vertical throughout its length.

The article was chiefly concerned with methods of computing safe loads on such dog-leg piles and, although it recommended the acceptance of such piles, suggested that lower bearing capacities be assigned. However, load tests, noted by the authors, indicated no reduction in bearing capacity; and tests by others, of record, also indicate that no reduction in bearing capacity actually occurs.

The procedure for inspecting hollow piles for misalignment, or other damage, depends largely on their length. Short piles can be checked by reflecting a beam of sunlight down the shaft with a mirror (not recommended during an Indian monsoon); longer piles, by lowering a light on the end of a cord of suitable length. To check the degree of tilt, batter, or misalignment, a plumb bob can be dropped down from the overhanging top edge, on a cord, until its tip touches the opposite side of the shell.

A device similar to that described in Sec. 1.14 for predicting landslides and there termed a tiltmeter has been used for measuring the inclination of tilted pile shells and is more precisely termed, in this connection, an inclinometer. Since both use a pendulum traversing a resistance-wound coil and employ the Wheatstone-bridge principle, the only substantial difference between the tiltmeter and the inclinometer is that the latter does not completely fill the pile shell. The inclinometer maintains contact with the wall of the shell at one point by means of a magnetized wheel assembly. It can record tilts of 1 in. in 40 ft but does not indicate the direction of tilt, a factor generally of no importance here.

The 110 piles found to be unacceptable at La Guardia Field were not extracted but were filled with concrete and ignored. The extraction of shell or composite piles is seldom practical. Designed, as has been described, for facility of driving with stepped or tapered mandrels, they do not have the requisite strength in tension to resist extraction stresses. Wood piles and steel H-piles or pipe piles, having welded connections, can be pulled more readily, but even here extraction will leave an undesirable void underground which cannot readily be filled and which may deflect adjacent piling still to be driven. In lieu of extraction, additional piles are usually required to be driven.

It is never desirable to fill a pile shell with concrete more than a few inches above cut-off level, but in the case of the airport, several *acceptable* pile shells were filled several feet too high. In cutting one of these down to grade, a substantial void in the concrete fill was disclosed. This disturbing disclosure immediately indicated the desirability of investigating the nature of the concrete fill in all the piles.

No acceptable method for determining the condition of the concrete in a filled pile has appeared. In the airport case, sonic devices were tried but indicated poor correlation with actual conditions and were abandoned. Drilling with wagon drills was attempted, but they had a tendency to drift out of vertical and plunge the drill bit through the shell wall. Hand drilling with jack hammers was finally resorted to since the operator could sense a change in consistency of the concrete by the speed of sinking. Holes were hand-drilled to a depth of about 15 ft; and where any indication of a change in density was apparent, 4-in. core borings followed the drilling.

Two types of defects were disclosed in 32 piles out of the 1500, generally at depths starting from 3 to 6 ft below cut-off elevation, involving either voids in the mass or segregation. The voids extended for distances of as much as 30 ft vertically with solid concrete both above and below them. The segregation consisted of stems of uncemented gravel for heights as great as 10 ft. Both these conditions suggest the difficulties involved in filling pile shells.

In the design of cast-in-place concrete piles, no bearing value is assigned to the shells, the concrete being presumed to carry the whole load. Consequently, on the airport job the value of all the 1500 piles was questioned when 2% were found to be voided or segregated. As a result it was decided to drive 917 additional piles, using a 10¾-in. o.d. pipe pile for the entire depth. In the case of pipe piles, of course, only the load value of the pipe wall is considered.

In placing concrete in cast-in-place concrete piles the following precautions should be observed.

1. Check the pile for water immediately before concreting. A light lowered into the shell should reflect this.

2. If the quantity of water is not too great, it may be blown out by air. A 1-in. pipe lowered along one side, hose connected to a compressor, can accomplish this. If infiltration is too rapid, the pile may have to be abandoned.

3. Provide absolute control of the concrete mix so that its consistency is neither too thick nor too thin. The placing slump should be 5 or 6 in. at least; and to maintain the strength, the cement content may have to be increased.

4. Place the concrete slowly but continuously. Do not stop the pour within a shell until it has been brought up to a level a few inches *above* cut-off level. There will be some shrinkage as the concrete sets, and these extra inches will compensate for the settling.

5. Vibration of concrete at great depths may be impossible. The head of an internal vibrator can become inextricably buried in the concrete, and the raising and lowering of 100 ft of flexible shaft can be arduous. Moreover, the continuously falling concrete can readily damage the vibrating head, while metal edges inside the shell can tear the rubber coating of the shaft.

6. When the concrete has reached a suitable level, however, some internal vibrating should be done, pouring a limited section and then vibrating.

7. Sonic vibration of the entire shell, with a device similar to the sonic hammer described in Sec. 5.18, has not been attempted. What effect this might have on the friction value of the pile shell is unknown, but it could readily serve to consolidate the concrete. Unfortunately, it could also serve to sink the shell deeper, resulting in a longer pile and creating confusion in the results previously obtained from the dynamic pile formula.

5.27 PILE ANCHORAGE

In discussing testing in Sec. 5.13, reference was made to uplift uses of piling to resist the flotation of hollow structures, to resist overturning, and to counteract the heaving action of frost or ice; but there are others. Overturning, generally thought of in relation to retaining walls and dams, may be caused by wind acting against tall structures, by earthquake forces which can produce a rocking action, or by intermittently applied forces such as those resulting from equipment with unbalanced loading, revolving on a platform (a dockside crane, for example).

It has been noted that in clays or similar soils, theory has indicated that 25% of the load on a pile is carried on the tip of a non-end-bearing pile, from which it might be concluded that uplift values are about 75% of the load-bearing value. Actually, tests indicate that the value in uplift is much less than this unless special precautions are taken to anchor the pile, such as spreading the tip with cast-in-place concrete piling.

The ordinary pile cap of concrete is designed primarily to transmit the structural load down onto the head of the pile, but consideration must be given to the strength of bond between pile and concrete, particularly where wood piling is involved. Hooked bars bolted through the pile butt have been used, as well as railroad spikes driven into the wood, but sufficient uplift force can tear these loose by ripping along the grain. Any type of composite pile must be carefully checked for uplift since the joint represents a particularly weak point in uplift.

An interesting possibility is the palm pile, a device developed by the Belgian, Jean Rooseu, and patented by him. Its latest use was in 1961 in Belgium, as a part of pile foundations for the Galerie Ravenstein in Brussels. The site of the Galerie was surrounded by such structures as the Albertine Library, the Fine Arts Palace, the Central Railroad Station, and the Bank of the Belgian Congo, some of which rested on simple spread footings. The driving of an obese pile would disturb these foundations, and the cross section of the pile was limited to one slightly under 14 in. square.

The palm pile is a cylindrical shell driven on top of a pre-cast concrete tip. Just above the tip, tucked into the shell, are four steel pivot plates. These plates lie within the shell during driving, but when bearing stratum is reached, the cylinder is slightly retracted, the plates (or palms) are mechanically thrust out into the bearing stratum, and the shell is then filled with concrete. Tests indicate that the palms double the bearing value of the pile, but it is apparent that such a device would also mightily increase the resistance to uplift.

The use of piling for anchorage was not entirely new when this method was considered for anchorıng a tunnel under New River at Fort Lauderdale, Florida, in 1959. Designed to be anchored into rock, it was found that the underlying limestone was poor, faulty, and riddled with solution holes and that, in fact, a sand layer of variable thickness underlay much of it, and so the 4-ft tremie slab was anchored into the sand.

A 7-in. oil-well casing was hammer-driven to the required depth with a plate sealed to its bottom not by welding but by wiring it to the bottom of the casing. Three radially placed steel plates were welded to the cap plate (see detail) to leave a 1½-in.-diameter void in the center. When sufficient penetration or suitable resistance to driving was reached, the depth of the casing was measured and a No. 11 reinforcing rod was fabricated with bottom stop plate and top anchorage fins. The end of the rod was dropped into the void between the radial plates, and pressure grouting of the casing followed.

A head of grout was developed in the casing, and then extraction of the casing was begun. To break the wires holding the bottom plate, holding pressure was applied to the No. 11 rod as extraction force was applied to the casing. Grouting was carried up to the underside of the proposed tremie slab. Each anchor was designed for an uplift pressure of 46,800 lb and was tested to 65,000 lb. Only 9 out of 2000 anchors failed at less than 65,000 lb.

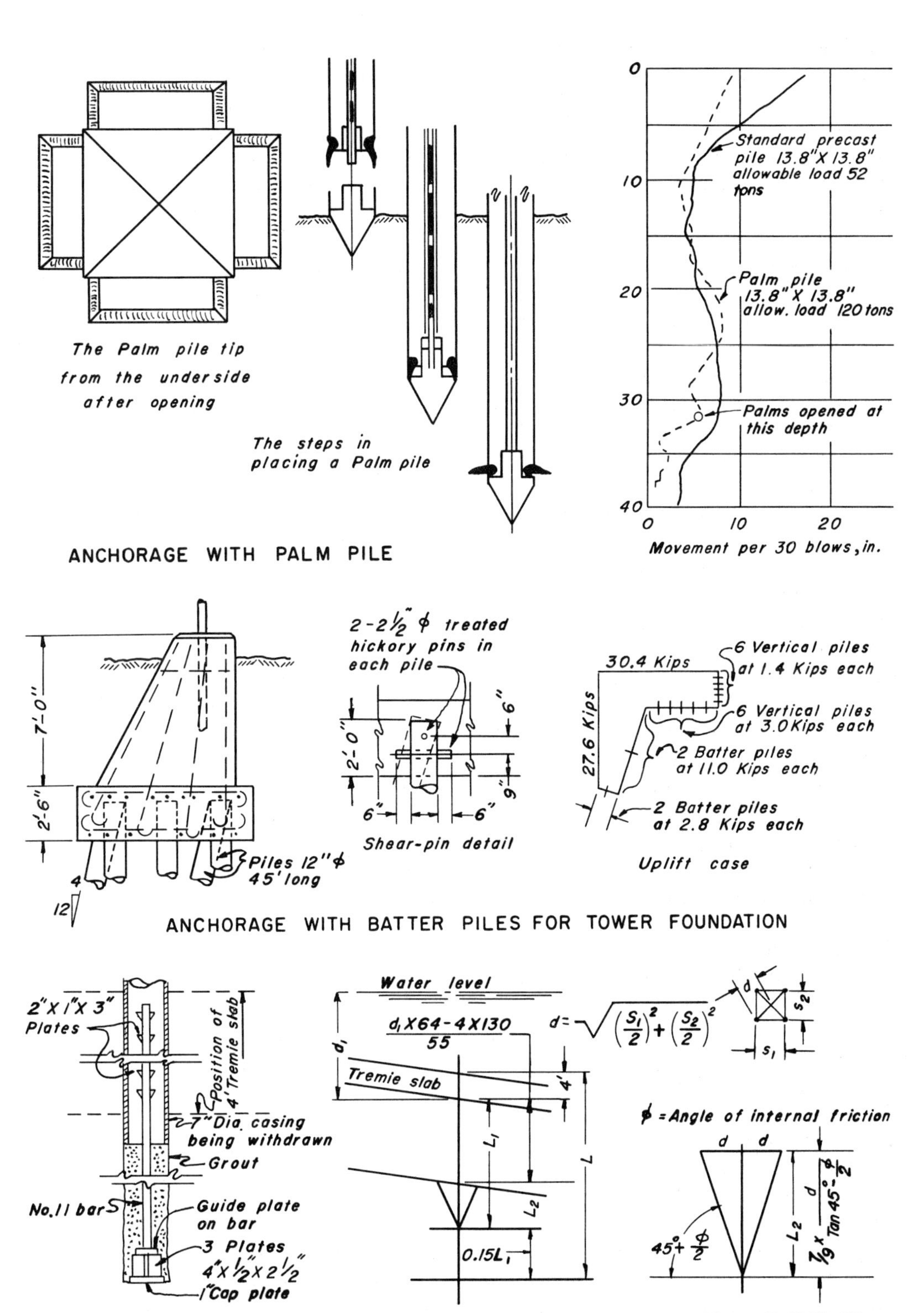
The Palm pile tip from the underside after opening
The steps in placing a Palm pile
Standard precast pile 13.8" X 13.8" allowable load 52 tons
Palm pile 13.8" X 13.8" allow. load 120 tons
Palms opened at this depth
Movement per 30 blows, in.
ANCHORAGE WITH PALM PILE
2-2½" φ treated hickory pins in each pile
Shear-pin detail
Piles 12" φ 45' long
30.4 Kips
27.6 Kips
6 Vertical piles at 1.4 Kips each
6 Vertical piles at 3.0 Kips each
2 Batter piles at 11.0 Kips each
2 Batter piles at 2.8 Kips each
Uplift case
ANCHORAGE WITH BATTER PILES FOR TOWER FOUNDATION
2" X 1" X 3" Plates
Position of 4' Tremie slab
7" Dia. casing being withdrawn
Grout
No. 11 bar
Guide plate on bar
3 Plates 4" X ½" X 2½"
1" Cap plate
Water level
d₁ X 64 - 4 X 130 / 55
Tremie slab
0.15L₁
φ = Angle of internal friction
ANCHORAGE OF TREMIE SLAB IN SAND WITH MAJOR DESIGN ELEMENTS

5.28 PILE CUT-OFFS AND CAPS

Piling used as a foundation is not complete until it has been adjusted to height, usually by cutting it off, and has been capped. The primary function of the pile cap is to transfer the structural load above onto the head of the pile; and, although the more customary method is to crown it with a substantial blob of concrete, other methods perform the same function.

Wood piles, particularly where cut off at or slightly above ground level, are conveniently sawed off by using a two-man saw, especially where the number of piles is small. For larger operations gasoline-driven chain saws or similar types of power saws are desirable. Care must be exercised in using power saws to be sure that the cut is truly horizontal (or if angled, as sometimes required on a batter pile, at the proper angle).

Wood piling can be cut under water by the use of a Wright air saw, which consists of dual reciprocating steel saw blades projecting from a 4-in. aluminum tube about 3 ft long. This is a one-man instrument, easily maneuverable and readily held in position. A special case is that of "removing" 18,000 piles under water, for which the Columbus Construction Company devised a steel frame with interior rails, in which a cutting blade, powered by a hydraulic ram, sliced the piles off, *à la guillotine*. The rig was simply lowered into position, its weight sinking it below the mud line where cut-off was made. This was a special case of cutting off old, water-soaked piles, where the exact cut-off elevation was irrelevant.

Where wood piles support timbering, opposite faces of the pile head are dabbed out to form seating for a pair of cleats. These are bolted to the tendon remaining of the pile head, using through bolts of galvanized steel or marine bronze with bolt heads and nuts held off the wood with ogee washers.

Steel H-piles and pipe piles are conveniently burned off by acetylene torch to the required height, and tubes for cast-in-place concrete piles are handled in the same way. Generally, pipe or tubes are adjusted to grade before filling with concrete; but occasionally, as is noted in Sec. 6.11, concrete is poured to the required grade, leaving a substantial section of pipe or tube hollow above it. Although convenient where excavation has not been completed to the desired level, the additional length of drop for the concrete may produce unsatisfactory fills.

In the past it has been common practice to weld a square steel plate, slightly larger than the outer dimensions of the section, over the top of steel H-piles, after cut-off, in order to reduce the "punching effect" of the web and flanges on the concrete of the cap. This plate is now frequently eliminated as unnecessary with the increase in concrete strengths to 4000 psi.

Pre-cast concrete piles are cut off by chipping away the concrete around their faces at cut-off level until the reinforcing steel is exposed. This can be done with hammer and chisel, but for larger projects air-driven chipping hammers are more desirable. The steel is burned away at the desired elevation; and the interior concrete in a solid pile can be snapped off, at its reduced section, by a sharp blow or tug of a crane hoisting line. It is frequently desirable to leave the reinforcing steel projecting for embedment in the cap, in which case the steel is not burned off but simply chipped free of the concrete.

Hollow pre-stressed (or pre-cast) concrete piles are generally handled somewhat differently, advantage being taken of the hollow core to provide a tie between pile and pile cap. Here cut-offs are made on a straight horizontal plane through concrete *and* steel by using a masonry saw, often hydraulically powered. Since hollow pre-stressed concrete piles are now so frequently used for bridge construction, requiring cut-offs over water, a rig is used consisting of a rail surrounding the pile on which the saw is mounted, with a standing platform about 3 ft below on which the operator stands. The rig is slipped over the pile head by crane suspension and temporarily hung on its top while the rig is bolted to the pile shell. A peripheral cut, not quite full depth, is made around the pile, and the last thin segment is broken off to prevent dropping the cut-off.

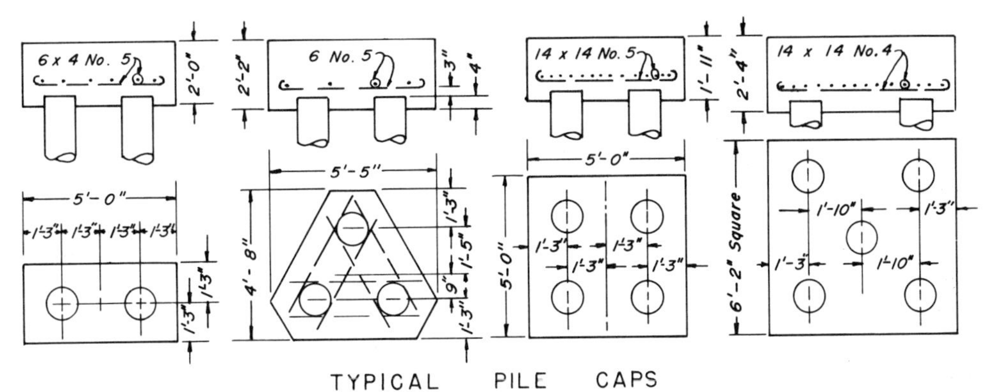

TYPICAL PILE CAPS

12" DIA. BUTTS 2'-6" SPACING – 40 TON LOADING – 3000 P.S.I CONCRETE

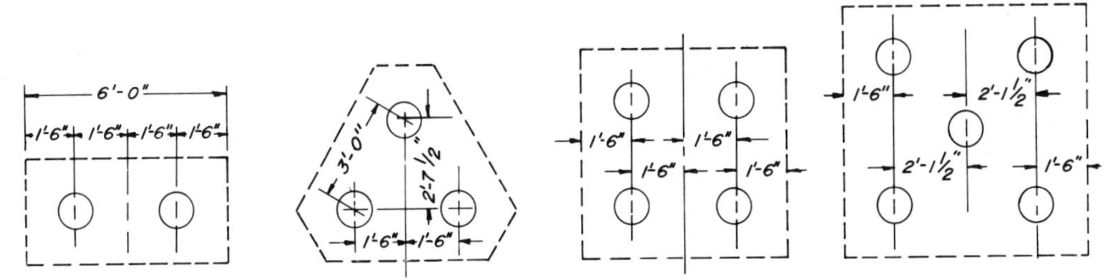

RELATIVE PILE CAP SIZES

USING 3'-0" SPACING OF PILES – 12" BUTTS

STEEL H-PILE PLATE CAPS

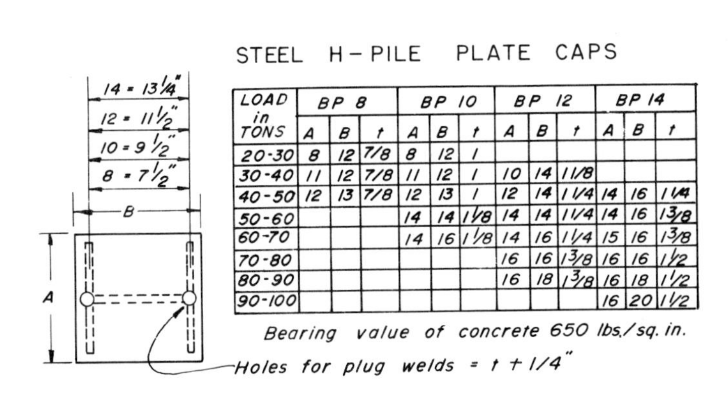

LOAD in TONS	BP 8			BP 10			BP 12			BP 14		
	A	B	t	A	B	t	A	B	t	A	B	t
20-30	8	12	7/8	8	12	1						
30-40	11	12	7/8	11	12	1	10	14	1 1/8			
40-50	12	13	7/8	12	13	1	12	14	1 1/4	14	16	1 1/4
50-60				14	14	1 1/8	14	14	1 1/4	14	16	1 3/8
60-70				14	16	1 1/8	14	16	1 1/4	15	16	1 3/8
70-80							16	16	1 3/8	16	16	1 1/2
80-90							16	18	1 3/8	16	18	1 1/2
90-100										16	20	1 1/2

Bearing value of concrete 650 lbs./sq. in.

Holes for plug welds = t + 1/4"

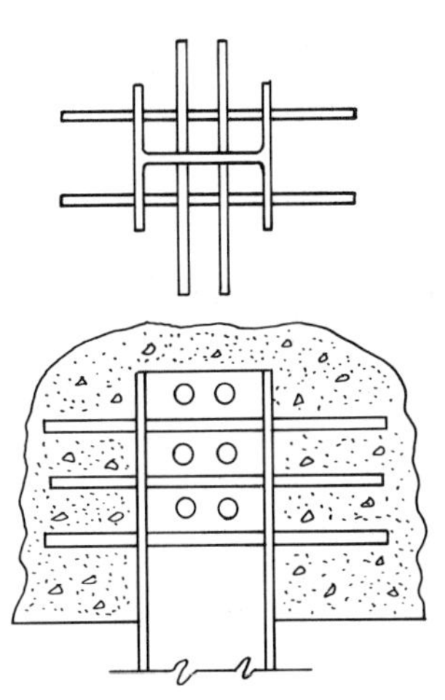

REINFORCING BAR CAP

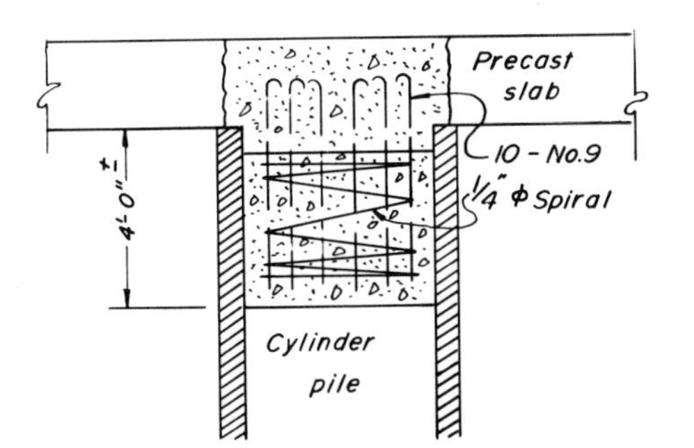

CAPPING A CYLINDRICAL PILE

1. *Hang disk form in pile, set steel cage and pour to within 6" of top.*
2. *Set precast slab with opening over pile, using temporary collar support on pile, if necessary.*
3. *Concrete remaining voids.*

5.29 PILING IN PERMAFROST

The principal and persistent difficulty, in any case where piling is intruded into permafrost, is that the pile, regardless of its material, will conduct heat. Since permafrost exists at temperatures only a degree or two below freezing, a slight rise in temperature can melt sufficient permafrost adhering to the pile to release pile friction. During the summer months under higher atmospheric temperatures sufficient heat may reach the point of the pile to relieve its value in end bearing.

The problem was studied at some length and extensive tests were conducted in 1950 in connection with the design of large tower foundations in permafrost areas. These tests (as well as some earlier Russian tests) showed a wide variation in the adhesiveness of frozen soils to steel pipe, depending upon moisture content, temperature, and density of the soil. Although values in excess of 100 psi of pile surface were recorded, a value of 20 psi seemed more aptly to fit the general condition, with a minimum embedment of 20 ft. (Bear in mind that we must be concerned with uplift possibilities.)

The design finally evolved in 1950 is shown opposite. Six 8-in. pipe piles are driven in a circle around the base of the tower, supporting radial spider beams. The purpose of this layout was to spread the heat flow from the tower itself to as wide an area as possible. A maximum rate of heat transmission, resulting from this design, was considered to be 1° per day per foot of pile during the summer months, and upon this basis a 30-ft-long pile was predicated. The pipe, after being placed, was filled with sand and plugged with concrete.

To additionally control tower and ground surface heat, a 2-ft-thick layer of sand encompassed the site, contained under a 6-in. concrete slab which in turn was surmounted by 18 in. of insulating material, generally the original tundra growth stripped from the site. Because there is a critical area immediately surrounding the tower, a ring of copper plate ⅝ in. thick was spread from the tower to a radius of 3 ft to supplement the conductivity of the concrete mat.

Another approach to the problem of piling in permafrost was employed in the construction of an Air Force facility near Bethel, Alaska. The structure to be supported was an all-purpose building about 250 ft square, designed to rest on 512 wood piles. Wood piling was selected as offering the least conductivity to heat passing down its length, but it was still recognized that degradation of the permafrost envelope would occur during the summer months.

The first step was the drilling of 18-in.-diameter holes to a depth of 30 ft by using a truck-mounted drill with a single-flight auger. The cutting edges of the auger were built up with boron carbide to resist the abrasiveness of the silty sand permafrost. Holes were completed at the rate of about one hole per hour.

With the hole completed, a creosoted wood pile with a 9-in. tip and a 12-in. butt was dropped into place *butt down* (to resist heaving). To each pile were strapped four evenly spaced 1-in. steel pipe loops extending for the full length of the pile, to act as a refrigeration coil.

The drill cuttings of permafrost had been carefully collected and were now shoveled into a ½-cu-yd concrete mixer where they were melted down with blast torches. The original water content of 25% was now increased to about 35% by the addition of hot water; and the resultant slurry, with a temperature of about 40°F, was poured into the drill hole around the positioned pile.

When a group of piles had been placed, braced, and slurried, their 1-in. pipe coils were connected to the refrigeration system and brine was passed through the lines to freeze the reconstituted slurry.

The decision to provide artificial refrigeration in the Arctic was based on three considerations: 1. The releasing of heat into the *undisturbed* permafrost could be counteracted. 2. The degree and the rate of natural freezing would be undeterminable and might not be adequate to support construction loads quickly. 3. During the summer months, artificial freezing could be employed to counteract heat transfer.

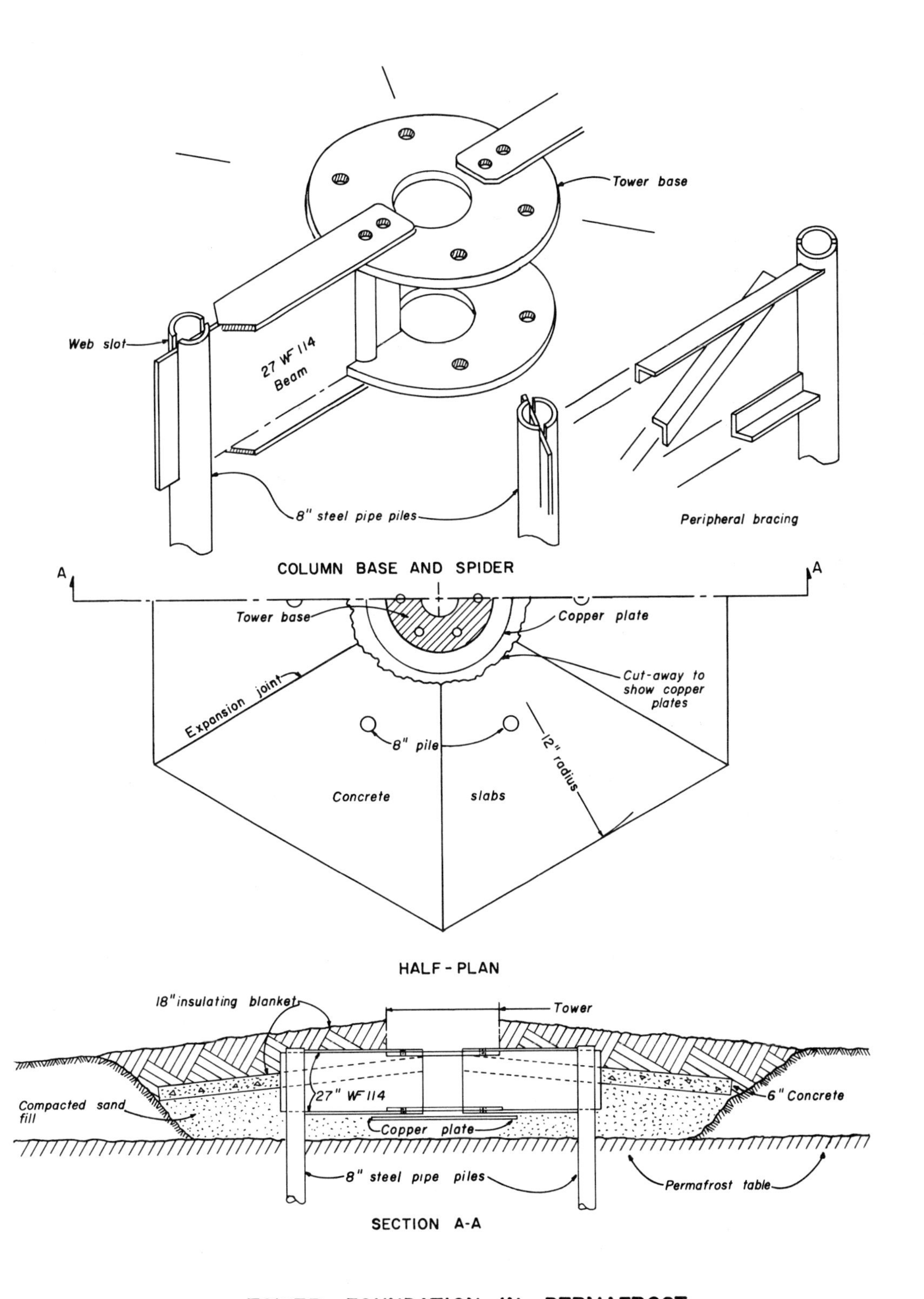

TOWER FOUNDATION IN PERMAFROST

5.30 DETERIORATION AND CORROSION PROTECTION OF PILING

The natural resistance of various types of piling to corrosion has been mentioned previously in dealing with each kind of piling. We are here more concerned with surface treatments to piling before it has been driven, or in the correction of limited corrosion after some years of service. The term corrosion is, perhaps, a misnomer since it is usually applied only to metals and we must here deal also with wood and concrete.

Refer, briefly, to the discussion of creosoted wood piles in Sec. 5.2. Creosoting resists virtually all types of marine attack or decay where properly applied; but, although timbers are pressure-creosoted, the complete penetration of the log by creosote seldom occurs. When a cut-off is made it is necessary to saturate the cut surface with hot creosote, which, since it penetrates but does not seal, should be covered with a heavy coat of coal-tar pitch, especially if it will be exposed. It would be desirable to do all timber boring prior to creosoting, but since this is not practical, all bolt holes should be treated with hot creosote, and a device is available which fits into one end of the hole while the other is plugged, to apply the creosote under pressure. Although creosote protection is greatest nearest the surface of the pile, scars or cuts into the pile during driving operations should be re-coated where possible.

Other types of treatments besides creosote have been used, such as zinc chloride, but without complete success. Perhaps the closest to creosote in effectiveness is pentachlorophenol, referred to as "penta," which is applied to the timber in the same way as creosote, with retentions varying between 10 and 16 lb per cu ft when combined with a petroleum solvent. Because it is colorless and odorless, timbering treated with penta is more easily and cleanly handled and is therefore preferred by workmen where extensive framing is required. The United States Navy generally permits the use of penta as a substitute for creosote on dock work.

Wood piling, not originally composite, is frequently concrete-encased by using gunite techniques within the zone of decay, and an epoxy resin has been developed that can be placed under water as a coating. The United States Navy and others have restored piers with damaged wood piling by cutting off the pile under water and removing the timbering up to the underside of the slab. A steel cylinder is fitted around the top of the cut pile extending up to the underside of the deck. Filled with gravel, the cylinder is intrusion-grouted under pressure, restoring full support for the slab and superstructure.

The corrosion, and resultant reduction in section of steel piling, has been noted. A slight increase in thickness of web and flanges of H-piles is one method of delaying failure, and the use of copper bearing steel further slows corrosion. Coatings of various kinds have been used, but all these are subject to damage in handling or driving. The most successful coating seems to be a bitumastic enamel, factory-applied after sandblasting the steel, in either two cold coats or one hot application. Although maximum corrosion generally occurs in the same aboveground zone as that for wood piling, and abrasion in driving can scrape off the enamel below ground, steel H-piles are generally coated throughout their full length.

Cathodic protection has been introduced as an additional safeguard in the protection of steel H-piles and pipe piles, as well as for steel sheet piling, where it is used for permanent bulkheads.

Remember your high school chemistry and the table headed "Electromotive Series"? If bars of any two metals thereon listed were placed in a suitable saline solution, ionization would occur, causing ions of the metal from one bar (termed the anode) to pass via the solution (termed the electrolyte) to the opposite bar (termed the cathode), where a feeble current would be released. The metal ionized might go into solution, releasing hydrogen to form on the surface of the cathode, or might be deposited on the cathode; but in any case the anode would be slowly eaten away. The table referred to showed, in effect, which metal of two would be the anode, the eaten-away metal, by its position in the table. Current would pass from that metal highest on the list to any metal below it.

Iron was well up on the list of the electromotive series; and although there were metals above it which would play anode to its cathode, there was a goodly list below it for which iron would be the anode and pass slowly away into solution in sea water and in many moist or damp soils. Corrosion of metals can now be defined, therefore, as the destruction of a metal by an electrochemical reaction in which a chemical change is accompanied by a transfer of electrical energy.

Two methods of counteracting corrosion due to electromotive forces operating in a solution are apparent. We can provide a "sacrificial" anode of sufficient strength to act as an anode, forcing the iron member involved into playing cathode. This was dramatically demonstrated some years ago by Dow Chemical, a leading producer of magnesium metal from sea water. They affixed blocks of magnesium to the steel hull of a seagoing vessel well below the Plimsoll line, completely relieving it from corrosion so that it needed painting only to keep it pretty. The magnesium blocks were, of course, expendable.

The second method that is apparent in reviewing earlier paragraphs is that of reversing the flow of current so that current of sufficient voltage will always be flowing *to* the iron member, as to a cathode, rather than away from it. Essentially this is what is meant by cathodic protection.

Cathodic protection was first developed for use in preventing the corrosion of buried pipelines, but was introduced in 1953 for the protection of steel piling in the New York Harbor by the Port of New York Authority and in the succeeding 10 years has been applied to an increasing number of wharf and pier structures involving steel H or pipe piles and steel sheet piling.

By the nature of the conditions, every case requires special design; but, in general, what is involved is feeding a direct current (usually obtained by rectifying alternating current) through a buried anode of sufficient strength to overpower any potential or stray currents developed in the sea-water medium. Anodes used in the New York installations have been graphite, with an expected life of 10 years, and Duriron, with a high silica content, for use in sea water. Where the anode must be placed in silt, Durichlor anodes are substituted for Duriron since Duriron is attacked by anodically produced free chlorine which cannot escape through the silt.

Whatever the type, anodes are placed strategically and in positions suitable for removal, sometimes suspended in the water from saran rope or hung in specially provided wells composed of porous concrete pipe. All the steel members, being protected, must be electrically bonded together, generally by means of continuous steel bars welded to each pile member near the upper end.

Concrete piles are often considered to be the most resistant to deterioration of any of the types of piling being discussed. This, however, is not necessarily so. Concrete itself is subject to chemical action under certain conditions; and even where such conditions do not exist, the manufacturing and driving conditions may introduce deteriorating factors.

In the manufacturing stage, the resistance of the concrete to attack by chemicals may be obtained through the use of the proper cement. Types II, IV, or V are frequently desirable in alkaline soils or in sea water. Air-entraining agents also provide greater resistance to corrosion. Careful manufacture, using dense mixes properly cured, is certainly the most vital aspect.

In driving or in position, physical damage resulting in the merest hair cracks can lead to deterioration when the reinforcing steel is reached by corrosive elements, and the resulting expanding rust further chips away the enveloping concrete.

Even under ideal conditions of protection, protective coatings do not adhere well to concrete and their effectiveness is sufficiently limited to make the cost unwarranted.

5.31 INSPECTION OF PILE DRIVING

The inspection of construction, generally, has never received the attention it deserves. In those instances where "difficulties" can arise in the inspection of wholly surface construction, consider the potentialities of inspection involving piling where the whole result is completely unobservable and the progress to that result is observed by an individual armed only with a notebook and pencil, a 6-ft rule, sometimes a transit, and often a jaundiced eye.

A program under way in 1962, conducted jointly by the Michigan State Highway Department and the U.S. Bureau of Public Roads, employed 12 engineers and technicians to observe and record pile-driving operations, and electronic computers to analyse the results. The instrumentation included a load cell to measure the hammer force, a penetrometer for measuring the dynamic deflection of the pile, an accelerometer to measure the acceleration of the pile top under impact, and high-speed camera equipment to replace the jaundiced eye. The observations of this equipment were automatically recorded by oscillographs producing pen-trace records for later contemplation. The study, sparked by the squabble over the energy rating of diesel hammers, may at least offer some better instrumentation aids to the inspector with the 6-ft rule.

One authority describes the inspector as follows: "A pile-driving inspector must be able to read plans, to keep neat and accurate records, and to understand the principles of the specification." This is certainly the understatement of the century, for an engineering degree or some natural intelligence or both are also very desirable—anything, in fact, that will induce the inspector to keep asking himself why!, why!!, why!!! when a pile does not go down in the same fashion as its predecessor. One case, now quite old but still pertinent, aptly illustrates this.

In the early 1930s the 30th Street viaduct was being constructed in Philadelphia between Market and Walnut Streets as part of an elevated street system to meet the level of contemplated bridges over the Schuylkill River, lying slightly over a block to the east of, and approximately parallel to, 30th Street. The area, once marshland, had been filled in to provide the lower street level; and the viaduct columns, resting on Gow caissons, were carried down through the fill and marsh 30 to 40 ft to sound sandstone below.

The Bridgeman Building was a three-story building, fronting on 30th Street between Chestnut and Walnut Streets, which had encountered trouble in construction when originally built in 1907 (see "Building Failures," by Thomas H. McKaig) and had more recently added a one-story warehouse at the rear. While the caissons were under construction, distress apparently due to settlement of the one-story addition was noted at the connection between the two sections of building. It was attributed to the dewatering of the area incident to the caisson construction, but the pile-driving records were consulted as part of an investigation.

The pile-driving records were "neat and accurate," the piles had been driven as per plans, and the final number of blows on each pile was in accord with the specifications. It was noted, however, that whereas most of the piling had pulled up at depths of from 17 to 20 ft, in one limited area adjacent to the older building they had consistently pulled up at only 9 to 10 ft.

By the sort of open-minded detective work for which the civil engineer should be more noted than he is, the reason was disclosed. The site of the Bridgeman Building, prior to a general filling of the area, had been occupied by the Wetherill Lead Company, who had maintained a barge slip, connecting to the river, precisely under the present building. It was shown that the short piles had pulled up on a sunken and abandoned barge, which was now slowly settling in the silt, either from vibration or from dewatering, incident to the caisson work.

The abbreviated form shown opposite for the use of the inspector should be kept "accurately and neatly" for every pile but should be questioned immediately by him upon finding any discrepancies.

PILE-DRIVING INSPECTION REPORT

Project ____________________

Engineer ____________________ Contractor ____________________ Date ____________

Pile No. ____________________ Location ____________________

The Pile	The Equipment	The Hammer
Wood: Tip ________ Butt ________	Basic Rig ____________	Make ____________
Pipe: O.D. ________ Shell ________	Leads? ____________	Model No. ____________
H. Beam: Sec. ______ Weight ______	Mandrel? ____________	Rated energy ____________
Shell: Tip ________ Butt ________	Follower? ____________	Weight of ram ____________
Concrete (note size below):	Cap: Type ______ Weight ______	Stroke ____________
Pre-cast ____________		Formula used ____________
Pre-stressed ____________	*Note:* Speed of hammer in blows per minute applies only to double and differential acting hammers.	Value of e? ______ of c? ______
Uncased ____________		Depth of cut-off ____________

Ft	No. of blows	Speed, blows per min	Ft	No. of blows	Speed, blows per min	Ft	No. of blows	Speed, blows per min	Ft	No. of blows	Speed, blows per min	Ft	No. of blows	Speed, blows per min	Ft	No. of blows	Speed, blows per min
1			21			41			61			81			101		
2			22			42			62			82			102		
3			23			43			63			83			103		
4			24			44			64			84			104		
5			25			45			65			85			105		
6			26			46			66			86			106		
7			27			47			67			87			107		
8			28			48			68			88			108		
9			29			49			69			89			109		
10			30			50			70			90			110		
11			31			51			71			91			111		
12			32			52			72			92			112		
13			33			53			73			93			113		
14			34			54			74			94			114		
15			35			55			75			95			115		
16			36			56			76			96			116		
17			37			57			77			97			117		
18			38			58			78			98			118		
19			39			59			79			99			119		
20			40			60			80			100			120		

Record number of blows for each 6 in., especially near limit of penetration. Remarks (note any delays or unusual aspects of pile being driven): ____________________

____

____

____

____

Inspector ____________________

5.32 PREPARING A BID ON PILING

The preparation of a bid on piling, while generally following the lines laid down for estimating in any category, has some peculiar facets of its own. By its nature, piling contains considerable uncertainty, and there is a tendency to accentuate this uncertainty by post-bid decisions on the part of the owner or engineer. It is true, of course, that contractors are not above a bit of chicanery here and there, but it is also true that what appears to the engineer as mere chicane is frequently simply an effort on the contractor's part to anticipate post-bid decisions that will be disadvantageous.

One such post-bid decision on piling was noted in Sec. 1.3 and occurred in the late 1930s. Another case concerning piling with which the author is familiar (having prepared studies and reports for the guidance of the attorneys in both instances) can be dated to the mid-1950s and, like that noted in Sec. 1.3, concerned piling supports for sewage-treatment-plant construction. The similarity of circumstances is not coincidental since sewage plants are frequently constructed on remote, unreclaimed marshland requiring piling. Furthermore, they are generally municipal structures controlled by local political bodies who look with disfavor on contractors who are not local boys or otherwise classified as being in a "most favored" category. (This is an aspect of bidding sedulously avoided by volumes on estimating and management.)

In this second case, the main structures for primary and secondary sedimentation and sludge digestion were set on cast-in-place concrete piles, but the digested sludge was to be dried in open lagoons. The six lagoons were formed by embankments of the native silt, partly obtained from material excavated from the plant site as well as by excavation at the lagoon site. They were fed by cast-iron pipelines specified to be carried on a series of two-pile bents of wood piling, and the diversion and outlet structures were similarly set on wood piling.

The contract called for a unit-price bid on the wood piling; but since the main portion of the plant was a lump-sum bid, the amount of the unit price for wood piling did not affect the determination of the low bidder. The contractor, himself an engineer, and aware (as he saw it) that piling would be mandatory under the lagoon structures, since it was clearly shown and specified, bid a substantial unit price for the work; indeed, a substantial portion of his profit was placed in this item. Consider then his chagrin and discomfiture when the engineers, having found the pile footage to be running some 50% over their estimate on the first few piling driven, ordered all the piling deleted from the lagoon lines and structures.

The contractor protested the elimination of the piling supports but proceeded to place the pipelines and structures as directed, proposing to sue for lost profits on the completion of the contract. However, as could have been predicted, the pipelines and structures, resting on a highly organic silt but a few feet above high tide adjacent to a tidal estuary, started to settle seriously before the contract had been completed. The farce continued, for the engineers condemned the work, intimating that the contractor was at fault for not having provided piling at his own expense to prevent subsidence. Here again a settlement was reached without litigation, but to no one's ultimate advantage.

On the opposite page is a checklist of items to be considered in preparing a bid on piling. These are all concerned with matters of fact, such as material and equipment, and are not concerned with intangibles. However, the contractor will do well to consider other aspects of the problem.

In the early 1960s the author encountered two cases where cast-in-place concrete piling, driven to end bearing on rock only 10 ft or so below foundation level, was specified. Efforts to drive the piling specified through a dense overburden were unsuccessful; the use of piling, in both cases, was abandoned and concrete footings were substituted. Anticipating this sort of possibility is one of the hidden hazards of bidding on piling.

Cost Factors in Piling Estimates

Type of piling: Friction? ________ End-bearing? ________ Kind of piling ________
Number of piles ________ Length of piles (estimated) ________ Total Pile footage ________
Driving conditions ________________
Estimated driving time ________ per pile (or) ________ per ft × count ________ = ________
Number of working days ________________ Months ________________
Test piles required ________________ @ $ ________ per pile = ________________

PILING MATERIALS

Wood piles: Class A ________ B ________ C ________ Butt dia. ________ Tip dia. ________
Treated piles ________ lineal ft @ $ ________ per ft . . . . $ ________
Untreated piles ________ lineal ft @ $ ________ per ft . . . . $ ________
Steel H piles ________ Size ________ Weight per ft ________
Total footage @ ________ lb/ft = ________ tons @ $ ________ per ton . . . . $ ________
Pipe piles ________ Diameter ________ Wall thickness ________
Total footage ________ @ $ ________ per ft . . . . $ ________
Shell piles ________ Type ________ Sections ________
Footage of section ________ = ________ @ $ ________ per ft . . . . $ ________
Footage of section ________ = ________ @ $ ________ per ft . . . . $ ________
Concrete fill: Pipe piles ________________ Shell piles ________________
Quantity in cu ft per ft ________ × Total length ________ = ________ cu ft
Cu ft of fill ÷ 27 = ________ cu yd @ $ ________ per cu yd . . . . . . $ ________
Pre-cast concrete piles ________ Size ________ Weight per ft ________ lbs
Total footage ________ @ $ ________ per lineal ft . . . . $ ________
Pre-stressed concrete piles ________ Size ________ Weight per ft ________ lb
Total footage ________ @ $ ________ per lineal ft. . . . . $ ________
Uncased concrete piles ________ Length ________ Cu ft per ft ________
Total footage
____ × ____ cu ft per ft ÷ 27 = ____ cu yd @ $ ________ per cu yd . . . . $ ________
Allowance for cut-off or other waste ________ ft per pile = ________
No. of piles ________ × Allowance per pile ________ = Total allowance ________
Pile cost $ ______ per ft + Fill cost $ ______ per ft = ______ × footage ______ . . . $ ______
Pile shoes ________ No. of piles ________ × $ ________ per unit . . . $ ________
Field pile splices ________ No. of splices ________ @ $ ________ each . . $ ________

EQUIPMENT

Driving rig ________________ Make ________________ Model ________________
Driving time per pile ________ × Pile count ________ = Use time ________
Rental rate or ownership cost ________ per month × ________ months . . $ ________
Power equipment for driving: Boiler ________________ Compressor ________________
Accessories: Hose ________ Control valves ________ Other ________
Cost of power equipment and accessories ________ per month × ________ months . . $ ________
Holding equipment: Leads ________ Skirts ________ Frames ________
Cost of holding equipment ________ per month × ________ months . . . $ ________
Pile hammer: Make ________ Model ________ Characteristics ________
Cost of pile hammer ________ per month × ________ months . . . . . $ ________
Expendables: Fuel ______ Oil ______ Cushions ______ Other ______ (lump sum) $ ______
Moving in and setting up (lump sum) . . . . $ ________
Dismantling and moving out (lump sum) . . . . $ ________

LABOR

Rig operator (rate per week) . . . . $ ________
Oiler (rate per week) . . . . $ ________
Fireman or compressor operator . . . . $ ________
Pile driving crew ________ men @ ________ per week
Total payroll per week . . . . $ ________
Estimated driving time ________ weeks @ ________ per week
Total labor . . . . $ ________

6 Cofferdams

The cofferdam is as ancient as any construction device. The Romans used it, as did the Chinese, who lashed bamboo mats to poles driven into the ground, in square patterns, with each square area filled with mud. Cofferdams are referred to in a contract dated A.D. 1421 for the construction of a bridge in Yorkshire, England. This contract provided that the owners were to furnish *brandereths* to permit the masons laying the stonework of the piers to work in the dry. These *brandereths* consisted of two rows of wooden stakes about 4 ft apart, with the intervening space filled with mud. The stakes or "brands" appear to have been carefully charred along opposite edges to create butting, flat faces, the overall resulting char producing a hardened stake, more resistant to the effect of water.

The term "cofferdam," like so many terms in the field of construction, derives from a French word, *coffre,* meaning a chest or casket. The method of constructing cofferdams has generally involved stakes or poles —"piling"—driven in a continuous line into the ground, although in Roman times cofferdams were built of woven baskets filled with mud and stacked progressively around the underwater site to be enclosed.

That pioneering unit of French military engineers, the Corps Royaux des Ponts et Chaussées, whose first duties involved the reviewing of all plans for roads, bridges, and canals built in France, followed this by providing inspection of construction. But their influence and the propagation of the French construction vocabulary may be more closely attributed to the founding of the first engineering school, the École des Ponts et Chaussées, from the military unit in 1760, under the initial direction of Leon Perronet, the designer and builder of the first flat-arched bridge at Nevilly.

The influence of the French engineering school on early American construction was direct and decisive. Claude Crozet, a graduate of that school, who had built roads for Napoleon on his way to Moscow and Waterloo, was appointed professor of engineering at the United States Military Academy at West Point in 1816, at the instigation of Albert Gallatin and Lafayette. Crozet introduced and used, during his 7-year tenure, a textbook which had been written by Joseph-

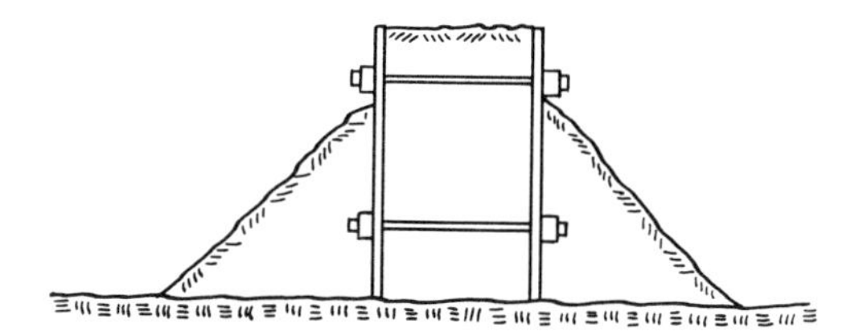

Mathieu Sganzin, former inspector-general of the Corps Royaux des Ponts et Chaussées. Military engineers trained in the French tradition were, therefore, ready for the first civilian assignment of the United States Corps of Engineers in 1824.

In April of 1824, Congress, as a result of a decision of the Supreme Court establishing its power to regulate navigation on the inland waterways, authorized the President to have surveys, plans, and estimates prepared covering those routes which could be considered to be of national importance. To accomplish this the President was to employ the officers of the Corps of Engineers and such other individuals as might be detailed to duty with the Corps. A month later the sum of $75,000 was appropriated for actual improvements on the Ohio and Mississippi Rivers, chiefly the removal of snags and sandbars.

The importance of the Ohio River as a navigable artery had long been recognized, and as early as 1811 a river packet, the *New Orleans,* had been making scheduled runs on it. During the succeeding 13 years traffic rapidly multiplied, so that by 1824 some 200 steamboats were plying the river.

Above Louisville, in 1824, the Ohio River dropped 28 ft in 3.2 miles over hard limestone outcrops, making navigation virtually impossible. To bypass this section of the river the Louisville and Portland Canal was begun in 1825. The construction included three locks with a total lift of 26 ft; and for lock construction then, as now, the use of cofferdams was required. Under the influence of the Corps of Engineers, a distinctive type of cofferdam was developed for this and later construction known ultimately as the Ohio River cofferdam. A double-wall timber cofferdam with earth slopes designed for underwater rock surfaces, it is still used to a limited extent.

Although the Ohio River type of cofferdam has given place to cofferdams constructed with steel sheet piling, the battle with the Ohio River continues. In 1950 the Ohio River Division of the United States Corps of Engineers began a 25-year program to increase the navigable depth of the river from 6 to 9 ft by a new series of locks and dams.

6.1 TYPES OF COFFERDAMS

The term cofferdam was applied initially to a very limited field, that of bridge pier construction, and could then be very clearly defined as a casket or containment used to hold out earth and water from the area of such construction. The water stopper was mud, and the casket or cofferdam retained the mud. This felicitous simplification no longer applies to the use of the word. While it still serves the primary purpose of restraining earth, water, or both, the aspect of the casket is no longer a necessary concomitant in the application of the term.

The elements of the modern cofferdam consist of sheets of wood, steel, or concrete—the sheathing—thrust or driven into the ground in various patterns. Horizontally, the pattern is maintained by walers; and the whole assembly is supported vertically by struts, shores, or rakers.

The original concept of two rows of sheathing cross-braced to each other and filled with soil to provide the sealant (as well as massive lateral stability), the basic principle of the Ohio River cofferdam, no longer meets current requirements in the use of the term. Early improvements in the shaping, grooving, or lapping of lumber; the introduction of steel sheet piling in the first half of the twentieth century; and the later introduction of concrete sheathing have led to the development of the single-wall cofferdam.

The single-wall cofferdam, no longer requiring the use of soils as sealant, is basically indistinguishable from those systems for supporting earth banks, more commonly termed "sheathing" or "shoring," in either case substituting the name of a part for the whole. Similarly, the aspect of casketry often disappears completely since these systems can be used in long thin lines, enclosing nothing.

The most familiar type of single-wall cofferdam is that employing walers and horizontal struts, the walers preserving the alignment of the sheathing horizontally while the struts preserve it vertically. This system implies two rows of sheathing, the one braced against the other.

When we eliminate one wall of sheathing, we no longer have anything to brace against, and new systems of support must be devised. The simplest alternative is that of driving the sheathing deeply into the ground to hold it rigidly erect by the reaction of the soil into which it is driven, coupled with the stiffness of the sheathing material which permits it to function as a cantilever.

In lieu of an opposing wall the sheathing may be supported by diagonal shores (hence the term "shoring") now more often termed rakers. The success of this method depends on the ability to secure the lower end of the raker, generally to a relatively immovable mass. Installation of diagonal rakers or shores requires careful planning to avoid the necessity of disturbing the members during subsequent construction.

The use of anchors behind the sheathing to which the wall was tied has long been a characteristic of permanent bulkhead construction. It is suitable for temporary use only when space permits and consequently has no universal application. A variation has appeared in recent years which is more practical, especially where rock or dense soil formations surround the site. Anchorage is secured by drilling, generally diagonally downward, into the adjoining formation and anchoring rods into the formation to support the sheathing. This is often termed a tie-back system.

Earth dikes are still used for cofferdams, now generally with a simple unsupported line of sheathing near their center to control the flow of water. The Ohio River type of cofferdam provides the additional stability of two rows of sheathing, but its primary function is similar to that of the modern cellular cofferdam.

The prime function of both the double-wall and the cellular cofferdam is to provide an independent structure in the water or adjacent thereto, where, because of rock or other subsoil conditions, no penetration of sheathing is possible. The cellular cofferdam, in particular, provides a string of islands linked together in a waterway area to form a continuous enclosure.

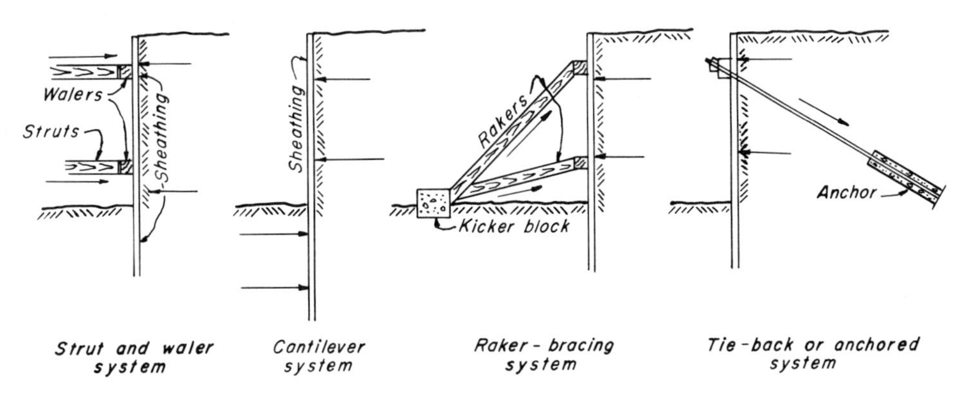

SINGLE WALL COFFERDAM SYSTEMS
IN ELEVATION

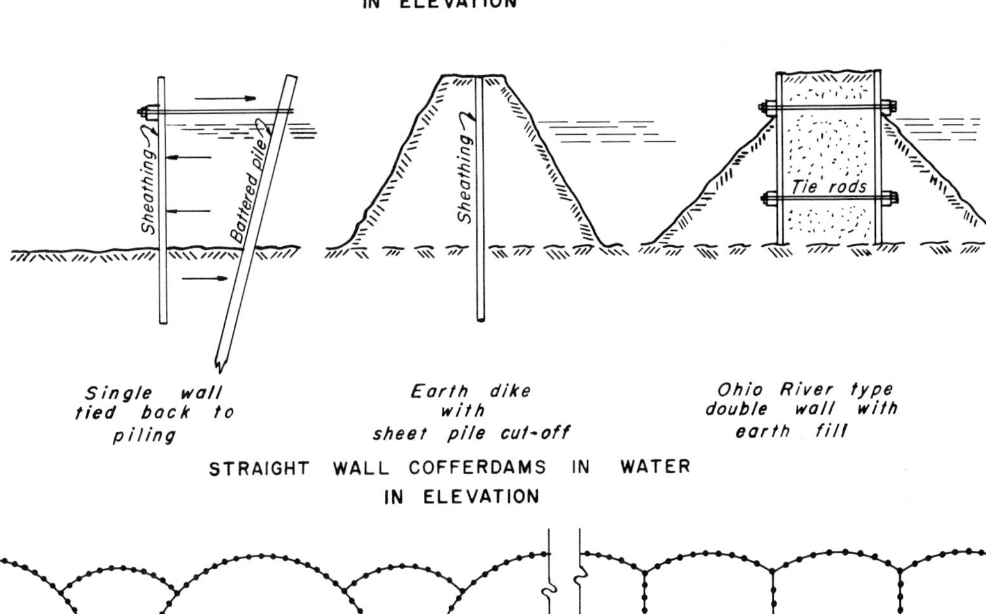

STRAIGHT WALL COFFERDAMS IN WATER
IN ELEVATION

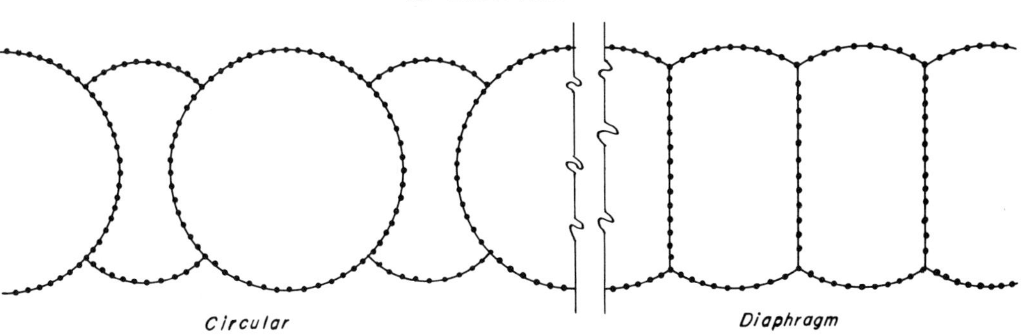

CELLULAR COFFERDAMS
IN PLAN

SINGLE AND DOUBLE WALL COFFERDAMS

6.2 BASIC SINGLE-WALL COFFERDAM DESIGN THEORY

The theoretical design of single-wall cofferdams is based on a number of factors, many of them not completely understood and some of them still a matter of dispute among the researchers in the field. To review these researches and the resultant theories from Coulomb to Rankine, to Terzaghi and Peck, to Tschebotarioff can serve no useful purpose here. All that can be attempted is a summary and simplification, maddening to the researcher with his plethora of suffixes, but of some practical use to the average contractor.

The materials to be restrained by a single-wall cofferdam will vary from a fluid condition to that of a rock wall, from a saturated silt or clay that, without restraint, will run off to a thin film, to soft rock formations chiefly subject to facial erosion. The degree or extent of the tendency to "run off to a thin film" will vary from 0% with rock to 100% with a true fluid, and the variation can be expressed by a coefficient K, "the active lateral earth pressure." The total pressure exerted by a true fluid is $\frac{1}{2}WH^2$, where W is the weight of the fluid and H is the height of the fluid to be supported. Under this condition the value of K is 1, which, added to the formula for fluid pressure, does not change the numerical value of the result.

Under any specific set of conditions the height H to be restrained by the cofferdam is known. The unit weight or density of the soil or soils is a laboratory determination from test-boring samples; and although the averaging of these, where a number of varying strata are involved, is complex, an approximation is generally possible. What, then, is the value of K?

In the early 1940s subways were under construction in Chicago, in some cases in open cuts. Terzaghi and Peck were consultants to the city on soil investigations. They introduced the use of the coefficient K and developed its value for clay from the equation shown opposite. The q is the unconfined compressive strength of the soil as determined by laboratory tests of undisturbed samples. Here again, however, the value of q will vary considerably with strata of soil, and the obtaining of a single usable value is an involved mathematical procedure.

Another advancement in theory occurred in the late 1940s when Tschebotarioff conducted a series of large-scale tests for the United States Navy at Princeton University, chiefly in sands and primarily to develop bulkhead designs. The field values of K in Chicago clay were found to range from 0.2 to 0.5, whereas Tschebotarioff suggests a single value for K of 0.5. His value of K for loose sand is 0.30 and for dense sand is 0.25. (It might be noted that in Chicago, some clay deposits had densities as high as 124.8 lb per cu ft, which is twice the density of water, so that for this particular density of soil, where $K = 0.50$, the total pressure would be the same as if water were being restrained.)

Since the single-wall cofferdam will require struts, rakers, or tie-backs, the distribution of this total load down the face of the cofferdam is important in designing these members. Terzaghi and Peck suggested the use of the trapezoidal form shown opposite.

This design method, which will be applied to a specific case in Sec. 6.3, is, presumably, the condition which will or can prevail when the cofferdam is completed, but as usual it is found that construction methods may alter this beautiful concept.

Terzaghi introduced the concept of yield, and its effect on the distribution of pressures is shown opposite. As can be seen, its effect will vary depending upon whether the contractor, being negligent in placing his bracing, permits the bank to yield, or deform, at either the top, the middle, or the bottom.

But the very act of placing the bracing can cause deformation without any negligence being involved. Struts must be wedged or pre-stressed into place to prevent yield, and the pressure thus applied to the bank's face can alter the nature of the soil, changing its W, q, and K, as the soil moves from a solid to a plastic state.

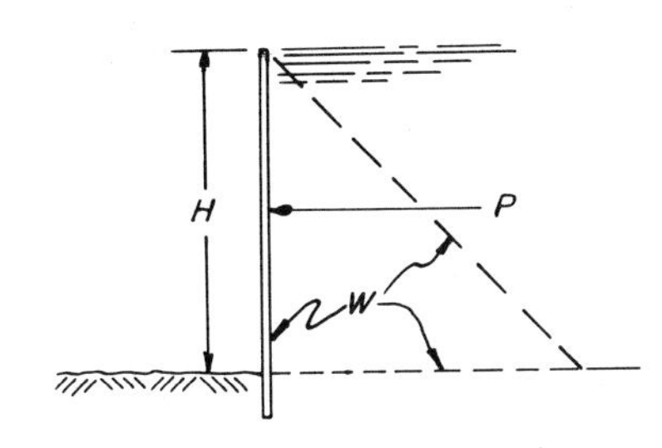

H = Height of fluid in feet.
W = Unit weight of fluid in lbs per cubic foot.
P = Total pressure on wall per lineal foot of cofferdam.

$$P = \tfrac{1}{2} WH^2$$

FLUID PRESSURE

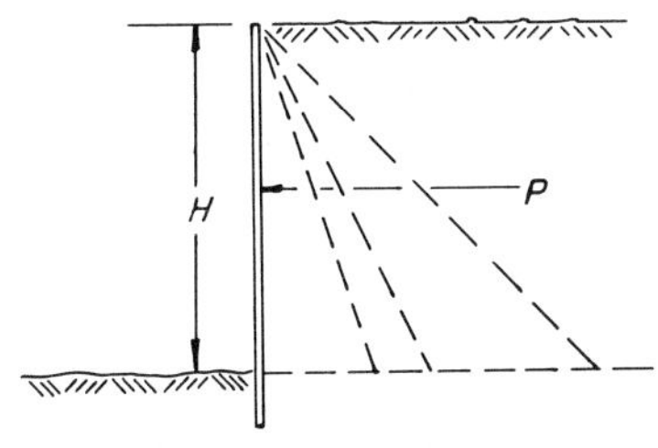

$K = K_A$ = Coefficient of active earth pressure.
$W = Y_a$ = Average density or unit weight of soil at depth H in lbs per cubic foot.
$q = q_u$ = Average unconfined compressive strength for depth H in lbs per square foot.
H = Depth below surface in feet.
P = Total pressure on wall per lineal foot of cofferdam.

EARTH PRESSURE ELEMENTS

$P = K \times \tfrac{1}{2} WH^2$ $\quad K = 1 - \frac{2q}{WH}$ $\quad$ Max H for placing first strut $= \frac{q}{W}$

RELATION OF PRESSURE ELEMENTS

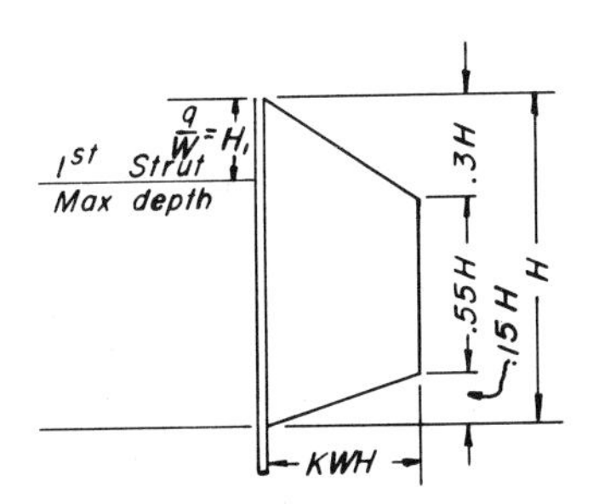

Maximum unit pressure per sq foot of wall = KWH
This extends from a point .3 of the total height from the surface to a point .15 of the height from the bottom - reducing to zero from these points.

TERZAGHI'S TRAPEZOIDAL DISTRIBUTION

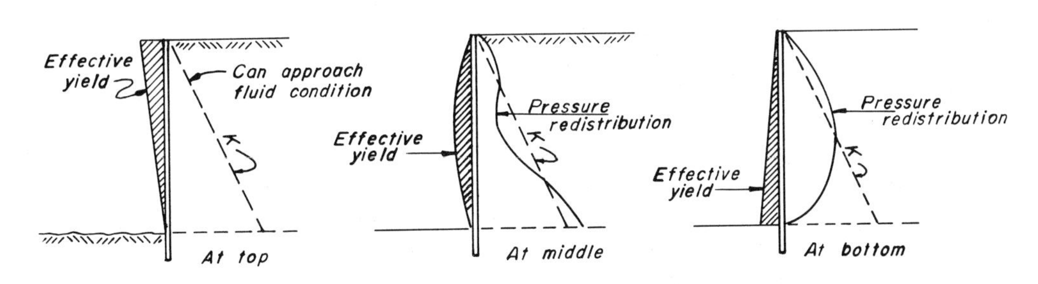

PRESSURE REDISTRIBUTION DUE TO YIELD OF THE BANK

THEORETICAL ASPECTS OF COFFERDAM DESIGN

6.3 A SINGLE-WALL COFFERDAM DESIGN

In Chicago, in 1958, it was necessary to enclose an area 99 by 195 ft with a single-wall cofferdam, to permit construction of the foundations for the Harris Trust Company Building. The site abutted on public highways on three sides, while an existing building occupied the fourth side. The contractor, the foundation firm of Spencer, White and Prentis, Inc., made a careful study of conditions and prepared a design for the cofferdam; then, as the cofferdam was being placed, checked actual loadings against design loadings at every step of the operation.

As was noted in Sec. 3.9, Chicago rests on a bed of clay up to 200 ft thick which varies, somewhat, in its characteristics, from place to place. Before undertaking a design, the contractors queried a number of engineers familiar with soil conditions in the area for their estimates of the total loading that might be expected on the cofferdam. The answers received ranged from 20 to 65 tons per lineal foot of wall for the necessary depth of 46.5 ft. Since these appraisals offered no precise basis for a design, the contractors used a figure of 30 tons per lineal foot, based on their own experiences with a comparable cofferdam, at a nearby site, the previous year.

In Sec. 6.2 it was noted that the theoretical approach started with the unconfined compressive strength, the density of the soil, and the height, and that loadings were developed from these data. The difficulty in using that approach here, as can be seen from the test-boring log, is that there are three strata of soil, each with a different unconfined compressive strength. To convert these to a single average requires assumptions equal in magnitude to that made, without reasonable means of verification.

The system of sheathing used consisted of soldier beams, 18-in. WF 96#, driven on 6-ft centers around the three sides of the perimeter, backed up with 3-in.-thick wood sheathing. The sheathing was set on a 3-in.-thick filler block behind the rear flange of the soldier beam. The soldier beams were intended to be cast in the finished structure, and the sheathing served as a form for the rear face of the concrete wall.

The soldiers were maintained in alignment by a 24-in. WF 160# waler, and the cross-lot horizontal struts were 14-in. BP 89# members spaced as shown in Sec. 6.4, with the spacing through the central area of the pit at 18 ft, providing a strut at every third soldier. Other aspects of this design will be discussed in Sec. 6.4, but here we are concerned only with the loadings on the struts whose vertical spacing is shown on the elevation illustrated on the opposite page.

The calculations required to convert the loading of 30 tons per lineal foot of wall to strut loadings are shown. The average density of the soil is assumed to be 120 lb per cu ft; and the value of H, the height, is that necessary for the structure to be built. From this, a value of 0.3 for K is obtained, which falls within the range of 14 values recorded by Terzaghi and Peck in their study of the Chicago clays of from 0.2 to 0.5. The maximum unit loading is now computed from KWH and found to be 1666 psf, which is rounded off to 1700 psf.

With this information, the loading on each strut is now computed for the areas shown shaded on the elevation opposite, on the assumption that the load of half the distance between each strut will be borne by any particular strut, with varying loadings vertically based on the trapezoidal form.

On this project, by means of strain gages, the actual strut loadings were recorded as excavation and the placing of struts progressed. (See Sec. 6.10 for methods.) As each level of struts was placed, the members were pre-stressed to the design loadings by means of calibrated hydraulic jacks, as shown opposite in detail. When all the cross-lot struts were in position, the strain gages recorded the actual loadings being received from the bank, and a comparison between design and the actual loads is shown. These indicated some variation, on two center tiers, from the design figures, and it was possible to connect the actual figures so as to show a parabolic distribution of pressures.

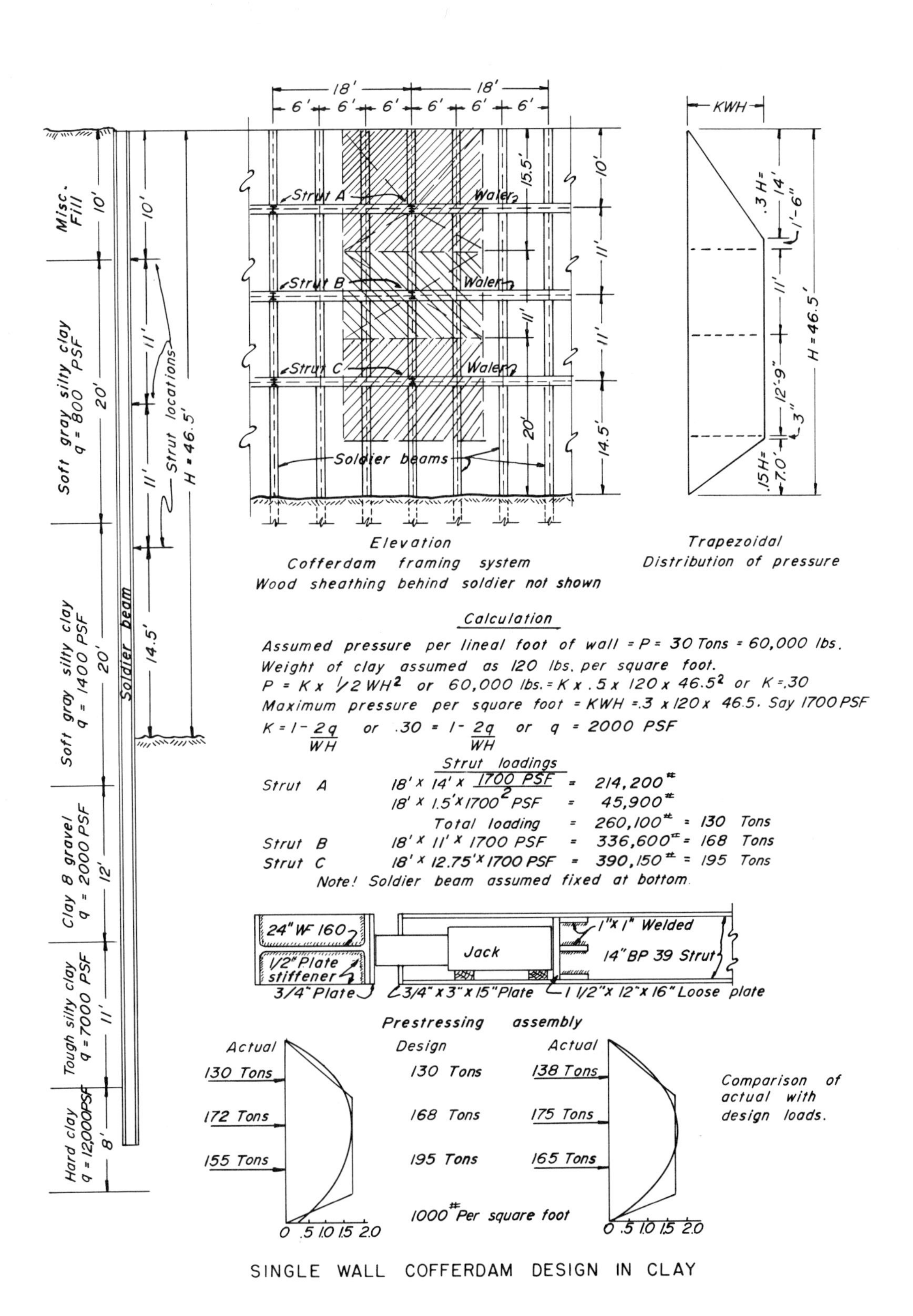

SINGLE WALL COFFERDAM DESIGN IN CLAY

6.4 STRUCTURAL DESIGN OF COFFERDAM MEMBERS

The design data in Sec. 6.3 does little more than establish the strut loadings. We still need to know the following:

1. The size of struts and their spacing
2. The size of walers (their spacing may depend on strut spacing or on sheathing characteristics)
3. The size of soldiers and their penetration into the soil below excavation level (if soldiers are to be used)
4. The kind and size of sheathing

Some of these factors depend solely upon the contractor. If he is habitually in the business of constructing cofferdams, he may have a stock of materials from which to provide struts, walers, soldiers, and sheathing. All that will be required will be some test calculations to check their suitability to a particular application. Some factors, such as strut spacing, may depend upon construction methods, involving considerations of the depth of successive layers of excavation or the positioning of struts in relation to subsequent construction of the foundation structure within the cofferdam.

Over all the contractor's thinking, like poisonous miasmata, swirl the conflicting mandates of cost. The construction cofferdam is a temporary structure (in contradistinction to the permanent bulkhead); and such matters as larger deflections of members, which do not imperil the structure, are permissible. Yet the placing of additional members to bolster a too closely calculated design not only can be costly but may invite danger to the whole system before installation can be completed. In Sec. 6.3 we have seen that theoretical approaches do not always represent actual results, and a realistic factor of safety must always be applied.

In the case discussed in Sec. 6.3, soldier beams and wood sheathing were selected as the cofferdam system to meet job conditions, specifically that they were to be left in place and that they were somewhat less expensive than steel sheet piling would have been.

The design of cofferdam systems brings up the old question as to which comes first, the chicken or the egg. In the discussion in Sec. 6.3 the strut spacing actually used was accepted, but we must consider how it was arrived at. To do this, let us start with the sheathing.

The 6-ft spacing of soldiers is quite standard where 3-in. wood sheathing is being used, since it provides a minimum of deflection in the wood sheathing. To avoid complex calculations and special conditions, the vertical spacing of struts should be in multiples of 6 ft. The 18-ft spacing used throughout the central area is arbitrarily chosen, and then the size of the strut is developed from this (see calculation opposite). The random spacings shown toward the ends of the pit were chosen to suit reduced estimated loadings or to meet foundation construction conditions or where it was necessary to vary the 6-ft spacing because of underground obstacles.

The vertical spacing of struts (and hence of walers) may be controlled by excavation procedures, particularly by the depth to which initial excavation can be carried before yield in the bank becomes serious. Ralph Peck has suggested the use of the average values of q/W as the maximum depth at which to place the top strut, in this case 2000/120, or 16 ft. However, in this case, the upper stratum of miscellaneous fill is so indeterminate in its characteristics that the value of 16 ft was reduced to 10 ft. In all cases the ability of the sheathing system to act as a cantilever above the top waler should be checked (see Sec. 6.6).

The position of the bottom waler may be determined by the requirements of the foundation construction below it, by limitations of strut loadings on members at a higher level, or by considerations of potential yield. It has been suggested that a spacing of $1.5q/W$ be used for the maximum distance above the bottom of the pit. In this case this is excessive since much larger struts would be required and the soldiers would need stiffening.

With the strut spacings determined, it is now possible to calculate the loading on the waler by simple beam formulas and to select a suitable size as shown.

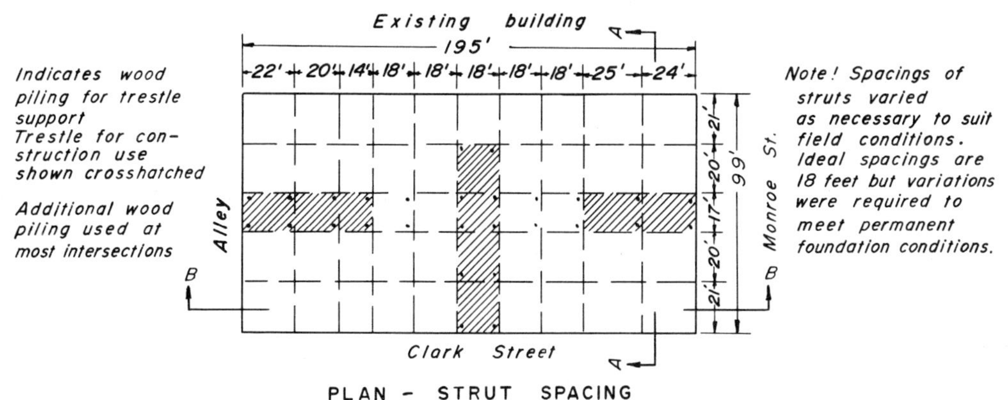

PLAN - STRUT SPACING
HARRIS TRUST COMPANY COFFERDAM

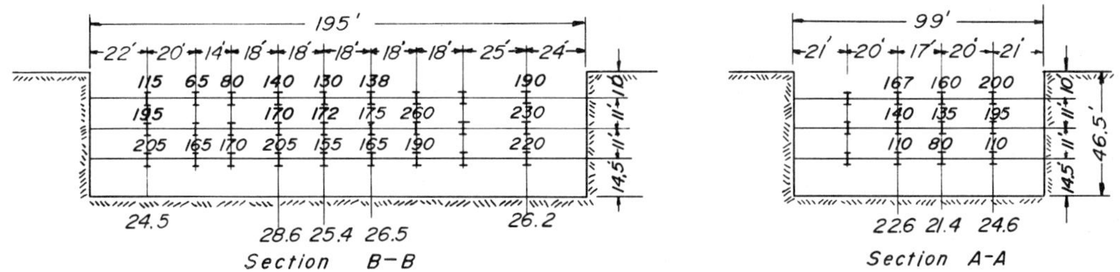

Figures above struts are actually measured loads. See 6.3 for designed loads
Figures below sections are measured wall loads in tons per lineal foot of wall. Design = 30 tons

Theoretical Design Check of Struts

From Sec. 6.3, maximum strut loading is 195 tons = 390,000 lb. f = allowable unit compressive stress = 18,000 psi. 390,000 lb ÷ 18,000 psi = 21.6 sq in. required.

Area of 14-in. BP 89# = 26.2 sq in., which is ample. Actual f = 390,000 lb ÷ 26.2 sq in. = 14,900 psi.

Considering the strut as a column, the longest unsupported span is 20 ft; l = length, r = radius of gyration.

$$L = 20 \text{ ft} \times 12 \text{ in.} = 240 \text{ in.} \qquad r = 3.53 \qquad 240/3.53 = 68$$

$18{,}000 - (1.485 \times 68^2) = 18{,}000 - 2240 = 15{,}760$ psi. This is a safe value.
Actual maximum loading was 230 tons, requiring 25.5 sq in.—still safe.

Theoretical Design Check of Walers

From Sec. 6.3, maximum loading occurs on bottom waler; investigate for 18-ft span.

65,000 lb 130,000 = F = 130,000 65,000 lb

Moment diagram ↓a = 6 ft ↓ a = 6 ft ↓a = 6 ft↓
↑R = 195,000 lb R = 195,000 lb↑

$$F = 6 \text{ ft} \times 12.75 \text{ (see 6.3)} \times 1700 \text{ psf} = 130{,}000 \text{ lb}$$

Maximum moment = Fa = 130,000 lb × 6 ft × 12 in. = 9,360,000 in.-lb.
f = allowable unit tensile stress = 24,000 psi.
Section modulus = 9,360,000 in.-lb ÷ 24,000 psi = 380.
Section modulus of 24-in. WF 160# = 413.5, which provides a safe value.
Actual values of F exceeded that above only slightly and over limited areas.

6.5 STEEL SHEET PILING

The discussion in Secs. 6.3 and 6.4 has been concerned with a particular system of single-wall cofferdamming consisting of vertical soldiers backed up with horizontal wood sheathing. It has been convenient to refer to this type of system here in order to compare theoretical design with the actual results obtained in a case where field tests were conducted. It is by no means the most important system of single-wall cofferdamming.

The earliest systems of cofferdams were constructed of wood sheathing, backed up with timber walers and struts. This system is still used extensively for relatively shallow earth cuts limited to about 20 ft in depth by the nature of the material. A brief résumé of data concerning timber systems is given in Sec. 6.19.

The use of concrete sheathing is limited to permanent installations or to uses where the sheathing becomes a part of the basic structure. A brief résumé of its characteristics will be found in Sec. 6.20.

Steel sheet piling has gradually superseded other types of sheathing for projects of substantial size or where conditions of driving are in any way critical. The nature of the material, steel, makes it possible to drive this sheathing to greater depths through more compacted soils and to provide a stronger wall with less walering and bracing. Its re-usability makes it economical for temporary construction purposes.

Steel sheet piling consists of rolled steel sections with interlocking edge joints produced in three general types, in graduated weights, to meet a variety of field conditions. Sections rolled in the United States are shown opposite.

Straight-web piling is designed primarily for applications where the tension value of the interlock is a prime consideration and consequently finds its principal use in cellular cofferdams. Sheets are available in two web thicknesses, ⅜ or ½ in., the choice depending upon the hardness of the driving conditions. Sheets employ the thumb-and-finger principle for interlocking, and the several sections will interlock with each other. Standard design tension in interlocks is 8000 lb per lineal inch; but in ordering, the design tension required should be specified since special rollings can provide tension stresses up to 16,000 lb per lineal inch. A deflection of up to 10 deg from a straight line, without loss of strength, is an additional interlock characteristic that has made the cellular cofferdam possible.

Arch-web piling is designed for conditions requiring a combination of beam strength and interlock tightness. These sections also employ the thumb-and-finger principle of interlocking, and all arch sections will interlock with each other. Arch-web sections are used for braced single-wall cofferdams, for light bulkheads, and for cut-off walls. Only the two flattest arched sections are used for circular cofferdams, the other sections having a tendency to flatten out in tension. The two flattest sections (MP-112 and MP-113 or SP-4 or SP-5) have a normal design tension in interlocks of only 3000 lb per lineal inch, although special rollings will provide strengths up to 12,000 lb per lineal inch.

The Z-pile is designed for high beam strength and offers the greatest possible ratio of beam strength to weight. The Z-pile differs from the other two types of sections in using a ball-and-socket interlock providing a tight joint but having little value of tension in interlocks. The shape of the pile also precludes its use in tension since, like the arch-web pile, it has a tendency to flatten out. The Z-pile is used chiefly in braced single-wall cofferdams or for permanent, filled bulkheads where high beam strength is required. Not all sections of Z-piling will interlock with each other, but all do interlock with the two flat arched sections.

Z-piling has a greater tendency to variation in driven length of wall than the other sections. It is, essentially, set in pairs, alternating between two faces pointing in and two faces pointing out. Since some clearance must be provided in the guides in which it is set, a slight twist of the pile will shorten the laying length.

STEEL-SHEET PILING SECTIONS ROLLED IN THE UNITED STATES

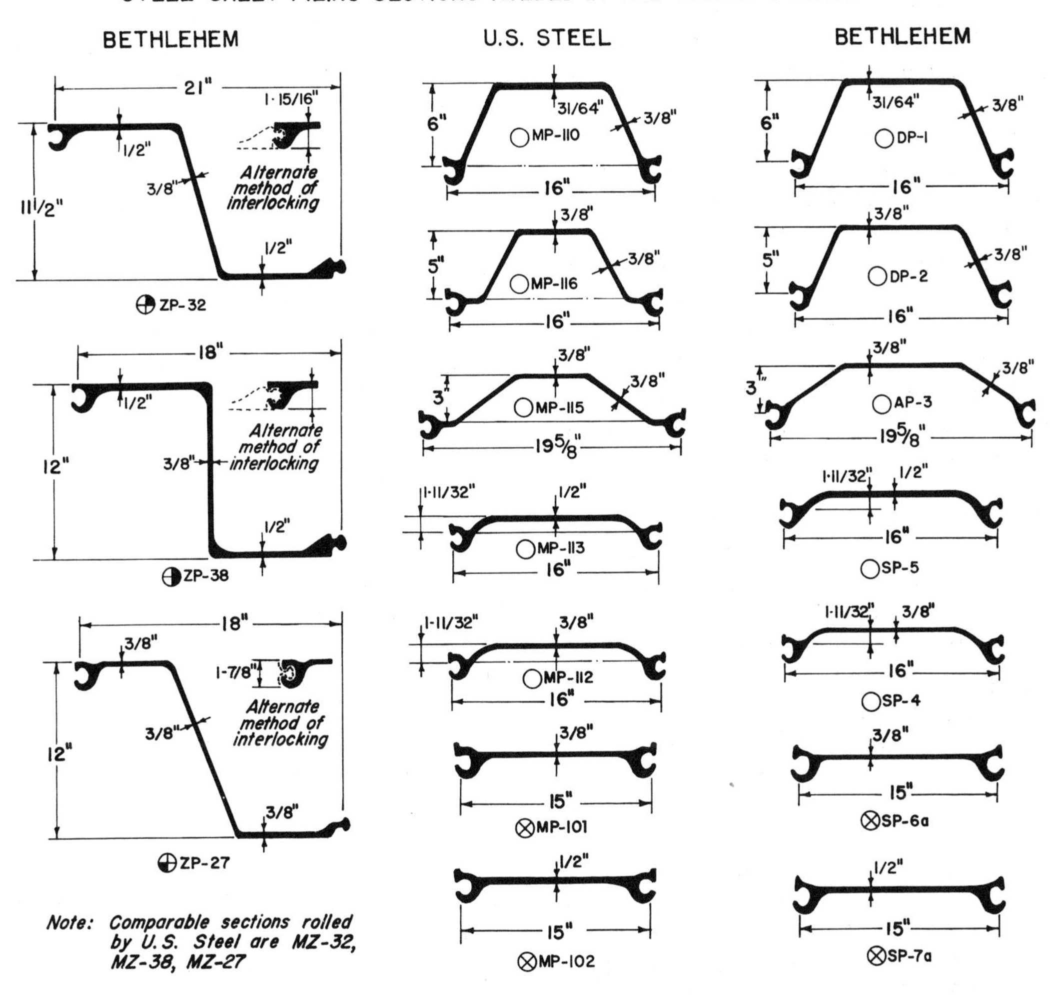

ROLLED CORNERS

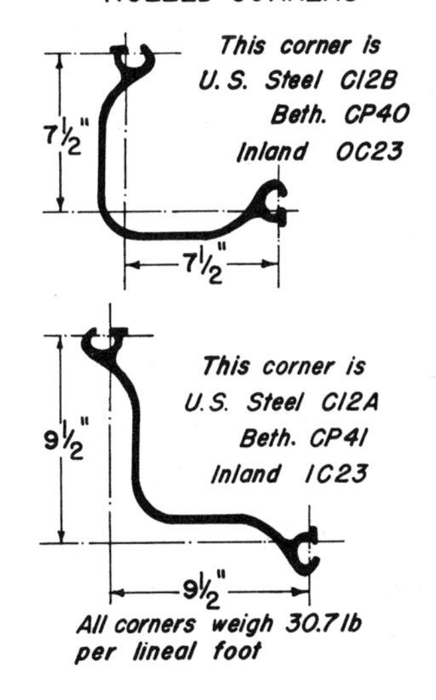

Characteristics of Steel-sheet Piling

U.S. Steel symbol	MP 110	MP 116	MP 115	MP 113	MP 112	MP 101	MP 102	MZ 32	MZ 38	MZ 27
Bethlehem symbol	DP 1	DP 2	AP 3	SP 5	SP 4	SP 6A	SP 7A	ZP 32	ZP 38	ZP 27
Inland Steel symbol	I 32	I 27	I 22	I 28	I 23	I 28				
Weight per lineal foot	42.7	36.0	36.0	37.3	30.7	35.0	40.0	56.0	57.0	40.5
Weight per square foot	32.0	27.0	22.0	28.0	23.0	28.0	32.0	32.0	38.0	27.0
Section modulus, in.^{3} per pile	20.4	14.3	8.8	3.3	3.2	2.4 B–3.0	2.4 B–3.0	67.0	70.2	45.3
Section modulus per lineal foot of wall	15.3	10.7	5.4	2.5	2.4	1.9 B–2.4	1.9 B–2.4	38.3	46.8	30.2

6.6 SELECTION OF STEEL-SHEET-PILING SECTIONS

It has been noted that the material which a contractor may have or can procure at least cost (generally this means secondhand material which has been used one or more times but which is structurally sound) will influence his decisions in the overall design of a temporary structure. In the case of single-wall cofferdams this will apply to the sheathing as well as to the struts or walers. Much lighter sections of steel sheet piling can be used than may be expected if water conditions can be controlled or if strut-and-waler spacing is judiciously arranged.

Steel sheet piling is purchased or rented by the ton f.o.b. site. New piling will cost about $175 per ton, varying somewhat depending upon location of the site, quantity, lengths, and other factors. Rented piling will cost about $70 per ton, with an additional charge of $2 per ton per month. Damaged or left-in-place rented piling must be replaced in kind, that is, with new piling, so that where considerable loss is to be expected or the time of use is lengthy, new piling is a better deal. It can readily be seen that the increased tonnage resulting from the use of a heavier-than-needed section can exceed the cost of additional supporting members, in many cases.

As has been noted, straight-web piling sections, because of their low beam strength, are seldom used in single-wall cofferdams; yet this use should not be entirely ruled out since in instances where loading is light or depths are shallow, straight-web sections can perform satisfactorily. Arched-web or Z-piling, either of which *do* develop substantial beam strength, *will* generally be used for single-wall cofferdams.

There are two loading conditions for steel sheet piling; the first is where it acts as a cantilever with one end rigidly held, and the other is where it acts as a simple beam under more or less uniform loading. Cantilevering can occur from the top strut to the surface of the ground, or it can exist where sheet piling is driven into the ground using the ground resistance for stability rather than struts. In this latter case it will generally be found that the critical condition is that of the resistance developed by the soil rather than the beam strength of the section of piling.

Terzaghi has pointed out that the resistance of a cohesionless soil (sand) against an outward movement of the buried portion of a steel sheet pile depends on the soil's effective unit weight and on its coefficient of passive earth pressure. For cohesive soils (silt or clay) the resistance depends only on its unconfined compressive strength. Values of these are tabulated opposite, and an example of their use is indicated.

On the opposing page, calculations are developed for selecting a sheet piling section for the loading conditions developed in Sec. 6.3. Calculations for alternative possibilities of strut loading are also investigated. To round out the picture, comparative costs between the necessary section of sheet piling and the method of soldiers with horizontal wood sheathing, which was actually used, are shown. It is quickly apparent why, for this permanent installation to be incorporated into the finished structure, the method used was selected.

Such alternatives as noted above are not always possible. The use of horizontal wood sheathing is not practical where water or liquid clay or silt combinations are expected to be encountered. And expected conditions may be altered by factors which have never been considered. In the case of the Harris Trust sheathing, considerable water intrusions were encountered in what should have been a relatively homogeneous clay bank due to leaking water mains and sewer lines, and the contemplated 3-in. wood sheathing had to be changed to 4-in.-thick wood sheathing at a depth of 35 ft to meet these local and unforeseeable soil conditions.

While it has been stated that greater deflections are permissible in the walls of temporary structures than can be tolerated in more permanent installations, bear in mind that deflection can produce yield in the bank, changing the distribution of earth pressures, sometimes disastrously.

Steel Sheet Piling Selection

The selection of a section of steel sheet piling, to be used in a particular instance, will be governed by the following factors:

(*a*) The basic characteristics of each type of section as noted in Section 6.5. Straight web sections will be used in tension; arch web sections will be used where some beam strength, as well as limited tension, is required; the Z section will be used where high beam strength is desired.

(*b*) The availability of a particular section. A surge of demand for certain sections may have depleted rental stocks. Some sections are not regularly rolled by manufacturers and an order of substantial quantity may be necessary to induce the mill to roll the desired section.

(*c*) The loading conditions of a soil bank are far from predictable and, moreover, are subject to change during the time interval of construction. Precise calculations frequently are unwarranted.

Let us take the loading conditions described in 6.3, where the maximum loading occurs in the third bay from the top and is 1700 psf, over a span of 11 ft. In this case, figured as a simple span:

$$M_{max} \text{ (at center)} = \frac{wl^2}{8} = \frac{1700 \times 11 \times 11 \times 12 \text{ in.}}{8} = 308{,}550 \text{ in.-lb}$$

$$\text{Section modulus} = \frac{308{,}550 \text{ in.-lb}}{20{,}000 \text{ psi}} = 15.45 \text{ in.}^3$$

If this is figured as a continuous beam it will be found that the section modulus is only 12.15 in.^{3}

It is apparent therefore, that an MP-110 section, with a modulus of 15.3 in.^{3} per ft of wall, will be adequate. However, the weight of this section, per square foot of wall, is 32 lb, whereas an MZ-27 section weighs only 27 lb per sq ft of wall, and that its section modulus is twice that of the MP-110.

The comparative cost of the *material* in the steel sheet piling system, versus the soldier beam system shown in 6.3, is:

Steel sheet piling: 18 ft $\times$ 11 ft $\times$ 27 psf = 2.673 tons @ \$175.00/ton = \$468.00, or \$2.36/sq ft.

Soldier beams: 3 pcs $\times$ 11 ft $\times$ 96 lb/ft = 1.584 tons @ \$175.00/ton = \$277.00
Sheathing: 18 ft $\times$ 11 ft $\times$ 3 in. = 600 FBM @ \$150.00/m = 90.00
Total = \$367.00

This is equivalent to \$1.90/sq. ft.

The relative costs of installation will vary with the type of project and its size. In both cases, equipment and crews for pile driving will be required and, under hard driving conditions, it is likely that the costs of installing the sheeting will be higher than the combined costs of the soldier system.

A check of the section above the top waler, as a cantilever, will show that the required section modulus is very low for the loading in 6.3.

The bottom section may be considered either as a cantilever, with zero soil resistance, or as a beam, with the reaction of the soil, in tons per square foot, as shown in the following table.

Unconfined compressive strength, tons/sq ft			Coefficients of earth pressure		
Consistency	Minimum	Maximum	Sand	K-active	Passive
Very soft	0	0.25	Clean dense	0.20	9.0
Soft	0.25	0.50	Clean medium	0.25	7.0
Medium	0.50	1.00	Clean loose	0.30	5.0
Stiff	1.00	2.00	Silty dense	0.25	7.0
Very stiff	2.00	4.00	Silty medium	0.30	5.0
Hard	4.00	∞	Silty loose	0.35	3.0

Steps Preliminary to the Driving of Steel Sheet Piling

A. *Initial Considerations*
   1. Selection of the proper piling section (see Sec. 6.6)
   2. The rental or purchase of the required steel sheet piling (see Secs. 6.6 and 6.23)
   3. Scheduling shipments
      a. It will seldom be practical to set up piling directly from a flatcar or trailer bed.
      b. Schedule shipments so that crane setting piling can unload each shipment without having to "walk" excessively.
      c. On large operations a separate crane should be deployed for unloading.
   4. Condition of received piling
      a. Inspect piling as it is unloaded and stacked.
      b. In shipments of random lengths see that each sheet is marked for length.
      c. Check piling to be sure it is straight.
      d. Check interlocks to be sure they are not damaged, that they are clean, and that they are free of obstructions.
   5. Handling
      a. Grabbing through handling hole is impractical for stacking except in the shorter lengths.
      b. Use bridled slings with tongs for long sections.
      c. Check for possible deflections in long lengths (using weight per foot and section modulus) in order to determine grab points. This avoids permanent bending of sheets.
   6. Storage
      a. Stack in nested lifts of about five sheets each, using oak blocks as lift separators at intervals of 10 to 15 ft.
      b. Stacks should be located so that the crane, when setting them up, can make grab with no more than a swing.
      c. Separate by lengths where depths of sheeting will vary throughout the length of the cofferdam wall.

B. *Field Preparations*
   1. The entire operation of setting steel sheet piling will depend on soil or water conditions at, or close to, the starting surface.
   2. Guides or templates (see Sec. 6.16) may be:
      a. Simple timbers staked to the ground surface
      b. Frames built on the surface or in partially excavated pits
      c. Templates on guide or spud piling
      d. Templates floating on water surfaces
   3. The use of leads does not replace guides or templates.
   4. A portion of the final bracing may be used as a driving guide. This possibility will be tied in with excavation procedures.
      a. Can any portion of the excavation be done before sheeting is placed?
      b. How closely will excavation follow sheathing?
      c. Is sheathing to be placed along the face of the existing soil bank?

C. *Equipment*
   1. Crane rigs, crawler- or truck-mounted (see Sec. 5.14)
   2. Pile-driving hammers (see Sec. 5.16)
   3. Selection of hammers (see Sec. 6.8)
   4. Leads (see Sec. 6.8 for use and Sec. 5.15 Piling, for description of)

6.7 SET-UP OF STEEL SHEET PILING

The preparatory steps incident to a set-up for driving steel sheet piling are tabulated on the opposing page and are fairly general in their application. It is now necessary to limit the description to single-wall cofferdams, either driven in a straight line or enclosing an area, generally rectangular. The basic procedures will also apply to circular cofferdams, although some special aspects of these will be discussed later.

With the guide frame in place, a start is made either at a corner of an enclosed area or in the middle or at the end of a straight run. Starting at a number of random locations will mean an equivalent number of special closures, which are expensive. Multiple theoretical spacings are difficult to maintain since there are clearances in the interlocks as well as mill rolling tolerances which may vary any length in actual practice.

Starting at a corner, the corner piece and the two adjoining straight pieces on either leg are set up. These are then plumbed carefully and braced to stakes driven diagonally into the ground or to members of the guide frame, using cable or rope.

Sections of sheet piling are raised by means of a screw-pin-anchored shackle 6 or 9 in. long hanging from a wedge socket on the end of the crane's hoisting line; a hook should never be used for this. For long sections a second lift may be required, using a choker at the sheet's midpoint to prevent excessive bending until the sheet is hanging vertically. A tag line is tied around the bottom of the pile to prevent it from swinging as it is raised and to guide it into position for "threading."

The "threader" or "pitchman" is atop the row of piles and is provided with a "saddle" and a set of steel "stirrups." He moves along the previously set piling by standing in the stirrups to slide the saddle along and then sits in the saddle to slide the stirrups up. With the bottom end of the sheet brought within his reach, he threads the thumb into the previously placed fingers and the sheet is lowered a foot or two. The threader then disengages the tag line, letting it fall, and the sheet is lowered to ground level.

The man on the ground thrusts the sheet against the guide frame and holds it there with a short length of 2 × 4 while it is raised a few feet and dropped. The operation at this point will depend on the nature of the soil. In soft soils where the weight of the pile will produce substantial penetration, the sheet may be raised and lowered several times in the slot produced by the first drop, to get it down to a uniform grade at the top. In firm soils it may be necessary to give the pile a few taps with a hammer to provide a reasonably level top for the threader to travel on and to provide retention for the bottom of the sheet.

In the usual case, one rig both sets and drives; the hammer, hanging slackly from one line, being set on the ground while a sheet is being placed. From this position it can be raised to provide initial penetration if need be. But before any penetration can be provided, each piece of piling should be checked for plumbness. Guide frames seldom touch the sheet at more than two points; and, unless plumb, the sheet can lean in or out during initial penetration. There is also a tendency of the pile to lean along the frame because of slack in the interlocks. Unless corrected as each pile is placed, this lean will become progressively greater.

A well-planned and well-placed sheet-pile enclosure should require no special pieces. Starting from one corner, setting progresses in each direction to the second and third corners (of a rectangle). If there has been any "walk" in the sheets due to play in the interlocks, strips placed horizontally along the guide frame on the last two sides can adjust the dimensions of the enclosure and thus avoid special closure pieces. At the final corner, usually the third side is completed and the corner piece set without penetration, closure being made in the fourth side. After the pieces at the corner have been threaded, two or three pieces can be raised together and then lowered to permit interlock adjustment. Some deflection at the final corner from a true right angle is seldom objectionable or dangerous.

6.8 DRIVING STEEL SHEET PILING

After all the sheets have been placed for a line or an enclosure, the assembly should be checked to be sure it is still plumb; and if necessary, guy lines should be used diagonally across the tops (or diagonally downward) at spaced piles to aid in maintaining the plumb condition during driving. Except on very short sections, driving should be done in increments of perhaps a third of the depth at a time. After each increment (or when adequate penetration has been secured), guy lines can be re-set. (They must be released when driving the pile to which they were attached.)

The selection of hammer will depend on the length and weight of pile, the depth of penetration, and the resistance of the soil. Generally a double-acting hammer is most advisable since it provides short, quick blows, but for driving heavy sections in tough ground a single-acting hammer is preferable. The hammer should be about 2½ times the weight of a sheet of piling.

Steel sheet piling is driven in pairs with the hammer centered over the intermediate interlock where the greatest section of metal is concentrated. Simple swinging pile leads should be used with single-acting hammers or drop hammers or where the pile is more than 60 ft in length. In these cases a cast steel driving cap, wedged to the top of the pile with a hardwood dolly, should also be used. The more customary method, however, is to bolt fishtailed "pants" to the bottom sides of the hammer to hold it in place on the top of the sheet. Since the operator must lower the hammer as the pile sinks, it is necessary, in soft ground, to limit the hammer to short, quick blows until substantial resistance develops, to maintain the hammer pants astride the sheeting. As a further precaution in these cases, a tag line secured to the hammer and controlled by a man on the ground can aid in re-positioning the hammer without undue loss of time if it does lose the pile.

Ground resistance must be carefully gaged by the man directing the driving operation. One type of resistance is that due to the plug of soil in the forward interlock which the thumb must extrude. This resistance can be decreased by jamming a bolt or plug into the bottom of the fingers before the pile is placed. The other type of special resistance is that due to underground obstacles, particularly boulders. The driving foreman must gage the amount of driving being required on any pair of piles to judge whether they have landed on a ledge, boulder, or other obstacle. Overdriving, in this case, can batter up either top or bottom of the pile or, if it is sliding down along a sloping face, can spring the interlocks, tearing the thumb out of the fingers.

Where apparent obstructions are encountered, leave the hung-up piling alone and continue driving along the line until piling on either side is well below the obstruction, then return and attempt to re-drive. This will indicate the extent of the obstruction and may provide some clue as to its nature. The deeper piling on either side may also act as a guide to the hung-up sheets. Where it is found that the sheet cannot be driven through the obstacle, it may be necessary to wait until excavation has reached that level and, by blasting or other means, break up the mass.

The correction of piling overdriven and forced out of interlock is a difficult and expensive operation. In constructing a portion of the Harvey Tunnel in greater New Orleans by open cut, in 1957, areas of I-23 sheet piling were driven between master piles set on 13 ft 4 in. centers. When excavation and dewatering began, extensive out-of-interlock conditions were encountered, some only at the bottom while others extended as much as 20 ft above the bottom. Some piling was extracted and re-driven with the same result; some was patched up with tapered sections; but for the most part the out-of-lock sections could not be pulled, and it was necessary to back them up with a concrete wall poured under water.

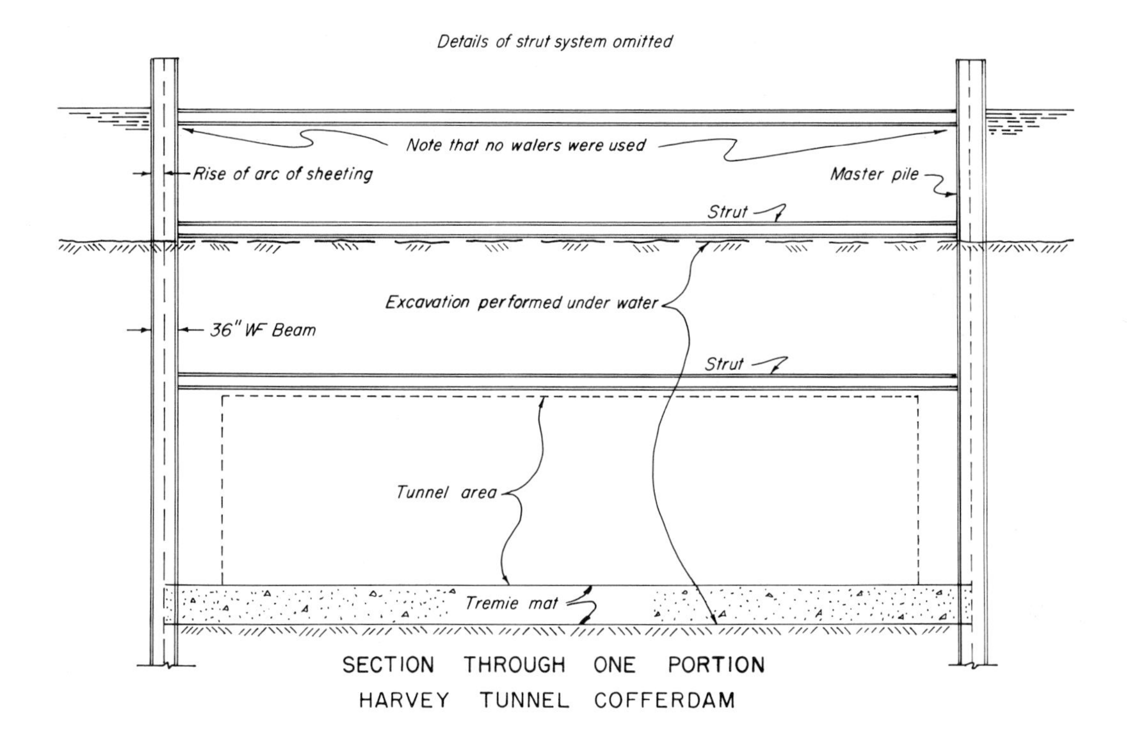

SECTION THROUGH ONE PORTION
HARVEY TUNNEL COFFERDAM

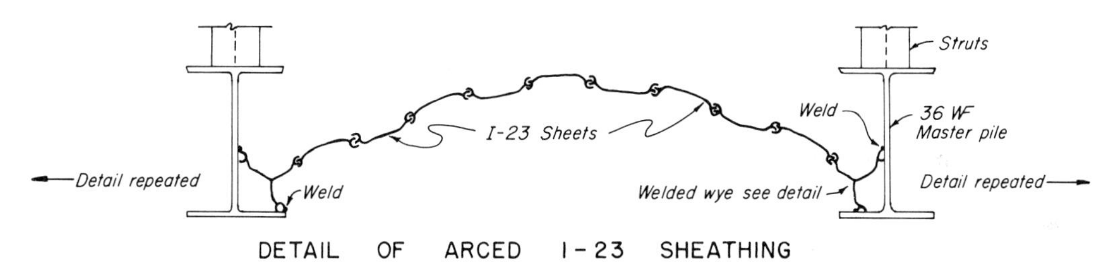

DETAIL OF ARCED I-23 SHEATHING

This is an adaptation of an arc buckstay wharf

1. *Master piles must resist - in bending - the entire lateral load.*
2. *A straight web sheathing should be used.*
3. *Even the flat arched section used caused the interlocks to flatten and twist under tension - throwing them out of interlock.*
4. *An odd number of sheets must be used between each pair of master piles.*
5. *Penetration of sheathing will be less than that of master piles.*

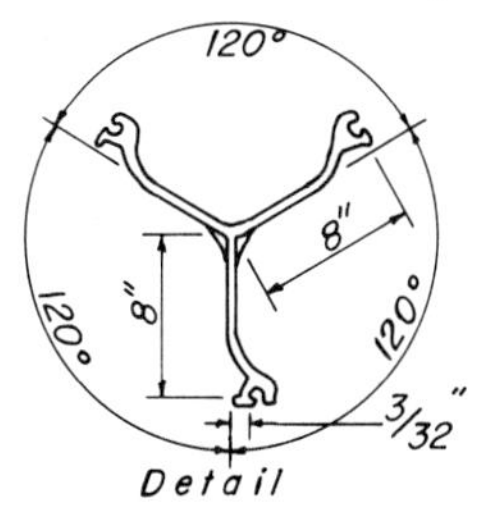

Detail

Standard welded wye

Remedial measures

A *Generally sheathing could not be pulled.*
B *When pulled and redriven it drove out of interlock though sheeting had 12,000 #/ lineal inch design tension.*
C *Tension rods 2 1/2" dia. were bent to a radius parallel to sheet piling arc.*
D *Threaded ends were thrust through webs of master piles and bolted in place.*
E *12" thick wall of concrete was poured around bolts on face of sheathing.*

6.9 A STRUTTED SINGLE-WALL COFFERDAM

Walered and strutted cofferdams have been referred to as though the struts were truly horizontal in all cases. While this is the desirable condition, it is by no means a limiting condition. Of more importance in defining this type of single-wall cofferdam is the fact that one cofferdam wall is braced against an opposing wall of sheathing. As a case in point let us consider the construction of a retaining wall 280 ft long and 38 ft wide along the edge of a traveled highway.

The highway had been built on a fill some 30 ft above the natural adjacent terrain; and, in the process of utilizing this ground level for expansion of an aluminum rolling mill, it was desired to carry construction up to the property line abutting on the highway. The bottom of the proposed retaining wall would be at a depth of some 10 ft below ground level, with the result that about 40 ft would exist as a difference in elevation between the two rows of sheathing (see opposite).

Alignment along the high side had to be precise, so a guide consisting of H-piles, 30 ft long, driven on 25-ft centers, with 10-in. H-beams tack-welded to brackets fastened to them was set up. Driven with leads, these piles would be removed as sheathing was placed.

MZ-38 sheathing 55 ft long was driven along the upper template and an MZ-27 section 25 ft long was driven along the lower template. With the sheathing in place, excavation between the two lines was carried down to a depth sufficient to place the top waler on each side, that on the lower side being just below ground level and that on the upper side being set just above. With the walers in place and welded to the brackets fastened to the sheathing, the struts—12-in. WF beams on 14-ft centers—were wedged into place and tacked to the walers.

Second-stage excavation between struts, by clamshell, now carried the level down to that of the second row of walers, where struts, now essentially horizontal, were placed. It should be noted here that although one condition is described there was considerable slope longitudinally along the road, so that as the difference in elevation decreased, it was possible to eliminate the bottom row of struts entirely.

With the sheathing in place and braced, with the excavation complete to foundation grade, the first step was construction of the vertical wall, 18 in. thick about 4 ft inside the sheathing on the high side. Where the struts projected through the wall, a box-out in the wall form left the struts free for later removal. Although the wall would rise above finished grade when completed, only enough to bring it above the upper waler line was poured in the first lift.

With a section of the wall poured, the slab section behind the wall was placed. When the concrete had obtained sufficient strength, it was possible to remove the lower strut, bracing the front waler against the wall with timber blocking. In the rear, the waler could be removed entirely, by substituting blocking directly against the sheathing if necessary. (Generally this was not necessary, since the single waler remaining would be sufficient to carry the short projection of the sheathing.)

With the initial phase completed and the boxed-out openings, where the lower struts pierced the wall, plugged with concrete, the remainder of the wall could be poured.

With the wall completed, the final step was that of transferring the loading of the sheathing on the high side from the struts and walers to the wall itself by the use of 12 × 12 in. timber struts. The WF struts and walers could now be removed and the upper holes in the wall plugged prior to the removal of the steel sheet piling. As the sheets were extracted or pulled, backfill followed closely, well compacted, to replace the timber struts, each of which was removed just prior to extraction of the sheet on which it provided bearing.

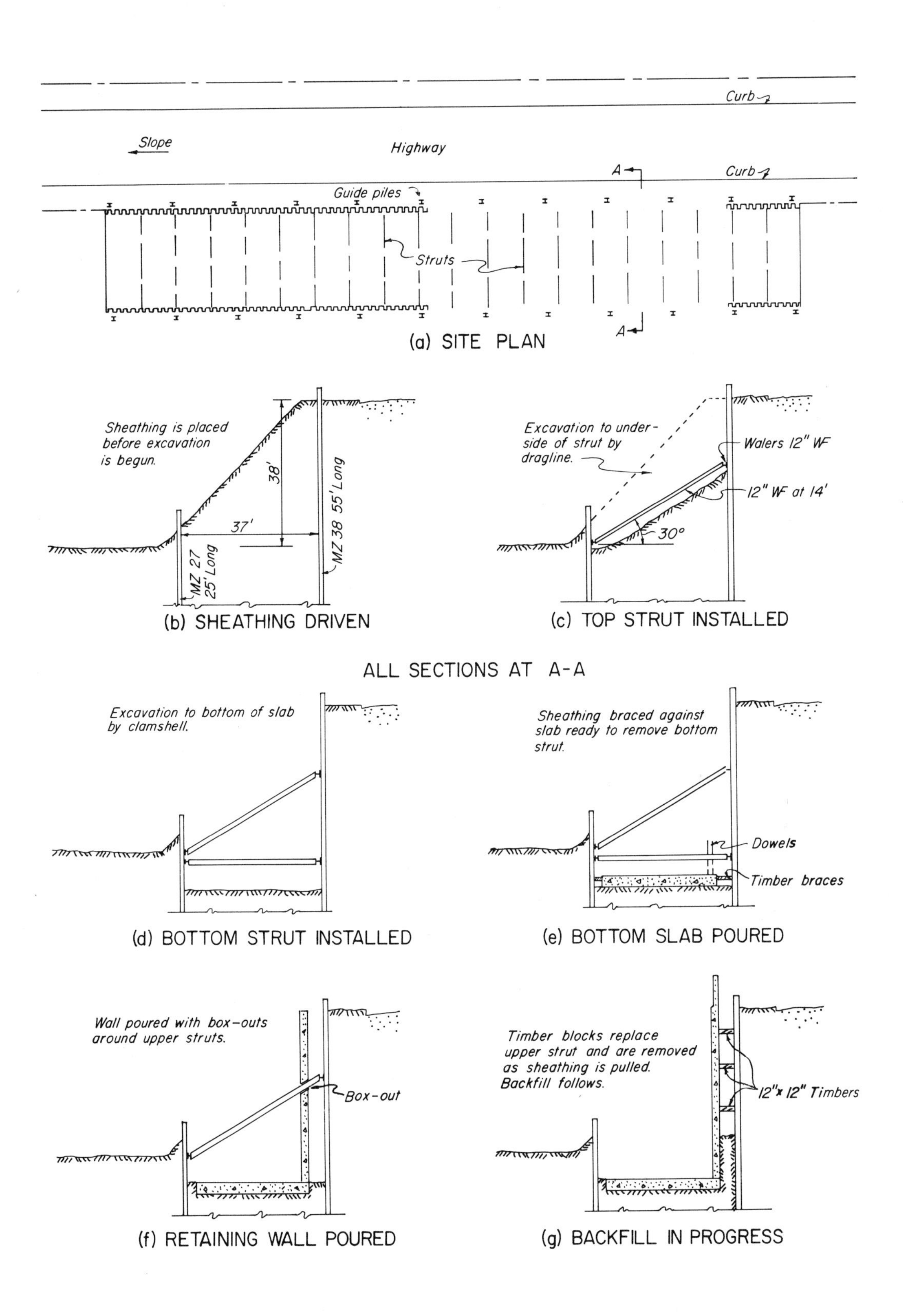
Curb
Slope
Highway
A
Curb
Guide piles
Struts
A
(a) SITE PLAN
Sheathing is placed before excavation is begun.
38'
37'
MZ 38 55' Long
MZ 27 25' Long
(b) SHEATHING DRIVEN
Excavation to underside of strut by dragline.
Walers 12" WF
12" WF at 14'
30°
(c) TOP STRUT INSTALLED
ALL SECTIONS AT A-A
Excavation to bottom of slab by clamshell.
(d) BOTTOM STRUT INSTALLED
Sheathing braced against slab ready to remove bottom strut.
Dowels
Timber braces
(e) BOTTOM SLAB POURED
Wall poured with box-outs around upper struts.
Box-out
(f) RETAINING WALL POURED
Timber blocks replace upper strut and are removed as sheathing is pulled. Backfill follows.
12"x 12" Timbers
(g) BACKFILL IN PROGRESS

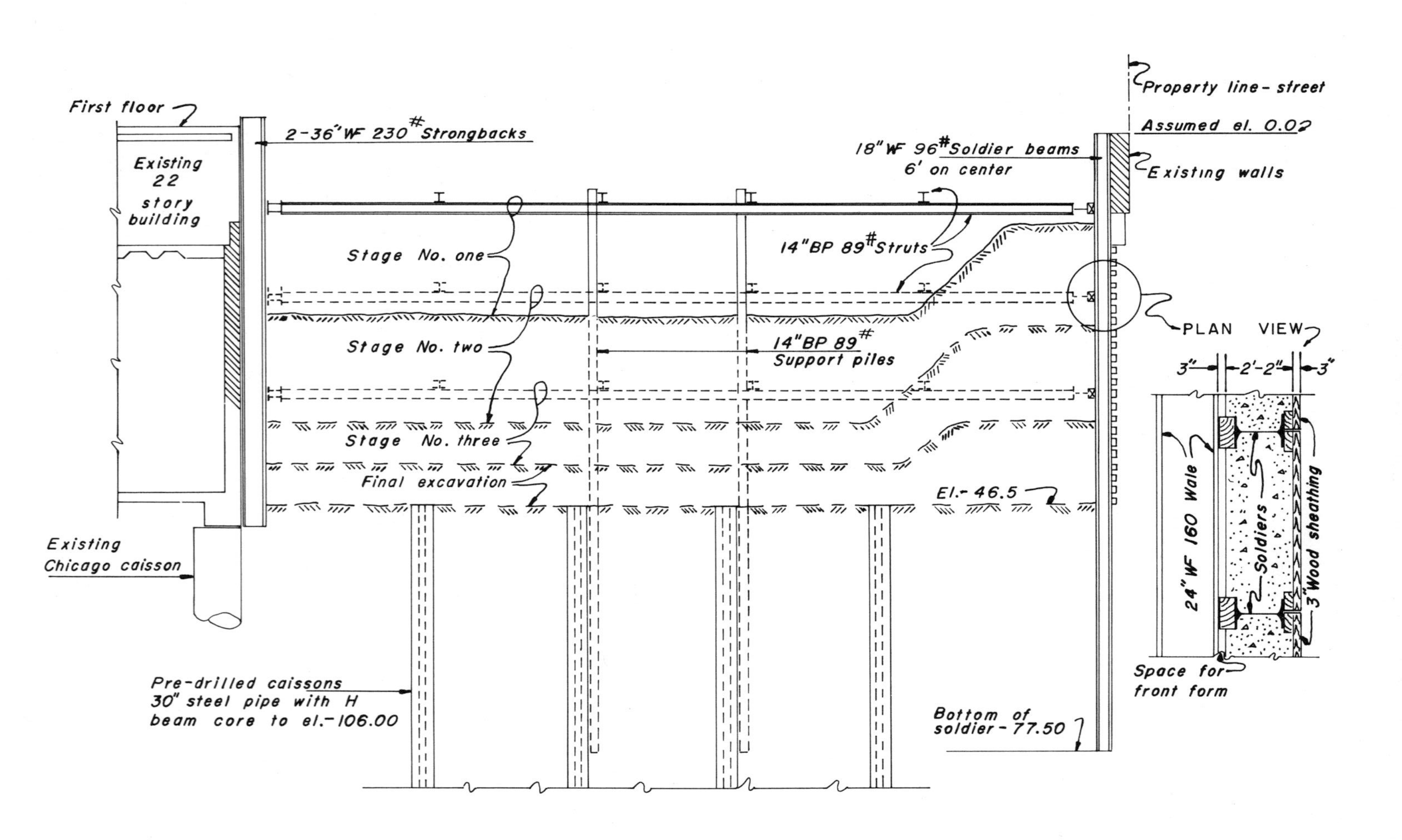

A HORIZONTALLY STRUTTED COFFERDAM

6.10 A HORIZONTALLY STRUTTED COFFERDAM

A prime example of the procedures involved in placing a horizontally strutted single-wall cofferdam is that required for the Harris Trust Company foundations, in Chicago, previously cited in Sec. 6.3.

The starting level at this site was 15 ft below street level, the basement depth of the pre-existing structures whose walls, together with berms, provided some temporary support around the perimeter of the site. From this depth, drilled-in caissons 90 ft long were placed to provide the main support for the interior columns of the new building.

Eighteen-inch WF 96# beams set on six-foot centers, the soldier beams for the exterior sheathing, were now driven around three sides of the perimeter using leads to maintain precise position and verticality since they were to be incorporated into the finished structure. A second pile-driving operation involved the driving of temporary wood piles within the area. These piles, cut off at street level, would provide support for three trestles constructed from the street out over the site, on which crawler cranes could be operated for excavation and the placing of the bracing system. These wood piles were so positioned that they could also support the struts at their intersections.

The fourth side of the perimeter was occupied by an existing building of varying heights and no great age, supported on Chicago caissons. The new structure was to abut the existing walls tightly, so that here the use of soldier beams was inappropriate. Since the cross-lot bracing would have to butt against the existing walls, precautions were required to prevent excessive induced stresses. The 6-ft spacings adopted on the other sides were not suitable here; instead, pairs of 36-in. WF 230# strongbacks were set vertically against the wall on the center lines of the interior columns, of sufficient length to span floor levels. These were then backed up with double 24-in. WF 160# walers set at the same level as the cross-lot struts.

Excavation proceeded in three steps as illustrated, each step being carried to just below respective strut levels, leaving a substantial berm with a 1:1 slope around three sides until the strut had been set and backup wood sheathing had been placed down to that level.

It is customary to place wood sheathing in soldier beams behind the forward flange, within the web area, requiring a minimum of excavation behind the face of the soldier. In this case the sheathing was placed not only behind the rear flange but behind a 3-in. filler strip required to obtain the desired wall thickness. The 24 in. of excavation resulting could not be performed by the clamshell traveling on the trestles; and a crawler-mounted Gradall, traveling on mats, was required to slice this out.

To set horizontal wood sheathing, cut away the soil (usually clay) behind two adjoining soldiers to the depth required and then shave off the soil between them. A length of wood sheathing, usually a 3 × 10, is thrust diagonally behind one soldier and is then rotated vertically until it is engaged behind the opposing beam. It is then tapped upward into horizontal position, beneath the piece previously placed. A spacer block 1 or 2 in. in thickness is provided between each piece of sheathing to permit bank drainage. Where a flow of water carries soil with it, salt hay or straw is used to plug the space, still permitting water to drain but restraining the soil. It can be seen at once that this type of sheathing is limited to banks of relatively firm soil.

Not only is considerable care and skill required in shaving the bank, but under the best conditions, full contact can not be expected. The pre-stressing of struts undertaken in this case may force the back face of the soldier tight against the bank but has little effect on the intervening planking.

In first starting this type of horizontal sheathing it is possible to stack a number of timbers, driving them down with a drop hammer with additional plank added to the top of the stack as required; but the depth to which this procedure can be performed is limited, and damage to the sheathing and vibratory damage to the bank face can nullify any advantage gained.

6.11 A RAKERED SINGLE-WALL COFFERDAM

A single-wall cofferdam used in 1961 to hold back the alluvial soils in New Orleans while the foundations for a 28-story office building were being built is of interest not only because it employed diagonal bracing rakers but also because these were constructed of pipe sections rather than the more usual H or WF structural steel sections.

The building site, 113 by 164 ft, is a corner lot with streets on two sides and buildings, if but rudimentary, on the other two sides. Two sizes of steel sheeting piling sections were decided upon for the 25-ft-deep excavation. Along the two street sides the sheathing could be pulled after completion of the foundation, and here an MZ-38 section was used. Along the building lot sides it would be necessary to leave the sheathing permanently in place, and here an MZ-32 section was used.

All sheet piling was driven from the street level of the cleared site by using two 14-in.-square timbers staked to the ground as guides. A foot of the 40-ft-long sheets was left projecting above ground level on the street sides to prevent street drainage from flooding the site.

Permanent support of the building was to be on 750 cast-in-place concrete piles 60 ft long. An initial excavation of the site to a depth of 6 ft was made, and the permanent piling placed from this level. Shells were left in place to the full height although concreted only to a depth 18 ft below this, their cut-off height.

At this point temporary wooden piling was driven to support trestles built out over the site, as described in Sec. 6.10, to carry excavating and concreting equipment.

The upper 6 ft of the steel sheet piling had been serving as a cantilever up to this time. Sixteen-inch WF 58# walers were now set in position around the periphery, hung from structural tee brackets welded to the sheathing, as shown.

Excavation to final depth now proceeded within the area of a central rectangle about 36 by 86 ft, leaving slopes of 1½:1 on all sides and substantial berms at the 6-ft level. To prevent the slope from creeping, lightweight sheathing (see Sec. 6.21) about 15 ft long was driven at the toes of slopes with air hammers.

Within this centrally excavated area, forms for the pile caps were now set and braced against the lightweight sheathing. The pile caps were poured with a wide beveled edge, or chamfer, as shown, and were tied together with 18-in.-wide grade beams for additional lateral stability.

With the footings in place, hand excavation through the berms and slopes was begun, to provide trenches in which to place the 16-in. o.d. pipe rakers. These had been pre-fabricated, as shown in the detail, to provide jacking lugs at the lower end. The top plates were field-welded to permit adjustment for length or angle if need be. Plate inserts to stiffen the waler flanges were also field-welded after the position of the raker was determined.

Rakers were set in opposing pairs so that loadings on pile caps would be balanced; and, having been temporarily wedged in place, the jacks were positioned and prestressing began. This operation required that the full pressure, which the waler would develop, should be transferred to the strut and then into the pile cap. The only useful criterion, therefore, was to apply pressure until the waler and sheathing began to move out of line. At this point jacking was terminated and the space between raker and concrete held with steel wedges and packed with a non-shrink grout.

Rakers were set on all four sides of the central group of pile caps at spacings which varied somewhat to suit pile-cap locations. With the rakers completed, excavation was continued outward to the face of the sheathing, and the remaining pile caps formed and poured.

The last areas to be excavated were the four corners. The placing of vertically diagonal rakers was considered to be impractical here, and the corners were braced by three horizontal pipe struts run diagonally across the corners. With these in place, final excavation and the remainder of the foundation construction could proceed.

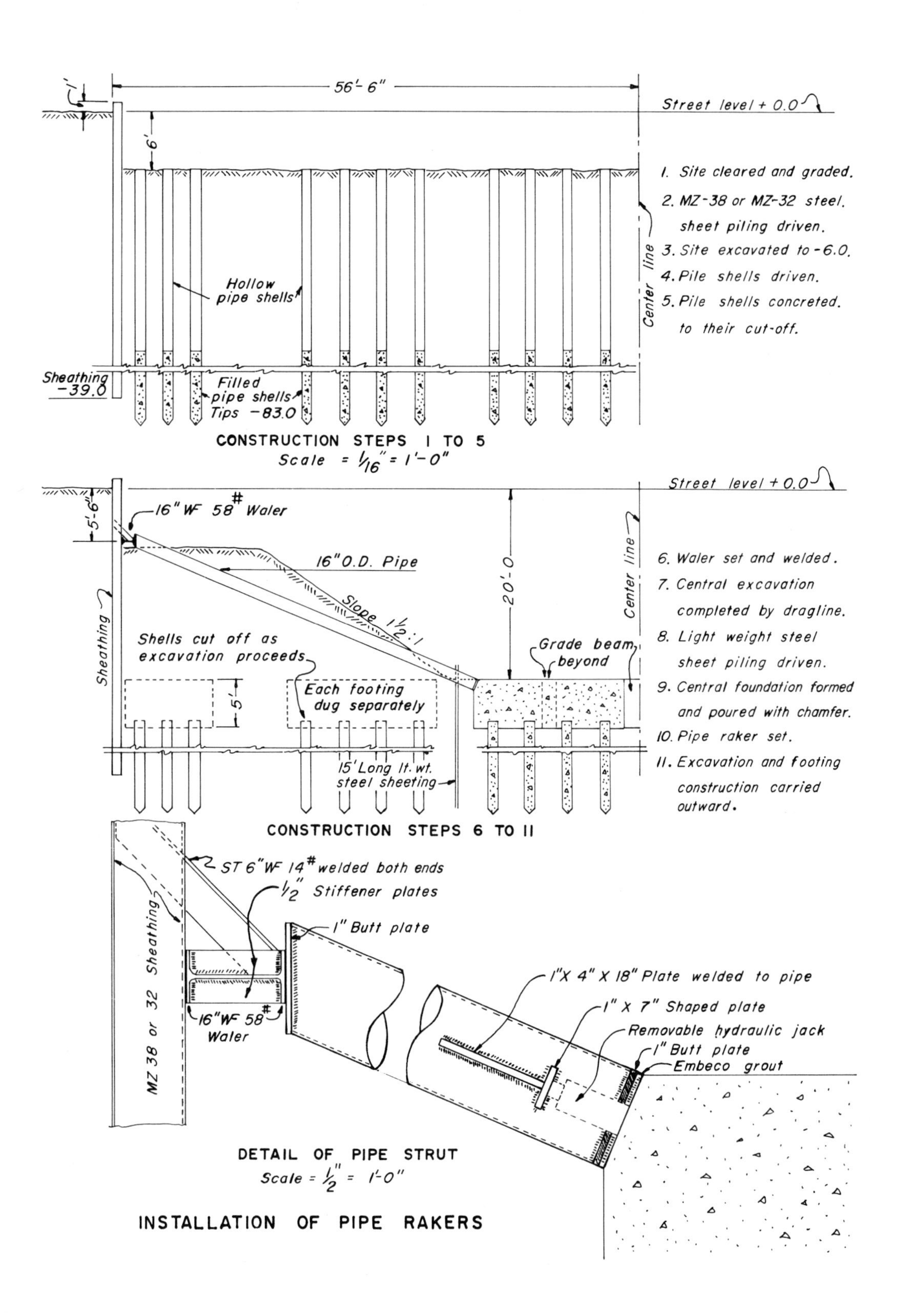

INSTALLATION OF PIPE RAKERS

6.12 TIED-BACK SINGLE-WALL COFFERDAMS

Any system of tied-back or anchored cofferdams depends on the existence of subsurface materials adequate to provide anchorage. Well-compacted soils with a high unconfined compressive strength may be suitable, but more generally the material will be rock of some sort, although it need not be continuous or solid. In general, the condition will be that of a heavy layer of soil overburden above a rock stratum into which excavation will also penetrate. The driving of steel sheet piling—or indeed of any vertical member—is impractical, and some means of avoiding the necessity must be devised.

Just such a condition existed in constructing foundations for the Pierre Laclede Building in Clayton, Missouri, in 1962. Here, a heavy clay deposit lay over a limestone bed of varying depth and density. The walls produced by excavation in the limestone required little or no support, whereas those in the clay demanded full support. For this condition a soldier beam and wood sheathing system was decided upon.

Fourteen-inch WF soldiers, thirty to sixty feet long, were set on eight-foot centers on three sides of the peripheral area fronting on streets or alleys. Eighteen-inch holes were drilled to a depth five feet below the bottom of excavation sub-grade through the clay and rock, the soldiers were dropped into place, and the bottom five feet secured by concreting.

The first 10 ft of excavation was performed by power shovel; and as it approached the face of the soldiers, 4 × 12 in. oak planks were set horizontally behind the forward flanges. In this case no care was taken in shaving the face of the clay, but the space between clay and plank was filled with lean grout to provide contact.

While excavation proceeded in the central area, a 15-ft-wide berm was left around the perimeter, below the 10-ft level. Operating on this berm, a drill rig with tilting mast drilled 8-in. holes on 4-ft centers at an angle of 30 deg into the bank—between soldiers. These holes penetrated from 20 to 42 ft into the bank, depending on the relative depth of the soldiers and the density and suitability of the subsurface material encountered. The lower ends were belled to a 3-ft diameter by means of an expandable bit, where rock seating was not available.

The anchors used were 1⅜-in.-diameter high-tensile-strength rods with a 6-in.-diameter steel plate on their lower ends, thrust down into the bottom of the hole. To prevent the rod from sagging against the side of the drill hole, a positioning guide was used consisting of a 4-ft length of 3-in. pipe with three radial fins of steel plate welded to provide an outside diameter of 8 in. The pipe was dropped down over the rod before it was placed in the drill hole. As the level of grout poured into the hole rose, the guide, to which a line was fastened, was pulled up to keep just ahead of the grout level.

Allowing 3 to 4 days for the grout to attain a satisfactory strength, 12-in. channels, back to back, were used to provide walers. These were backed up against the soldiers with tapered washers while an ogee washer, under a nut, held the rod against the waler. A projecting section of the anchor rod held a second washer and nut to which hydraulic jack loading was applied. This not only pre-stressed the rods but tested the holding power of the tie-back. The bottom nut was snugged up when full stress had been obtained.

Because of the considerable slope of the original ground surface, walers were not carried continuously but in 32-ft sections, maintaining the 10-ft depth from the top but stepped up or down horizontally as ground surface varied.

This operation was repeated when the excavation had reached a depth in excess of 25 ft, to provide a 15-ft vertical spacing for walers. The horizontal wood sheathing was carried down to the surface of rock, below which it was no longer needed.

Also illustrated is an alternative tied-back system, at another site, in which steel piling, driven on a batter, was used.

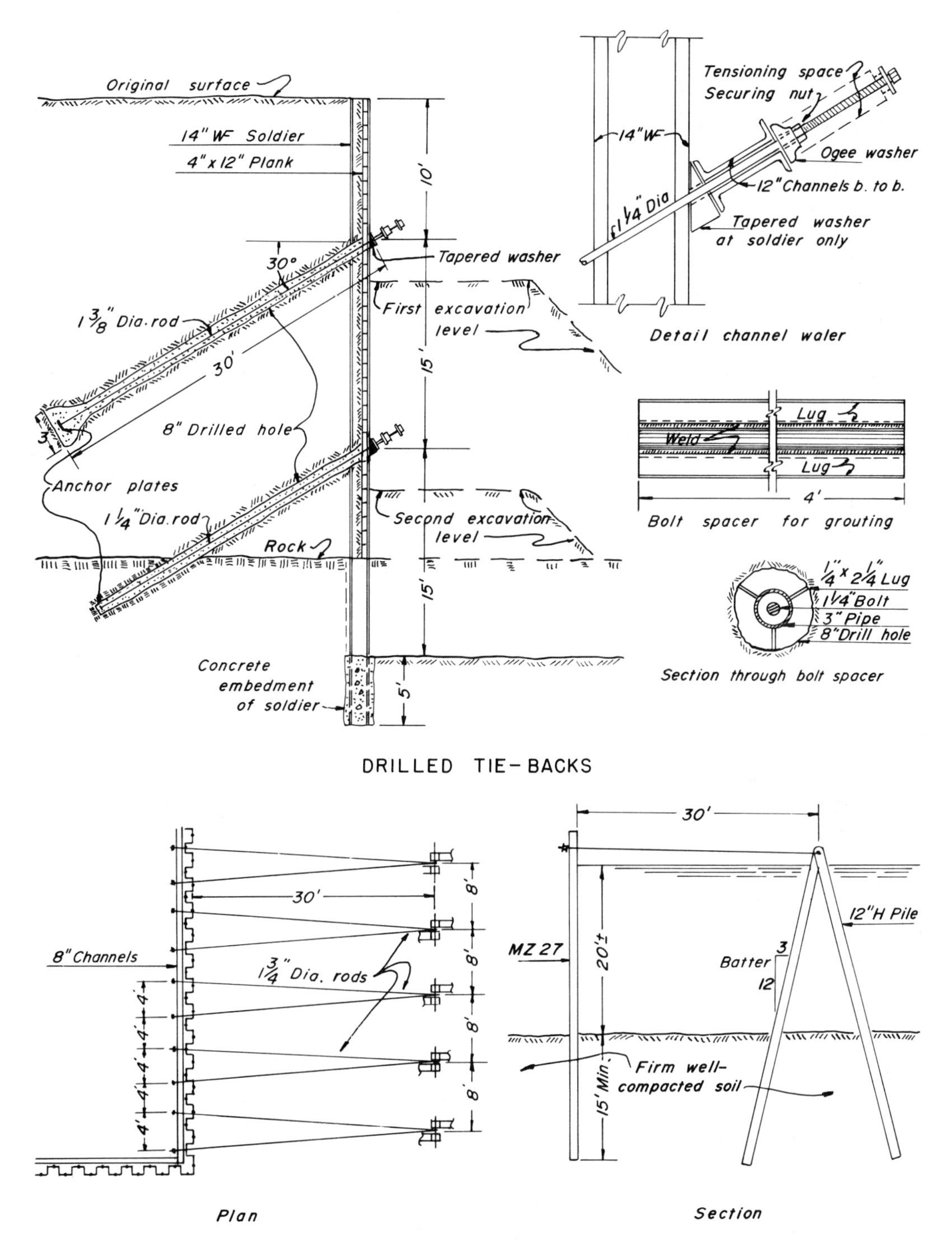

DRILLED TIE-BACKS

BATTER-PILE TIE-BACKS

6.13 A CASE OF A DOUBLE-WALL COFFERDAM

It is very seldom indeed that a single subsurface construction method can be isolated and treated singly without extraneous factors making it a special case. Ohio River type double-wall cofferdams are now less often used, partly because steel sheet piling and cellular cofferdams have taken their place and partly because heights and depths of the structures to be built behind them have increased tremendously. The principle of the method remains, however, even when steel sheet piling is substituted for wood sheathing.

An adaptation of a double-wall cofferdam was used in 1961 in constructing a tunnel in open cut under the Intra-coastal Highway Canal at Houma, Louisiana, although in this case one significant aspect of double-wall cofferdams was not used; the space within was not filled with soil; and the walls, consequently, were not tied together, but braced apart.

Half of this canal crossing was constructed at a time, permitting maintenance of canal traffic as well as allowing construction operations to progress from the shore outward. The tunnel itself was designed to be built within the space provided by two double-wall cofferdams paralleling the area on either side.

To provide a guide for the cofferdam walls, a double row of wood piles, 10 ft apart on 12.5-ft centers, were driven from the shore outward. Each pair was tied together by 6 × 12 in. timbers bolted to the tops of the piles. These rows of piles provided a guide and support for the movable template needed for driving the MZ-38 steel sheet piling. The template, a light wooden frame, was moved forward by crane lift as the sheathing was driven.

With the sheathing driven to depth, a top row of walers and horizontal struts was placed above water level, the temporary wood piling being withdrawn as the work progressed. To maintain spacing between the two cofferdams, walers and struts were also placed, at the same level, between them.

Excavation was now undertaken by clamshell under water, within each cofferdam, to a depth sufficient to place the bottom waler. The walers for the lower level had been temporarily hung in place before the upper struts had been set. These were now lowered into position on posts welded to the top surface of the lower waler, and the top of the posts were then welded to the underside of the upper waler, the posts providing the proper vertical spacing. The bottom struts were wedged into place by divers and tack-welded.

Clamshell excavation of the central area lowered the bottom sufficiently to place the waler-and-strut system at the 18-ft depth. Because of the span it was necessary to truss the struts, and the truss sections were now substituted for the single strut originally placed. The lower waler was again placed under water with divers.

With the truss sections installed, excavation under water continued until the level of the third strut line was reached. Water level in the three compartments was lowered—in the outer bays to elevation +90.00, in the inner bay to elevation −70.00—and this row of struts was placed. Further lowering of soil and water to elevation +62.00 permitted placing of the lowest strut.

Final excavation was completed, support piling driven, and the tremie mat of concrete placed with water level at elevation +93.00. After this mat had been placed, it acted as a strut to support the sheathing at its lowest point; and complete dewatering for the tunnel construction could progress in the central bay, with the water level lowered to elevation +80.00 in the two double-walled cofferdams.

After completion of the first half of the tunnel and its backfill to canal bottom level, sheathing was pulled and re-used in a similar double-walled cofferdam placed from the opposite bank. Tunnel closure, at midstream, was provided by a third cell, 65 ft long, with double walls on 4 sides, the bottoms of the cross-wall sheathing welded to a ¼-in. steel skin plate, sheathing the tunnel concrete, where necessary.

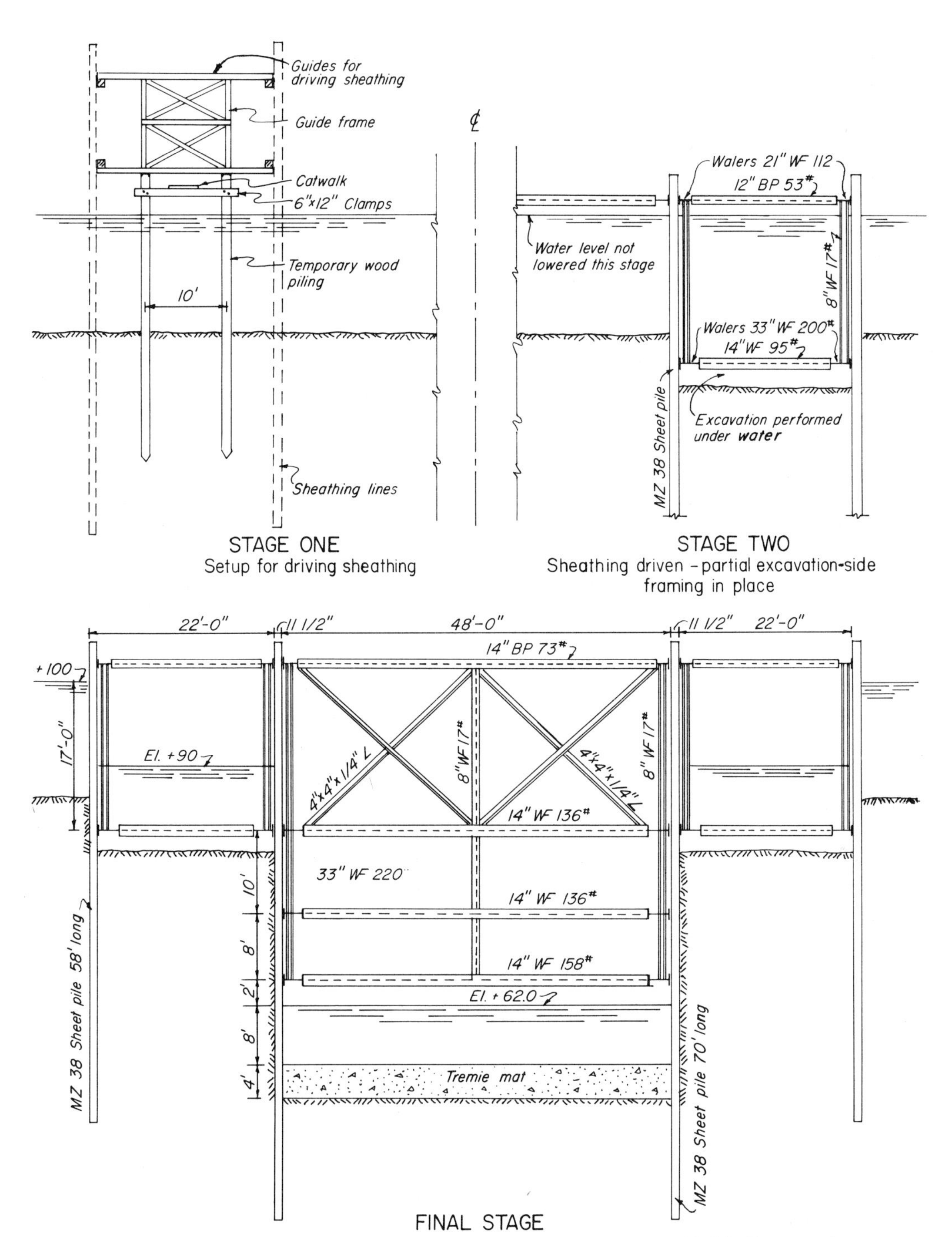

FINAL STAGE

Central section excavated under water. Water level lowered to elevation +62.00. Tremie mat, as strut and sealant, placed

A DOUBLE-WALLED COFFERDAM

6.14 CELLULAR COFFERDAMS

On April 22, the President of the United States ordered a naval blockade of Cuba, not in 1962, 1963, or 1964 but in 1898. The President then was McKinley, and the occasion stemmed from the sinking of the United States battleship *Maine* in Havana Harbor on the previous February 15. The *Maine* had been sent to Cuba to protect American "life and property" during a rebellion on the island against Spanish rule. A Naval court of inquiry, with the aid of divers, had reported that the sinking had been due to an explosion of a submarine mine attributable to "a person or persons unknown." The declaration of war on April 20 had included an ultimatum to Spain to quit the island and grant Cuba independence. One of the earliest cellular cofferdams to be built was constructed around the sunken *Maine* in 1910, 12 years later.

The Spanish-American War went very well with the aid of Teddy Roosevelt and the little torpedo that wasn't there in Manila Bay, but the cofferdam gave its builders considerable trouble. It was built on the soft, compressible, harbor-bottom soil and was filled with a heavy clay; and these two features almost led to failure, but also provided valuable information on the do's and don't's of cellular-cofferdam construction.

Today, cellular cofferdams, employing steel sheet piling, are used as temporary devices in the construction of bridge pier foundations, dams, river locks, dry docks, and other marine facilities. The principles of cellular cofferdams have been incorporated into permanent structures such as docking facilities, retaining walls, breakwaters, and mooring dolphins or have been used simply to provide foundation islands in a watery expanse.

The two chief types of cellular cofferdams are the circular cell and the diaphragm cell, each of which has certain merits and demerits in particular applications.

The circular-cell cofferdam consists of a series of spaced circles enclosed with steel sheet piling, generally a straight-web section, filled with sand. These are tied together on both sides by short arcs of sheet piling, also filled, to provide a continuous wall. Connections between arc and circle are accomplished by substituting a tee section, at this intersection, for the usual straight-web piece. The effective width of such a cofferdam is considered to be the width of a rectangular wall having an equal resistance against overturning.

The diaphragm-cell cofferdam consists of a series of straight transverse walls joined at opposing ends by circular arcs. These arcs spring from a common wye section at each end of the straight walls.

The selection of one or the other type of cofferdam will depend not only on site conditions but on construction problems, and consequently the circular cell is more commonly used. The circular cell is used for the greater depths and particularly where the cofferdam is to be built in swiftly flowing water. Each cell can be filled independently, without endangering the structure, and can then serve as a base for further construction operations. This is not true of the diaphragm cofferdam, whose cells must be filled in stages to prevent distorting the straight walls between cells. Generally one cell is not filled more than 5 ft above the level of the sand in adjoining cells, so that a substantial number of cells must be completed before filling can be attempted.

For shallow cofferdams the diaphragm type requires less tonnage of steel sheet piling per foot of cofferdam; but as depth increases, closer spacing of cross walls is required to keep interlock tension within permissible limits, so that for deeper cofferdams the cost of steel per foot of dam is less for circular cells. This will be true regardless of the radius of the basic cells.

When completed and filled, either type of cofferdam provides a working base on the surface of its fill for construction operations, but it is advisable to relieve it of heavy, fixed surcharge loads. Where a whirley, for instance, is to be set permanently (for the duration of construction) on a particular cell, an independent platform constructed on spud piles driven down into the supporting soil below the cofferdam is advisable.

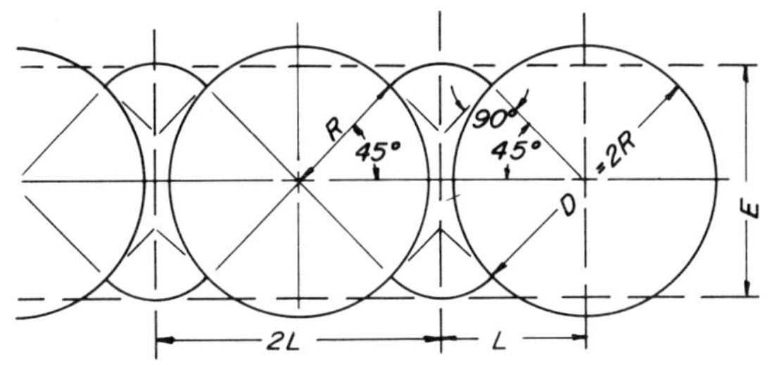

1. *Decide on width "E". Minimum is that equivalent to rectangular wall that will resist overturning.*
2. *E = 1.7 R or 2R = D = 1.13 E*
3. *Spacing = 2L = 1.3 E*
4. *Connecting arcs tie into tees on full circle so arc radius must be tangent to circle.*

Theoretical relationships must be adapted to width of sheet to be used (See F-15)

THEORETICAL RELATIONSHIPS
CIRCULAR COFFERDAM

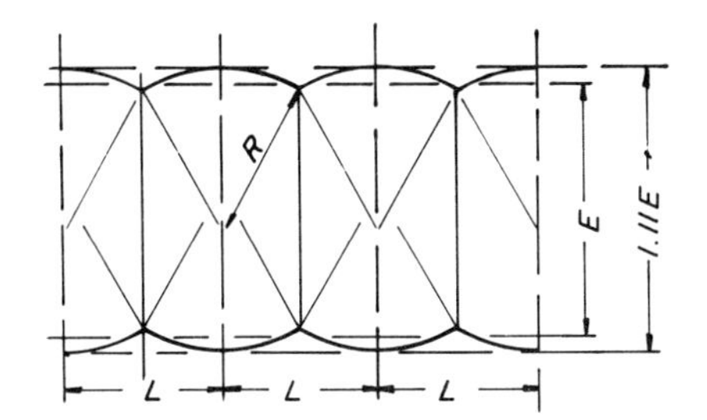

1. *Select width "E" as above.*
2. *Dimension for overall width will be 1.11 x E.*
3. *Arcs spring from 120° wyes.*
4. *$R = \frac{1}{2}L \div$ Sin. 30° = L.*
5. *L can vary but is limited by interlock tension.*

DIAPHRAGM TYPE COFFERDAM

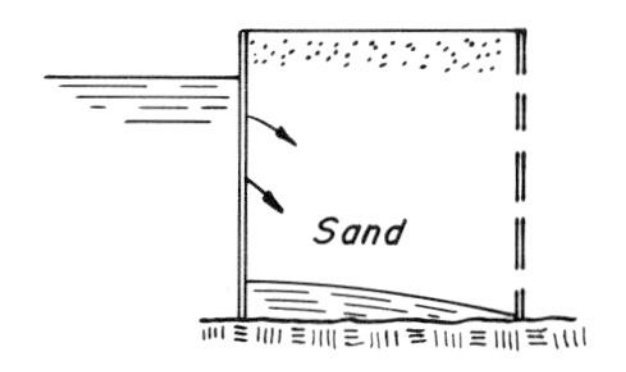

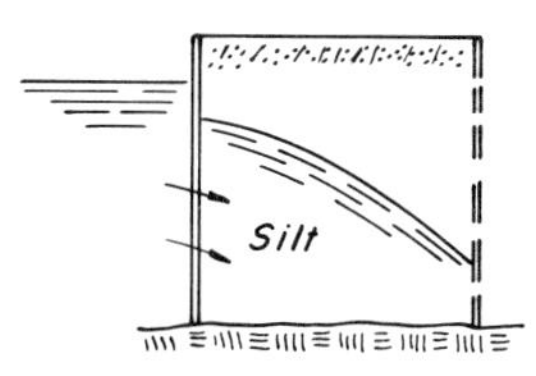

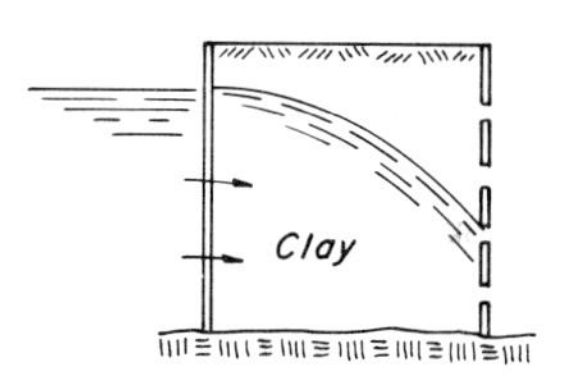

Effect of type of fill on flow through cofferdams on impermeable rock

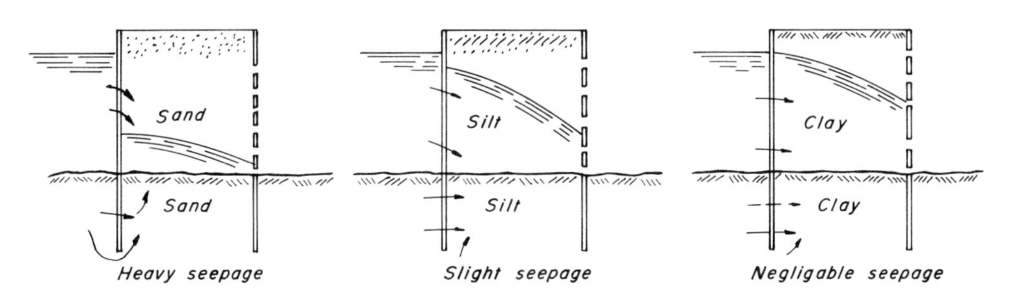

Effect of type of fill on flow through cofferdams on similar soil foundations

6.15 CELLULAR COFFERDAM DESIGN

"Remember the *Maine!*" was a slogan that helped to rally the nation and led to a quick and satisfactory settlement of the Spanish-American War, but it can also serve as a warning cry in the construction of cellular cofferdams.

The cofferdam constructed around the *Maine* in Havana Harbor, for the ultimate purpose of raising the sunken vessel, consisted of 20 circular cells, each with a radius of 25 ft. The steel sheet piling was driven through 30 ft of a soft red loam, abounding in shells, and several feet into an underlying blue clay stratum. The cells were filled by dredging—with clay.

When dewatering had lowered the water level about 20 ft inside the cofferdam, an inward tilt developed of such alarming proportions that pumping was halted while a berm of clay was placed around the inside toe of the cofferdam. This palliative was only partially successful, and in the final stages of dewatering it was deemed expedient to brace the cofferdam against the hull of the *Maine*. Settlement had been observed during the process of filling the cofferdam; and this tendency continued, accelerated by the inward tilt, and ultimately amounted to 4 or 5 ft.

Although considerable study has been given to the factors affecting the design of cofferdams, the approach still remains largely empirical, based on previous experience with successful cofferdams. The major controlling element is the clear height required, the difference in elevation between the interior bottom desired—which is fixed by the nature of the construction operation—and the elevation of the surrounding medium, usually water. Here the question of how high flood tides or floodwaters will rise is important, but a sufficient height to hold out the rarer, highest levels is seldom used. Generally, provisions are made in estimates to cover the costs of cofferdams being flooded once or twice during construction. The cost of the increased height is generally greater than the damage and dewatering resulting from the flooding of the cofferdam.

With the height established, an empirical formula is used to determine the width of the cofferdam. The effective width E is selected as $0.85H$, the clear height *under the selected flood or maximum-water-level conditions* on the outside of the cofferdam. Where only soil is involved there is but the one outer elevation. In shallow cofferdams, this width may be increased to provide working space on the top. (For instance, a height of 20 ft requires an effective width of only 17 ft, but this may not provide a working plateau of sufficient width for the operation contemplated.) From this point on, manufacturers' tables will provide all the information required to lay out the cofferdam.

Cellular cofferdams are subject to possible failure from four basic causes: (1) failure of their foundations; (2) shear deformation due to an inward tilt or producing such a tilt; (3) sliding on their base; (4) bursting of their cells due to failure in interlock tension. These causes can be controlled as follows:

1. Do not construct cofferdams on soft compressible soils. "Remember the *Maine!*" In sands, control the flow of water under and into the cofferdam from the bottom. (See Sec. 6.17.)

2. The tendency of a cofferdam to deform is resisted partly by the shearing value of the fill soil and partly by frictional resistance in the interlocks. Do not fill the cells with clay ("Remember the *Maine!*") but with sand or sand and gravel. Provide drainage through the cofferdam and weep holes on the interior face. Do not lubricate the interlocks. This makes the sheets easier to pull but reduces the friction needed.

3. There are no cases of failure by sliding even on smooth rock surfaces; but to avoid providing a case history, use seating precautions discussed in Sec. 6.18.

4. Bursting of cells results from careless driving or overdriving, in which one sheet is forced out of interlock with its mate. Sheets should be driven plumb against a secure template, and any unusual resistance to driving should be investigated. Failure has also resulted from poorly fabricated closure pieces or improperly field-welded or bolted tees or wyes.

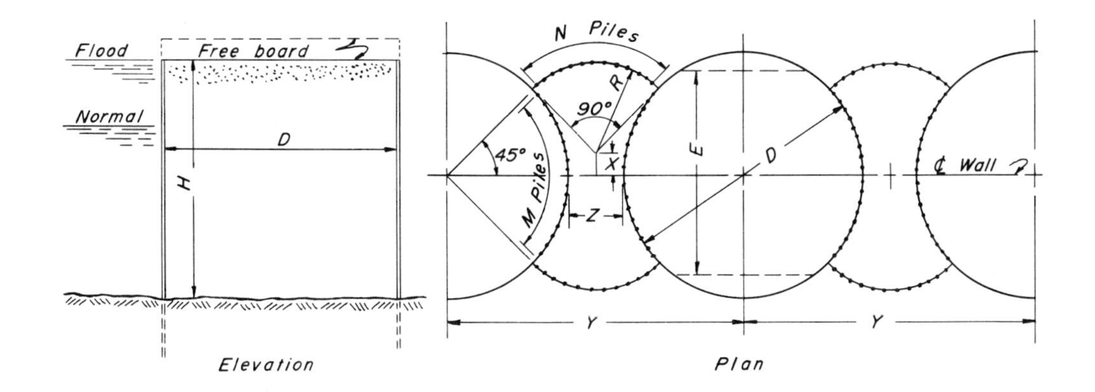

Selected Dimensions For Circular Cellular Cofferdams

Using Sections MP101, MP102

Dimensions in feet								Pile count		These figures for length = Y			
H	E′	E	D	Y	Z	X	R	M	N	Total piles	Area, sq ft	Fill, cu yd	Sand, tons
20	17.0	17.57	20.69	25.97	5.28	1.66	8.01	12	9	70	478	390	530
25	21.3	21.52	25.46	31.61	6.15	2.23	9.60	15	11	86	715	720	970
30	25.5	26.64	31.83	37.19	5.36	3.90	10.39	19	12	104	1042	1230	1670
35	29.8	30.60	36.61	41.71	5.10	5.03	11.19	22	13	118	1342	1840	2490
40	34.0	34.10	41.38	45.10	3.72	6.72	11.19	25	13	130	1636	2540	3440
45	38.3	38.37	46.15	49.62	3.47	7.86	11.99	28	14	144	2007	3500	4720
50	42.5	43.66	52.52	56.32	3.80	8.96	13.58	32	16	164	2589	5000	6730
55	46.8	47.62	57.30	60.83	3.53	10.10	14.37	35	17	178	3050	6450	8690
60	51.0	51.72	62.07	65.36	3.29	11.23	15.17	38	18	192	3551	8170	11,010

Using Sections MP112, MP113

H	E′	E	D	Y	Z	X	R	M	N	Total piles	Area, sq ft	Fill, cu yd	Sand, tons
20	17.0	17.40	20.36	26.47	6.11	1.16	8.54	11	9	66	482	390	530
25	21.3	21.50	25.46	31.28	5.82	2.36	9.39	14	10	80	706	750	950
30	25.5	25.70	30.55	36.15	5.60	3.60	10.24	17	11	94	977	1160	1560
35	29.8	29.95	35.65	40.88	5.23	4.77	11.08	20	12	108	1282	1750	2370
40	34.0	33.94	40.74	44.48	3.74	6.57	11.08	23	12	120	1588	2470	3340
45	38.3	39.51	47.52	51.68	4.16	7.76	12.78	27	14	140	2152	3750	5060
50	42.5	43.77	52.62	56.48	3.86	8.97	13.63	30	15	154	2603	5020	6770
55	46.8	48.00	57.71	61.28	3.57	10.17	14.48	33	16	168	3096	6550	8820
60	51.0	52.21	62.80	66.08	3.28	11.37	15.33	36	17	182	3631	8340	11,260

E' = theoretical width for height H where $E = 0.85\,H$.

E = theoretical width of a rectangular wall having a resistance against overturning equal to that of the cellular wall.

Total piles includes M and N piles plus 4 tee piles.

Yardage and tonnage figured for 2-ft freeboard.

6.16 COFFERDAM GUIDES, TEMPLATES, AND DEFLECTORS

Guides for straight-wall cofferdams of steel sheet piling are a relatively simple matter, and several methods of providing guides have already been cited. In Sec. 6.9 a row of steel H-piles, pre-driven on 25-ft centers with horizontal H-beams secured to one face, provided guidance. In Sec. 6.11 two timbers staked to the ground sufficed. In Sec. 6.3 a double row of temporary wood piles supporting a movable frame turned the trick.

In general it will be found desirable to provide a guide frame or template which is independent of the bracing frame, although in some instances the bracing has been used for guidance. The bracing frame of modern cofferdams will usually consist of steel members which are less adaptable, by cutting, than wood. Also it will seldom be practical to do much excavation until the sheet piling has been placed. The best procedure is, therefore, to construct a wood frame on the ground surface or on a subsurface level 5 ft or so below ground level, drive the sheeting, and then place the bracing members individually as excavation proceeds. It is seldom practical to fabricate a bracing frame and drop it into position. The walers should have uniform bearing on the whole width of every other sheet, which a pre-built frame can seldom provide since the frame will need to be a few inches short of sheeting dimensions for clearance in placing.

For circular cofferdams more precise templating must be provided, and these are basically of two types: a single inside template and a double inside-and-outside template.

The single inside template is generally fixed in position on spud H-piles or pipe piles, driven in at the four corners of a square, and cross braced to prevent sway. The template, of light framing members, is then dropped over the spuds and secured to them by bolting or wedging. A light exterior guide member may be fastened to the frame and made removable to provide a slot for guide positioning of the sheets. For large cells, templates must be made up in sections to permit handling with available rigs, and they should always incorporate provisions to permit slight size reduction in case binding occurs during removal.

Where underwater overburden above rock does not provide sufficient penetration to support spud piles, a floating template may be necessary. Floating templates are generally provided for both the inside and the outside of the cell being constructed and are made up in sections that can be readily disassembled for lifting from the inside and floating off from the outside. Since the first of a string of cells will generally start at the land end and be constructed progressively outward, such templates can be anchored to the previously placed cell. However, in cases where there is considerable current, due either to the tide or to stream flow, maintaining the template in position may require the additional security of independent concrete anchors to which the template can be tied by tensionable cables.

In turbulent waters or those with a high velocity of flow, additional problems present themselves. These begin with the driving of the sheeting and require an independent anchorage to prevent the sheet from being driven out of plumb. An independent dolphin consisting of a cluster of piles can be used, not only to maintain the sheet plumb during driving but to prevent tilt in the template from developing as additional sheets are added.

A deflector has been used, as detailed, for providing quiet water within which to set up the template and drive the piles. This one was developed by the Dravo Corporation for constructing circular cofferdams in the Long Sault Rapids in connection with power development of the St. Lawrence River.

A frame secured to the previously driven cell (four cells were constructed before reaching the rapids section of the river) supported guides into which pre-cast concrete slabs were dropped. This device worked satisfactorily in water with a surface velocity of 13 fps striking the bulkhead at an angle of 45 deg.

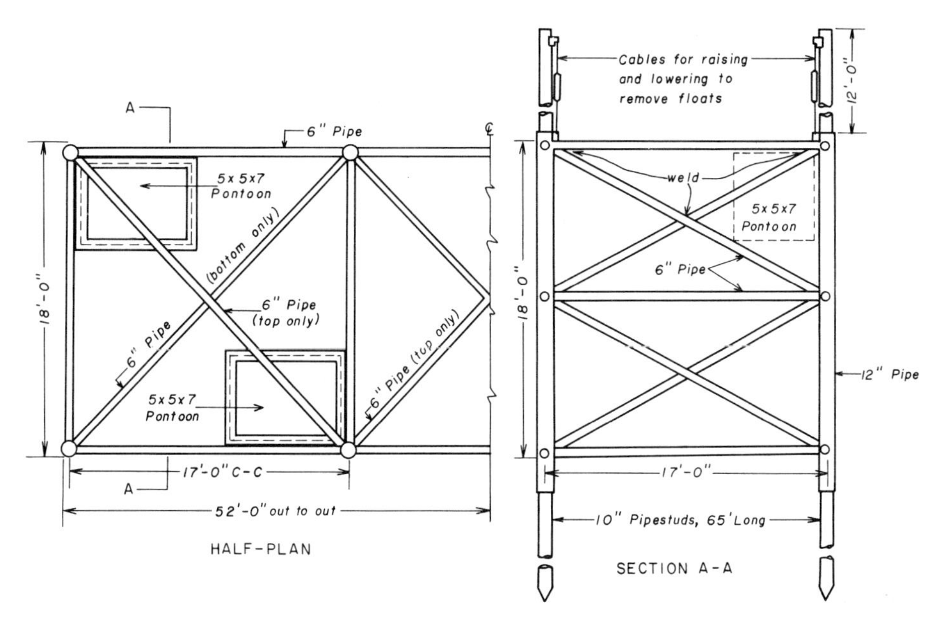

A FLOATABLE COFFERDAM TEMPLATE

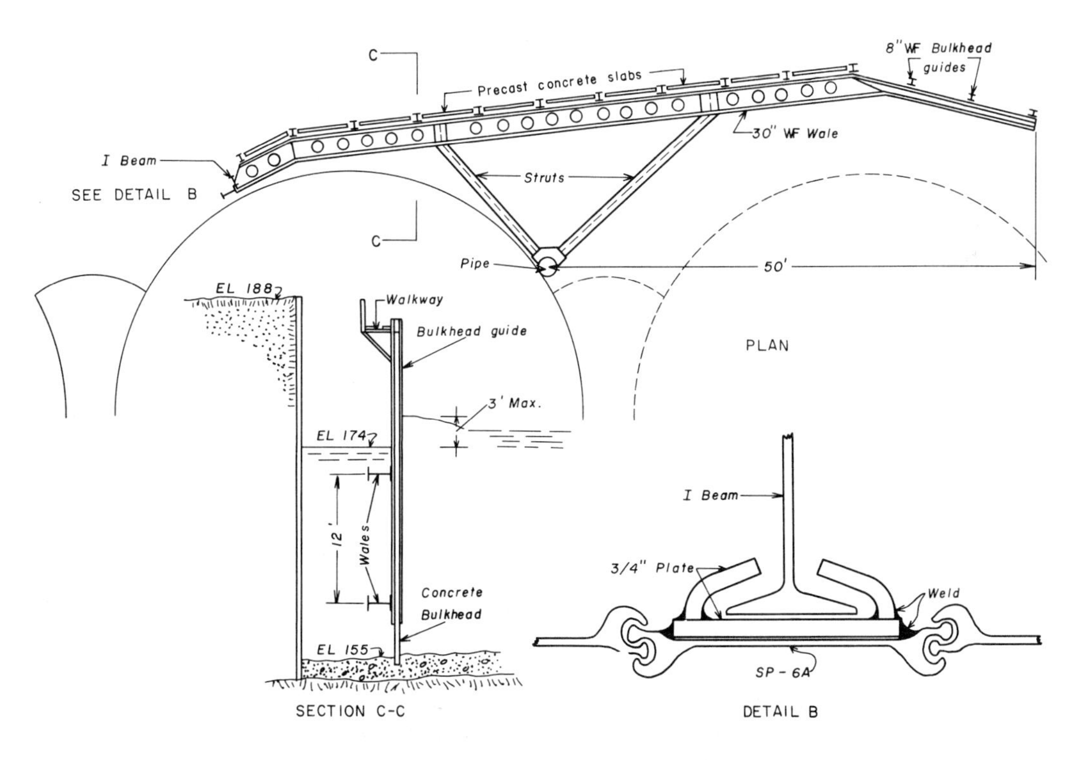

CANTILEVERED DEFLECTOR SHIELD

6.17 CELLULAR COFFERDAMS IN SAND

Cellular (or single-wall) cofferdams are commonly used to restrain substantial heads of water; and although sands can provide a stable base, the flow of water under the cofferdam must be controlled to maintain that stability. The driving of sheeting into dense sands can be difficult, but this can be overcome by jetting (see Sec. 5.25) if necessary. It is in dewatering the cofferdam, that is, in lowering the level of water on the inside of the structure while the level of water on the outside remains fixed, that we encounter difficulty.

Sand is a porous medium containing a substantial percentage of voids, the percentage varying directly with the size of the sand particles. Before dewatering starts, the sand voids are filled with water, but as the head differential between interior and exterior water levels increases, water begins to flow through these voids in the direction of the side with the lower head. The rate or velocity of flow increases in direct proportion to the increase in the head differential. At some point the velocity of flow becomes critical for each size of sand particle, in that it becomes sufficient to counteract the weight of the particle and toss it upward. This, of course, will occur not with a single particle but with all the particles in the area where sufficient velocity exists and leads to what are known as cofferdam "boils." In this condition the sand becomes a "quick" sand and is no longer of value in supporting the sheathing.

Considerable study has been given to this problem in an attempt to pre-determine the possibility of boiling, but these attempts have been largely frustrated by the reluctance of natural sands to occur in beds of uniform consistency. Out of these studies has come the concept of the "flow net."

The flow net is, basically, a depiction of the path flowing water will take under a cofferdam. It can be developed theoretically from the head differential and the porosity of the soil for any specific cofferdam structure, or it can be determined experimentally by models simulating field conditions. Neither method will be detailed here, only the general conclusion to be drawn from such data.

In a copyrighted graph published in 1953, the Engineering Division of the Moretrench Corporation (well-point suppliers) showed safe depths for open pumping in cofferdams based on a typical flow net and varying with the differential head of water, the depth of cofferdam sheathing penetration, and the effective weight of the soil. However, like many such attempts, it was necessary to make specific assumptions which tended to limit its field use. The soil was assumed to be "homogeneous, cohesionless and granular," and the determination of its effective weight was presumed to be practical by undisturbed sampling procedures. These and other assumptions limited the graph's applicability to a narrow range of possibilities.

The flow-net concept has been more valuable in its general implications than in the specific data that it could furnish. This general implication is that if we lengthen the path of water flowing through porous soils, due to a specific velocity-producing head, the resultant velocity can be reduced to a value below the critical velocity that could result in boils.

If the construction cofferdam is to function for the purpose intended, the head differential cannot be controlled; it is fixed by the depth of the structure to be built within it, in relation to the surrounding water level. It is possible, however, to lengthen the flow path by driving the sheathing deeper. It is possible to approximate the additional depth beyond that needed for cofferdam stability at which no boiling can result, but the cost of this added sheathing may be prohibitive.

The alternative methods of lengthening the flow path are shown opposite. These consist in the first case of placing an impervious clay blanket along the *exterior* face of the cofferdam or providing a berm along the *interior* face of the cofferdam. The interior berm may provide lengthening; but if its material is a crushed rock, it may additionally prevent boiling by holding down the sand.

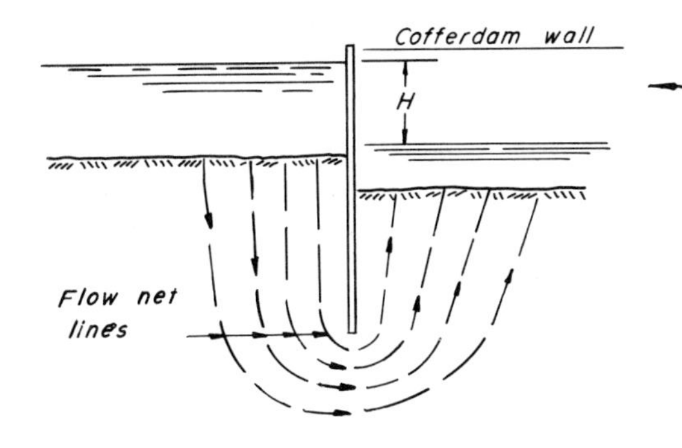

(*a*) A TYPICAL FLOW NET

In porous soils, such as sand, a flow net can be plotted, or developed experimentally. Its shape and egress points will depend on the degree of porosity of the soil and the amount of head, H, the velocity producing factor.

(*b*) COFFERDAM BOILS

Here, the quick condition in the interior soil is induced by the high velocity of water flowing upward from under the cofferdam. Driving the sheathing deeper will lengthen the path of flow and may stop the boiling.

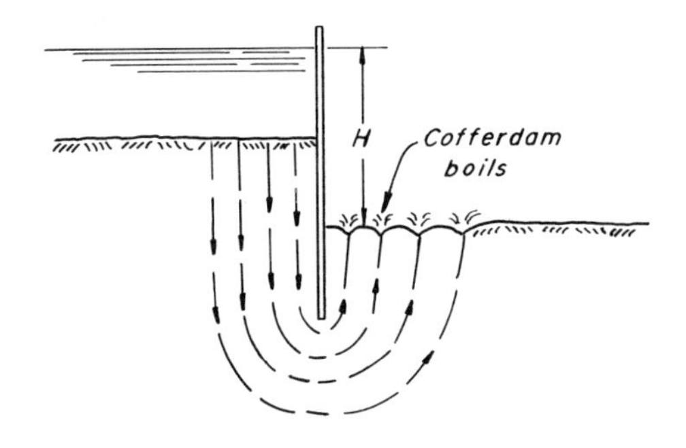

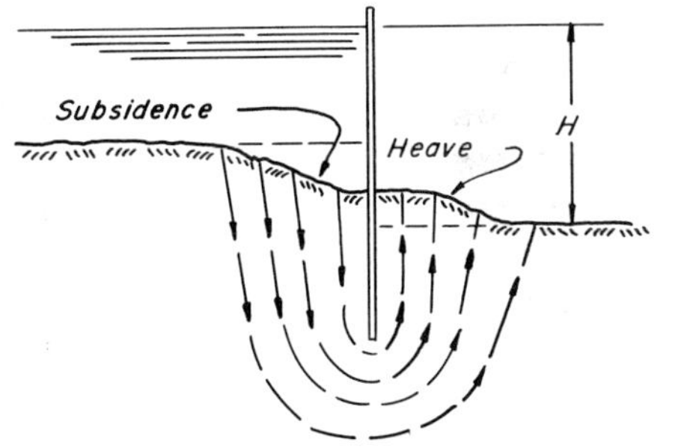

(*c*) COFFERDAM BOTTOM HEAVE

This may be the end result of the boiling shown in b . The settlement on the exterior is caused not only by velocity, but by the void produced on the interior. This can also occur where the interior soil is dense, preventing suitable egress.

(*d*) CLAY BLANKET SEAL

A clay blanket seal on the exterior, laid down over more porous soils, forces the ingress points of the flow net farther away from the cofferdam wall, lengthening the path of the flow net lines and reducing their velocity.

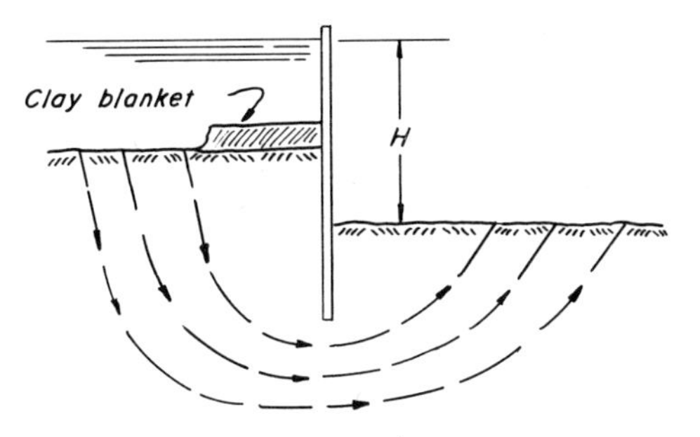

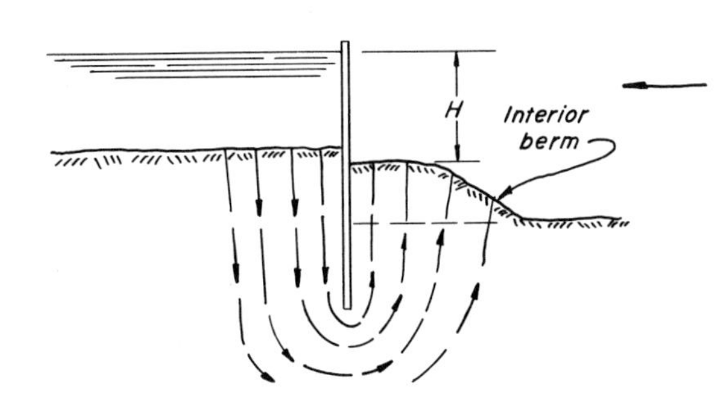

(*e*) COFFERDAM INTERIOR BERM

An interior berm, even of porous material, may decrease the net head, increase the length of flow path and prevent boiling. The disadvantage is that a larger cofferdam area may be required.

6.18 CELLULAR COFFERDAMS ON ROCK

The original Ohio River cofferdams were intended primarily to rest on rock, and their construction was based on that premise. Frames made up in sections connected by articulated joints were placed progressively from a barge. Sheathing was set on both sides of the frame, driven to rock through overburden, if any existed, or simply secured to the frame by through tie rods where the rock surface had been swept clean. Fill progressed within and without the sheathing walls simultaneously, forming, in effect, an earth dike with two parallel cut-off walls.

Today the preponderance of cellular cofferdams are still built on rock surfaces, with varying overburdens, for the simple reason that most of the foundations to be built within them are carried up from rock surfaces. The design considerations discussed in Sec. 6.15 are valid here with the exception that (1) failure of their foundations is unlikely; (2) inward tilt is still of primary importance and has been discussed; (3) despite the fact that cellular cofferdams have been built on relatively slick rock surfaces, there are no known cases where sliding has occurred. Friction of the steel sheet piling as well as of the fill seems always to have prevented sliding. Bursting of the cells is a very real hazard, although control is not difficult.

The varieties of rock and their bed conditions are very extensive: they may range from hard granite to soft coral; they may have surface layers that have decomposed; their surface may be tilted extravagantly or be fairly level; they may be relatively smooth-surfaced or extremely rough and pitted; or the upper layers may have broken up into boulders.

Although sliding is not considered a factor in failures, it is advisable to drive sheathing into soft rock or soft surface layers to the extent that this can be done without mushrooming the top of the sheet. In harder rocks it may pay to have a shallow trench blasted under water into which the sheathing can set, particularly on very smooth surfaces. In these cases actual driving becomes an academic term better described as "setting up" the sheathing. In some cases, to provide a toe-in, a 5-ft layer of clay has been barge-dropped over the site into which the sheets are penetrated. The chief purpose of this is to seal the bottom of the cofferdam against the infiltration of water. In this case exterior or interior (to the cofferdam) berms of soil are added after the sheathing has been placed.

The real problems of cellular cofferdams on rock arise where the surface is pitted, contains boulders, or is highly irregular. It is in these cases that the possibility of forcing the sheets out of interlock is greatest, and consequently the resulting possibility of bursting the cell. Pre-blasting, under water, of the rock surface must be considered; but the breakage, in certain types of rock (limestone, for instance) may simply produce a new configuration of hills and valleys. The only possible solution is to avoid overdriving, even to the extent of underdriving, and then take measures to seal the perimeter.

Sealing techniques (applicable as well to single-wall cofferdams) may consist of interior or exterior berms set after the cell has been filled, but before dewatering. In some cases, more particularly in single-wall cofferdams, a band of tremie concrete is placed on the outside of the cofferdam sheets. The dimensions of such a band, generally square in section, will depend on the variations in the rock surface but should lap the highest hung-up sheet by about 2 ft.

Another problem in setting cellular cofferdams on rock is control of water, not under the toe of the sheathing or through the interlocks but through the rock underneath the cell. Many rocks are extremely porous or contain substantial channels which, while not dangerous to the cofferdam, demand control to prevent excessive pumping. In some cases, as in coral formations where voids can run as high as 20% of the mass, intrusion grouting has successfully controlled this factor; but, unfortunately, in the majority of cases control measures can be undertaken only when the source is located, after dewatering is substantially under way. (See Sec. 3.18 for methods.)

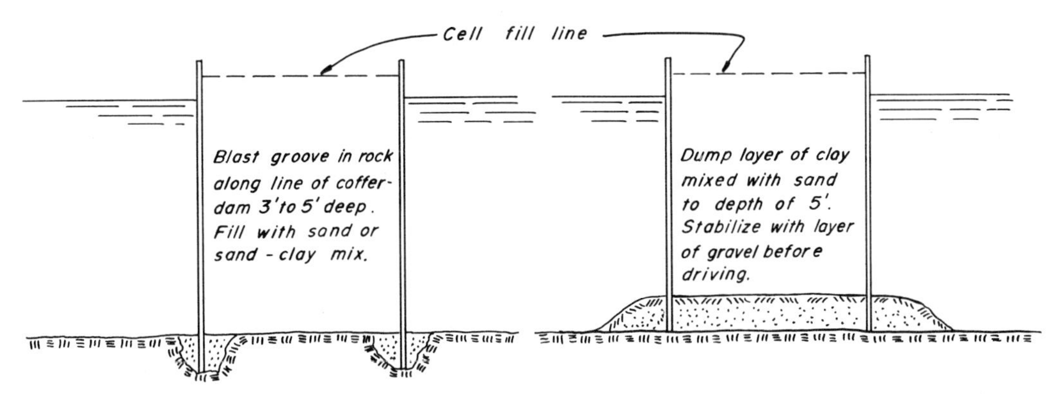

BLASTED-IN

FILLED - UP

COFFERDAMS ON SMOOTH ROCK SURFACES

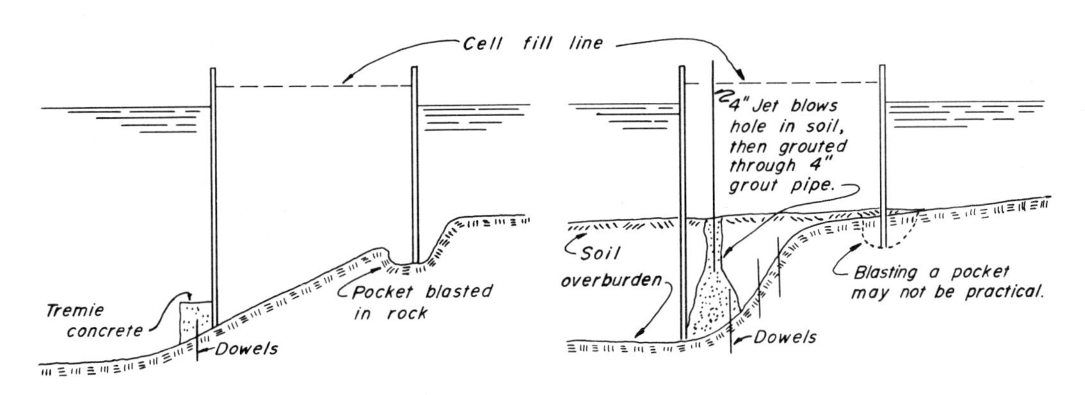

SLOPING SURFACE

DIPPING SURFACE

COFFERDAMS ON IRREGULAR ROCK SURFACES

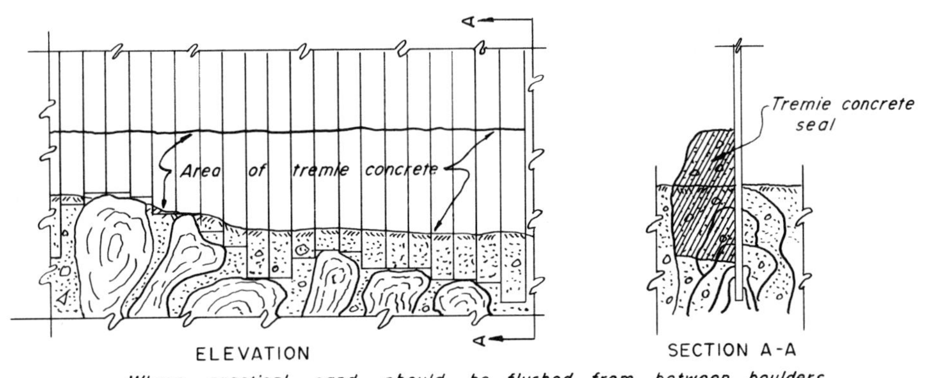

Where practical sand should be flushed from between boulders

COFFERDAMS ON BOULDEROUS FORMATIONS

6.19 TIMBER COFFERDAMS

Comparable to steel sheet piling, in wood, is tongue-and-grooved sheathing. This was the earliest material used for cofferdams, and steel sheet piling was a successful effort to duplicate and exceed the capabilities of timber sheathing. Tongue-and-grooved plank for sheathing is still used for bracing banks of limited depths and is, possibly, more effective in restraining the infiltration of water than steel sheet piling since the water-soaked tongue and groove tends to expand and form a tight seal; but for substantial depths, over 20 ft, it has disadvantages which make it impractical, costly, or both.

The section modulus of wood is considerably less, for a given thickness, than that of steel, so that a greater thickness of wood is required to reach the same strength in bending as that of steel. In the days prior to steel sheet piling this thickness was achieved by the use of what is known as Wakefield sheathing, consisting of square-edged plank in three 2-in. thicknesses, the center plank offset to form a groove on one edge and a tongue on the other. The three planks were bolted together before driving. Additional layers of thicker planks were also used, as circumstances required, and it is notable that Roebling used 9 *feet* of timbering for the walls of his Brooklyn Bridge caisson.

The alternative to increasing the thickness of wood sheathing is to decrease the spacing of struts and walers. Aside from the increased cost of the material, the increased cost of the labor required in this operation is considerable. When the construction of cofferdams was a laborer's operation and laborers received 25 cents an hour, the additional labor cost was still not excessive; but when the setting of wood sheathing became a carpenter's function and rates climbed to from $3 to $4 an hour, the additional cost became impressive.

But there are other disadvantages to wood cofferdams. Wood is a much softer material than steel and cannot be successfully driven through compact soil layers. As a result, it is necessary, in many cases, to pre-excavate along the interior face of the sheathing and, in effect, to drive the sheathing into the void created. In order to drive wood sheathing, whether it be square-edged or tongue-and-grooved, it is necessary to chamfer the long bottom edge and taper the narrow edge in order, in the one case, to keep it tight against the guide frame and, in the other case, to keep it tight against the plank previously driven. This results in a thinned edge that can mushroom or split upon encountering rather small boulders or variations in soil density.

The use of hardwoods such as oak, rather than softwoods such as pine, would provide some amelioration of the difficulties were it not for the fact that hardwoods tend to warp and twist excessively in large sections and tend to split when nailing is required.

Although, for small excavations of limited depths, what we have been describing as cofferdams are still frequently built of timber, and timber struts, for instance, are still frequently used to brace steel walers on steel sheet piling, the most important use of timber in supporting excavation is in a more limited area.

Shoring, as the term is generally used in reference to the bracing of banks resulting from excavation, consists in supporting the soil face as a unit rather than as an aggregate of grains. Well-compacted clay strata, for instance, will stand with minimal support for depths of 8 to 10 ft. Any tendency of such banks to collapse will arise from breaks in the continuity of the strata. Bracing of such banks, particularly in trenches, is necessary but need be done only at intervals of 8 ft or so, the cohesion of the soil being depended upon to support the intervening space.

In shoring, the timber plank—generally square-edged—spreads the reaction of the struts—which may be trench jacks—sufficiently to prevent deformation of the bank face. Shoring of portions of open excavation banks is often resorted to where a single area in the pit wall is unstable. Here, a continuous face of planking aligned by walers and supported by rakers may be employed, with characteristics no different from those for any other cofferdam.

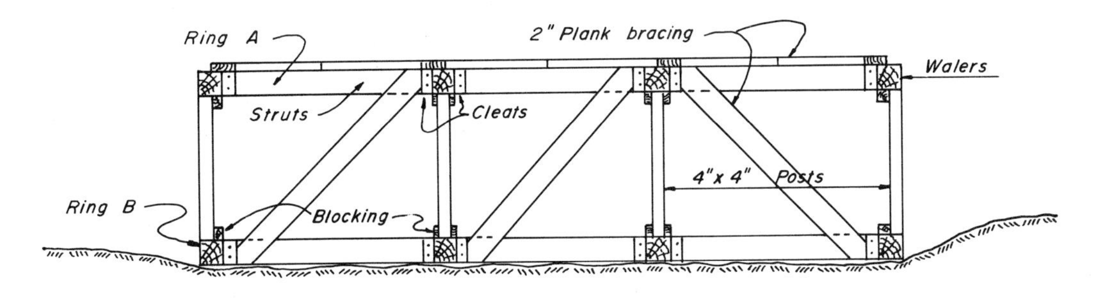

SECTION THROUGH TIMBER FRAMING

SET-UP FOR COFFERDAM CONSTRUCTION ON SOFT GROUND

Sizes of members and their spacing will depend on spans and on soil loadings

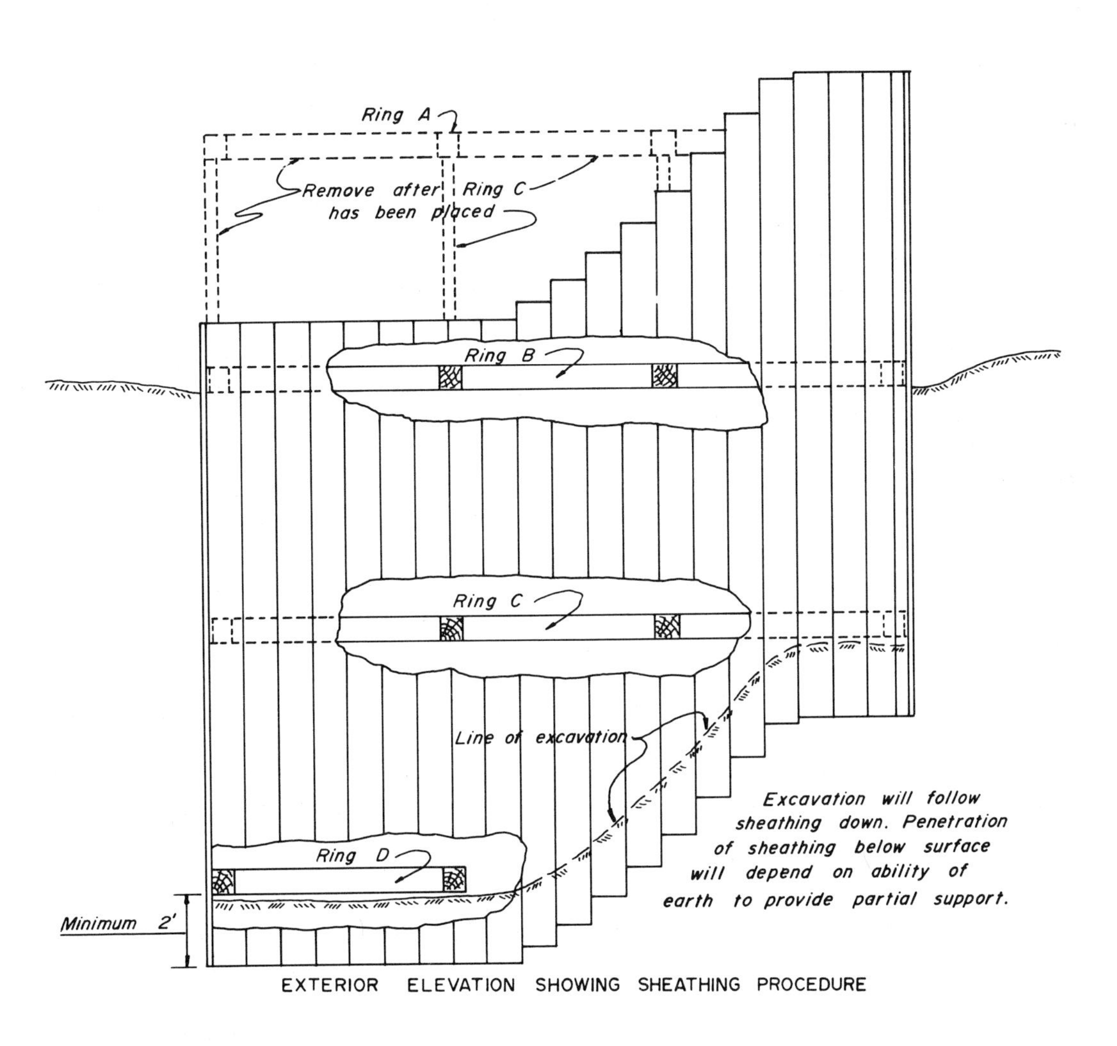

EXTERIOR ELEVATION SHOWING SHEATHING PROCEDURE

6.20 CONCRETE COFFERDAMS

As has been said previously, concrete cofferdams are seldom used for any but permanent installations, and in these cases chiefly as bulkheads of relatively low height. In tropical and semitropical waters they have the advantage of resistance to corrosion as well as to marine life when they are properly made and placed. For waterfront work in low-lying terrain where residential development is the aim of separating sea and land, they present a far more satisfactory appearance when new, and retain their appearance longer, than creosoted timber or steel sheet piling. As a result, concrete sheathing for low bulkheads is extensively used in Florida, for instance.

The construction problems involved in handling and placing concrete sheathing are similar to those discussed under concrete piling (see Sec. 5.6) and arise from the rigid nature of the material. Concrete shapes do not adapt themselves as readily to field conditions as do steel or wood, and far more careful planning is necessary in their use.

As with concrete piling, the use of pre-stressing techniques has been found to be particularly suited to concrete-sheathing production; and except for special shapes where pre-stressing is impractical, most concrete sheathing is pre-stressed. A typical set of shapes with details of pre-stressing wires is shown opposite.

Concrete sheathing, since it is used generally for permanent bulkheads, is designed for the specific job at hand, both as to thickness and as to length, with special sections for closures rather than field fitting. In the majority of cases, these bulkheads do not depend on depth of strata for their length, but are designed so that a certain portion of the *weight* lies below ground surface. (Minimum penetration would be 0.6 of the height above ground in good soil.) They are generally held in position or alignment by concrete caps poured in place and anchored by tie-backs to piling or deadmen.

Although concrete sheet piles are designed to be driven, driving can be difficult because of the dead weight of the sheet itself; and the more desirable method, particularly in granular soils, is to jet them into position, resorting to driving only where soils cannot be suitably eroded by jet action. However, if the material is dense, pre-punching or the creation of trenches by blasting may be required.

Since concrete sheathing is used for bulkheads to a large extent, they are subject to considerable wave action and, where voids in the sheathing occur, can lead to progressive loss of soil behind the bulkhead. One design of concrete sheathing, extensively used in Florida, has one edge grooved from top to bottom of the sheet while the other edge is grooved from the top down to a point which will be 4 or 5 ft below water surface after the sheet is driven. Below this the sheet has a tongue.

Whether there is a partial tongue, as noted, or a continuous tongue, it is impossible to maintain one tongue in the adjoining groove tightly enough to prevent a void. With the partial tongue indicated, a slot remains at the top which can be filled with grout for the purpose of sealing. Here again the possibility of the grout's leaking out during filling is considerable, especially in the section under water.

One solution to sealing the slot seems to have proved satisfactory. A thin polyethylene tube is slipped over the bottom end of the grout pipe and carefully lowered down in the slot. The grout pressure forces this tube against the edges of the piles—sealing the void. Although no bond is provided in this procedure, between grout and sheet or between sheets, bond is not considered necessary in a uniformly loaded wall.

With the wall in place and grouted, a concrete cap poured in conventional forms is placed along the top of the sheathing in lieu of a waler. Pre-cast tie beams are cast into this at the intervals required by anchorage conditions, and the inner end of the tie beam is similarly cast into the anchor or a pile cap set over the anchorage pile.

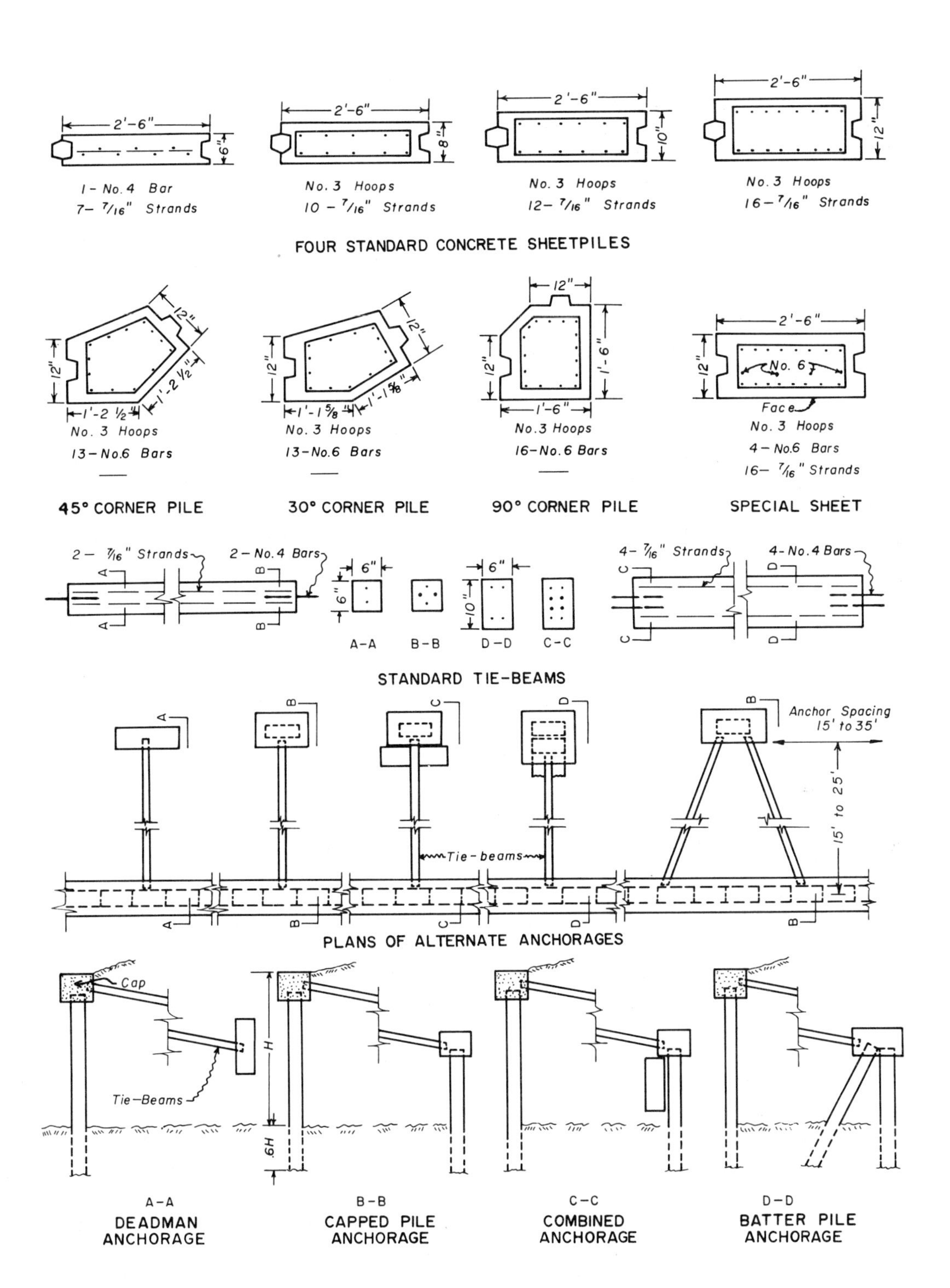
2'-6"
6"
1 - No. 4 Bar
7 - 7/16" Strands
2'-6"
8"
No. 3 Hoops
10 - 7/16" Strands
2'-6"
10"
No. 3 Hoops
12 - 7/16" Strands
2'-6"
12"
No. 3 Hoops
16 - 7/16" Strands
FOUR STANDARD CONCRETE SHEETPILES
12"
12"
1'-2 1/2"
1'-2 1/2"
No. 3 Hoops
13 - No.6 Bars
12"
12"
1'-1 5/8"
1'-1 5/8"
No. 3 Hoops
13 - No.6 Bars
12"
12"
1'-6"
1'-6"
No.3 Hoops
16 - No. 6 Bars
2'-6"
12"
No. 6
Face
No. 3 Hoops
4 - No.6 Bars
16 - 7/16" Strands
45° CORNER PILE
30° CORNER PILE
90° CORNER PILE
SPECIAL SHEET
2 - 7/16" Strands
2 - No. 4 Bars
6"
6"
6"
10"
A-A
B-B
D-D
C-C
4 - 7/16" Strands
4 - No. 4 Bars
STANDARD TIE-BEAMS
Anchor Spacing
15' to 35'
15' to 25'
Tie-beams
PLANS OF ALTERNATE ANCHORAGES
Cap
Tie-Beams
H
.6H
A-A
DEADMAN ANCHORAGE
B-B
CAPPED PILE ANCHORAGE
C-C
COMBINED ANCHORAGE
D-D
BATTER PILE ANCHORAGE

6.21 MISCELLANEOUS SHORING SYSTEMS

The emphasis, in preceding sections of this chapter, has been on the larger, more difficult types of cofferdamming since these require more care and judgment in constructing to avoid catastrophic consequences. Still, from a numerical standpoint, the number of jobs of modest size with modest restraint requirements greatly exceeds those larger, special, more flamboyant exercises in human ingenuity. For these modest exercises wood or timber remains the most commonly used material, but it has important drawbacks. No longer as cheap as it once was, it is easily damaged either in driving or in removing, and patching it or rehabilitation of the members is generally in the range from difficult to impossible. Some type of sheathing intermediate between wood and steel sheet piling with substantial re-usability therefore offers advantages.

The L. B. Foster Company, one of the principal suppliers of rented steel sheet piling, has developed and now markets a lightweight steel piling for limited earth support. Essentially a corrugated sheet similar to steel decking, its section modulus is from one-half to three-fourths of that for straight-web sheet piling; but its weight per square foot is only about one-third. Even though its cost per ton is somewhat higher than that of steel sheet piling, it still represents a considerable saving in material cost. Handling and driving costs will also be lower because of the lighter weight; and, if it is properly handled, the re-use potential is high. Details are shown opposite, and its use in an instance where it was left in place is mentioned in Sec. 6.11.

Lightweight steel piling is driven with a sheathing driver or converted paving breaker using a special driving head grooved to fit the corrugations of the sheathing. The nature of the interlock permits two sections to be folded together in lieu of "threading" before driving. The driving required is limited to soft uniform soils where battering of the thin metal is unlikely and where obstacles will not force sheets out of interlock.

An ingenious sheathing known as Telesheeting has been developed by a New York contractor, primarily to avoid the necessity of *driving* sheeting. Telesheeting consists of three tubular sections of ⅛-in. steel plate, two of them telescoping within the basic 4 × 12 in. × 6-ft-long section. Excavation to an approximate depth of 6 ft is first performed. Across the gouged trench at 20-ft intervals, 6-in. H-beams are placed. To the underside of these are bolted two pairs of 4-in. H-beams, one pair along either wall of the intended ditch. Spaced 4¼ in. apart, these pairs of 4-in. H-beams form at once a slot into which the Telesheeting is dropped and the top waler of the finished system. The transverse 6-in. H-beams, left in place, provide a top strut.

Plates welded to the face of the sheathing indicate spacing for the waler at the bottom of the 6-ft section; and struts used here are timber, wedged into place after being hung on top scabs. These struts are on 10-ft centers. The area outside the sheathing can now be backfilled. Excavation proceeds to the next level some 5 ft lower with the next telescopic section dropping down, as hand excavation against the bank follows machine excavation in the center of the trench.

This system, originally designed for a specific depth and set of conditions (an 8-ft-wide trench 16 ft deep) does not appear to have sufficient flexibility for general use.

The effort to reduce labor costs in the shoring of trenches has led to a number of ingenious devices, one of which is the Saf-T-Jax sold by the Sigma Engineering Corporation of California. A unit consists of two 7-ft-long shoring pads of aluminum with hydraulic cylinders fixed to these near the top and bottom on swivel connections. The collapsed unit with jacks retracted can be dropped vertically into the trench, positioned against the banks and stressed against the earth walls by jacking, actuated by a remote quick-coupled bucket pump. Intended to replace wood planking jacked apart by trench jacks, it would seem to employ mechanical devices less easily handled and more subject to failure or damage with time, without any real decrease in overall cost.

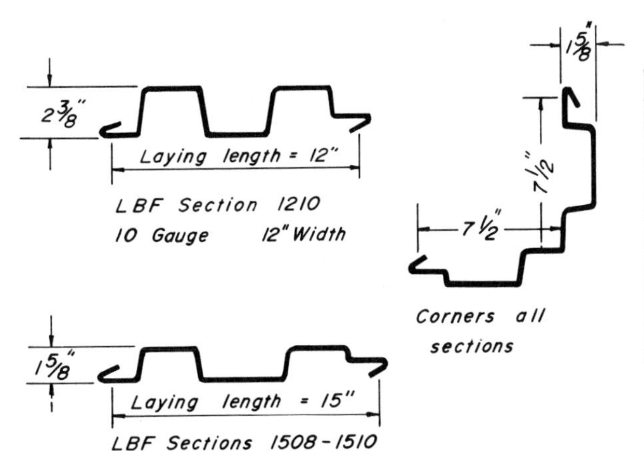

LBF Section	1508	1510	1210
Thickness in.	0.165	0.135	0.135
Laying width in.	15	15	12
Lbs. per sq. ft.	10.5	8.6	10.6
Lbs. per lin. ft.	13.13	10.75	10.6
Area sq. in.	3.1	2.5	3.1
Moment of In. in.^{4}	1.18	0.96	2.38
Sec. Modulus in.^{3}	1.36	1.14	1.97
Rad. of Gyration in.	0.62	0.62	0.88

L.B. FOSTER LIGHTWEIGHT STEEL PILING

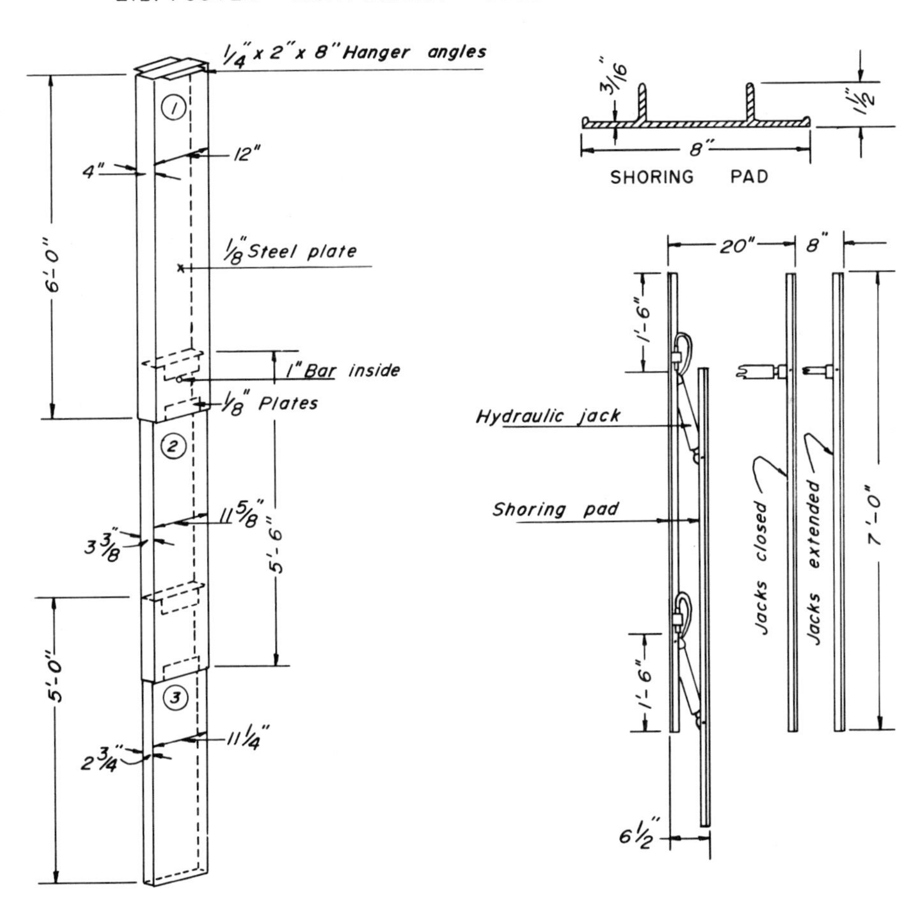

TELESCOPIC SHEETING

SAF-T-JAX SHORES
(SIGMA ENGINEERING CORPORATION)

6.22 DEWATERING COFFERDAMS

Of equal, or sometimes greater, importance than the construction of cofferdams is their dewatering. (The term "unwatering" has come into use in recent years. From an examination of dictionary definitions of the two prefixes, there seems to be little choice so far as meaning is concerned. The author will, therefore, continue to use the term "dewatering.")

The chief problem in dewatering is that of determining how much water must be pumped, and this applies equally with the single-wall, double-wall, or circular cofferdam. The volume contained within a cofferdam enclosure is a simple mathematical calculation, but the rate of flow into the cofferdam is a more complex matter. There are three basic conditions of flow into a cofferdam.

The first condition is that in which the sheathing is driven deeply into a dense stratum of incompressible clay. This is a very effective sealant for the bottom, permitting only leakage through the interlocks above clay level. Methods of sealing interlocks with fly-ash, cements, or other fines have never proved satisfactory, particularly where this work must be done under water. Circular cofferdams produce relatively less leakage because tension in interlocks presses metal firmly against metal, but this fails to form a complete barrier. The sand fill in circular or double-wall cofferdams further decreases the rate of flow but does not eliminate it.

The second condition, where a cofferdam is driven to a seat on pervious soils, has been discussed in Sec. 6.17. Here, stability considerations require that the rate of flow be reduced or spread over a wide area, but the volume of water to be handled can still be prodigious. Bear in mind too that the quantity of flow will increase as the head differential between water levels inside and outside increases.

The third condition is that of flow upward from the bottom of the cofferdam. This may be only a symptom of the second condition, or it may occur where the cofferdam lies over a sunken aquifer or where the cofferdam is seated on a porous rock stratum. Although this inherent condition may be revealed by test borings, the quantity of flow is very difficult to estimate realistically.

Several empirical formulas have been suggested for determining the quantity of water likely to require pumping under a specific set of conditions; but none of these has proved particularly satisfactory for general use, and they are occasionally misleading. The most satisfactory expedient is to provide a dewatering system with considerable flexibility, adding to or subtracting from it as conditions reveal themselves.

Alternative systems involve open pumping with centrifugal pumps, well-point systems, submersible or deep-well installations, or combinations of these.

Open pumping with centrifugal pumps is limited by the potential depth of suction, not over 20 ft, so that for deeper depths, cofferdam pumps must be lowered with the water level. This can be accomplished by setting up the pumps on a float or barge, inside the cofferdam, which floats down on the lowering water level.

The use of well-point systems is limited to those cases where a berm of pervious soil exists into which the well point can be penetrated. For many cofferdams, therefore, they are impractical.

Dewatering by submersible or deep-well pumps is likewise limited to conditions where wells can be drilled. (True submersibles can be used with open pumping, but their capacities are limited and for substantial flows may be inadequate.)

Shown opposite are two types of dewatering used under two sets of conditions on the same project. The Almendares River Tunnel was completed in February of 1953 (well in advance of the Revolution) to connect the Miramar area with the Vedado area of Havana, Cuba. Constructed in open cut, it used both single-wall and double-wall cofferdams, depending upon depth and water conditions, and employed a three-stage well-point system in one area while reducing the open pumping requirements in the other area by the grouting of porous rock.

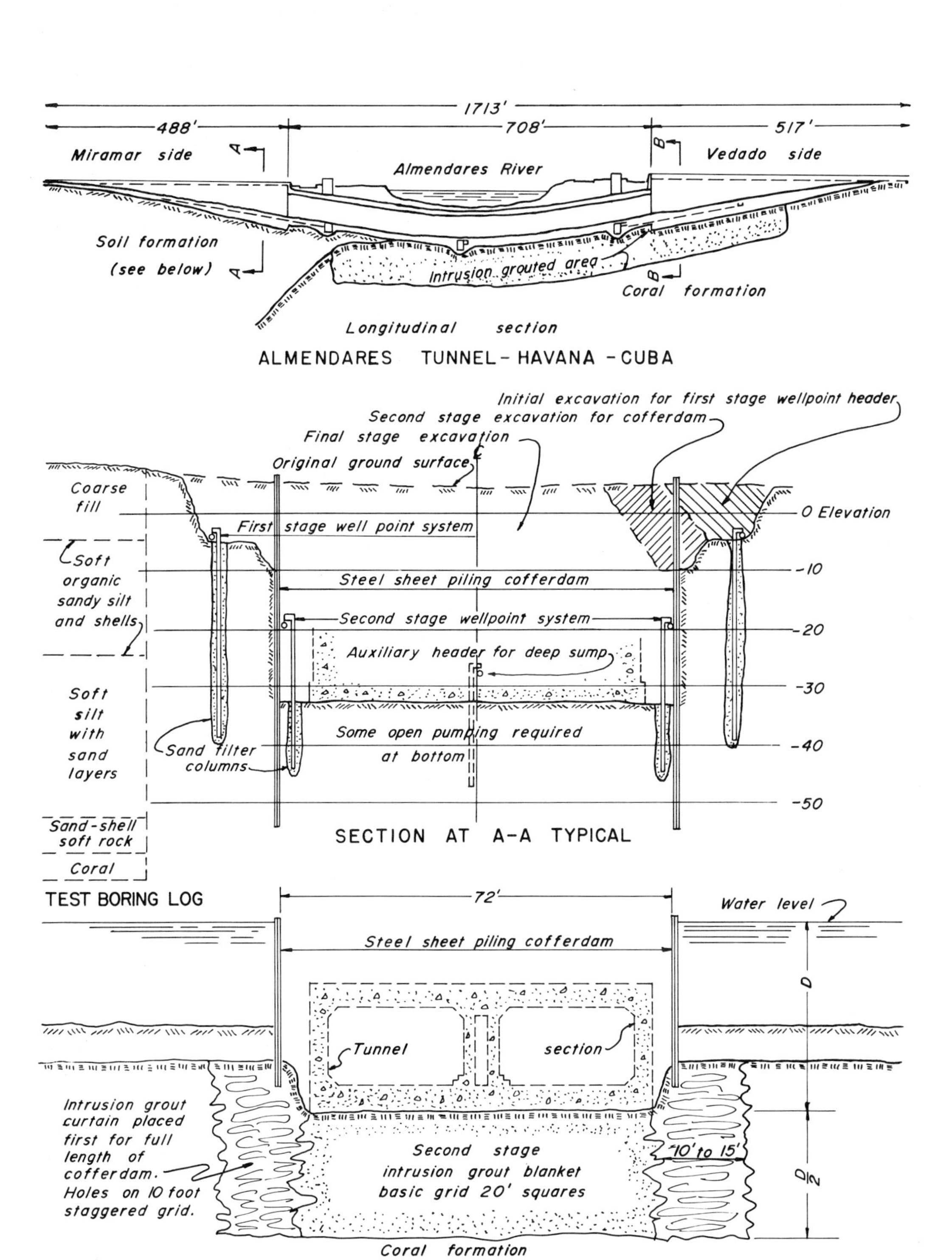

1713'
488'
708'
517'
Miramar side
Vedado side
Almendares River
Soil formation
(see below)
Intrusion grouted area
Coral formation
Longitudinal section
ALMENDARES TUNNEL - HAVANA - CUBA
Initial excavation for first stage wellpoint header
Second stage excavation for cofferdam
Final stage excavation
Original ground surface
Coarse fill
O Elevation
First stage well point system
Soft organic sandy silt and shells
-10
Steel sheet piling cofferdam
Second stage wellpoint system
-20
Auxiliary header for deep sump
-30
Soft silt with sand layers
Some open pumping required at bottom
Sand filter columns
-40
-50
Sand-shell soft rock
Coral
SECTION AT A-A TYPICAL
TEST BORING LOG
72'
Water level
Steel sheet piling cofferdam
D
Tunnel section
Intrusion grout curtain placed first for full length of cofferdam. Holes on 10 foot staggered grid.
Second stage intrusion grout blanket basic grid 20' squares
10' to 15'
D/2
Coral formation
SECTION AT B-B TYPICAL

6.23 EXTRACTION OF SHEATHING AND REMOVAL OF COFFERDAMS

What ever happened to the Trojan Horse? We can, if we wish, learn how the Greeks planned and built it and how they towed it into the citadel and debouched from it, but then Odysseus wandered off and Homer (like his literary descendants when writing of the construction of cofferdams) left it standing in the center of the captured city of Troy, a useless monstrosity.

The case is similar with cofferdams and other temporary construction devices, many pages being devoted to how they should be built and few, if any, providing instructions on how to be rid of them. But turn back to Sec. 6.8 and review the difficulty of extracting the sheathing in that case, and then consider how you would have removed the circular arcs after they had been faced with a foot of concrete—under water.

Ordinary demolition techniques (which, of course, are not being discussed here) are, in most cases, unsuitable since not only is damage to the structure within the cofferdam possible, but the vibrations of explosives, for example, can produce unpredictable stresses.

The extraction of supporting piling is a relatively rare occurrence, whereas steel sheet piling is driven, for the most part, to be extracted. The force needed for extraction of steel sheet piling cannot be adequately pre-determined and may bear no relation to the force required to drive it. The length of pile and consequently its weight, together with the cohesiveness of the soil into which driven, are primary factors; but there are a number of secondary factors of equal importance.

Under almost any conditions, fines of clay or silt particles will drift into and plug the interlocks, thus increasing the friction. The length of time the sheeting has been in the ground can greatly increase the cohesiveness of the soil against the faces and hence the frictional resistance. Piling that has become bent or tilted, no longer permitting a straight vertical pull, can be as difficult to extract as sheathing driven out of interlock, as in the case already cited in Sec. 6.8.

In cases such as that noted in Sec. 6.9, where blocking is substituted for the original strut and waler system, if the sheathing is permitted to tilt out of plumb, the pull along the line of the tilted sheathing may be directly into the structure just completed; this is one reason for allowing a sufficient space between the sheathing and the structure when originally planning and driving.

In cellular cofferdams or other cofferdams where sheathing is in interlock tension, the force required to pull the first sheet and break the tension can be considerable, after which the remaining sheets may pull with relative ease.

With cellular cofferdams, shall we remove the sand fill first or pull the sheathing first? Here, as in other cases, unbalanced pressures against the cofferdam must first be relieved, not only for ease of extraction but to prevent distortion of the sheathing structure. Generally the head differential between the inside and the outside of the cofferdam can be relieved. Backfill may be required around the structure just completed inside the cofferdam; and, in any case, water levels can be restored. These factors, however, will not necessarily permit sheathing extraction without excavation of the cell fill.

Generally a substantial portion of the contents of a cellular cofferdam should be removed before extraction is begun; and where there is little penetration into the subsoil, all the fill may require removal. The methods of accomplishing this may or may not be the reversal of the filling process. Dredging methods used for filling will seldom be practical for removal; and although the air lift used for caisson excavation can be adapted to this use, it is rather slow and expensive. A return to clamshell excavation will permit barge loading of the excavated material for transportation away from the site, for it will be a rare instance where much of the cell fill will be required within the area surrounding the completed foundation structure.

Sheathing extractors generally resemble driving hammers except that they are designed to deliver upward rather than downward blows; and, in fact, a double-acting hammer can be hung upside down from a crane hook and cable-rigged to serve as an extractor. In either case a pair of drilled double plates slide down along the faces of the web, and the three thicknesses of metal are fastened together with a short pin or bolt 1½ to 2 in. in diameter. (A handling hole is a single hole 4 in. from the top of the sheet; but where difficult extraction is anticipated, two pulling holes 4 and 9 in., respectively, from the top of the sheet should be provided.) Pin assemblies should have a tight fit to prevent tearing the web.

In some cases a few taps downward are beneficial in breaking the initial frictional resistance; but where sheathing has been driven to substantial refusal at the beginning, this recourse may have no effect. Since strut-and-waler systems are replaced with backfill to equalize pressures, swaying the sheathing to relieve facial tension is generally impractical.

Even when an extractor is used, a straight, steady, upward pull will be required, often provided by a multiple-reeved crane hoisting line, but fixed frames supporting large block-and-tackle assemblies have been used (see illustration of Sec. 7.18). Where satisfactory support is available, rigs employing hydraulic jacks can be devised, especially where only a few piles are expected to offer massive resistance.

This discussion of sheathing extraction presumes that the contractor, like Old King Cole—"he called for his pipe, and he called for his bowl, and he called for his fiddlers three!"—got what he needed in the way of equipment with little effort. He may, indeed, get what he wants with no more effort than a phone call; but the resulting costs, frequently unanticipated in the original proposal, may be quite considerable.

Another example of unanticipated costs in sheathing removal is that a void is left which may require filling. While sheathing is often left in place to avoid displacements due to the resulting void of extraction, where pulled, jetting equipment may be required to wash fill into the void, and sand may have to be brought in for this purpose.

Even with the sheathing extracted and neatly piled ready for shipment the contractor has not reached the end of his costs, particularly if it is rented sheathing. He calls for the rental agency's truck to haul it away but receives a visit from their inspector instead. Those holes burned through the web to install tie-backs must be plugged; those battered tops and bottoms must be neatly cut off; those bent sheets must be straightened; where battered top cut-offs are too close to the handling hole, new handling holes must be provided. Some of the interlocks are ripped? Cut the section off the sheet and scrap it; indeed, if the sheet has been reduced to less than 20 ft in length, scrap the whole sheet.

Now the mud must be washed off the sheets (rust can remain) and the interlocks must be cleaned out, and then the remaining sheets must be loaded on the rental agency's truck. At their yard the steel sheet piling is weighed, and thc difference in tonnage under that shipped is billed to the contractor.

Now all these items of costs are set forth in the fine print of the rental agreement; and even though the contractor dons a powerful pair of reading glasses to scrutinize the terms, how evaluate them when bidding? How long will it take a welder to patch and trim the sheathing, plus crane and labor time for handling and cleaning? True, the rental agency will do all this for the contractor for a price per ton, but will this be more or less than it will cost him to do it?

The contingent possibilities incident to the removal of cofferdam systems are enormous and are not, sad to relate, subject to any very rational system of evaluation. Perhaps this is why Homer left the Trojan Horse, all woebegone, standing in the market place of the city of Troy.

7 General Foundation Construction

What is a foundation? In construction parlance the term is equally that segment of the earth's subsurface on which a structure is placed and the lowest or contact portion of that structure itself; illustrating, once again, that chronic imprecision in construction terminology repeatedly noted in earlier chapters.

The comment has been made that, since a volume on subsurface construction methods will deal heavily in foundations, would not simple logic dictate that it should open its first chapter with a discussion of "simple" foundations, progressing thence to the more complex and esoteric varieties? Unfortunately, experience has shown that the only thing "simple" about simple foundations is the simpleton who fondly assumes them to be simple. Indeed, what constitutes a simple foundation?

The leaning tower of Pisa, 150 ft high, has stood for several centuries leaning 14.7 ft out of plumb although it rests on ordinary spread footings; while the Campanile, at Venice, resting on closely spaced wood piles, driven solidly into sand and clay layers, collapsed in a cloud of dust in 1902 although it was a mere 2.6 ft out of plumb in its 330 ft height. So-o-o! What else is simple in foundations?

The *very* ancients tell us little about the foundations of their structures in that portion of their documents which has been preserved, and this vacuum continues into the eighteenth century with only an occasional glimpse at foundation strategy.

In 1400 the Milan Cathedral, then being built with local Lombard craftsmen, was encountering some difficulty with foundation problems. The Church authorities brought in various French and German architects (here often referred to as engineers) for consultation. On February 21, 1400, three French engineers were asked to determine ". . . if it seemed to them that this church was adequately founded to sustain and carry the weight. . . ."

In due course the engineers reported: "We the aforesaid engineers and masons say that . . . we have the foundations of two piers exposed, which two piers should sustain and abut the apse . . . , and [they] are inadequately and poorly founded. And one of these is more than a foot at fault inside the work, and of poor material."

They then recommended: "All the piers

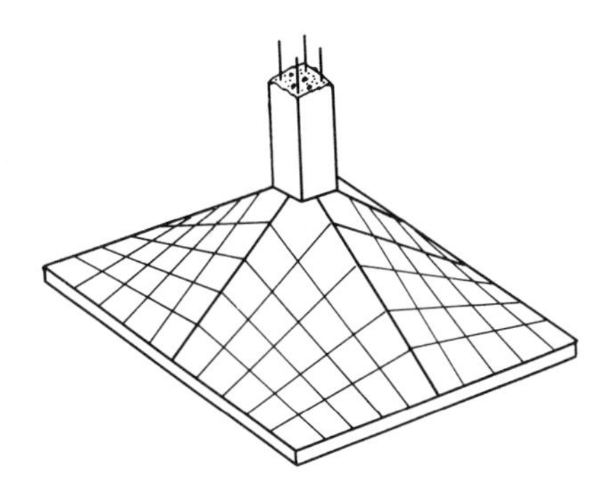

of said Church . . . are to be reviewed down to the lowest base and all those which were badly founded . . . are to be refounded of large blocks of well-laid stone, and their bedding should be well leveled and planed and joined, and buttressed by dovetailing into the other foundations . . . and built in with a mortar bath. These foundations should be made two braccia [about an arm's length] or more beyond the plumb line of the bases of the piers, coming to one braccio at the surface by a set back."

Foundations continued to be built in this general fashion until, in the dawn of the nineteenth century, this approach was deemed inadequate. Soils which had once seemed sufficient for support now appeared questionable, and attention turned from the structural foundation to the soil foundation. This led to a series of values being established as "allowable bearing values" of soils. Once properly established, these seemed to end further question on the matter.

The assumption of safe bearing values for soils has worked well in the majority of cases; but the exceptions have been notable and, as the twentieth century progressed, it was belatedly realized that soils are destructible and do not necessarily retain their earlier characteristics.

Modern foundation experts have paraphrased an old saying ("Where rape is inevitable, relax and enjoy it") to read "Since settlement is inevitable, relax and control it." Settlement of structures (even those founded on rock) being expected, it can be made predictable, within limits, the only cautionary measures being those to prevent unequal settlement.

Foundations for important structures, bridges, skyscrapers, and the like will usually have been designed by a coterie of experts diligently probing at the earth's bosom prior to producing a design; but thousands of lesser structures are built annually which do not warrant such extensive investigation and certainly do not receive it.

Many of these lesser structures reach their final construction stage and then begin to show cracks in the walls or other signs of distress. Although there are many reasons besides foundation difficulties which can produce trouble, it is the foundations, now decently interred in backfill, toward which the accusing finger points. All too often it is indeed the foundation which is at fault.

7.1 THEORETICAL ASPECTS OF FOUNDATIONS

It is neither practical nor desirable, here, to discuss thoroughly the many and devious theories of soil mechanics, as they relate to foundations, nor to discuss the design of foundations in relation to the soils on which they may be placed. The depth and size of footings have usually been provided, for the contractor, by the architect or engineer, and he is inhibited from changing them, first by the terms of his contract and secondly (where he would improve upon them) by the costs incident to such action. Yet the contractor should be alert to the general possibilities inherent in improperly designed or constructed footings since he is frequently the scapegoat who must pay the bill for corrective measures. He may not possess the personal dynamism needed to deter a determined designer; but if he has gone on record at the appropriate time, it may later save him a stack of pennies in remedial work.

When we place a load on a soil mass in the form of a square, the pressure resulting from that load is transmitted downward and outward through the soil so that at any particular depth the effect of that pressure will have spread out to form the base of a truncated pyramid of which the top surface is the square base of the load. (The same effect will, of course, occur with any shape of load although producing different configurations and distributions of loading.)

If there were no restraining influences, this pyramidal distribution would extend indefinitely either until the area of distribution became so wide that the loading effect was negligible or until it encountered a solid resisting medium such as rock. In fact, however, soil itself will resist this effect in both the horizontal and the vertical direction to an extent depending upon its characteristics. The cohesiveness of clay will provide resistance as will the angularity of sand particles; and the voids, moisture, and density of any type of soil will further condition the influence of the foundation pressure.

The net result of the initial, spreading pressure distribution, curbed by the soil, is to produce a pressure zone in the shape of a bulb. The illustration shows such a pressure bulb with the general effect of pressure distribution indicated by influence lines. Such a regular bulb, as shown, presupposes a homogeneous soil of indefinite extent.

In nonhomogeneous soils the pressure bulb can be extensively distorted or twisted askew, resulting in an excessive concentration of the superimposed load in the weakest portion of the soil. Where a soft stratum lies under a harder stratum, the distribution of loading mushrooms in the softer layer to a point where it may exceed the supporting capability of the soil. A harder underlying stratum may effectively dampen the distribution, and there is no danger of failure here, unless the harder stratum exists on a steep tilt. Some of the possibilities are illustrated.

The pressure bulb indicates a fact which has been checked by experiment, that the ground at the center of a uniformly loaded footing will settle more than the soil at the edges. This leads to the conclusion that, unless the foundation is flexible, a condition which exists only in special cases, a higher loading than anticipated can occur on the soil at the edges of the footing. If one edge of the footing soil is "softer" than others, a tilt in the foundation can develop.

Also apparent from the pressure-bulb illustration is the fact that there is a tendency to produce a settlement crater extending beyond the edge of the footing, which can induce uneven settling in a structure built immediately adjoining that footing. In other words, influence lines from two adjoining buildings meet to produce a higher load concentration than originally intended for either; or, in yet other terms, the first footing produces a crater which robs the support from the second footing.

Not so evident from the chart, but well established nevertheless, is the fact that a larger footing with the same unit loading will settle more than a smaller footing. Thus a 4×4 ft footing with a load of 16 tons will settle more than a 2×2 ft footing with a load of 4 tons although each will have an evenly distributed load of 1 ton per sq ft.

Vertical Stresses Under Square Footing

The pressure bulb at left and the charts at the bottom show the distribution of pressures under a square footing resting on a soil which is a semi-infinite elastic solid, that is, where the soil is a homogeneous clay stratum of indefinite extent.

The data presented here shows only relative magnitude and this too will vary depending upon the physical characteristics of the underlying soil.

The two part-sections at the left indicate two possible effects where the soil is stratified.

To Use Charts at Bottom

Determine the vertical stress P_H at a depth of 20 ft, where the footing is 8 ft square and the uniformly distributed lead P = 400 kips (1000 lb).

Enter Chart II at the 8-ft footing width and, horizontally, find intersection with P = 400 kips.

Enter Chart III at the 8-ft footing width and, vertically, find intersection with H = 20 ft.

Extend intersection on II vertically and on III horizontally to find loading PH = 0.45 kips/sq ft.

The pressure bulb homogeneous soil

Effect of hard stratum

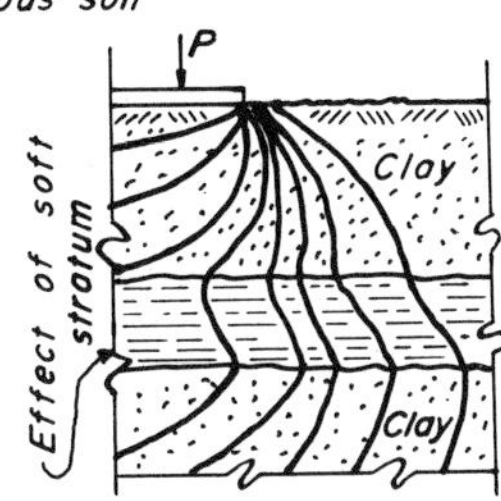

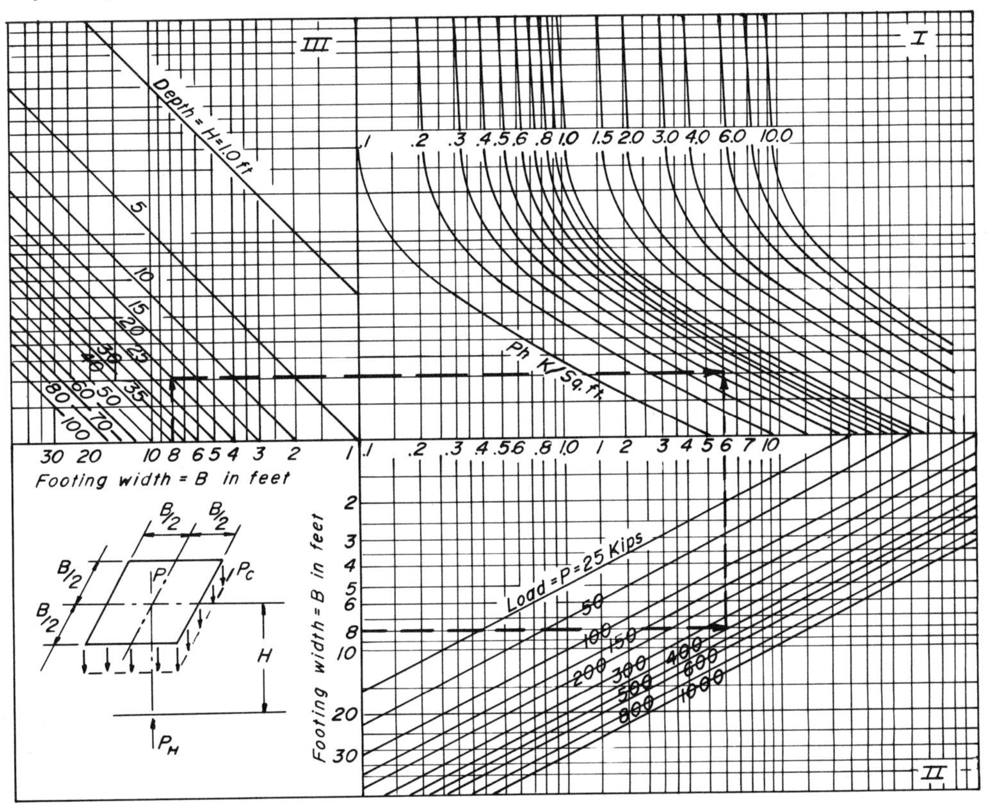

7.2 PREPARATION OF FOUNDATION-BEARING SURFACES

The treatment of the soil surface on which foundations are to be placed has been discussed in various aspects in Chap. 3, and the reader ought now to review that chapter before proceeding. Special attention should be paid to those aspects to be found in Secs. 3.3, 3.6, 3.9, 3.11, and 3.16. These sections cover general aspects of many specific soils; but the potential variations are virtually inexhaustible, the more so as new construction sites appear and new types of structures are required. Have you ever built the foundations for a variable-angle launcher, for instance?

The variable-angle launcher was a United States Naval Ordnance test facility constructed at the Morris Dam Torpedo Range near Azusa, California. Its purpose was to launch various types of projectiles into the air and observe the effects of their entering the water at variable angles and at high velocities. A complex construction project, it was built on the two sides and peak of a peninsula jutting out into the Morris Dam Reservoir, the body of water which was to receive the projectiles.

The project required a rock slope of 30 deg on the launcher side and 45 deg on the counterweight side. Test borings showed the site to consist of highly fractured metamorphic and igneous rock sliced by substantial fault zones.

The slopes had a landslide potential, and slides had occurred in the general area. To control the slide possibility, all soil and loose rock was stripped from the two slopes, much of it being piled at the toes of the slopes to further resist their sliding. The quantity of rock removed, 9000 cu yd on one side and 15,000 cu yd on the other, also reduced the loading which could contribute to slides.

The second step in treatment was deep pressure grouting to seal deep voids, fissures, and faults, followed by surface grouting to seal open crevices. When this work had been completed, foundation construction could proceed.

Or consider the foundation preparation for the South Bay Power Plant at Chula Vista, California, where not only was the structure placed on compacted fill, but the fill was a *dune* sand. Here the site consisted of 10 to 20 ft of soft alluvial soils lying over firm sandy clay. Water was to be found about 10 ft below ground surface. Rock lay more than 200 ft down, and alternative foundation systems were deemed too expensive, so the fill method was chosen. An average settlement of 1½ in. was anticipated and a differential maximum settlement of ¾ in. It was believed that this settlement would be completed prior to the time turbines and other equipment would be set.

Some details of the operation are shown opposite. Drainage ditches were dug around the site and backfilled with ¾- to 1½-in. crushed stone. These ditches drained to 42-in. concrete standpipes at the corners, acting as sumps—from which the collected groundwater was pumped.

The site was excavated by scraper down to within 12 to 18 in. of final grade. To prevent disturbing the final sandy clay layer, a backhoe excavated the remainder by backing over the unexcavated sub-grade using a sharp blade fastened across the bucket lip to shave the ground down to grade. Over the final grade a 12-in. layer of ¾- to 1½-in. crushed stone was placed by using a light bulldozer pushing from a continuously augmented pile and traveling only over the stone previously placed. There was then placed a 2-in. layer of pea gravel, and the foundation area was topped off with 6 in. of fine to coarse sand. The placing of dune sand fill could now begin.

The fill sand, known locally as Coronado Fine Sand, was placed in 12-in. layers, moistened, and compacted by three passes of a type B vibratory compactor. Tests showed that the highest relative density of the sand was reached under conditions of substantial saturation. To maintain the water level 3 in. below the top of the sand, the flow in the underdrainage system was reversed by pumping water *into* the 42-in. pipe sumps.

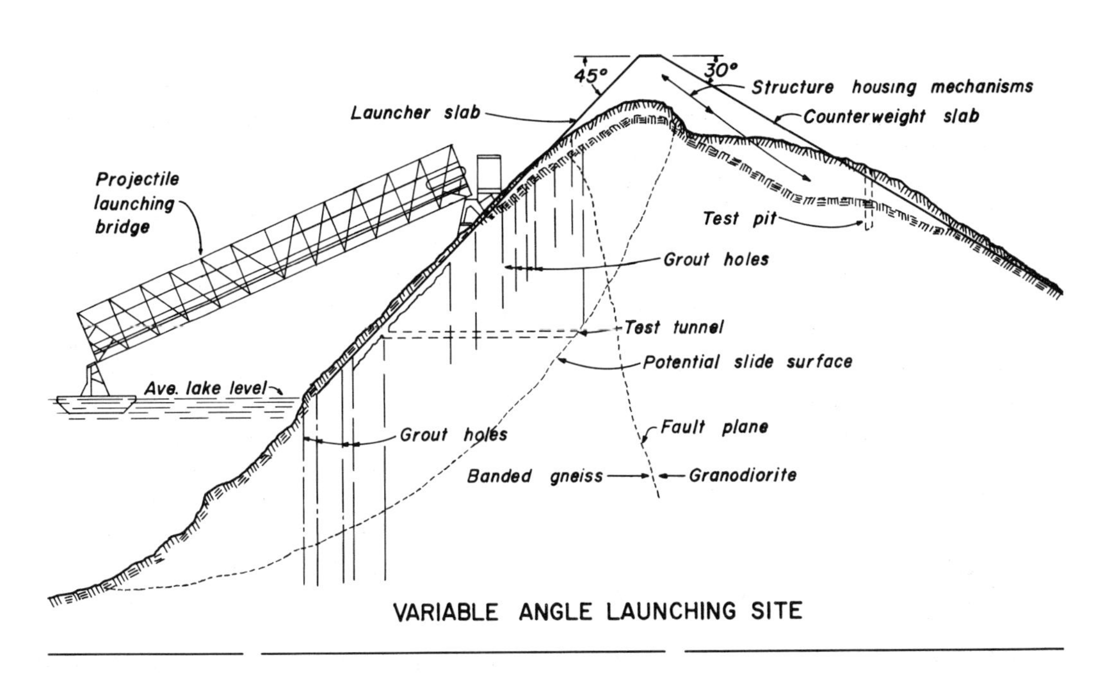

VARIABLE ANGLE LAUNCHING SITE

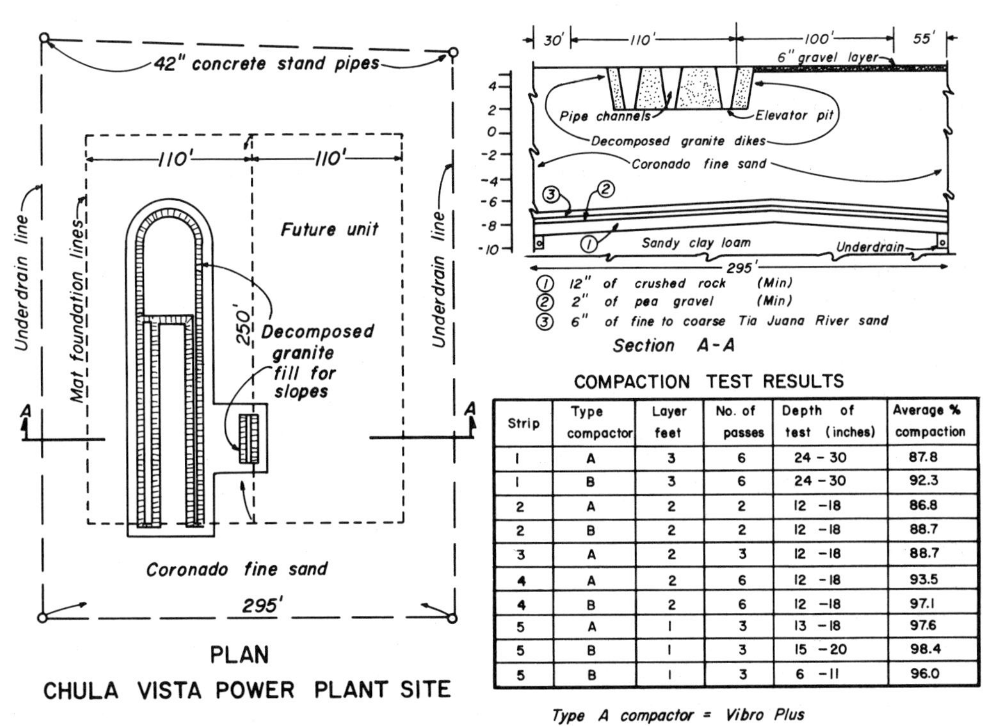

PLAN
CHULA VISTA POWER PLANT SITE

COMPACTION TEST RESULTS

Strip	Type compactor	Layer feet	No. of passes	Depth of test (inches)	Average % compaction
1	A	3	6	24 - 30	87.8
1	B	3	6	24 - 30	92.3
2	A	2	2	12 - 18	86.8
2	B	2	2	12 - 18	88.7
3	A	2	3	12 - 18	88.7
4	A	2	6	12 - 18	93.5
4	B	2	6	12 - 18	97.1
5	A	1	3	13 - 18	97.6
5	B	1	3	15 - 20	98.4
5	B	1	3	6 - 11	96.0

Type A compactor = Vibro Plus
Type B compactor = Essick

7.3 SPREAD FOOTINGS

The spread footing discussed here might be termed a "simple" footing in that it is essentially simply a pad of concrete placed at foundation level, on subsurface soil, to spread the loading of the wall above it. (Although the footing of a column or pier is also generally a spread footing, there are often variations of sufficient importance to warrant making a distinction here. See Sec. 7.6.) This pad, which is today constructed of concrete, may be from 1 to 2 ft wider than the wall above it; and the wall, although generally of concrete too, may be of brick, stone, or concrete block construction. Where materials other than concrete are used for the wall, the spread footing serves simply as a leveling pad to start the masonry coursing. Where concrete is involved in the wall above, the concrete of the spread footing may serve other purposes.

The depth to the bottom of the spread footing, where no basement or similar subsurface structure is involved, will be a minimum of from 3 to 5 ft below the *finished* ground surface around the exterior of the building. In cold climates the depth is kept a minimum of 4 to 5 ft below surface because of possible frost action, but in any climate the possibility of erosion of the soil around the outside walls must be considered. In sloping terrain the bottom of the footing is generally stepped down in elevation to maintain the required depth below the original ground surface.

The means employed to construct spread footings will depend, to a large extent, on the reinforcing-steel requirements. Pads for masonry walls, of any type, frequently have no reinforcing or, at the most, only light longitudinal bars classed as "temperature" steel. Where concrete walls are used, however, the spread footing may be designed as an integral part of the wall, with rods in two directions horizontally and dowel bars projecting upward into either or both faces of the wall above. In some cases only longitudinal bars are used or even none at all, with a minimum of straight-ended dowels as a tie between wall and footing.

Where only longitudinal bars are required, they are generally provided for the upper surface of the footing and can be forced down into the concrete after it has been placed. The same thing can be done with straight-end vertical dowels. Where either of these reinforcing methods is employed or where no reinforcing is used, it may be possible to dispense with forms altogether. In wet conditions, or where sand is encountered, a form must be provided to retain the soils; but in many cases soils are self-supporting, for limited depths, and it appears to the author that forms should be dispensed with wherever practical.

Where forms are not used, the excavation for the footing should be carried about 2 in. wider on both sides. The cost of the additional 4 in. of concrete is negligible; the extra width provides a better footing; and the problems of backfilling the 2-in. slot left by the removal of the forms is avoided.

Footing forms will be required not only where ground conditions dictate their use but where mats of steel or dowels, requiring precise positioning, are used. Footing forms are generally constructed of 2-in. plank 8, 10, or 12 in. wide, often unfinished, and battened together to obtain the required height. These are supported by 1×4 in. stakes driven flush with and flat against the outer face of the form and are spiked to the plank. Steel bars ¾ in. in diameter can serve as stakes and are secured to the plank with a partially driven 8d nail, bent over.

Footing forms can also be set using a measured length of 1×4 between planks at their bottom, as a spreader, and a 1×4 in. cleat is spiked across the plank tops. The spreader is removed as concreting proceeds.

Either type of form construction will provide sufficient support for the reinforcing steel. Although often supported by blocks off the ground surface, it is more desirable to hang reinforcing bars, by wire, from the cleats; or separate short bars can be laid across the top of the form to provide support. Hooked or bent dowel bars are generally tied into the mats and can also be hung or blocked up where their projection is limited to the usual 25 diameters.

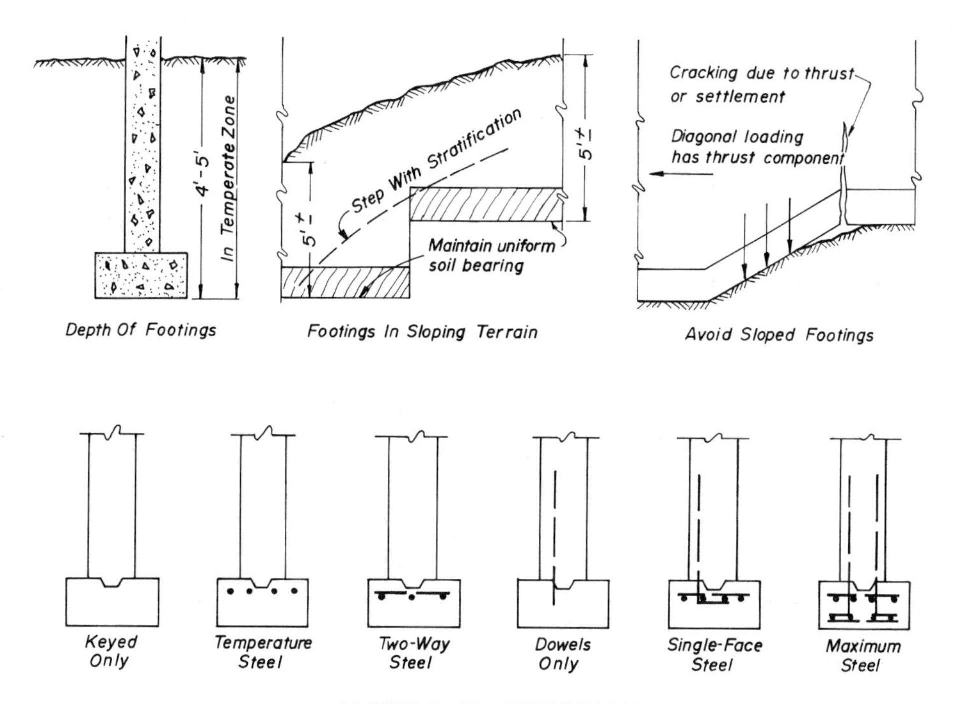

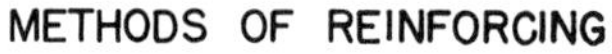

METHODS OF REINFORCING

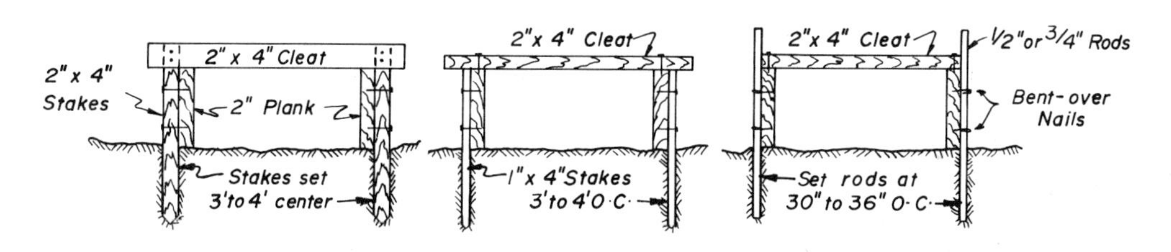

WALL FOOTING FORMS ON CLAY OR OTHER FIRM SOILS

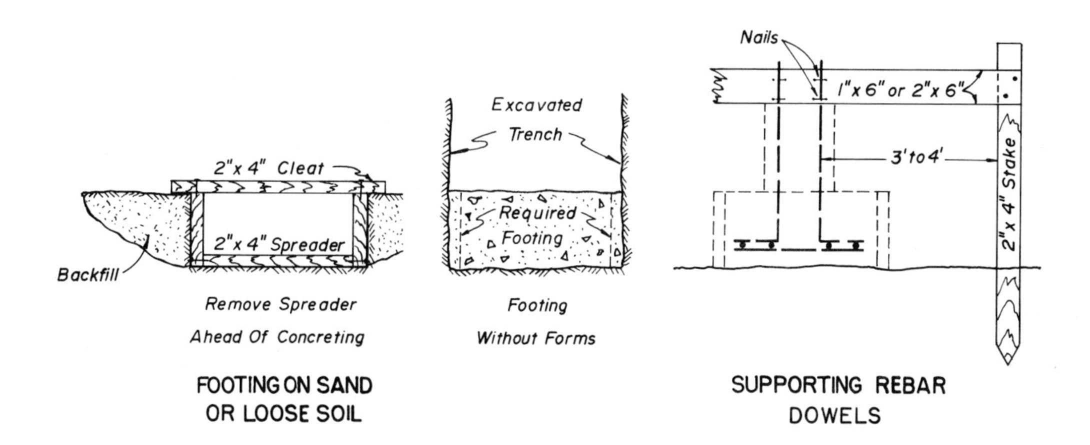

Remove Spreader Ahead Of Concreting

FOOTING ON SAND OR LOOSE SOIL

Footing Without Forms

SUPPORTING REBAR DOWELS

7.4 FOUNDATION WALLS

When the extensive use of concrete construction mushroomed during the first three decades of the twentieth century, little attention was paid, at first, to formwork construction. For the most part, capable laborers were selected from the gang and, termed "hatchet and saw men," were employed to build forms. Designated as "form builders" and paid a rate between that of laborer and carpenter (in one instance labor was $0.25 per hr, form builders $0.45, and carpenters $0.65), they performd the function adequately. Carpenters, then trained in the cabinetmaker's tradition, disdained the work. With the advent of the depression of the 1930s, the carpenters changed their mind, claimed the work, tucked the form builders under their organized wing and, after multiple strikes on early P.W.A. projects, brought the form builders' rate up to that of carpenters—prior to assigning their own men to the work.

The character of foundation forms, once slapped up with rough scantlings, gradually changed with the times and the rising labor costs. Other factors were at work too. The Portland Cement Association began an extensive educational campaign in the 1930s to promote the use of concrete. Fewer foundations walls were merely re-buried in the ground; more of them formed basement walls with smooth interior surfaces; waterproofing came into widespread use, requiring smooth *exterior* surfaces for application.

There has been a persistent trend toward the use of manufactured or pre-fabricated (patented) form systems. Many contractors continuously engaged in concrete construction own the elements of one or more of these systems. In most cases, for a single use, such systems can be rented. Because the manufacturer also furnishes form design as well as materials, they are particularly valuable in cases requiring special form designs, or where the class of labor available is presumed insufficiently trained to build adequate forms at the site. Such systems, whether faced with metal, plywood, or compo board or whether framed on steel members or magnesium metal frames (to reduce weight), are expensive in first cost, and the economics of their use depends on the number of potential re-uses for individual members. Careful study of conditions should be made before adopting any system, and this should include salvage value.

Formwork consists of sheathing, studs, walers, strongbacks, and bracing; and in many cases it will be advantageous to fabricate forms from standard materials at the site. Patented formwork will generally cover only sheathing and studs (as well as ties and spreaders) but will seldom include the necessary walers, strongbacks, and bracing. Foundation-wall forms can be designed for each job or use, based on concrete pressures and other variables. Conscientious design, here, can lead to a variety of odd sizes for members providing limited re-usability, and for foundation walls is usually uneconomic.

Concrete, freshly mixed, exerts a liquid pressure against the form, essentially a hydrostatic pressure, which varies with the height of the mass, the temperature, and the "slump" or consistency of the concrete. This condition prevails until it has obtained its initial set, a time varying from 1 hr at 70 to 80°F to 1¾ hr at 40 to 50°F. The mere act of setting does not provide any great strength for the concrete but does eliminate the earlier liquid pressure.

Today's specifications generally limit the slump of concrete to 3 or 4 in., and the tendency is toward still lower slumps. (When the first pre-stressed concrete bridge beams were built in the United States for the Walnut Lane Bridge in Philadelphia, the Belgian designer called for a 1-in.-slump concrete, which the workmen could not be trained to place.) The lower the slump, of course, the lower the liquid pressure exerted.

The principal form of control to be exercised by the contractor is that of liquid depth. In other words, if he is to use a standard form, involving the least practical cost, he must control the rate at which concrete is placed in the form. Rates of rise of 1 to 2 ft per hr are adequate to permit setting up of the previously poured layer without overstressing the form.

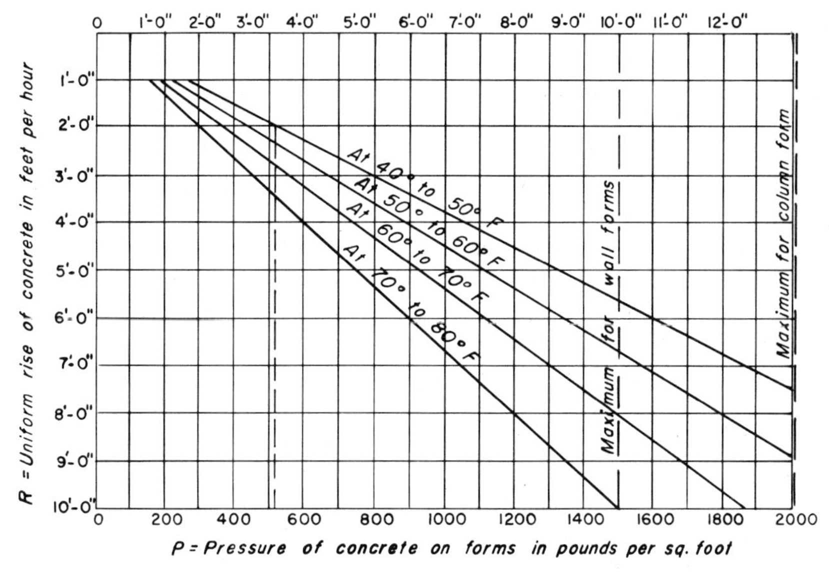

FORM PRESSURE CHART FOR CONCRETE

Limited head is the maximum effective height of liquid concrete for a given rise per hour, based on setting times assumed as 1¾ hr for 40° to 50°, 1½ hr for 50° to 60°, 1¼ hr for 60° to 70° and 1 hr for 70° to 80°.

Many other factors affect the result such as the use of accelerators and retardants.

Sheathing thickness must be adequate to prevent excessive deflection between studs. Stud spacing can be reduced or sheathing can be increased to ¾ or 1 in. thick for higher rates of pour. ⅛-in. deflection between studs is usually considered a maximum. Exposure of wall face will determine permissible deflection.

Studs must be checked for their loading between walers. It will generally be cheaper to go to 3 in. × 4 in. or 2 in. × 6 in. studs rather than decrease the spacing.

Walers are always figured as being a double member. Their spacing must be balanced with tie strength since we are concerned with the total area loading.

Form ties for walls are available in strengths of 3000 and 5000 psi in snapties. Other tying devices are available which extend the limit to 9000 psi. However, all elements of the form must be increased to match the size used, with increase in cost throughout.

TABLE OF FORM SPACINGS

Concrete pressure, psf	Uniform rise of concrete in ft per hr for temperature F				Form member sizes			Tie spacings		
	40° to 50°	50° to 60°	60° to 70°	70° to 80°	Sheathing	Studs	Walers	Hor.	Vert.	Areas, sq ft
450	1'9"	2'0"	2'6"	3'0"	⅝" plywood 1" sheathing	2" × 4" @ 16" o.c. 2" × 4" @ 19" o.c.	Double 2" × 4"	2'8"	2'6"	6⅔
525	2'0"	2'3"	2'9"	3'6"	Same	2" × 4" @ 15" o.c. 2" × 4" @ 18" o.c.	Same	2'6"	2'4"	5¾
600	2'3"	2'9"	3'3"	4'0"	Same	2" × 4" @ 14" o.c. 2" × 4" @ 17" o.c.	Same	2'0"	2'6"	5
675	2'6"	3'0"	3'6"	4'6"	Same	2" × 4" @ 14" o.c. 2" × 4" @ 17" o.c.	Same	2'0"	2'3"	4½
750	2'9"	3'3"	4'0"	5'0"	Same	2" × 4" @ 13" o.c. 2" × 4" @ 16" o.c.	Same	2'0"	2'0"	4

7.5 STANDARD CONCRETE WALL FORMS

Concrete wall forms consist of two faces of sheathing held to the proper dimensions by means of wall ties and spreaders, backed up with studs (to keep the sheathing thickness to a minimum); held in horizontal alignment by means of walers; supported vertically by means of strongbacks braced to outside objects as required.

Standard formwork frequently used today, where patented form systems are not used, consists of ⅝-in. plywood, held apart and pulled together by rod wall ties, laid over 2×4 in. studs with pairs of 2×4 in. walers, backed up by strongbacks with a minimum dimension of 4×6 in. and braced by 1- or 2-in. plank.

Plywood sheathing has replaced 1-in. sheathing boards for several reasons. A tight form (to prevent leakage of mortar) would require tongue-and-grooved sheathing, which cannot be salvaged in stripping the form. Either tongue-and-grooved or square-edged sheathing will leave form marks in the concrete. One-inch boards of standard (six, eight, or ten in.) widths are more costly to nail to the studding. Five-eighths-inch plywood is the minimum thickness that can be used without deflection between studs on 16-in. centers for concrete rises of 1 to 2 ft per hr. Metal sheathing in the thicknesses required (at least 11-gage) is heavy to handle, cannot readily be cut for wall inserts, is subject to rust and consequent wall discoloration, and leaves a glazed surface on the concrete which makes finishing of exposed faces difficult.

The studs, to which the plywood is nailed, are set vertically on 16-in. centers providing even spacing on a 4×8 ft sheet of plywood. Since there is no pull between plywood and stud, 6d nails on 12-in. centers are adequate. End studs may be set flush with the edge of a sheet of plywood, or the plywood sheets can meet at the center of a single stud. Studs used are 2×4s and, in most cases, a 2×4 in. plate, also set on edge, is used at top and bottom of an 8-ft sheet of plywood.

Walers are set as pairs of 2×4 in. timbers with a space of about ¾ in. between the members of each pair. Generally, the bottom waler will be set about 6 in. above the bottom of the wall form with another waler set a similar distance below the top of the form. Maximum spacing of walers is about 30 in. center to center, although this dimension will often be reduced to 24 in. toward the bottom of the wall where maximum pressure can occur.

The wall ties used with these materials are a standard manufactured product consisting of a steel rod to which two disks have been welded at the spacing required for the thickness of the wall. The overall length of rod is sufficient to leave the ends, terminating in a metal button, protruding beyond the outer face of the walers. Wall ties are spaced on line with the center of each waler and in a vertical line alongside every other stud.

Usually one side of the form is erected with the plywood drilled and the wall ties inserted. The second side is then brought up into position, its plywood similarly drilled, and threaded over the opposite end of the wall tie. Walers are lightly toenailed in place and "hairpins" or other similar devices are slipped over the wire tie between the waler and the terminal button, where, by means of wedging action, they draw the form up so that the plywood is forced against the spacing disks.

Strongbacks are now set against the tier of walers, held in place by temporary toenailing, and bracing is set as conditions warrant, to align the wall.

Standard wall ties are made for the dimension lumber noted. Five-eighths-inch sheathing can be increased to three-quarter-inch without change in wall ties, but if 2×6 in. studs or walers are to be used, the standard projection on the wall tie must be increased. The wall ties mentioned here are generally termed "snapties" since a nick is provided in the rod which permits them to be snapped off inside the face of the concrete. Normally this snap-off point is only ¼ in. inside the disk, but 1-in. snap backs can also be obtained. The deeper snap back is desirable in an exposed wall, subject to moisture, to prevent rust from showing through.

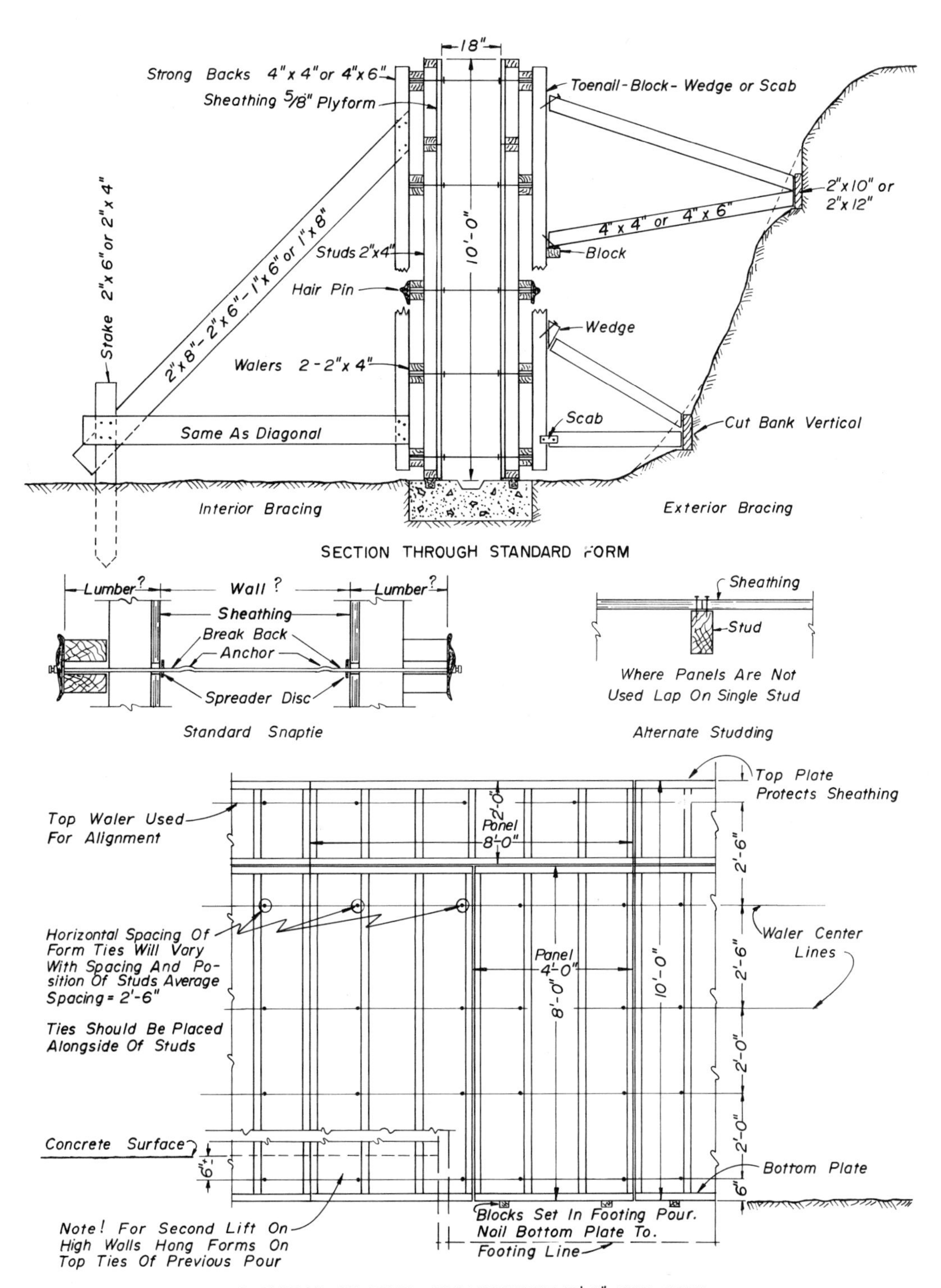
18"
Strong Backs 4"x 4" or 4"x 6"
Sheathing 5/8" Plyform
Toenail-Block-Wedge or Scab
2"x 10" or 2"x 12"
4"x 4" or 4"x 6"
Block
Stake 2"x 6" or 2"x 4"
2"x 8"-2"x 6"-1"x 6" or 1"x 8"
Studs 2"x4"
10'-0"
Hair Pin
Wedge
Walers 2-2"x 4"
Scab
Cut Bank Vertical
Same As Diagonal
Interior Bracing
Exterior Bracing
SECTION THROUGH STANDARD FORM
Lumber ?
Wall ?
Lumber ?
Sheathing
Break Back
Anchor
Spreader Disc
Standard Snaptie
Sheathing
Stud
Where Panels Are Not Used Lap On Single Stud
Alternate Studding
Top Plate Protects Sheathing
Top Waler Used For Alignment
2'-0"
Panel 8'-0"
2'-6"
Waler Center Lines
Horizontal Spacing Of Form Ties Will Vary With Spacing And Position Of Studs Average Spacing = 2'-6"
Panel 4'-0"
8'-0"
10'-0"
2'-6"
Ties Should Be Placed Alongside Of Studs
2'-0"
Concrete Surface
6"±
2'-0"
Bottom Plate
6"
Note! For Second Lift On High Walls Hang Forms On Top Ties Of Previous Pour
Blocks Set In Footing Pour. Nail Bottom Plate To.
Footing Line
ELEVATION OF PANEL ARRANGEMENT 10'-0" HIGH POUR

7.6 COLUMN AND PIER FOOTINGS

Spread footings for columns and piers cover a wide range of possibilities and vary somewhat from those discussed in Sec. 7.2 for walls. Once again we are chiefly concerned with footings for concrete columns and piers, although stone, brick, and block columns are occasionally used. Special and extensive pier footings will be discussed under separate headings, and it should be recognized that spread footings, as discussed here, frequently are not used under columns, as in caisson construction, for example

Although spread footings for columns can employ essentially the same form materials as those used for wall footings, there are some important differences. Because of the column punching potential, spread footings for columns are generally thicker than those for walls, are more heavily reinforced, involve proportionally more corners in the formwork, and require a greater precision in the support of projecting dowels.

Two-inch plank is generally used for column spread footings in convenient standard widths of 6, 8, or 10 in., battened together as required. At the corners, the simple lapping and spiking of one battened plank form over the end of another is not adequate since nails are easily forced out of the end grain. Cleats spiked diagonally across the tops of the planks at the corners may be adequate unless the form has considerable height, but backing up the form with soil alone is seldom satisfactory.

The most desirable method is to extend one form beyond the end of the adjacent right-angle form some 6 to 8 in., set a 2 × 4 or 4 × 4 in the interior corner resulting, on the exterior of the form, and spike both forms into the vertical member.

An alternative method, on deeper footings, is to carry walers over the face of the plank battens, of sufficient length to lap each other at the corners, with alternate walers offset so that the top member of one lapping pair fits between the two members of the opposite pair (see detail). Walers can be spiked together, held by a 4 × 4 in the corner, or bolts can be dropped through matching holes drilled in the ends. Bolts permit multiple uses without damage to the walers.

Another method of supporting column footing forms is to drive 2 × 4 in. pointed stakes behind the form, bracing them back diagonally to stakes driven 3 to 4 ft behind the form, as might be done with wall footing forms. This requires a homogeneous, fine-grained soil which will permit fairly precise driving.

Horizontal reinforcing steel can be prematted and supported off ground level, as for slabs, where it is solely in the bottom of the footer. Where a mat also occurs in the top portion, or only in the top portion, it must be hung in position. This is generally done by placing plank on edge over the top of the form, the mat hanging from wires wrapped around the plank. The hanging plank should be toenailed to the top of the form to prevent horizontal movement (see detail). If properly positioned, these hanging planks can also support the dowel steel.

Spread footings for columns frequently occur in large numbers of similar size. Rather than invest in form material for all footings, proper organization can make do with a limited number of forms. Assuming that we have 36 footings of comparable size and complexity, we might have laborers preparing the ground surface for six footers and carpenters setting the forms for an additional six footers. Rod setters are placing rods in the forms completed yesterday while concrete is being poured in forms completed by the rod setters and a separate crew of laborers is stripping, cleaning, and moving forms previously poured.

Such programming requires careful planning. Basic excavation must have been carried to a point where a limited group of laborers can fine-grade six footings a day. Forms must have been prepared so that only assembly, in place, is necessary. The footer mats and dowel cages should previously have been assembled and adequately tied for handling as a unit. A crane rig should be available for setting these mats and cages. Carpenters must be available to provide supports or place hanging planks.

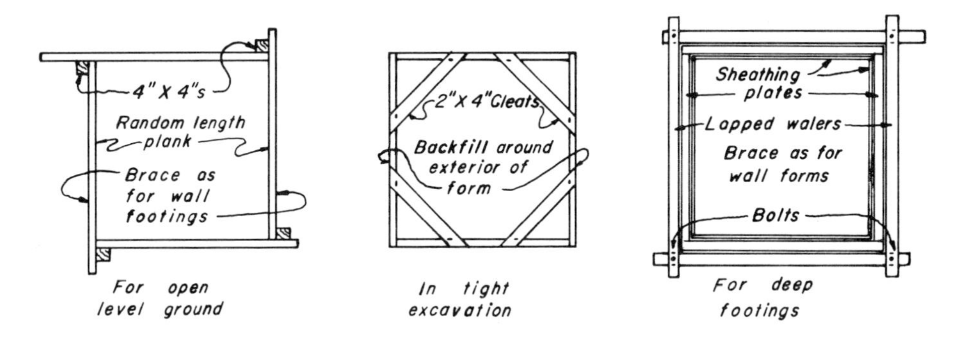

COLUMN FOOTING FORMS

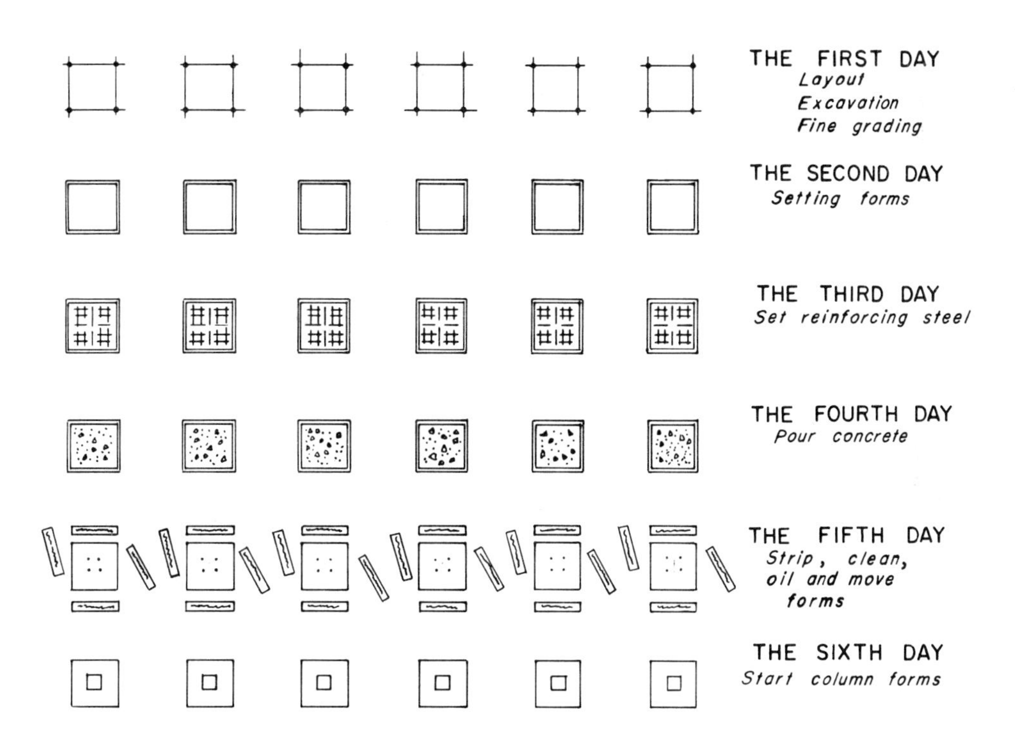

ON THE SIXTH DAY

A FOUNDATION CONSTRUCTION PROGRAM

Some operations may be combined and done in a single day. Some days you won't make a nickel. It will rain or snow; the temperature will be too high or too low; or everyone may knock off to go hunting.

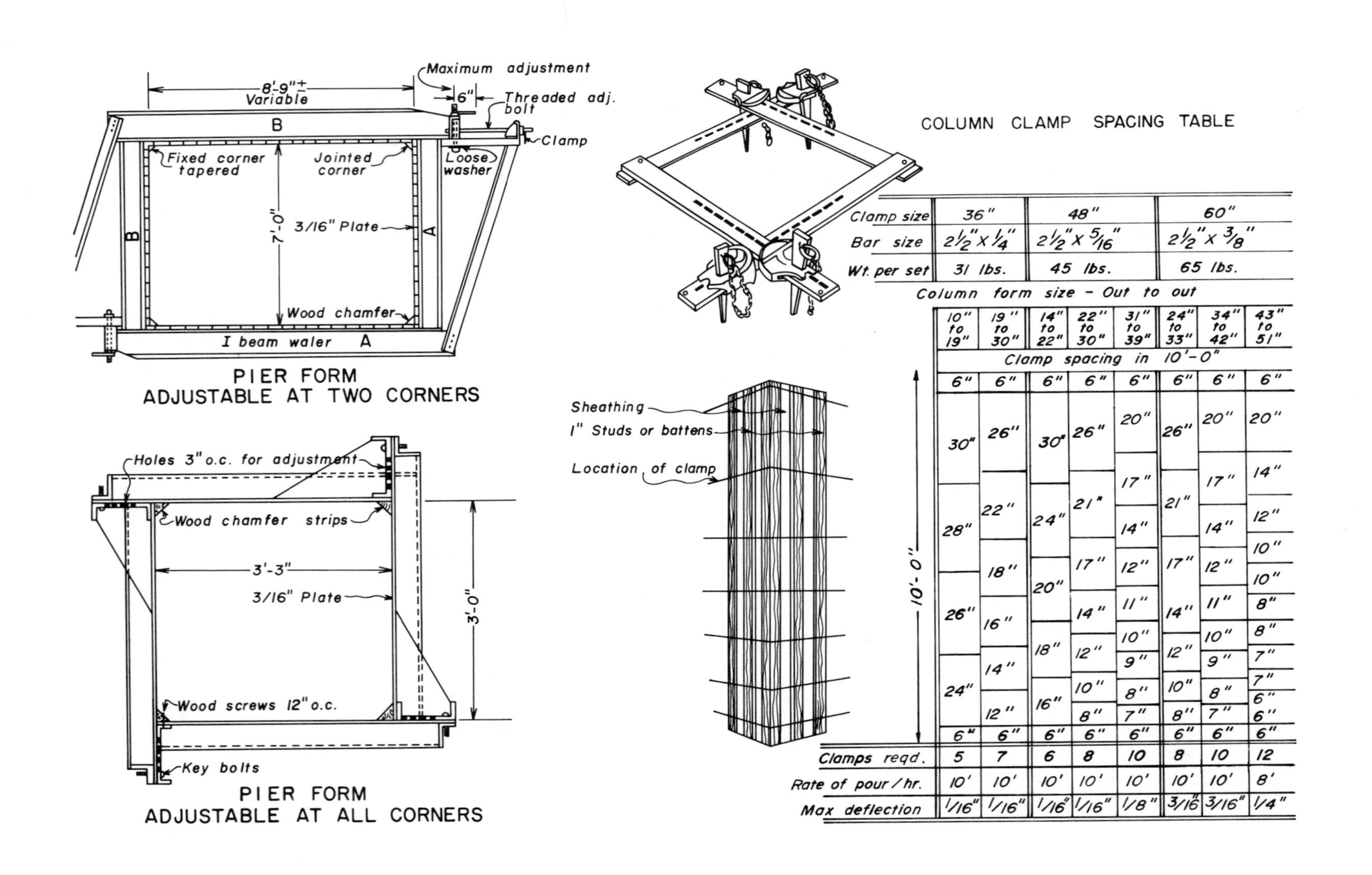

PIER FORM
ADJUSTABLE AT TWO CORNERS

PIER FORM
ADJUSTABLE AT ALL CORNERS

COLUMN CLAMP SPACING TABLE

Clamp size	36"		48"			60"		
Bar size	2½" x ¼"		2½" x 5/16"			2½" x 3/8"		
Wt. per set	31 lbs.		45 lbs.			65 lbs.		
Column form size – Out to out	10" to 19"	19" to 30"	14" to 22"	22" to 30"	31" to 39"	24" to 33"	34" to 42"	43" to 51"
Clamp spacing in 10'-0"								
	6"	6"	6"	6"	6"	6"	6"	6"
	30"	26"	30"	26"	20"	26"	20"	20"
	28"	22"	24"	21"	17"	21"	17"	14"
	26"	18"	20"	17"	14"	17"	14"	12"
	24"	16"	18"	14"	12"	14"	12"	10"
	6"	14"	16"	12"	11"	12"	11"	10"
		12"	6"	10"	10"	10"	10"	8"
		6"		8"	9"	8"	9"	8"
				6"	8"	6"	8"	7"
					7"		7"	7"
					6"		6"	6"
								6"
								6"
Clamps reqd.	5	7	6	8	10	8	10	12
Rate of pour / hr.	10'	10'	10'	10'	10'	10'	10'	8'
Max deflection	1/16"	1/16"	1/16"	1/16"	1/8"	3/16"	3/16"	1/4"

7.7 COLUMNS AND PIERS

The construction of concrete columns and piers varies considerably according to whether they are, in fact, columns or whether they are piers, with the particular demands of the design and with the class of contractor performing the work. (Building contractors and bridge contractors have somewhat different approaches to their trade.)

The term column is usually applied to a vertical member supporting a portion of a building. It is frequently concealed or covered by a finishing material or, below ground, is re-buried in the soil.

Piers, however, such as those used in the support of bridges, although much of them may lie below ground, often have substantial projections above the surface; and this portion, formed and poured as a continuation of the below-ground section, being exposed to weathering, cannot be successfully patched or mended without time revealing this meddling.

The construction of forms for columns and piers is influenced by the method of placing concrete. Concrete is, normally, said to be *placed* in wall forms and *dumped* in column forms, the distinction being chiefly the rate of rise of the fluid concrete. Unless a large number of columns is involved, or a pier of great obesity, the quantity of concrete needed cannot be poured into them at a controlled rate of rise. Not only is such a method uneconomic, but arrangements for small pours, lost time, and the possibility of "cold" joints add to the difficulties. The result is that column and pier forms must withstand considerably more pressure than wall forms.

For columns, we may start with the same materials used for wall forms, ripping the plywood into strips of the width required and nailing battens of 1 by 4 or 1 by 6 in. vertically to the exterior face. The four panels (for a rectangle) are stood in place and temporarily tacked together. Column clamps, of ⅜-in. steel flats, available in pairs, are now set and adjusted to precise dimensions by means of attached wedges and slotted holes (see table for spacing).

An alternative method employs 2 × 6 in. walers set singly on the same centers, of sufficient length to lap at the corners, where they are spiked or bolted together.

To eliminate the battens, 2-in. planks placed vertically are sometimes used (although these too will require some limited battening for handling).

Column forms set up in rows will be cross-braced to each other by diagonals nailed to yokes of 2 by 4 in. encircling the column form on 4 to 6 ft centers. Diagonal bracing in the right-angle direction will be placed to ground stakes, previously poured columns, or walls, or as circumstances indicate.

Circular columns may employ pre-fabricated steel forms which are field-assembled and braced before pouring. Cardboard tubes such as those developed by Sonoco require yokes and bracing before pouring. Such paper tubes are not re-usable.

Pier forms frequently require more elaborate formwork since, instead of straight sides, these may be tapered, bowed, or stepped, whether they be square, rectangular, or circular. These require a careful layout and generally some millwork.

The critical portion of piers, that part above ground, will be laid out on paper to a large scale (3 in. = 1 ft or even full size) where complexities warrant. Templates to be cut from 2-in. plank are then laid out to partially encircle the pier. These are then mill-cut to the required shape of the exterior face of the form. In the field these templates are set in position on the ground and temporarily secured together while the sheathing is nailed to them. Sheathing may be plywood up to the limit of its ability to be bent; and beyond this, 1-in. strips of 1-in. plank can be used, each strip nailed to each template.

The sections of sheathed template are then stood in place, and the templates are pulled together and held by scabs and bracing, as for columns.

Opposite are details of the brackets or column clamps for two sets of pier forms. Although of simple rectangular shape, the first of these is for a pier tapered on two faces; the second is for a stepped pier.

7.8 SLAB FOUNDATIONS

The consideration of construction methods for slab foundations fall into two categories. The first is that where the slab is indeed a portion of the foundation, while the second is that basically of a free slab on ground. In the first instance we are dealing with a slab tied into the walls so that we have a monolithic structure, and in the second case we are dealing with an independent slab separate from the foundation walls. The second case, being the simplest of the two conditions, can be disposed of first.

Where ground conditions are dry and the bearing soil is firm and not unusual, as is the case in many one-level basements on relatively high ground, a slab is poured inside the footing walls. This slab may be poured up to the edge of the footing and separated by an expansion joint (a condition not noted in Sec. 7.3, where not only a form but careful finishing of the surface of the footer is necessary), it may be poured to rest on the footer (a rare condition since a sliding surface must be provided between the slab and footer surface), or it may be poured at a higher level than the footer with backfill of soil between the two.

Slabs of this kind are seldom bearing slabs, do not exceed 8 in. in thickness, are lightly reinforced (often with only one layer of mesh), are placed under cover after the first floor slab or roof is placed, and seldom contain anything, such as conduit, within their body, although pipe projections from the underside frequently must be handled.

Because this slab is not involved in completion of the superstructure, all the underground work can be completed and backfilled, and grading can progress without weather interruptions. Slabs up to 20 ft in one dimension will have an expansion joint, varying from ¼ to 1 in. in thickness, against the wall only. Pre-molded expansion-joint material, the full depth of the slab, can be nailed to the wall by using special concrete nails or may be held in place by small sticks driven in front of it which are removed as concrete is placed.

In order to grade the surface of the slab, screeds will be necessary. These must provide a level top edge and be firm enough to support the operation of screeding. There are several metal screed assemblies manufactured, but the simplest method is to set a 2×4 on edge on 1-in.-thick stakes driven into the ground below the slab level. The 2×4 is set in position, the 1-in. stake is driven down along its face (alternate stakes on alternate faces), 8d nails are driven through the 1-in. board into the 2×4, and any stake projection is cut off flush with the top of the 2×4. The spacing of screeds is seldom less than 6 ft and may be as much as 10 ft, so that only one would be needed in the center of a 20-ft slab. However, an additional screed at each end of the 20-ft slab would also be needed. If this screed is set against a wall, it should be kept away from the face of the wall by about a foot to provide freedom for the screed board, which will be run back and forth over the top of the screed.

After the poured concrete section has received its initial set and is firm enough to support the weight of the finisher, kneeling on short pieces of plank, the screed boards and stakes are removed and the space filled with concrete, which is then troweled into the finished slab surface.

Large exterior slabs are generally broken into sections separated by expansion joints, to provide movement for wide temperature variations; but interior slabs of some size, while poured in sections, are considered to exist in a sufficiently uniform temperature milieu to eliminate intermediate expansion strips. There is a wide variation in specification practice, however, as to whether adjoining sections should be bonded to each other or left as a cut joint with no reinforcing carried through. Although temperature variations will be limited on interior slabs, once the building is fully enclosed and heated, there is an interval of construction time during which such slabs may expand and contract excessively. With cut joints, cracking of the concrete will be concentrated at these points; with continuously reinforced slabs, cracks may develop in unsightly patterns across the floor.

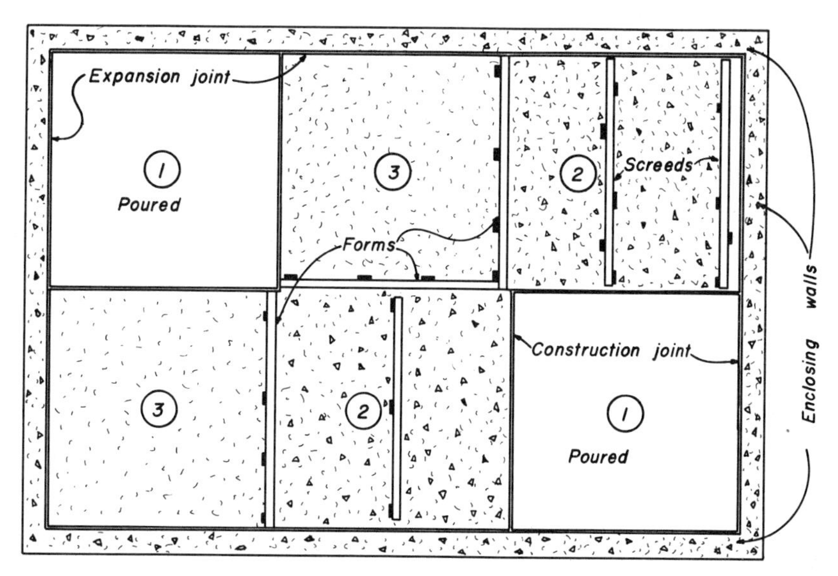

ROTATION OF POURS – INTERIOR SLAB
(Where joint strips must be set)

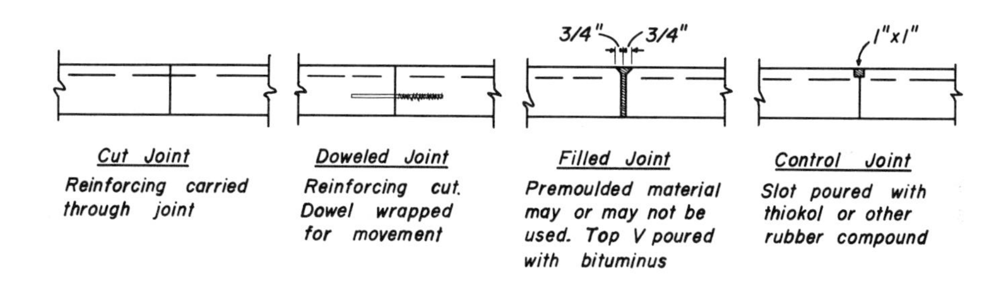

TYPES OF CONSTRUCTION JOINTS

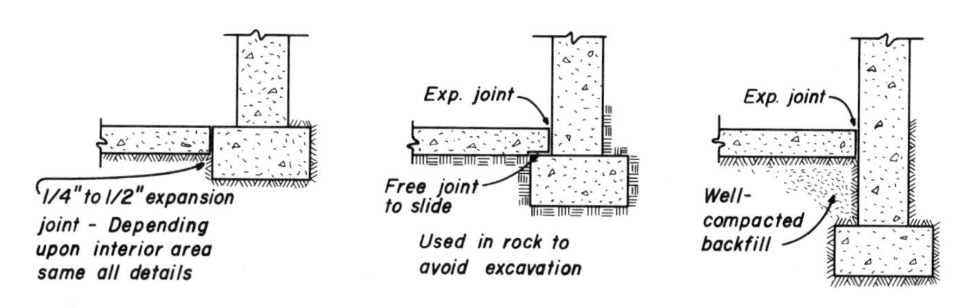

ISOLATED SLABS VERSUS FOOTING DEPTHS

7.9 RAFT-TYPE SLAB FOUNDATIONS

Strictly speaking, a raft foundation is one which floats. It is a foundation resting on a subsurface stratum at sufficient depth so that the weight of the soil overburden, removed to that depth, is approximately equal to the weight of the structure resting on that foundation. In sensitive clays, some heaving of the bottom and sides of the excavation pit can be expected, as was shown in Sec. 3.11; and unless such heaving is controlled, structural settlement equal to the amount of the heave can be expected.

The raft foundation is only an extreme case of the mat foundation, more frequently encountered, but the construction of the two presents similar problems. Mat foundations are extensively used for deep structures intended to be watertight, such as pumping stations, sewage-treatment plants, power plants, apartment houses, and the like, where foundations will lie below the water table and must resist hydrostatic pressure and uplift as well as substantial foundation loadings. Such foundation slabs are designed integrally with the walls and involve a thicker slab, heavily reinforced. They contain inclusions of electric conduit, bolts, and drain lines cast into them.

Mat foundations are seldom less than 12 in. thick and more often 2 to 3 ft thick. They must be placed before any other portion of the structure, consequently, any protection from the elements can be provided only by temporary construction coverings, always expensive and sometimes impractical.

After the exterior slab forms have been positioned (see Sec. 7.3, Spread Footings) and any stone placed, the lower mat of reinforcing steel is set, usually 3 in. above the bottom concrete surface. This mat can be supported on pre-cast concrete blocks or on brick (2¾ in. thick) without difficulty.

Once any interior piping has been completed, the top mat of steel, from 6 to 12 in. or more above the bottom mat, is placed. It will generally be necessary to set this by tying it to pins driven or drilled into the underlying foundation soil. As the top mat is set, the trampled bottom mat is lifted and re-set, now possibly also tied to the pins and to the upper mat. With the reinforcing steel in place, the carpenters now invade the field to set screeds, perhaps to set anchor bolts, and, where cross walls are involved, to set stub forms, support dowel steel, or set up water stops. After these items are completed, the rod setters return to re-adjust the steel before the concrete crew takes over. All in all, it is like trying to develop a good turf on a football field between Saturday afternoon games or grow a field of corn between visitations of herds of stampeding buffalo.

Subject to site limitations on space, concreting will generally start along one wall or exterior form, progressing across the slab for the full width until it reaches the opposite side. The progress of concreting should parallel the direction of the screeds. If care is exercised in placing, the concrete crew can work on that concrete already in position which, even when liquid, still provides some support for the enveloped steel and other insertions.

With slabs of this thickness of the wall-bearing or raft type, certain conditions must be met that the simpler slabs, described in Sec. 7.8, do not encounter. Slabs of greater thickness may take longer to set at finish level than those which are thinner; screeds cannot be pulled as soon, and partial set plus the filling of a deep crevice may produce a joint subject to leaking. Here, metal screeds are more desirable, often set on pins which are not pulled and are so designed as to provide a cut-off for seepage.

The method employed for placing concrete in mat- or raft-type foundations will have a bearing on the overall operation. Where concrete is placed by drop-button concrete buckets, suspended from a crane-boom hoisting line, greater speed and less disturbance to the setup results. Where site conditions make this impossible, concrete may have to be wheeled in Georgia buggies from a receiving hopper. Runaways supported on bracketed legs will be installed as a final preparatory operation, and concreting will start at the center of the slab and move outward to the perimeter form.

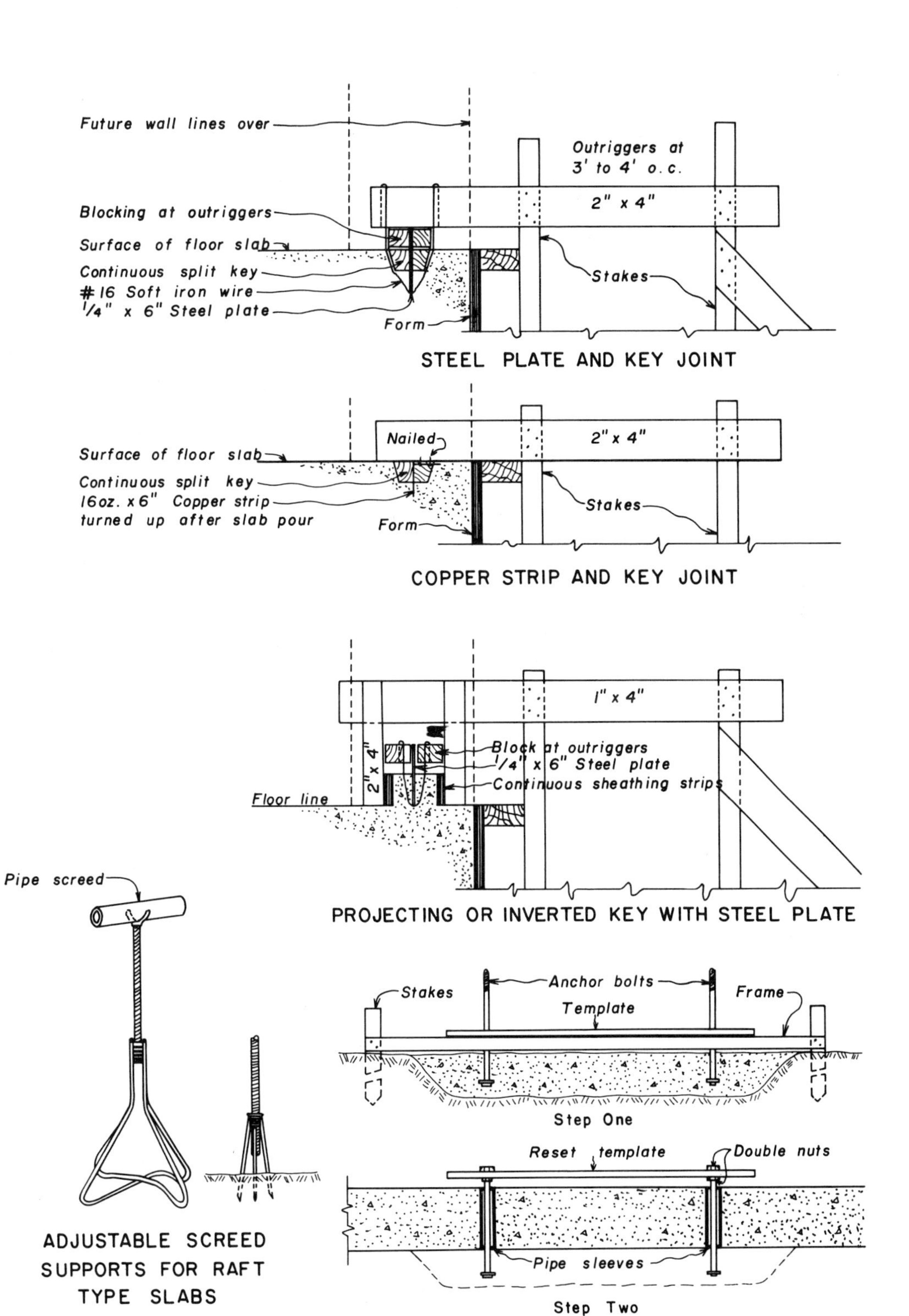

ADJUSTABLE SCREED SUPPORTS FOR RAFT TYPE SLABS

SETTING ANCHOR BOLTS IN A RAFT SLAB

7.10 THE FOUNDATION PROBLEM IN MEXICO CITY

The mat foundation discussed in Sec. 7.9 is a not uncommon form; the true raft, designed to float a structure on sub-grade, is more rare and its use is warranted only in quite special conditions. Such a site, with possibly the most special conditions in the world, is to be found in Mexico City. Here, in 1961 and 1962 the United States was engaged upon the construction of a new Embassy Building.

Mexico City lies in a large flat valley ringed with mountains. It is the site of an ancient lake whose glassy surface once reflected the glow of the active volcanoes surrounding it and which was gradually filled to the surface with volcanic debris. This volcanic ash has produced a spongy clay some 100 ft in depth which still contains water to the extent of 500 to 600% of the weight of its dry solids. At about the 100-ft depth there is a 15-ft-thick stratum of dense sand, with more clay below this.

A clay with this high a water content will consolidate under overburden pressure if the water is removed, and such a removal has been taking place over a period of some years. The withdrawal originally began when a drainage tunnel was pushed through the surrounding mountains; but the effect has been heightened, in recent years, by increased pumping of groundwater from the basin. Subsidence amounts to from 8 to 12 in. a year in many areas.

To further complicate the picture, Mexico City lies in an area subject to earthquakes, probably a heritage of its ancient, now quiescent, volcanoes. A shock, of a Mercalli intensity of 7½, was experienced on July 28, 1957, causing extensive damage precisely in the area where the greatest number of modern buildings exist.

A number of types of foundations, designed to meet this congeries of conditions, have been used for substantial buildings in Mexico City. They are illustrated opposite.

The first of these methods is a solid concrete mat, often 8 to 10 ft thick, heavily stiffened and set near ground surface. On such a mat sits the Palace of Fine Arts building. Here the mat is 8 ft 9 in. thick and laced with steel beams and grillages. The entire building, begun in 1904 and finished in 1934, has now settled 10 ft below the surrounding street levels.

A second method sets a substantial concrete mat on sectional wood piling extending down to the sand layer. These sections are supposed to deflect with settlement, but the concept has not worked and this method has been abandoned.

Some modern structures have been set on concrete piles, either pre-cast or cast-in-place, extending down to the sand layer. These foundations have successfully resisted earthquake and settlement, but subsidence in the area around them has left them sitting up in the air.

A fourth method also employs concrete piles, but here a mat is pierced by the piles, from the tops of which the mat is hung by yokes and screw jacks. The building can thus be lowered as surrounding subsidence occurs. This method seems to transmit a considerable portion of the earthquake shock to the building.

Another method, which is of the raft type, uses transverse inverted barrel shells to increase the contact area between the soil and the concrete, cast on the clay between heavy exterior walls designed to contain the perimeter banks. This is a new concept on which no results are available.

A sixth method, and that selected by the designers of the United States Embassy building, is that type termed a compensated raft foundation. Instead of the solid slab of fixed weight, the compensated raft is a hollow slab consisting of closed boxes. This foundation can be ballasted like a cargo ship and can be trimmed to maintain a level floor slab. It can be loaded or unloaded to suit settlement conditions. The design is based on the premise that the weight of the whole structure will equal the weight of the soil displaced, plus the uplift of the liquid soil beneath it.

In Sec. 7.11 details of construction of the compensated raft foundation will be discussed.

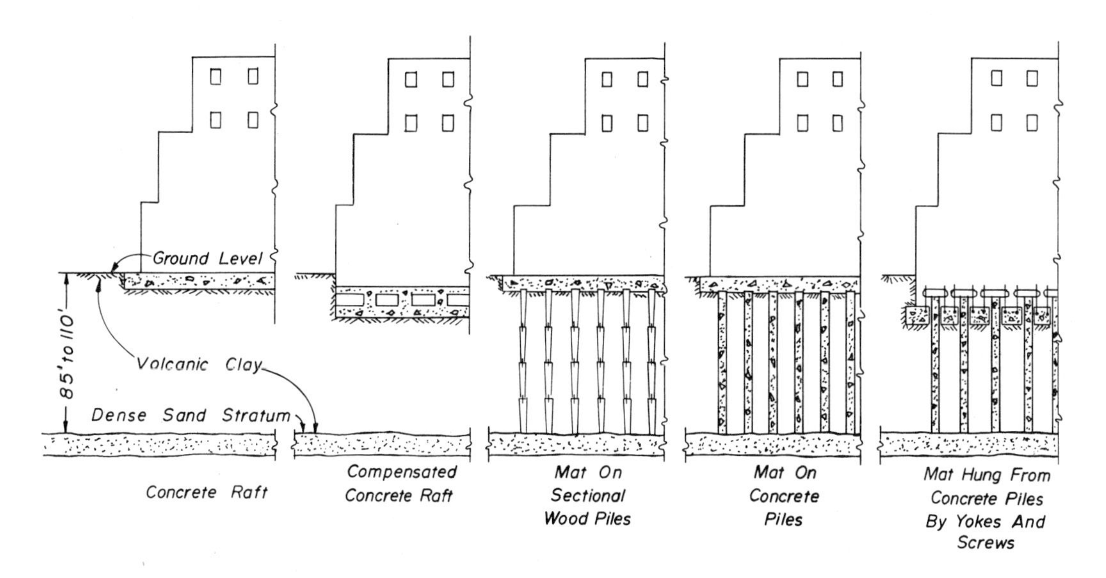

EARLIER TYPES OF MODERN FOUNDATIONS

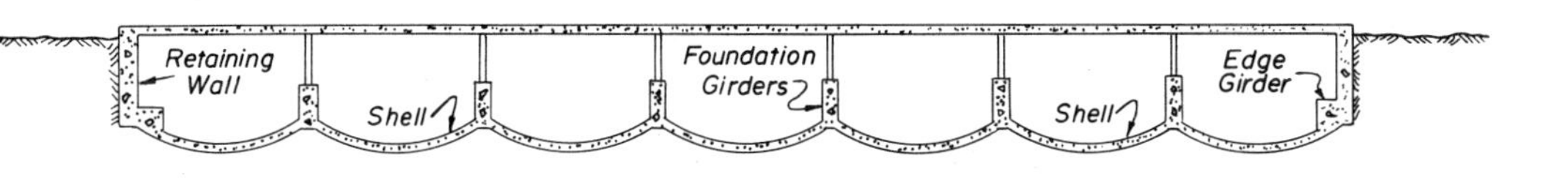

BARREL SHELL FOUNDATIONS
INTRODUCED IN THE NINETEEN SIXTIES

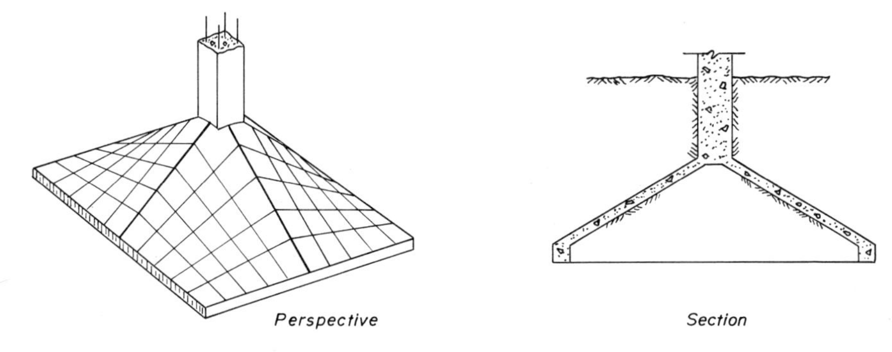

Perspective

Section

THE INVERTED HYPAR FOUNDATION
USED IN PLACE OF BARREL SHELLS
ON THE NONO ALCO-TLALTELOLCO PROJECT

FOUNDATIONS USED IN MEXICO CITY

The first step in the construction of the foundation for the United States Embassy building in Mexico City, referred to in Sec. 7.10, was to excavate, by power shovel, over the whole site to a depth of 13 ft, a depth requiring no dewatering. At the same time, a modified Wakefield sheathing was driven around the perimeter of the 180 × 240 ft site to a depth of 50 ft. Constructed of four layers of 2 × 12 in. timbers, bolted together, with offsets, to provide tongue and grooves, this sheathing would support the surrounding banks until walers and struts could be installed.

The initial construction involved a pattern of 3-ft-wide by 8-ft-deep concrete beams set, in both directions, on 30-ft centers. Near the center of the site, within one of the resulting 30-ft squares, a cluster of piling was driven to temporarily support a tower crane to be used in excavating for, and the concreting of, the beams.

The 3 × 8 ft beams spanning the site were to be built in 10-ft-wide trenches, and the excavation for these was begun at the center of the site area. As excavation proceeded outward and downward, steel beam struts were set, starting at the 13-ft level, on centers ranging from 6.5 to 3.5 ft vertically. The struts were placed along one side of the 10-ft-wide trenches and terminated at walers along the perimeter wall sheathing.

This beam trench-excavation procedure left a 20-ft-square mound of earth bounded by trenches, each of which was enclosed in 2-in. sheathing and banded by steel walers tied together at the corners.

To control the water level in the trenches, as hand excavation proceeded, wells had been driven in alternate squares; and well points, packed in sand, had been set in these. To avoid overpumping in the area, the water table was lowered just ahead of excavation and only in those areas where work was actually in progress.

The excavation was carried a foot below beam bottoms, and in this space a perforated pipe was laid in sand. These underdrains were tied into headers carried along the perimeter sheathing, and their sand surface was covered by 6 in. of lean concrete on which beam construction was started.

Forms and reinforcing were now positioned for the beams with dowels provided at intersections. Although a top slab could be placed over the tops of the beams, a special tie-in for the bottom slab was obtained by welding a heavy projecting angle to the bottom of the reinforcing steel cage. Beams in the four center bays were poured first, and concreting progressed outward.

Around the inside of the perimeter sheathing a wider trench had been excavated in which a 15-ft-wide foundation slab was poured monolithically with a 14-in.-thick perimeter wall, with the 3 × 8 ft beams tied in to wall and slab.

Three-foot-square columns, rising from the intersections of the foundation beams, were next poured up to the underside of the main floor slab, four feet above grade; and a depressed central plaza floor and mezzanine floor were cast.

The structure now had sufficient weight to permit excavation of the 20-ft mounds of earth, one by one, in a staggered pattern; and as excavation in each was completed, a 2-ft-thick foundation slab was placed at the bottoms of the beams, its reinforcing welded into the angles previously placed. The basement floor slab was now poured on top of the beams.

Five of the resulting boxes, at the center, would be used for the permanent storage of 30,000 gal of water. Eight exterior chambers were filled with sand to provide the desired ballasting, and other chambers could be filled in the future, if desired.

During foundation construction, groundwater level in the area had been controlled by means of the wells noted above, the beam underdrains and the peripheral headers. These headers had drained to sumps where water was collected and transferred by pumping to a similar header laid down around the *outside* of the sheathing. (See Sec. 7.12 for a discussion of groundwater recharge.) Nine of the wells were retained, to permit recharging from the 30,000 gal of storage.

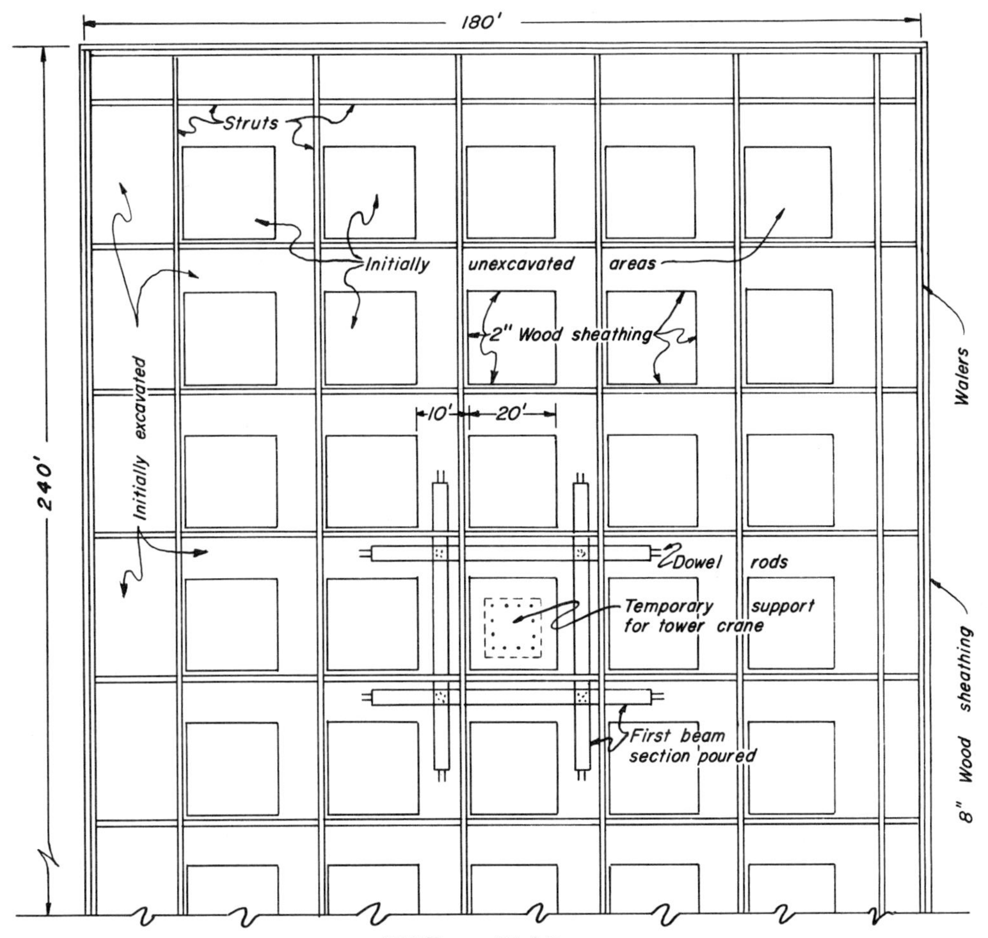

PART PLAN
U.S. EMBASSY FOUNDATIONS - MEXICO CITY
IN EARLY STAGES OF CONSTRUCTION
SCALE 1" = 40'

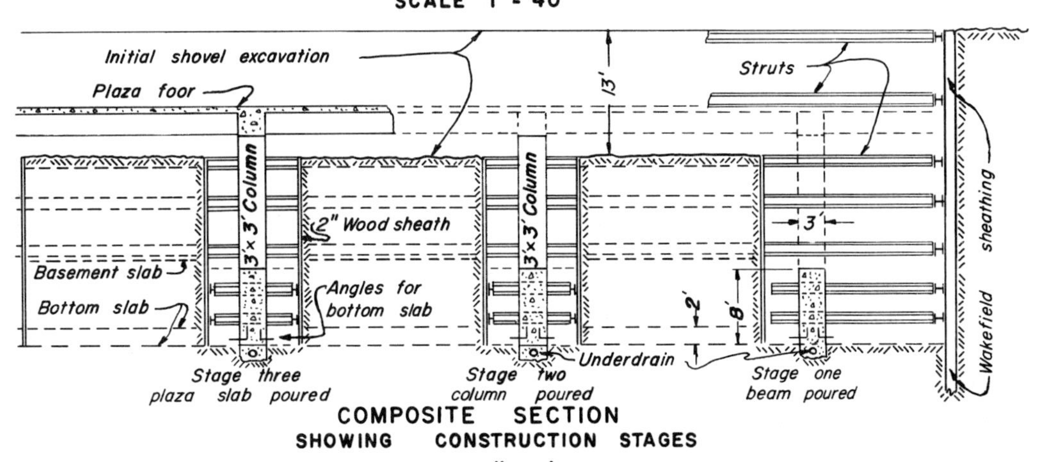

COMPOSITE SECTION
SHOWING CONSTRUCTION STAGES
SCALE 1" = 20'

7.12 GROUNDWATER RECHARGING

It is not necessary to go to Mexico City to study unusual foundation operations unless you want also to pick up a quick Mexican divorce while you're there. You can go to Coney Island instead, where, between rides on the roller coaster and trying to thrust your toe into the ocean through massed bathers, you can study the methods used to construct raft foundations where control of groundwater levels was vital to the operation.

In Coney Island, behind the Boardwalk and before Surf Avenue, between West 29th Street and West 32nd Street, in 1954, the New York City Housing Authority proposed the construction of five middle-income 14-story apartment houses. At the site lay dune sand to a depth of 61 ft, pierced by lenses of organic silt.

Raft foundations set at elevation −10.00, or 9 ft below mean sea level, could be floated on the saturated dune sand. However, lowering the water table in the area from −1.00 to −10.00, during construction, could produce disastrous settlements in surrounding structures and disturb the neighbors living across the streets. The problem thus became one of lowering the water table at the construction site while continuing to maintain it in the surrounding area.

The method decided upon consisted of a well-point system for dewatering, laid down on the *near* side of each street, with a diffusion system that would restore subsurface water as fast as it was withdrawn, laid down on the *far* side of each street.

A 200-ft test section was set up in 29th Street and operated for 2 weeks. The level of drawdown and reconstitution of groundwater level were checked from observation wells and settlement levels on the over-street buildings. One important difficulty was disclosed in this test section. After a time the diffusion piping became clogged with a gelatinous ferric hydroxide; and the need for cathodic protection, as well as filtering or screening, became apparent.

The well-point dewatering system was installed in the usual way using a continuous 10-in. pipe header around the site. Two pumping stations were set up, one on 32nd Street for normal operation and one on 29th Street as a standby. Discharge from the well-point pumps was fed into a centrally located surge tank, 10 ft 6 in. in diameter and 25 ft high, which maintained a constant head pressure on the diffusion system and, by overflow, permitted excess water to flow away to a nearby sewer.

The diffusion system used specially designed diffusion points set on 4-ft centers along 8-in. pipe headers. The diffusion points themselves were similar to the well points, having a cap on the bottom instead of the customary jet holes and ball check valve. An 8 ft 3 in. length of screened area replaced the customary 3-ft screen of the well points. The diffusion points were jetted into position with a separate jet so that their screens were between elevations −1.00 and −10.00. The well-point screens, on the other hand, were set between elevations −24.00 and −27.00, although maximum drawdown never exceeded −12.50.

In operation the system started out withdrawing water at a rate of 6000 gpm, which eventually dropped to 4500 gpm. Of this, about 3000 gpm was sent through the diffusion piping, the remainder being wasted to sewer. The surge tank provided a pressure of 10 psi, 7 psi of which was lost in the piping system, leaving a 3-psi pressure at the diffusion points. Vented overflow loops at the ends of lines maintained the 3-psi pressure.

The cathodic protection consisted of 1-in. zinc rods set at 4- to 6-ft centers along the diffusion headers. The rate of ion exchange was checked by potentiometer readings on reference electrodes spotted along the header route. In addition, the piping of the diffusion system was coated inside with a vinyl paint, and the points were galvanized. One-hundred-mesh screens were installed in each of the swing joints between headers and diffusion points, and fifty-mesh screens were installed at the surge tank.

Despite the anti-corrosion provisions, gelatinous formations continued to cause trouble, requiring frequent cleaning of the screens.

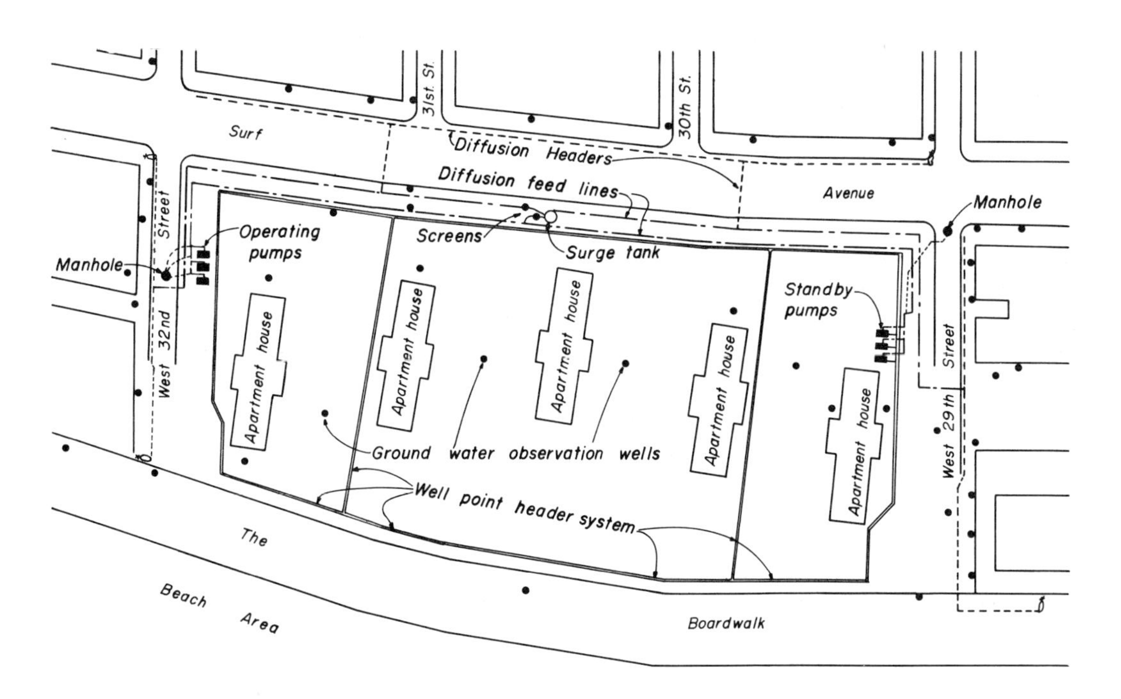

THE SITE
ON THE BEACH AT CONEY ISLAND

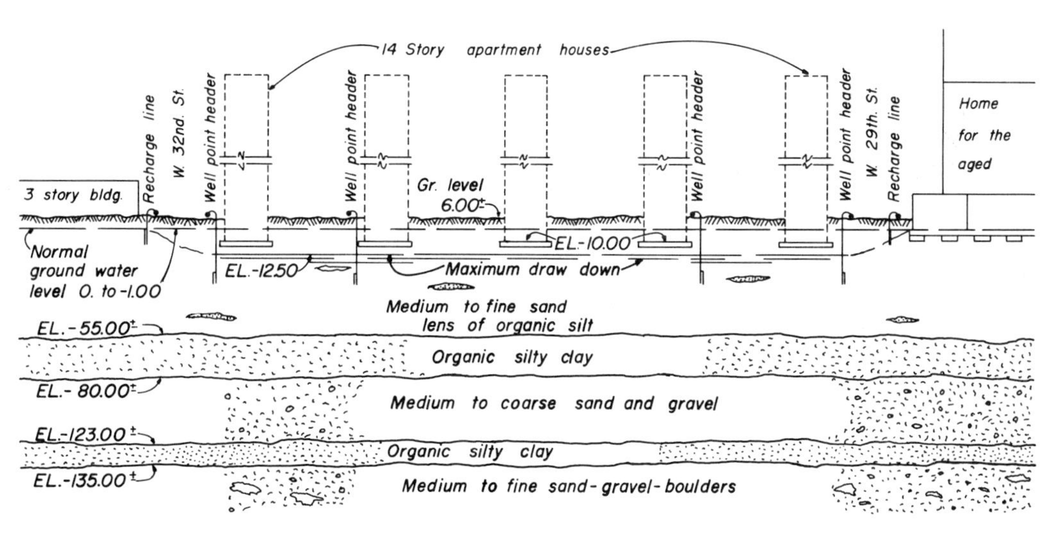

Longitudinal Section

A GROUND WATER RECHARGE SYSTEM

7.13 DISHED AND SLOPING FOUNDATION SLABS

Many types of special structures have what are essentially mat foundations but, instead of being level or lightly graded, are built on substantial slopes. In circular tanks, the bottom may be dished or concave. These types of slabs may be found in sewage and industrial waste treatment plants, and many industrial processes require concrete vats with sharply sloping bottoms to permit the concentration of sedimentary deposits. The inverted barrel slabs used in Mexico City, and noted in Sec. 7.10, are a special case in point.

The first decision to be made in dealing with dished or sloping slabs is whether to form them or to slope-pave them. The maximum slope on which paving should be attempted is a 1:1 or 45-deg slope; but unless necessity compels, the more practical limit is a 30-deg slope. Relatively small pits with sloping sides, often termed "hopper bottoms," interject the problem of working space. Unless there is adequate space for the concreting and finishing crews to work, forms must be used and the degree of slope on the sides is irrelevant.

Forms for hopper bottoms must be designed to resist uplift, the tendency of the concrete to float the form upward. The form must be held down, and the method used will depend on the nature of the underlying soil and the size of the hopper bottom.

In rock formations sloping forms can be held down with anchors drilled and grouted into the rock. Small hoppers, inverted truncated pyramids, may be held down simply by anchors tying down the bottom slab form. Larger sloped areas may require tie-downs at 30- to 36-in. centers.

In soils, sufficient anchorage for small hoppers may be obtained by a concrete block cast into the soil below slab level, with anchor bolts projecting. Larger hoppers or sloping slabs must be held down by weighting. Since direct weighting of the form will generally interfere with concreting, timber or steel beams are carried over the form and secured by weights at their ends; the beams can be diagonally staked, where practical, or can be wedged against soil banks or other objects.

The placing of concrete in sloping forms must be done through windows cut into the sheathing between studs, in alternate panels, at spaced intervals along the form. Concrete is placed, generally by chuting, up to the bottom of a window, the section of panel is replaced, and concreting is then carried up to the level of the next opening.

Neither formwork nor slope paving is practical on soft or sandy soils since, even if their angle of repose is adequate, the soil can be mixed into the concrete. It may be practical to stabilize the surface with a thin layer of concrete or by means of guniting or surface grouting, but placement of full slabs by this method is not often satisfactory.

On the 30-deg slope of the variable-angle launcher referred to in Sec. 7.2, a base slab 18 in. thick was placed by guniting. Side forms for this base, on which 90-lb rails would later be set for the counterweight carriage, supported templated anchor bolts, and there was considerable reinforcing steel. The grout was a 1:4.5 cement-sand mix sprayed on at nozzle pressures of from 45 to 74 psi.

Guniting was started at the top and placed in successive layers to control rebound, but grout accumulated and had to be leveled by screeding. The entire process was expensive and the results were unsatisfactory. Although initial tests had indicated a potential strength of 3840 psi, test cores developed only 2100 psi, because of lack of bond between layers. The bond between steel and grout was also deficient.

Concrete slabs placed on slopes require screeds which should be placed parallel with the slope. Screeding should be done *up* the slope, using a low-slump concrete which should be tamped rather than vibrated.

For smaller dished slopes a pipe sleeve can be anchored in concrete at the center. A mast, or pipe, is set in the sleeve with a screed hung from it, as shown. This can then be turned through the full arc to control the grade of the concrete surface.

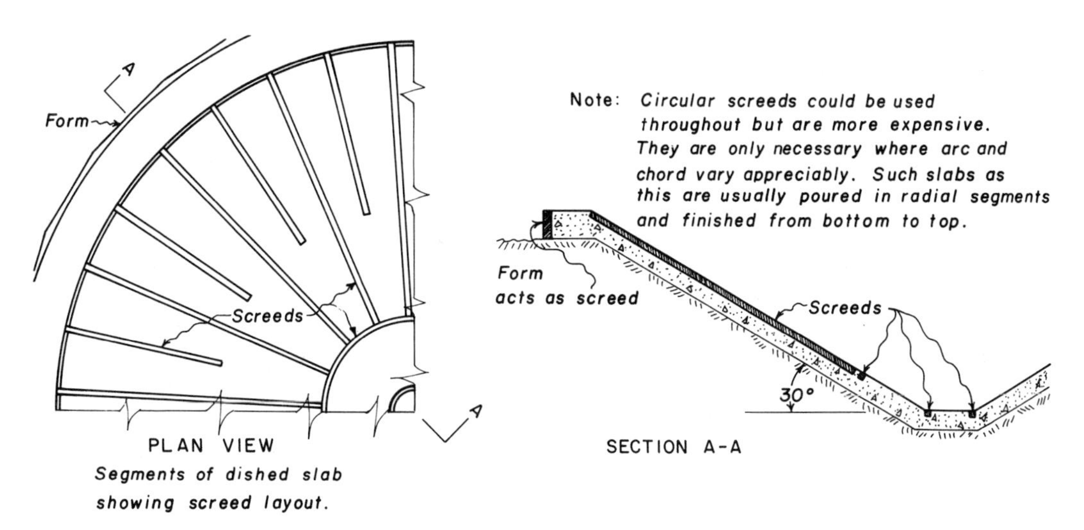

PLAN VIEW
Segments of dished slab showing screed layout.

SECTION A-A

SCREEDS FOR DISHED SLAB

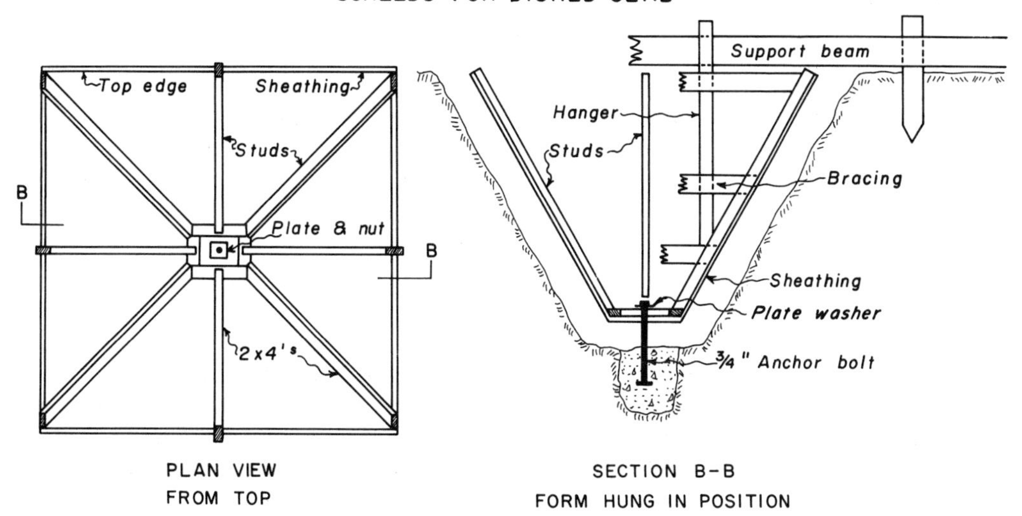

PLAN VIEW
FROM TOP

SECTION B-B
FORM HUNG IN POSITION

A HOPPER BOTTOM FORM

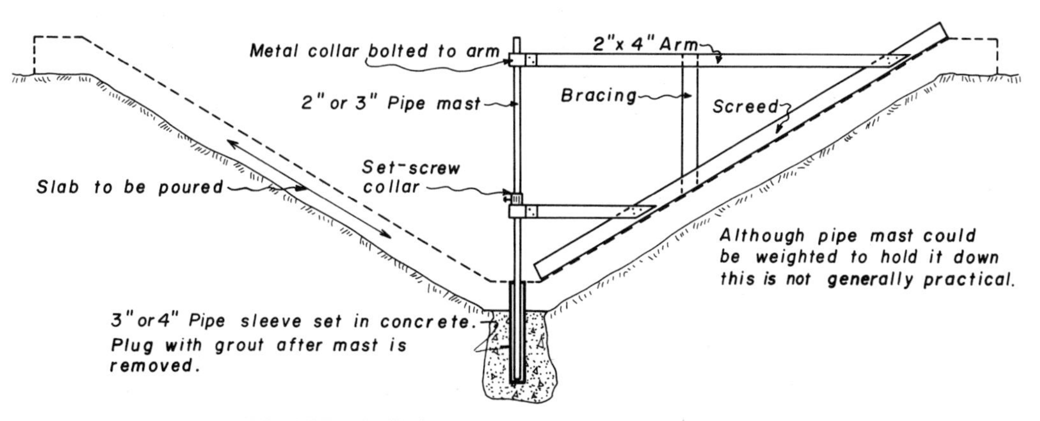

SCREED SET-UP FOR SMALL CIRCULAR DISHED TANK

7.14 WATERTIGHT CONCRETE FOUNDATION WALLS

Many foundation walls serve the dual purpose of supporting a structure and at the same time enclosing a usable space, not always merely a basement. In these cases it becomes necessary to ensure that these walls are watertight or, at the least, that they are dampproof, since even when they are not placed below a water table the surrounding soils may be damp much of the time.

Concrete walls can be made watertight if properly designed and constructed. The concrete should be a designed mix having a suitable water-cement ratio to provide a compressive strength in the 3000- to 3500-psi range and should be placed with a low slump. Although finely divided admixtures are available as densifiers, they accomplish no more than can be obtained by increasing the cement content by from 5 to 10%.

Proper mixing and placing, using controlled vibration (that is, the concrete should be neither excessively vibrated nor insufficiently vibrated), should produce the proper density to control moisture. The chief sources of leakage or of moisture penetration are segregation of aggregates and "cold" joints where bonding between successive pours has not occurred.

Walls designed to be watertight should have a substantial thickness. Although thickness alone will resist some moisture penetration, the chief value of a thicker wall is that concrete can be placed and worked more effectively than in a narrow wall.

Careless handling of concrete is the greatest foe to producing a watertight wall, assuming that batching and mixing have been adequate, since this leads to segregation. Practical considerations will generally limit the height of a wall pour to 12 ft, but this is an excessive depth for free-falling concrete. Elephant trunks hung in the forms control segregation, as does the placing of concrete by chute in stepped windows cut into the form. A steep, wide chute will produce segregation more readily than a flatter, narrower chute in which a depth of concrete can be maintained. Troughed belt conveyors can produce segregation since the coarse aggregate is often cast off separately at a higher velocity than the accompanying paste.

Higher walls will be poured in lifts requiring horizontal construction joints. Where watertightness is of no interest, a shear key is the only device, in addition to dowels, that is needed. For tight walls a water stop will be required, generally in addition to the keyway. Sixteen-ounce copper strips, dumbbell-shaped plastic or rubber, and steel plate have all been used successfully. Copper can be bent flat in placing the first lift of concrete, and there is a possibility that the same thing will happen with plastic or rubber strips. One-quarter-inch steel plate has the necessary rigidity to prevent its being bent. Moreover, steel plate has a tendency to be self-sealing since, if moisture should reach it, the steel will rust and the rust, in expanding, will tend to seal the moisture-producing void.

Horizontal joints, whether consisting only of a shear key or including water stops, have a tendency to accumulate sawdust, shavings, or dirt from the upper form construction. The joint surface should be cleaned with an air blast before concreting. As a further precaution, 4 to 6 in. of grout should first be placed over the joint surface. This grout should be the concrete mix specified, less the coarse aggregate.

Except for very small structures whose entire perimeter can be placed in one pour, vertical joints will frequently be needed. Although reinforcing in the wall may be designed to prevent expansion and contraction at this joint, it is safer, for watertightness, to use an expansion-type vertical joint than to rely on the steel-concrete bond. Details of this type of joint consisting of a V-shaped copper strip filled with mastic, are shown opposite. Dowels are necessary to maintain alignment of the two sections of wall but must be permitted to move horizontally.

In all concrete walls which it is desired to have watertight, vertical expansion joints should be provided at intervals not greater than 40 ft. Expansion and contraction in walls with fixed ends, such as corners, will occur with temperature changes; and reinforcing is often insufficient to resist the development of cracks.

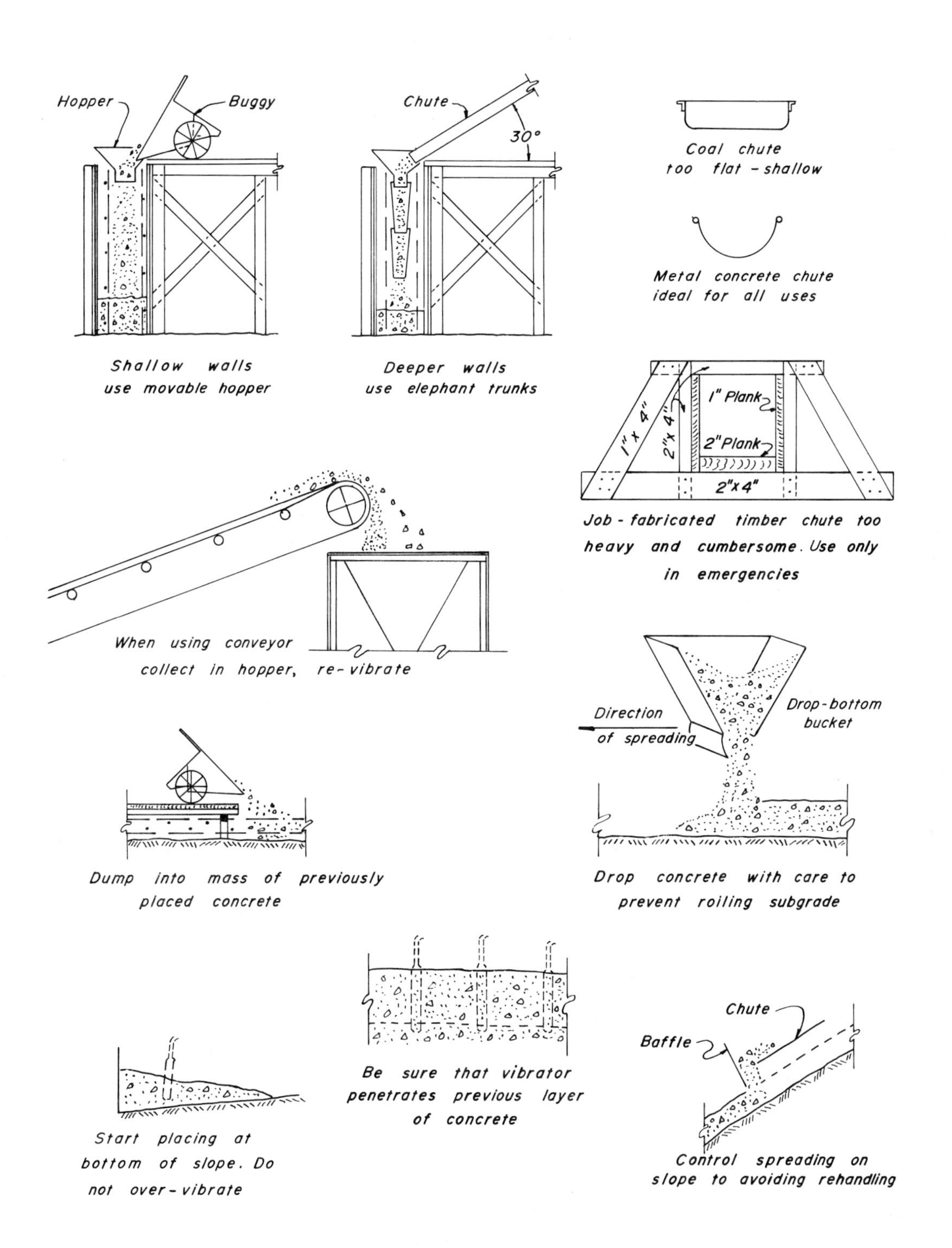

SUGGESTIONS FOR PLACING CONCRETE

7.15 WATERPROOFING CONCRETE FOUNDATION WALLS

Although many concrete structures almost wholly immersed in water have been made watertight without any special treatment, there are many cases where such treatment seems desirable or is even necessary because of uncertainties as to the final quality of the concrete or simply as a precaution against unforeseen contingencies.

The simplest situation occurs where no actual water is present—only damp soil. The degree of dampness will vary depending on the season, the climate, and the frequency of rainfall, but at most the soil is saturated only for short periods of time. For this condition, dampproofing alone may be sufficient.

Dampproofing consists of from one to three coats of asphalt or coal-tar pitch applied hot or one or more coats of an asphaltic emulsion applied cold to the exterior face of the wall, each coating spaced 24 hr apart. Since dampproofing can be mopped or brushed on, it is suitable for use on block or masonry walls or on relatively rough-surface concrete walls. It cannot be placed too thickly since, being viscous, it can sag downward on the surface.

Where consistently water-soaked soils will lie against the exterior face of a foundation wall, a membrane waterproofing may be used. Membrane waterproofings consist of two or three layers of fabric held to the wall by successive moppings of asphalt or coal-tar pitch.

Fabric may be cotton cloth saturated with asphalt or coal-tar pitch, may be rag felt similarly saturated, or may be asbestos felt saturated with asphalt. Two or three layers of cotton fabric are used with three or four moppings. Two layers of felt are generally applied with a third layer of cotton fabric between the two, each layer mopped in.

The surface of the concrete wall should be reasonably smooth, without pits or sharp projections, and should be free of surface moisture. An asphalt or creosote primer is first applied to the wall surface. After priming, a first bituminous coating is placed and a layer of felt or fabric is pressed into this coating. The final layer is surface mopped.

Membrane waterproofing must be protected from damage which could tear and puncture it, not only during the backfill period but at a later date should ground movements occur. A 4-in.-thick course of brickwork has been used for this purpose; but this is rather expensive, as would also be a 4-in.-thick concrete wall poured against the membrane. A plaster coat of sand-cement mortar has been used, from ½ to 1 in. thick, sometimes incorporating an integral waterproofing compound such as Ironite. Although admixtures may inhibit shrinkage, there is still danger of cracking of these plaster coats. Since the coating adheres to the mopping, this may also result in the cracking of the membrane.

A more common and less costly method of protecting membrane waterproofing is to adhere a ½-in. thickness of a Celotex board to the top mopping, butting the sheet edges to form a uniform surface.

With dampproofing, with membrane waterproofing, or with no surface coating at all, backfill procedures have a great influence on the result. A 12- to 24-in.-wide backfill of clay rammed into place at optimum moisture content not only prevents damage to coatings but is itself an effective waterproofing means. Certainly a selected material, free of stones and debris, should be used for backfill adjacent to foundation walls. Care in the handling of tamping tools is necessary.

Waterproofing membranes are also placed under floor slabs in saturated soils or wet conditions. The slab forms are set and a 2- or 3-in. layer of concrete is first poured and troweled smooth. The membrane is then applied to this slab with the fabric being turned up on the edge forms, to which it is temporarily tacked. When the wall is placed, these fabric layers are then interleaved with those on the wall surface.

The effectiveness of underslab waterproofing is questionable. The main concrete slab is poured over the waterproofing, whose surface has not only hardened, to a degree, but has usually been trampled on in the process of preparing the slab for pouring.

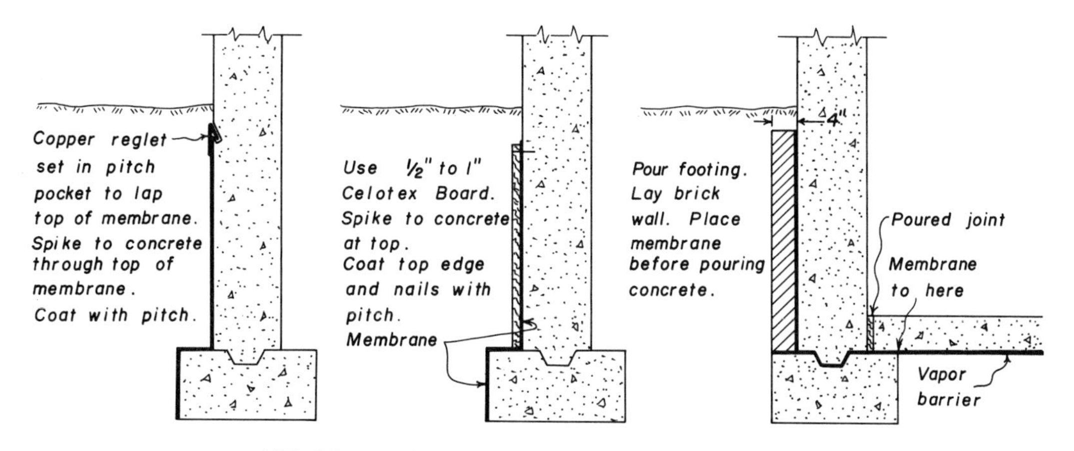

MEMBRANE WATER PROOFING VARIATIONS

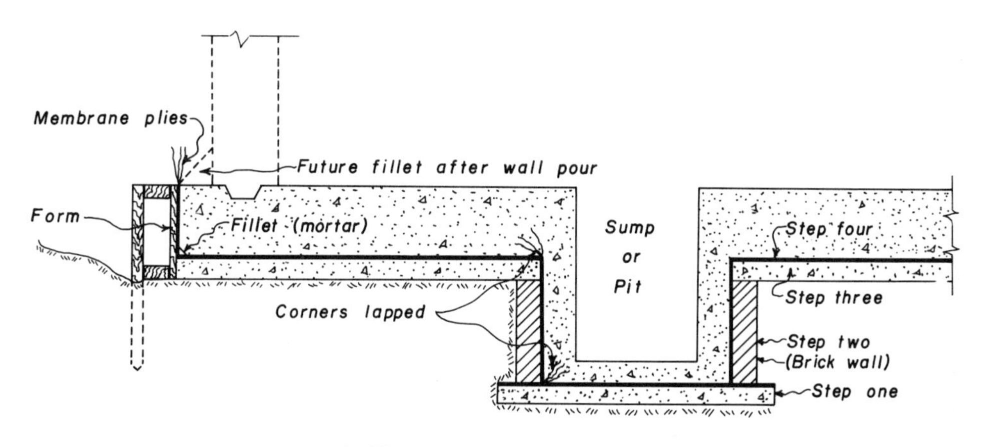

WATER PROOFING RAFT SLABS

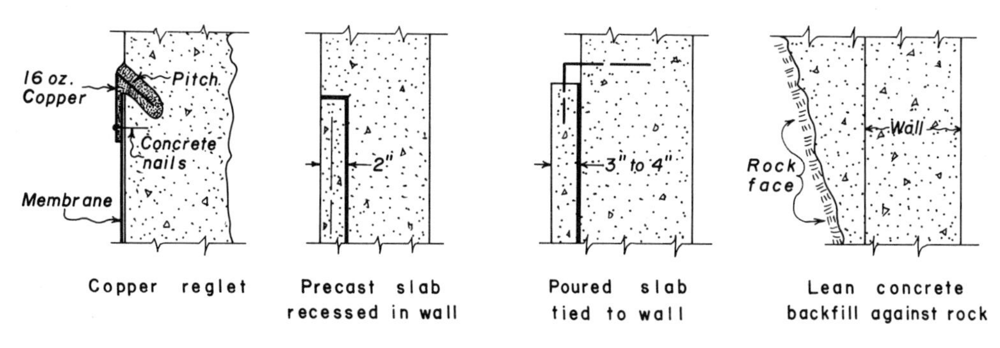

Copper reglet

Precast slab recessed in wall

Poured slab tied to wall

Lean concrete backfill against rock

MISCELLANEOUS DETAILS

7.16 HOT-WEATHER CONCRETING

In Chap. 1 there were some general references to foundation construction in the tropics, in the desert, and in the arctic, with some casual references to concreting problems. But in temperate climates we get intermittent temperature extremes which, precisely because they are not prevailing, can lead to unwitting damage.

It is true that foundations are usually substantial masses of concrete seemingly immune to the effects which weather can produce on more fragile sections high in the superstructure, but there are enough cases of record to prove the fallacy of such facile assumptions. Here was a building, virtually complete, where belated cores showed the foundation concrete to be far below design strength. The building was shored up in increments while the foundations were cut out and re-poured. There was a case where concrete, placed on a frozen sub-grade, itself froze. When the spring thaw commenced, the foundation not only settled but partially disintegrated.

For concrete operations in *hot* weather the following suggestions are offered.

1. Forms, reinforcing steel, and inserts should be liberally sprinkled with cool water. The same procedure should be followed with foundation soils or rock, except that care must be taken to prevent turning clay surfaces into mud. Wetting down the area outside the form will cool the air and increase the humidity.

2. Fog nozzles should be used for wetting down sub-grades rather than open hoses. Where hot, dry winds prevail, wind breaks should be set up.

3. The temperatures of the cement and aggregates must be controlled. Cement at temperatures close to 170°F can be flash-setting and should be mixed only with cool water.

4. Aggregate stockpiles should be protected from excessive heat and from drying winds. They should be kept moist by sprinkling during mixing operations.

5. Cool mixing water is essential. Pipelines and storage tanks should be buried, insulated, or painted white to reflect the sun's rays. Crushed ice can be used but should be thoroughly melted down before any mixing water containing it is used.

6. Admixtures which tend to retard the setting time of concrete are useful, particularly where the concrete must be re-worked, screeded, or finished, as in slabs. Prior to using any retardant, experiments should be conducted to determine its possible effect on the concrete. Variations in the characteristics of cement, sand, or stone may vary the effect of retardants.

7. The time between mixing and placing must be carefully controlled. Times, often specified as limiting, range from ½ to 1 hr, but in hot weather the time must be figured from the beginning of mixing to the time when the last of the batch is placed. Where the slump drops, the concrete should be discarded and should not be re-tempered by the addition of water. Where known time delays may occur, it may be desirable to increase the quantity of the initial mixing water, but this requires careful judgment. It is not the quantity of water or slump that is important but the time required for initial set of the concrete.

8. An important aspect of temperature control lies in the placing operation. In hot weather it is desirable to have two points at which concrete can be placed, so that if delay occurs at one site, the batch may be diverted to another.

9. The continuity of placing should be maintained to prevent "cold" joints, where the surface of the previously poured batch has become excessively dry or where partial set has taken place. Scheduling of concrete deliveries is extremely important in hot weather.

10. After placing concrete in hot weather, all exposed surfaces should be kept wet down continuously for the first 24 hr. Not until the day following the pour should curing compounds, plastic membranes, or sealing papers be applied. In the wetting-down operation the exterior of the forms should be kept just as wet as the exposed concrete surfaces.

7.17 COLD-WEATHER CONCRETING

It is generally true that the placing of concrete in low temperatures is more dangerous to the structural result than concreting in high temperatures. Although exposed slabs may be more difficult to finish in hot weather because of their rapid set, the finishing of slabs in cold weather is more costly, not only because of the precautions needed but because the slow rate of set increases the labor cost by the waiting time.

1. The most serious initial problem is that of obtaining a foundation sub-grade free of frost. In temperate climates an overnight freeze can develop frost several inches thick, and, of course, freezing soils will expand. During the winter months it is desirable, therefore, to provide an enclosure of the foundation area. Tarpaulins hung on a wooden frame with some slight heating maintained may be adequate.

2. Wetting down either forms or soil in cold weather can be dangerous, unless this is done within just the limited area where concrete is to be placed and warm water is used.

3. On large projects, aggregates can be pre-treated in revolving drums to bring their temperature above freezing. On smaller jobs, heating coils or fire drums can be set into the aggregate piles.

4. Hot water should be used, but the temperature should be kept to a maximum of 150°F, and this limit may have to be lowered if the aggregate temperature is high. However, it is the effect of the hot water on the cement that is critical, and hot water should be added only in revolving drums.

5. In cold weather, admixtures which tend to accelerate the setting time and control the water content are often used. The effect of these at specific temperatures should be pre-determined. Their use in the coldest weather may be satisfactory, but if the temperature rises rapidly during the pour, their use may have to be discontinued.

6. Some building codes and specifications prohibit the placing of concrete at ambient air temperatures under 50°F. However, lower temperatures, above freezing, are not dangerous if the proper precautions are exercised. Since these precautions are costly, the contractor must decide whether the additional expense is warranted.

7. Concrete cannot be wet down in cold weather. Slabs should be covered with a layer of paper as soon as sufficient set has been obtained, and 4 in. of salt hay should be placed over the paper. The heat of hydration of the cement serves a useful purpose so long as it is kept within the concrete; and it is common, on exposed walls, to place bat insulation between the studs for heat-containment purposes.

8. The heating of areas in which concrete has been placed may present more difficulties where foundations are involved than would be true in superstructure construction. Although foundations are frequently constructed in pits which can be covered over for protection, this is not, by any means, always the case, and exposed slabs in shallow excavations often require substantial enclosures.

9. The use of coke-burning salamanders has resulted in so many fires started by blowing tarpaulins, or windblown hot ashes, that they have largely been abandoned in favor of portable burning units. These units burn gasoline or kerosene under containment and are provided with built-in fans which produce a blast of hot air. The problem with any kind of heating unit is that of placing it so that hot air is supplied at the lower level of an enclosure without subjecting a portion of the concrete to excessive heating or drying. Just as salamanders should be set outside of slab areas, so blowers should be hung from an enclosure with their blast of air directed several feet above slab level. Unless the foundation area poured is very large, walls can usually be protected by limiting the applied heat to the exterior faces, covering over the interior so that the self-generated heat maintains the temperature above freezing.

10. Foundation forms can usually be stripped much sooner than those for suspended slabs or thin sections, but in inclement weather the protection offered the concrete by the form, against freezing rain or snow, should not be overlooked.

7.18 PIER CAISSONS

In Sec. 7.6, the discussion of foundation piers had to do with those constructed in an open excavation employing adaptations of standard forming techniques in which to place the concrete. In contrast, a pier caisson is a concrete pier constructed inside a casing which has been dug or driven into the soil. A still more recent method employs drilling to sink the excavation, and this method will be discussed in subsequent sections. Despite the increased use of drilled caissons which, for deep foundations, may be considerably cheaper, there are still occasions where pier caissons must be dug in.

Dug-in pier caissons are generally limited to 40 or 50 ft in depth and to 6 ft in maximum diameter since, beyond these limitations, costs increase disproportionately to the results. Although their use in the past was quite general, they are now used only where working space is confined or in special cases. Two methods are employed, the Gow method and the Chicago open well method.

The Gow method, introduced by the Gow Foundation Company, a corporate ancestor of Raymond Concrete Pile Company, employs steel shells 7 or 8 ft long in sinking the caisson. A shallow starting pit is excavated at ground surface, and the first length of shell is set upright in the pit. This shell, of ¼ or ⅜ in. steel plate, is first plumbed and then backfilled around the exterior. Hand excavation now proceeds inside the shell, at the bottom of the pit; and, as excavation is removed from beneath the bottom edge of the shell, the casing is driven into the resulting void.

A sheave, hung from the apex of a tripod 20 to 30 ft in vertical height and centered over the shell, carries the rope lifting the caisson bucket, a bucket suspended in a yoke which, when unlocked, can be tipped in either of two directions. The rope may be hauled by hand or powered winch. The haul rope is also used for raising and lowering a drop weight for driving the casing down. Shells are limited to 7 or 8 ft in length since the exterior wall friction makes the driving of longer lengths difficult.

When the top of the first shell has reached ground surface, a second shell 2 in. smaller in diameter is dropped down inside the first one, similarly excavated, and driven down until its top is just above the bottom of the first shell. Additional shells are added until the desired depth is reached.

The Gow caisson may be driven down to a seating on rock, in which case the desired bearing area there will determine the diameter of the starting shell. Where seating is on firm soil such as hardpan, the final shell may be held several feet above founding level while a tapered or bell section is underreamed to extend the bearing area. The belling potential will depend upon the soil's ability to stand firm until it is concreted.

The Chicago open well method is essentially similar except that it employs wood lagging in lieu of the steel shell. The lagging, usually 2 × 6 in. tongue-and-grooved plank, is set around the perimeter of circular steel channels which act as walers. The preshaped channel may be in one split length or may be composed of two pieces which are tied together by bolting when the ring has been positioned.

The individual pieces of lagging are driven down as excavation proceeds, the channel frames, set at from 3 to 6 ft intervals, acting as interior guides. As a length of lagging reaches ground surface, an additional length is set on top of it, held by the tongue and groove of the adjoining pieces.

With the Gow-type caisson the steel shell is pulled as the concreting operation progresses upward, but with the Chicago method the casing is left in place. Since the critical shaft diameter is either at the top of the bell or at the surface of the rock, the Gow type will require greater quantities of hand excavation and concrete than the Chicago type. For considerable depth of caisson, therefore, the Chicago method will generally be found to be cheaper.

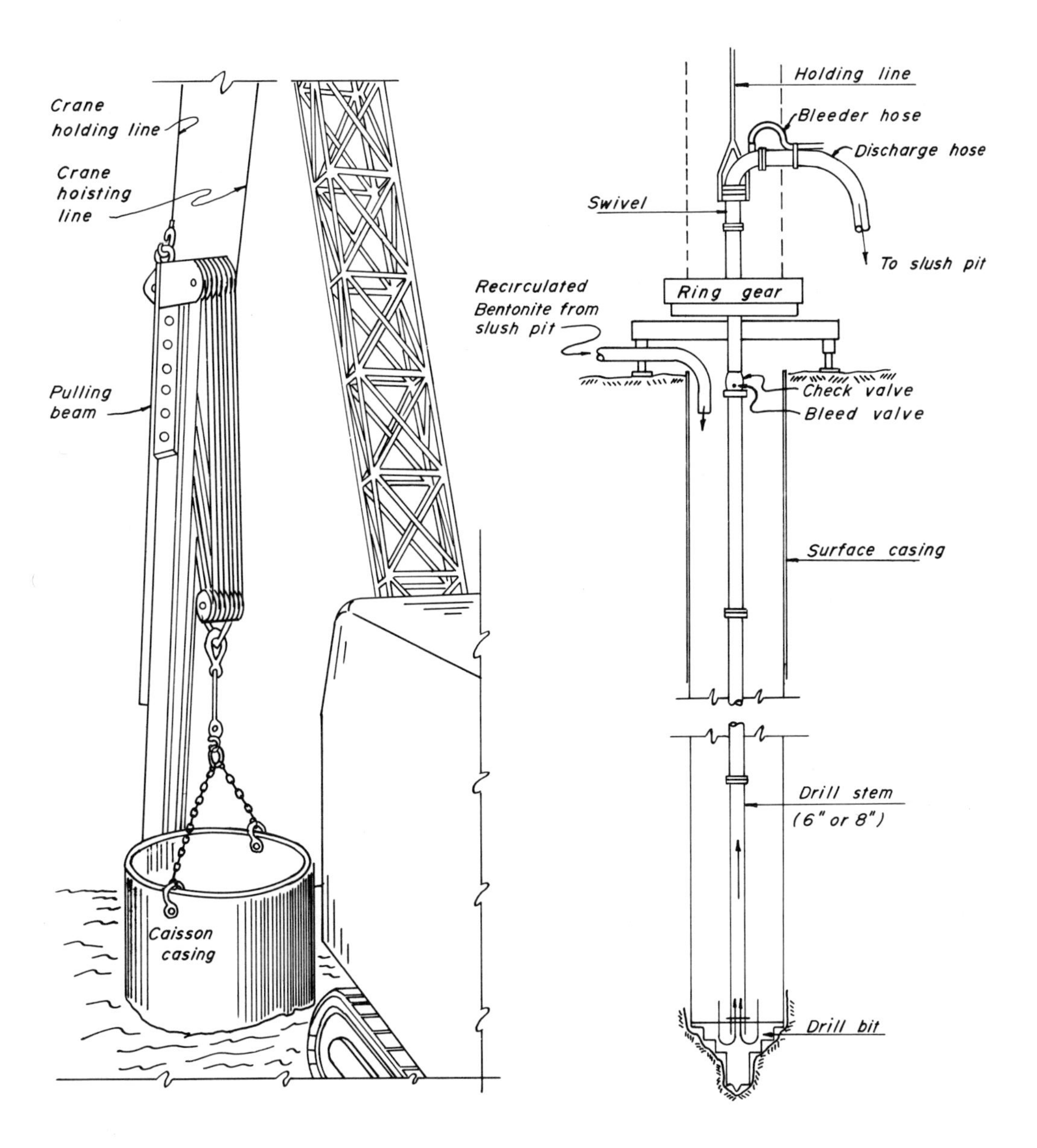

PULLING BEAM ASSEMBLY
FOR EXTRACTING CAISSON CASING
SEE 7·22 FOR DETAILS

REVERSE CIRCULATION SYSTEM
FOR BENTONITE SLURRIES
SEE 7·23 FOR DETAILS

7.19 A CASE OF A DUG-IN CAISSON

The foundation contractor is not unlike an opera singer; the more tricks he has in his repertoire the more valuable he becomes, so that, although the dug pier caisson has been replaced to a very great extent by the drilled caisson, a situation may arise where the older technique is useful. Such a situation did arise in 1960 in Chicago, where 36 caissons were required, not for a new structure but to add 22 stories on top of an existing four-story garage.

It may be argued, with some justice, that this sounds like a tale of underpinning, but here the trick was to leave the existing 9 × 9 ft spread footings undisturbed while putting caissons down between them for independent support of the higher structure.

The contractor, Case Construction Company, would normally have used a rotating caisson bucket (see Secs. 7.20 and 7.21) for the job, but the headroom available on the first floor was only 13 ft, insufficient for this type of drill, and the contractor decided to use Chicago open well caissons.

The first floor was opened up at each caisson site, and a hole was hand-excavated to a depth of 6 or 7 ft. Within this hole 10-ft lengths of 2 × 6 in. lagging were set up around the perimeter of 4-in. channel iron rings. Hand excavation progressed within the lagging, which, driven by a sheathing hammer, followed the excavation down. When the top of the lagging reached floor level, an additional 5 ft was added to provide a caisson with an inside diameter of 5 ft 2 in. and a depth of 15 ft. It was at this level that a stratum of water-bearing sand was encountered.

Excavation in this sand layer would have required dewatering, but to lower the water table at this depth might result in settlement of the spread footings, and a steel casing of the Gow type was now reverted to.

To shove this steel casing down through the sand layer, the first expedient tried was that of jacking it down from the second-floor beam system. These beams could withstand a 30-ton loading without damage, but this was inadequate for the purpose. The second expedient tried was that of churning bentonite (see Sec. 7.23) into the water-bearing sand and driving the casing down with a pile hammer, but this required piercing of the second floor for the churn drill mast, and the process was slow.

The final solution proved to be that of driving the casing down by using two pile hammers operating simultaneously. To do this it was necessary to reinforce the upper edge of the ¼-in. shell with a ½-in. curved plate, welded to the inside face of the casing. Lugs were then welded to the ½-in. plate to support a 10-in. WF beam which took the direct impact of the pile hammers. In the same fashion a 6-ft extension was added to the 11-ft section, by welding, to provide a total shell depth of 16 ft.

For the lower portion of the caissons, which were now carried down to hardpan through solid clay, the contractor returned to the Chicago open well method, setting the lagging inside the steel casing to provide a caisson diameter of 3 ft 5 in. Leakage of water at the joint between lagging and casing was temporarily removed by a Flygt submersible pump until the caisson had been bottomed. This joint was then sealed with dry-pack rammed into the void.

The caisson diameter had now been reduced from 5 ft 2 in. to 3 ft 5 in., so it was necessary to bell out the bottom, when hardpan had been reached, to provide the starting area of the 5 ft 2 in. diameter. Since the material at bell level was a hard clay, no difficulty was encountered in holding this slope until concreting was completed.

Concreting was carried up from the bottom, leaving both lagging and casing in place, to a point 5 ft below floor level, where structural steel framing would begin. A cage of reinforcing was provided, full height.

All the equipment for this project was air-operated: the pile-driving hammers, sheathing hammers, and clay spades. The use of air was also extended to include the winch hoist on the tripod and a small sludge pump used in handling sand and water during the steel-casing stage. A Lamb air mover, also air-operated, was used for circulating air to and from the bottom of the caisson.

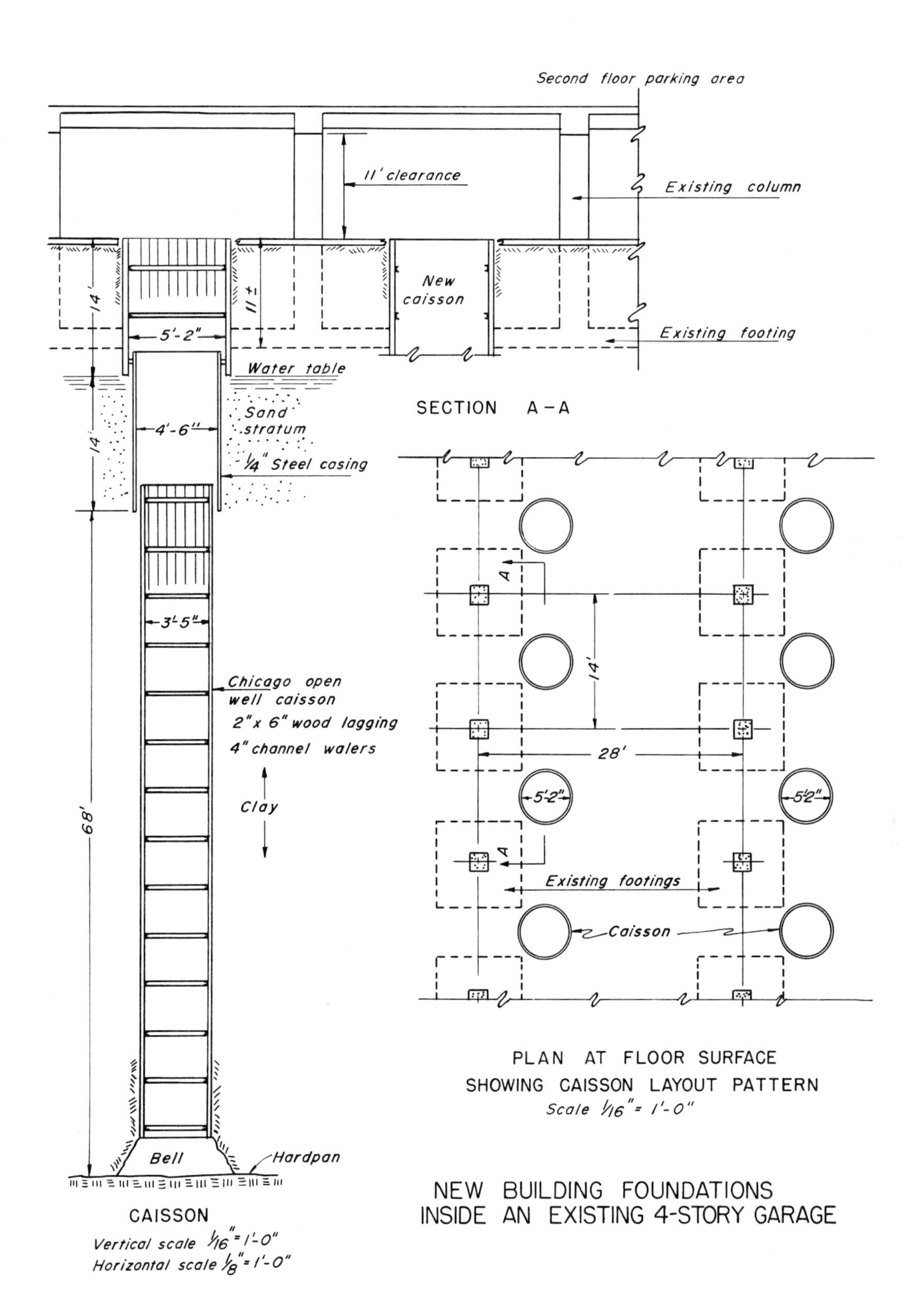

NEW BUILDING FOUNDATIONS
INSIDE AN EXISTING 4-STORY GARAGE

7.20 EARTH DRILLING RIGS

Although we generally think of drilling in terms of a rotating mechanism, the act of drilling is simply that of piercing; and the earliest drilling, in search of water, was performed by punching a hole in the earth. The practice survives, for the same purpose, in the percussion or churn drill. With this type of drill the soil or rock is pounded to bits, water is added, and the resulting slurry is scooped out—like getting to the bottom of a chocolate parfait.

The percussion drill proved to be too slow for the drillers of the petroleum industry when they went in search of oil at thousand-foot depths instead of the usual hundred-foot depths, and their hunger for oil led to the development of rotating drills, first only for rock piercing and then, finally, for general drilling.

Once we have solved the problems of applying power so as to rotate a vertical shaft and hold it in alignment, the rest follows. For the cutting end of the drill shaft there was, first, the solid steel bar with projecting ridges; then a smooth steel face grinding on crushed steel shot; or various steel faces, diamond-studded. The first led to the construction-drill bit, its ridges now built up by an insertion of tungsten carbide. The second led to the Calyx or shot core drill and the third to the diamond drill, both of which permitted the removal of cores, when required, as well as providing a drill hole.

For earth drilling, an auger was developed resembling the simple wood auger. For firm soils a single-flight auger is used, since this requires the least power. For softer soils, continuous-flight augers have been found to be practical, the available power, no longer needed to cut through the soil, being applied to a larger load of soil at each bite. It now became important, however, to find means of quickly raising the loaded auger and of disposing of the soil accumulation without letting it fall back into the drill hole.

The next logical step was the development of a vertical digging bucket which would carry an entire load of soil to the surface and which could be swung free of alignment for unloading. One of the earliest to develop such a bucket, and such a means, was Calweld of California, applying their knowledge of oil-well drilling rigs to the special problems of the digging bucket.

The earliest rig consisted of a truck-mounted mast for supporting the "Kelly" bar, or shaft, vertically, designed to tilt into horizontal position when the truck was being moved. The Kelly bar, from which the digging bucket was hung, was a square bar piercing a square hole in a steel yoke. The yoke, in turn, was set into a ring-and-pinion gear assembly, power-driven, which rotated the yoke and the bar with it. The Kelly bar moved vertically on a rope hung over a sheave at masthead and was powered by a truck-mounted winch. When the digging bucket was filled, the Kelly was raised until the bucket reached the yoke, when bucket and yoke were both lifted through the ring gear. The assembly was swung aside, on the rope support for the Kelly, to be unloaded.

Today's drill rig remains much the same in principle but with some significant improvements. Originally the weight of the bar and bucket, plus the engagement of the tilted bucket lip into the soil, was depended upon for downward thrust. Recent models include a hydraulic crowd mechanism to force the bucket into the soil. Digging depths of the original models were limited by the length, or height, of the Kelly bar. A telescopic bar has now been introduced to permit digging to depths of as much as 150 ft.

The original assemblage, once only truck-mounted, has now been mounted on a crawler tractor for greater mobility over soft terrain otherwise impassable for truck tires. A separate unit has been developed that is an attachment for a standard crane. Hung from the boom foot pins and controlled, for positioning, by hydraulically operated bars tied back to the crane's gantry frame, it utilizes the crane boom as a mast. Since the crane rig can swing on its revolving base, it is no longer necessary to lift the bucket through the ring gear, and the yoke operates in a fixed position, rotated by a power source built integrally into the attachment.

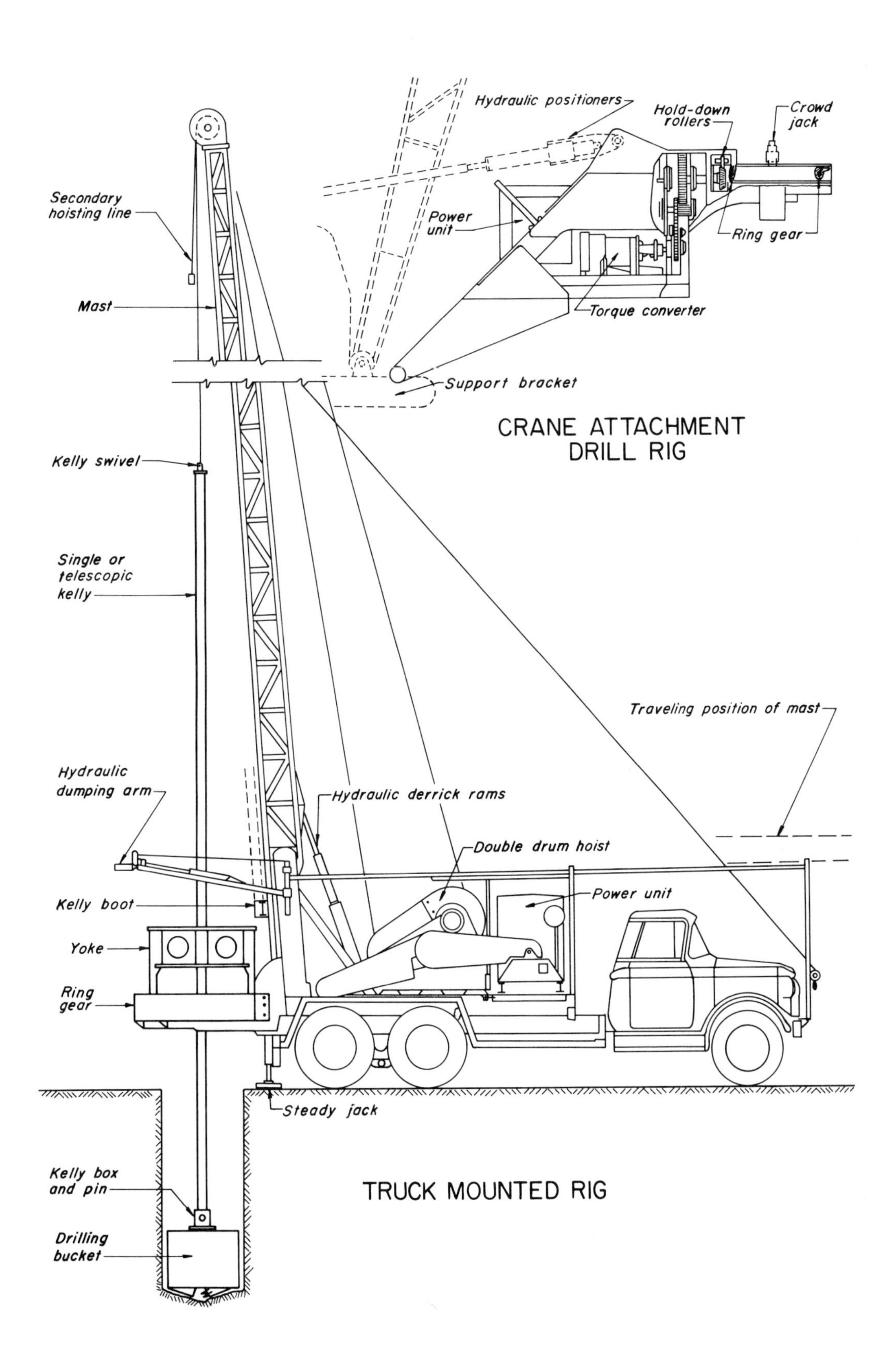
Hydraulic positioners
Hold-down rollers
Crowd jack
Power unit
Ring gear
Torque converter
Support bracket
CRANE ATTACHMENT DRILL RIG
Secondary hoisting line
Mast
Kelly swivel
Single or telescopic kelly
Traveling position of mast
Hydraulic dumping arm
Hydraulic derrick rams
Double drum hoist
Power unit
Kelly boot
Yoke
Ring gear
Steady jack
Kelly box and pin
Drilling bucket
TRUCK MOUNTED RIG

7.21 DRILLING TOOLS

Foundations are drilled into or through soil formations by using either augers or drilling buckets as noted in Sec. 7.20 and, whether a casing is used or not, are generally termed caissons. But augers or buckets are not sufficient to cope with every underground condition encountered and must be supplemented by other drilling tools.

Neither augers nor buckets will cut through boulders or through thin layers of sedimentary rock which may be encountered in otherwise continuous soil formations. In these cases a core barrel is used not only to cut through the obstruction (which may be buried timber or roots as well as rock) but also to provide the equivalent of a casing to support the walls of the hole.

The core barrel consists of a section of steel tube or casing to the bottom of which a heavy ring of tool steel has been welded. The bottom of the ring, of slightly greater diameter than the exterior of the casing, is provided with cutting projections which are reinforced with tungsten carbide or boron carbide to resist abrasion. The top of the core barrel is suspended from a Kelly bar; and the entire unit is rotated in the hole, cutting a core from the rock or obstacle, which is then broken up and removed. The core barrel is also used where it is desired to socket a caisson into a rock stratum.

The rooter is a device for breaking up the core cut by a core barrel, for mincing roots or debris to a consistency where an auger or bucket can dig them loose, or for clearing a hole of loose stones which could jam an auger or bucket. A heavy point set into a flanged plate of steel, the rooter is bolted to the bottom of the Kelly bar and alternately dropped and rotated.

A muck bucket must be used where soft wet soils are to be lifted from a hole, since they will not stay between the flights of an auger and will run out of the bottom of a drilling bucket. The muck bucket scoops up wet soil by rotary action; but rubber flap valves, built into it, retain the material until it is lifted from the hole. With its load brought topside and swung away from the hole, a trigger mechanism opens the flap valves and permits the contents to spill out.

For more liquid conditions or where water alone must be handled, a bailing bucket is used. This is lowered into the hole until it rests on the bottom. The weight of the bucket thrusts a foot piece upward, raising a valve head from its seat and permitting the liquid to flow into the bucket. As the bucket is raised, the valve drops into place on the seat, holding the contents in until they are lifted and dumped.

A belling bucket has now taken the place of hand excavation to provide a bell at the bottom of a caisson. This bucket, suspended from the Kelly, opens when the bar is pressed downward into a wedging slot, forcing two blades, with toothed edges, diagonally outward. The bar is then rotated to scrape away the earth in a bell shape, while continued downward pressure forces the teeth into the soil and gradually widens the bell. The belling bucket, despite its name, does not lift the loosened soil, which must be removed by other means.

The choice between auger and drilling bucket for general caisson excavation depends on a number of factors. Augers turned through gravelly soil tend to unravel the sides of the hole, while stones tend to jam in the flight, obstructing further movement of earth. It is difficult to maintain the vertical alignment of single-flight augers; and continuous-flight augers, which are bolted together in sections while being turned down, cannot rapidly be lifted for cleaning. A single-flight auger or one with two or three flights can be readily lifted, rotated at high speed, and its contents thrown off in a windrow around the hole.

Digging buckets, once started plumb, tend to maintain alignment as they move downward. In loose soils, such as sand, they will bring a larger load to the surface than can be held on an auger.

Continuous-flight augers are used only for smaller holes, whereas single-flight or limited-flight augers can cut holes of considerable size. Digging buckets have been used up to 6 ft diameters, but are more apt to bind in tight clays than augers.

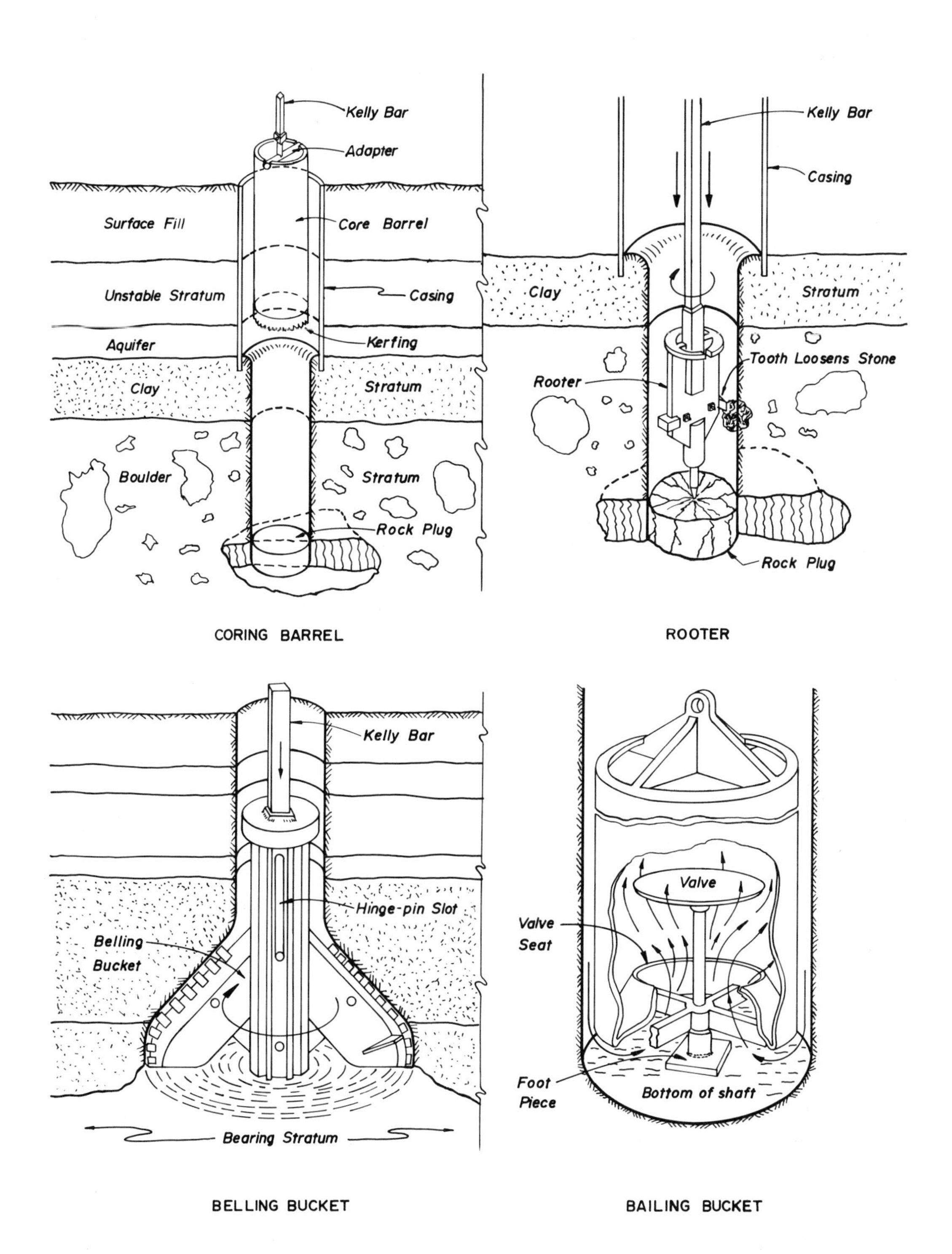

DRILLING TOOLS

7.22 CASINGS FOR CAISSONS

The development of drilling equipment and drilling techniques has led to the extensive use of caissons in lieu of piling and to the use of drilling instead of driving methods for placing piling. A cast-in-place concrete pile is a shell driven into the earth and filled with concrete; and a caisson, in most cases, is a shell or casing driven into a pre-drilled hole and filled with concrete. Large pre-cast concrete piles are frequently dropped in a drilled hole and driven only a few feet. This differs from a large-size caisson only in that the bottom or bearing surface can be examined before concrete is poured, whereas with the pile, bearing surface is judged by resistance to driving.

It is only in those cases where no satisfactory bearing stratum can be reached and a friction pile is needed that the pendulum swings in favor of piling. Against this is the fact that caissons, capable of being belled at the bottom, can utilize weaker strata. The only way that a sound comparative judgment can be reached is to compare estimated costs for a whole foundation, since fewer caissons than piling are generally needed, and the ground-level structural design may be somewhat different.

Although some soils will stand from the surface downward without casings, this is a relatively rare condition, and even here the top few feet of the hole will be cased to avoid breaking down the walls because of drilling operations.

Generally a section of hole is drilled and the casing dropped, pushed, or driven into the resulting void. The diameter of casing provided should be 1 to 1½ in. larger than the specified diameter of the finished caisson, and the hole that is drilled should be that much larger again. This means that the starting hole should be at least 2 to 3 in. greater than the specified shaft diameter.

The length of casing is limited to the depth of hole that can be drilled before there is danger of caving the walls, plus the additional aboveground length which will fit under the rotary table of the drill. The Kelly bar may be telescopic and extendable, but the rotary table (that is, the ring-and-pinion gear assembly) is fixed and there must be clearance between its underside and the top of the casing. Although it is desirable to have the bottom of the casing section come to rest on firm material below the unstable stratum, this is not always practical.

Once placed, the length of casing should be driven down as far as possible before drilling is resumed. Where driving is easy, an additional section can be welded or bolted to the top of the casing; but where resistance is encountered in driving, it may be desirable to drop a second section down inside the first or earlier one. The second section should be about 1 in. smaller in diameter than the previously placed section. When drilling is resumed, it will be necessary to adjust the drill-tool diameter.

Where considerable depths of casing are involved, and a number of telescoped sections are required, the initial hole diameter must be large enough to provide the minimum specified shaft diameter at the bottom. This contingency must be taken into account when estimating the cost of the operation, and the resulting increase in the quantity of concrete must be provided for.

When the casing is concreted, the concrete level should be brought 1 to 2 ft above the bottom edge of a section of casing before pulling is started. If concrete is carried too far up within the shell, the resulting skin friction can lift the concrete and cause "necking"—a reduction in shaft diameter.

The pulling of casing should be done with a slow even lift utilizing a pulling beam such as that shown in Sec. 7.18, to provide sufficient power. Direct pulls by hoisting lines are not generally practical. Once begun, pulling should be continuous, and the placing of concrete should follow the casing up, keeping always above the bottom edge.

Where the pulling is difficult, a delay in pouring, or necking, may result; and in these cases it may be necessary to leave the casing in place. In estimating casing requirements on a job of substantial size the contractor should allow 10 to 20% for casing left in place.

7.23 BENTONITE OR DRILLER'S MUD

Bentonite is a natural clay product found extensively in the western part of the United States, particularly in such states as Montana, South Dakota, Wyoming, and Colorado. Bentonite is essentially a montmorillonite clay (see Sec. 3.8) formed naturally by the chemical decomposition of certain igneous rocks such as volcanic ash. Unlike the pure montmorillonite described in Sec. 3.8, bentonite contains inclusions of sodium or calcium, in this respect resembling illite, and the particular inclusion affects its properties.

Sodium bentonite, found in Montana, South Dakota, and Wyoming, has a high degree of swell, sometimes of as much as 1500%. Calcium bentonites, such as those found in Colorado, have a lower swell but have been found to be useful in sealing canal banks and similar earth structures.

The product known as driller's mud, a bentonite found chiefly in Wyoming, was first exploited by the petroleum industry in oil-well drilling operations and there first acquired its name. A commercial product, dried, ground, and distributed in paper bags containing 100 lb each, it is sold under such trade names as Volclay and Aquagel. When drilling spread to the construction industry, driller's mud or bentonite was adapted to perform the same function it had served in oil wells, chiefly that of retaining the banks of drill holes without the use of a casing.

Bentonite, mixed with water, will remain in suspension, forming a homogeneous mud, up until it reaches a ratio of 3.5 lb per gal of water. Beyond that concentration, separation and settling will occur. Not only will the resulting mud expand considerably in the mixing process, but the weight of the mixture, at maximum concentration, will increase from the weight of water of 62.4 lb per cu ft to a weight of mud of about 91.5 lb per cu ft.

The principal function of bentonite in a drill hole is to control the flow of water into that hole. Most caving results from water flowing into the void. However, if we fill that void with a fluid which is heavier than water, the flow will be reversed and the bentonite slurry will tend to flow into the walls.

Bentonite is usually mixed in the drill hole. Compute the capacity of the drill hole when filled with water. (Bentonite has little value near the surface, and the first 4 or 5 ft of hole should be cased.) Add bentonite at the rate of 3.5 lb per gal of water. (Smaller ratios are, of course, usable; but since it is difficult to assess the precise density required, using the maximum is the safest policy.) Lower the drill being used into the mixture, and turn the drill up to its highest speed of rotation while alternately raising and lowering it through the slurry.

The process of drilling will produce cuttings of rock or soil particles which will float or settle in the slurry, depending upon their relative densities. In some cases, these fines may so nearly approximate the density of the slurry as to remain in suspension, but this will not destroy its efficacy.

To remove particles resulting from drilling while at the same time maintaining the liquidity and density of the slurry, a slush pit can be excavated adjoining the site of drilling. This should be dug in a long rectangle with one end deeper than the other, the deeper end adjoining the drill hole. The capacity should be at least twice as great as the ultimate volume of the drill hole.

The loaded slurry is raised by pumping, by an air lift, or by being forced up the hollow shaft of the drill rig, and is then piped to the shallow end of the slush pit. As the slurry flows through the pit, settlement occurs, leaving the heavier cuttings in the pit and permitting the remaining mud to flow back into the drill hole after passing over a weir at the deep end of the slush pit. In the exceptional case where the soil intrusions are lighter than the slurry, a 2×10 in. baffle board, placed with half its depth below the surface, will collect floating particles which can then be skimmed off.

It is possible to re-use bentonite slurries several times, but the costs of handling will seldom warrant this. Sodium bentonites cost about $40 per ton, and this quantity will fill a hole 12 in. in diameter and over 700 ft deep even if we disregard the swell potential.

Marina City, a five-building complex in central Chicago noted for the circular design of its structures completed in 1963, is supported on caissons, a method selected after spread footings and driven piles had been tossed out the design window. Involved were 151 caissons, ranging in diameter from 24 to 60 in. and drilled 110 to 115 ft in depth, first through a 20-ft layer of rough and miscellaneous fill, then through 70 to 80 ft of Chicago clay, and finally punched through a 5-ft stratum of water-bearing gravel and boulders to bed on sound limestone.

For the 60-in. caissons, drilling was begun with a 78-in. limited-flight auger, and 20 ft of 72-in. steel casing was placed. The clay was then drilled through to within 5 ft of the gravel layer, and a 66-in. casing was installed. The gravel layer was under artesian pressure which could force water 55 ft up into the caisson, and the 5 ft of clay was left to retain this pressure until a core barrel could be placed.

The remainder of the drilling was done with a core barrel 65 ft in length and 60 in. in diameter, of rolled steel plate ½ in. thick. As described in Sec. 7.21, the bottom of the core barrel was reinforced with a kerfed steel plate with carboloy tungsten slugs, silver-soldered to it. The top rim of the barrel was also reinforced and provided with four shoe-shaped slots at the quarter points of its periphery. The slots engaged four 3-in. steel pins projecting from the plate of a spinner attachment secured to the end of a 4½-in.-square Kelly bar. The core barrel was reinforced on 3- to 4-ft centers by steel rings welded around the interior surface.

Coring was carried on continuously through the clay and gravel layers and several feet into the limestone, effectively sealing off the water pressure and permitting dewatering with a Flygt submersible pump. A laborer was sent to the bottom, where he loaded out the fractured material in a caisson bucket and broke up the limestone core with a paving breaker. Finally, an inspector was gently lowered in a boatswain's chair to pass upon the bottom before concreting began.

Concreting was carried up to the top of the left-in-place core barrel before a cage of reinforcing steel was set, and concreting proceeded to the bottom of the 72-in. shell. At this point a light 60-in.-diameter shell was set, centered, and concreted to foundation surface, before withdrawal of the original 78-in. shell.

There were contretemps, of course. In one caisson a core barrel sheared off in the gravel layer and the piece could not be removed. Here, it was necessary to drive two additional caissons down, one on either side of the damaged unit, and to span the faulty caisson with a beam on which the column could be supported. One caisson was forced out of plumb by a recalcitrant boulder, and the eccentricity was compensated for by loading the bottom of the shaft with reinforcing steel.

On the Martinez-Benecia bridge, along San Francisco Bay, eight caissons 130 ft deep were required under each pier foundation where the bridge spanned the Sacramento River. For each caisson an 80-in.-diameter steel casing of ½-in. plate was lowered through the water and pushed through the bottom mud to resistance. Through this casing, a 78-in. drill bit, specially designed as an adaptation of a dredge's cutterhead, its teeth star-studded with tungsten carbide chips, drilled into the river bottom. The Kelly bar, providing torque, was specially designed, using a 6-in. steel pipe to which four continuous angles were welded to provide a square exterior.

The cuttings and river water combined to form a slurry which was forced up through the hollow-shaft Kelly by a combination of air pressure at the bottom and pump suction at the top. The slurry was drained off through a swiveled slot in the Kelly and dumped into the water outside the caisson.

At bedrock, a 60-in. core barrel drilled 5 ft into the rock; a 72-in. caisson shell was dropped into place, followed by a steel cage; and concrete was placed by tremie. The 80-in. shell was withdrawn and re-used.

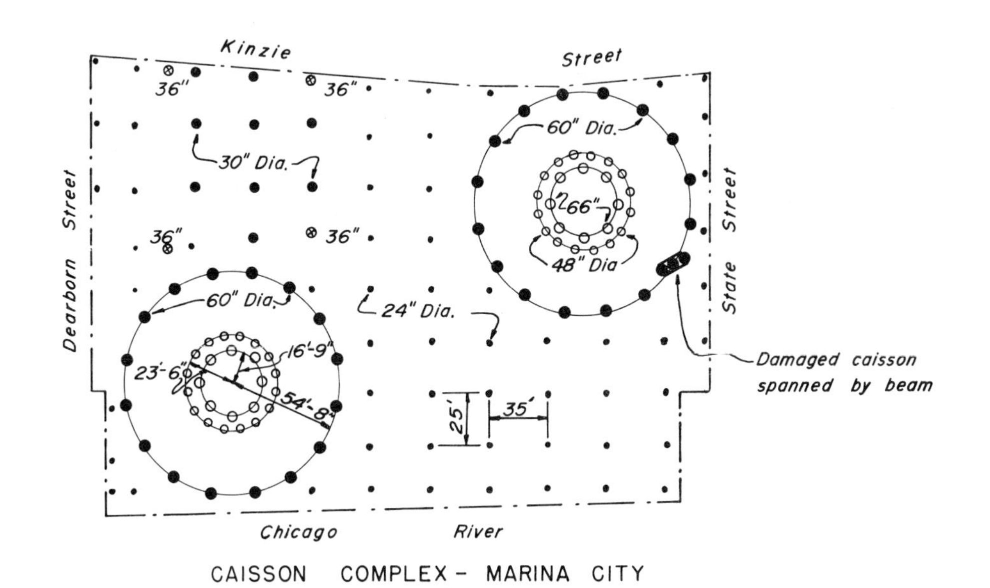

CAISSON COMPLEX – MARINA CITY

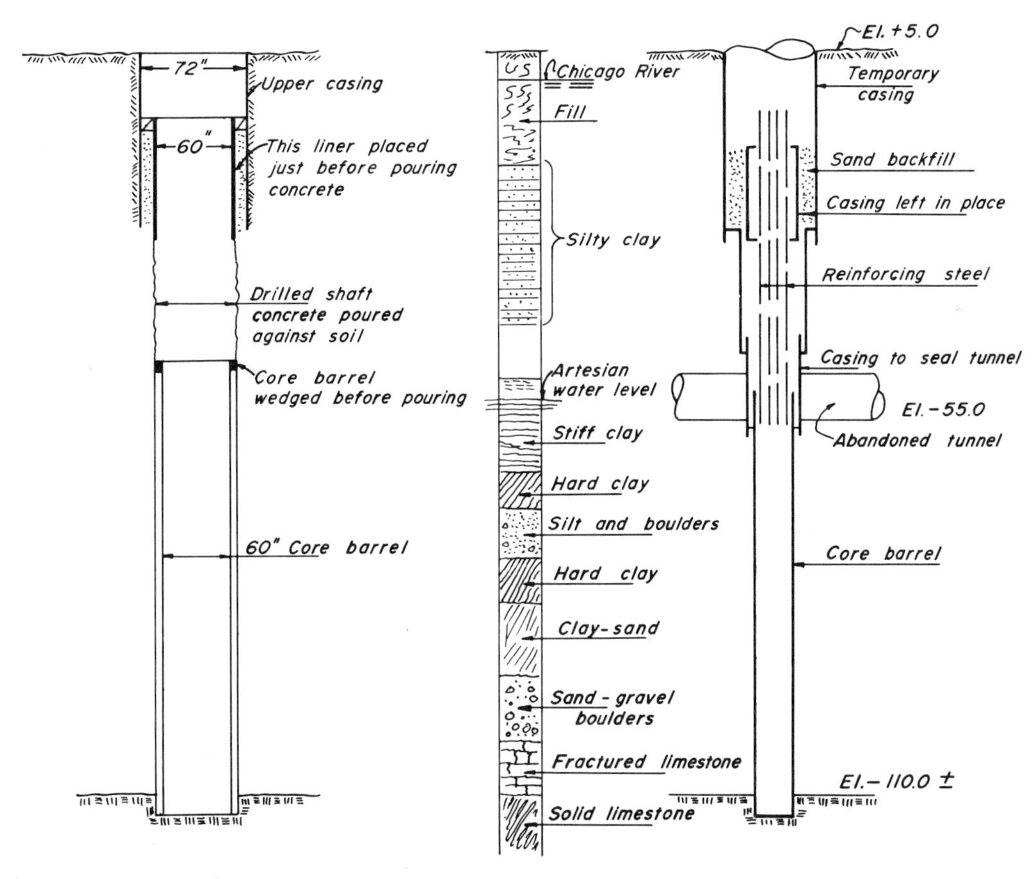

STANDARD 60" CAISSON CORE BORINGS SPECIAL CASE

In 1959, the Western Foundation Corporation was engaged in placing 144 drilled-in caissons for the foundations of the Prudential Tower in the heart of Boston. These were 130-ft-deep caissons traversing fill, silt, and clay and socketed 15 to 23 ft into the underlying limestone. There was water all the way and, because of older surrounding structures, the water table could not be substantially lowered.

A 65-ft length of 30-in. steel pipe, ½ in. thick, provided with a cutting edge of rolled steel plate 15 in. wide and 1 in. thick, was driven to its full length by using a Vulcan "O" pile-driving hammer. A second section, also 65 ft long, was then welded to the top through a standard coupling, and the combined sections were driven to rock.

Excavation, which proceeded with the driving, was by jet blast. A 1½-in. jet nozzle, fed by a 6-in. riser, was forced down along the inside edge of the casing to a depth below the cutting edge. The jet was left in place and the hole around it was plugged with sand. Water at 400 psi was now fed to the jet, from a high-pressure pump, and a plug of clay was blown out of the upper end of the casing. When the clay became too liquid, a bailer, or muck bucket, was used for cleanout.

At rock level a 5-point star drill weighing 6500 lb, standing 8 ft high, and with a 29-in. outer diameter was raised and dropped from a spudding arm to cut up the rock and reduce it to a slurry. The slurry was removed by bailer; and when the desired depth of penetration had been reached, the shell was driven from 1 to 4 ft into the resulting socket.

It was not possible to dry these caissons up for personal inspection, and a closed-circuit television camera was used. The fines were first removed by means of a submersible pump, and the camera, in its watertight case and with a built-in light source, was lowered and rotated slowly. Results were cabled to a screen in the field office.

The final step in completing this caisson was the placing of a 50 lb per ft H-beam for the full length of the caisson. This was sealed into the rock socket by tremied grout. The grout sealed off the water entrance, the caisson was pumped out, and the remaining concrete was placed in the dry.

In 1960, bids were called for the foundations of a grain elevator on the Montreal waterfront, in an area of unstable dredged fill laid down over a rock mattress. The original design called for a cluster of nine piles at each of 600 column centers. The contractor, Petrifond Foundation Company, Limited, suggested a 39-in. caisson in lieu of the nine piles, and this alternative was accepted.

Petrifond imported two Benoto rigs from France, each weighing 30 tons, to put down the caissons. Shells were placed in 20-ft sections, the bottom section having a cutting edge. Instead of driving, however, these sections were forced down by using a battery of hydraulic jacks acting vertically, while a vibratory mechanism, acting through a collar clamped to the top section, vibrated them horizontally (a variation of the vibratory hammer discussed in Sec. 5.18). Excavation progressed concomitantly, using a specially designed 4-cu-ft grab bucket, dropped into the casing after the jacking frame had been swung aside.

When rock or other obstructions were encountered, the Benoto rig was temporarily moved to a new location and a crawler-mounted churn drill took over. From the spudding arm of the churn drill a 6-point star drill, 36 in. in diameter and 8 ft long, was hung to chop up the obstruction. The same unit was used, when final rock surface was reached, to provide a 42-in.-deep socket into the rock. The slurries resulting from the rock-drilling operation were removed by a standard bailer.

No steel core was used, and the reinforcing cage was limited to the top 10 ft of the caisson. Concrete was placed by tremie in 90 ft of water, and the casing was removed as concrete was placed. In removing the casing, the jacking operation was reversed, the rim of the casing being gripped and jacked upward while the horizontal vibratory action was maintained.

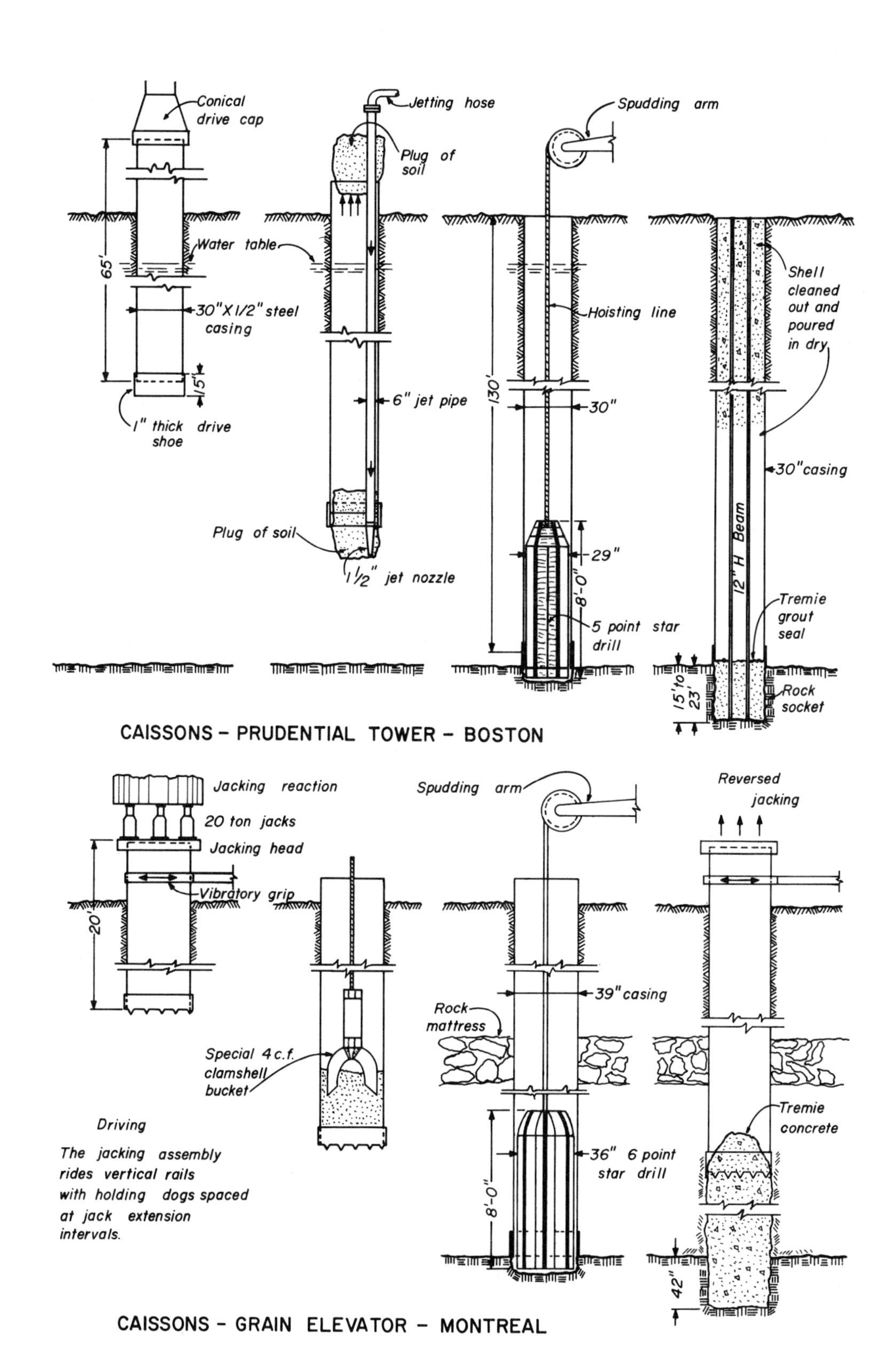
Conical drive cap
Jetting hose
Spudding arm
Plug of soil
Water table
65'
30"X1/2" steel casing
15'
1" thick drive shoe
6" jet pipe
Plug of soil
1 1/2" jet nozzle
Hoisting line
130'
30"
29"
8'-0"
5 point star drill
Shell cleaned out and poured in dry
30"casing
12" H Beam
Tremie grout seal
15' to 23'
Rock socket
CAISSONS - PRUDENTIAL TOWER - BOSTON
Jacking reaction
20 ton jacks
Jacking head
Vibratory grip
20'
Spudding arm
Reversed jacking
39" casing
Rock mattress
Special 4 c.f. clamshell bucket
Driving
The jacking assembly rides vertical rails with holding dogs spaced at jack extension intervals.
8'-0"
36" 6 point star drill
Tremie concrete
42"
CAISSONS - GRAIN ELEVATOR - MONTREAL

The use of drilling equipment has been adapted to a wide variety of foundation problems, many of which lie on the boundary between caissons, piling, and more ordinary types of excavated footings and are, in fact, "neither flesh nor fowl nor good red herring."

On the section of the Illinois Toll Highway sweeping around Chicago, a number of bridge foundations were to be supported on 36-in.-diameter pre-stressed cylindrical piles. The overburden, here, consisted of 20 to 30 ft of sand and clay separated from a suitable bearing stratum of hardpan by a shallower layer of gravel and boulders. The danger of damage to or deflection of the 36-in. piles was considerable, and a drill rig was employed to provide a pilot hole into which the pile could be dropped.

A 36-in. drilling bucket was used which, in the overburden, could be filled to its capacity of 37.5 cu ft in 8 sec. When the gravel-boulder stratum was reached, the material was loosened by dropping a cutting bucket, an open-end shell with teeth set around its perimeter, and the loosened gravel was then removed with the digging bucket. Where larger boulders protruded into the pit, a core barrel was employed to cut them off to pit size.

When excavation reached the hardpan layer, the 36-in. pile was dropped into place and driven the remainder of the distance required to provide adequate bearing.

On the Romona Freeway in Los Angeles, cast-in-place concrete piles were placed by drilling. Although ground conditions varied somewhat, from place to place, all were firm soil formations. The drilling tool was a 16-in.-diameter continuous-flight auger, the flights hardfaced with Stellite. The flights were constructed around a 4-in. hollow shaft 50 ft long, the unit guided, during drilling, in 60-ft-high pile leads.

After centering and plumbing the auger, drilling progressed until resistance indicated that the flights were filled. The auger was raised; plywood was laid over the hole; and, with the direction of rotation reversed, the auger was spun to clear the soil off the flights. In the final increment of drilling to required depth, the auger remained in the hole with loaded flights.

A pressure pot, its outlet closely coupled to the 4-in. hollow shaft, hung from the leads above the auger during drilling. With the auger at full penetration, concrete from the previously loaded pot was forced down the shaft under 30 psi pressure while the auger was slowly raised, rotating. The bottom of the auger thus maintained pressure on the concrete while the loaded flights prevented hole collapse above concreting level.

When concreting by this method had reached a point 12 ft from the surface, the auger was removed, a cage of reinforcing was inserted, and the remainder of the pile was filled by chuting.

Rotary drills were used to sink 200 missile shafts or silos at the Minuteman Missile Base near Cheyenne, Wyoming. Although there was considerable variation in soil conditions, requiring special treatments, the basic unit of excavation was a single-flight auger, 14 ft 8 in. in diameter.

The shafts were 95 ft in depth; but, because of the requirements of auxiliary construction, the top 30 ft of soil was first removed by scrapers, leaving a 65-ft-deep hole to be drilled. A pilot hole was first dug by using a 6-ft-diameter limited-flight auger; then, with the pilot hole as a guide, the 14 ft 8 in. diameter single-flight auger reamed the hole to full size. The single-flight auger had a very flat pitched blade with a depressed cutting edge armed with teeth. Turned through a transmission having a ratio of 300:1, it rotated at 5 rpm during cutting and, when loaded and lifted, could be spun at 30 rpm to toss the excavated material away from the hole. The windrow of tossed soil was picked up by front-end loader.

Where harder drilling was encountered, the 6-ft hole was first reamed by a 10-ft limited-flight auger before the final blade was used. Where sand or running soils existed, liner plates were placed for support, after shaving the bank wall, using the single-flight blade as a working platform.

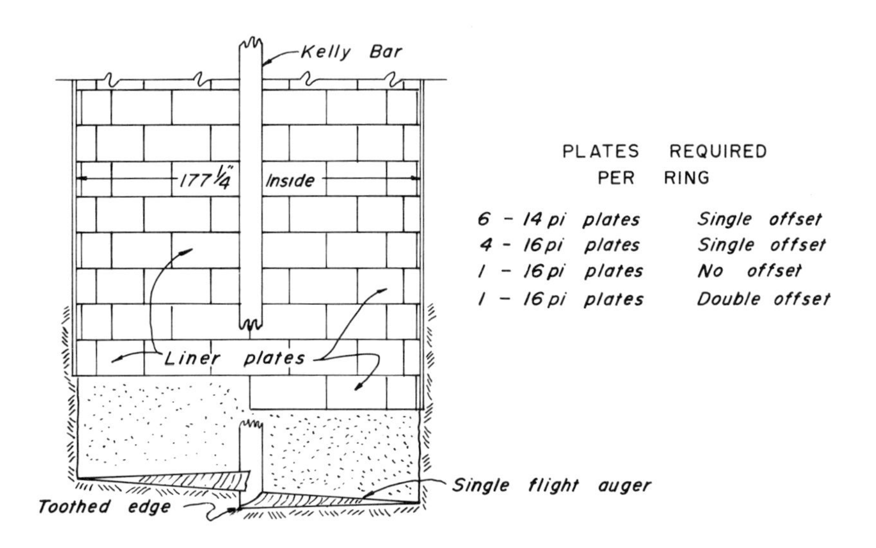

PLATES REQUIRED PER RING

6 – 14 pi plates	Single offset
4 – 16 pi plates	Single offset
1 – 16 pi plates	No offset
1 – 16 pi plates	Double offset

SETTING LINER PLATES IN MISSILE BASE SHAFT

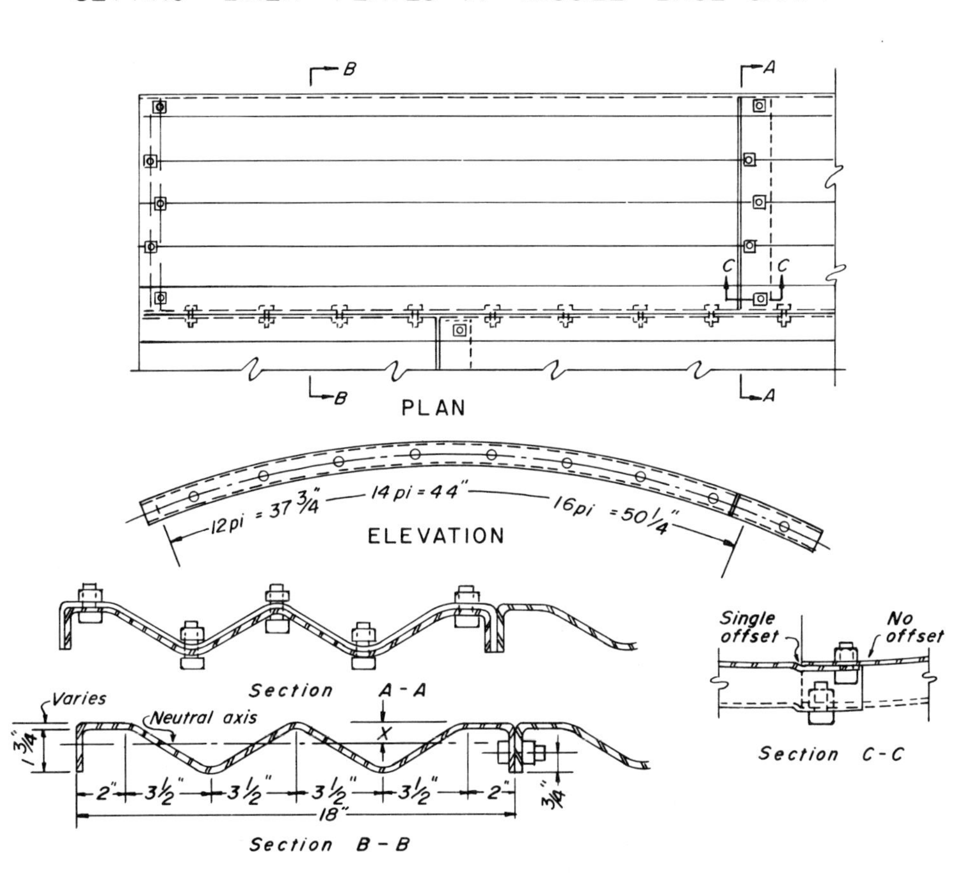

DETAILS ARMCO LINER PLATE

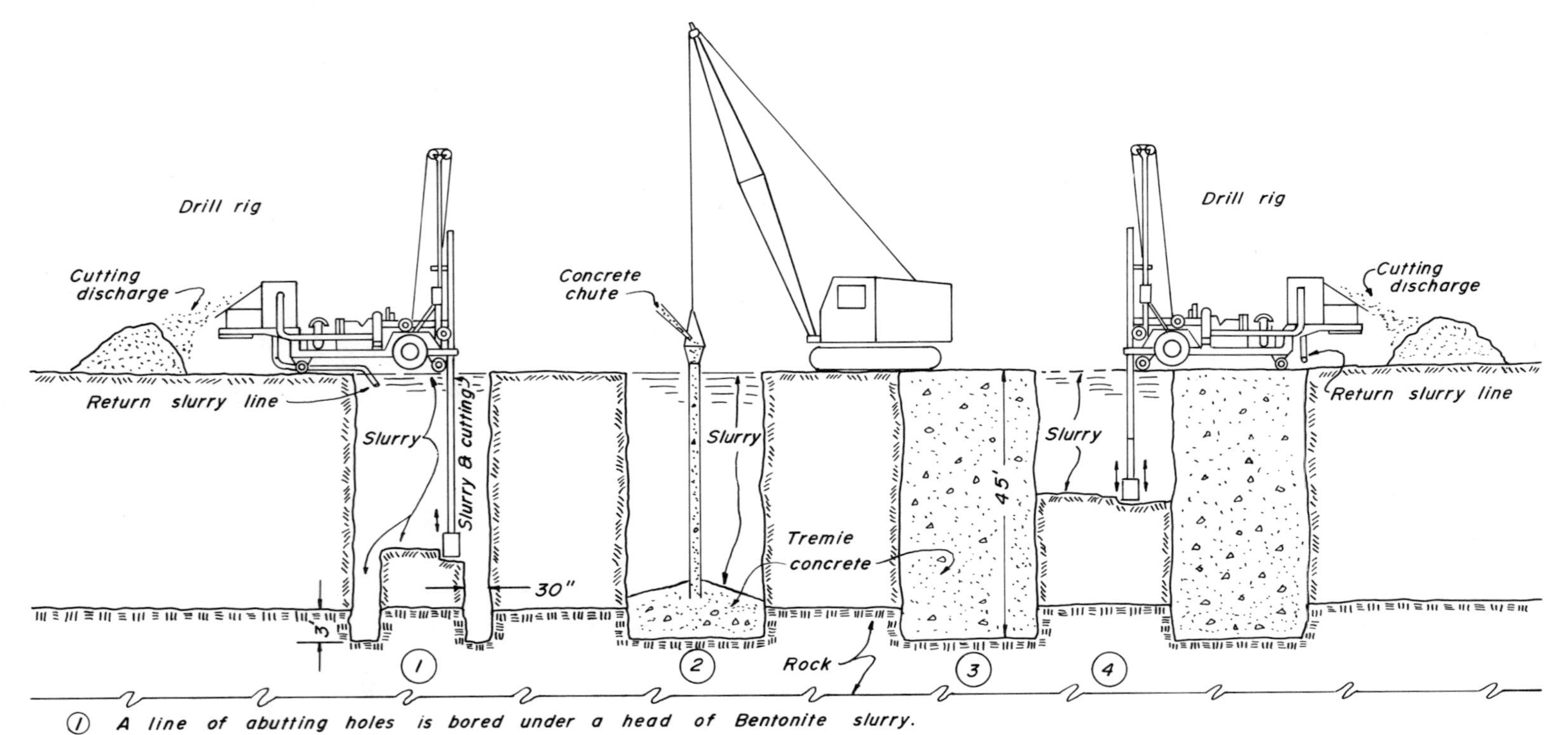

① A line of abutting holes is bored under a head of Bentonite slurry.

② The resulting slurry – filled trench, 7' to 12' long, is tremie concreted. (Reinforcing steel not shown)

③ Concrete wall displaces slurry which is collected and cleaned for re-use. (Collection plant not shown)

④ After alternate sections of wall are cast, Intervening sections are drilled and poured.

Note! All rigs are turned 90° from their normal working positions. They would work facing the trench and might be supported on planking spanning the trench. (Planking not shown)

FOUNDATIONS BUILT UNDER SLURRY

The Wanapum Dam, in the state of Washington, is an earth dam which was under construction in 1959. The specifications called for the cut-off wall beneath it to be constructed using a bentonite slurry. Also required was a 4000-ft-long cellular cofferdam enclosing the outlet works of the dam during construction. The contractor, Grant County Constructors, a joint venture sponsored by Morrison-Knudsen Company, conceived the idea (or had it sold to them by Cronese Inc., who controlled certain patents on the equipment involved) of applying the bentonite treatment to seal off the cofferdam.

Stratigraphically, the cofferdam site consisted of silt over gravel, over rock. Driving steel sheet piling through the boulder-filled gravel involved the possibility of twisting some of it out of interlock.

The specifications called for a 12-ft-wide trench as much as 80 ft deep for the cut-off wall; for his cellular cofferdam, the contractor used a 7-ft-wide trench of somewhat less depth, laid out along the center line.

The trenches were excavated by dragline to a depth somewhat below that of the water table in the area and were then filled with the bentonite slurry. The slurry came from three portable mixing plants each having a capacity of 150 gal of slurry per minute. Each plant had a mixing chamber, with attached hopper, into which the dry bentonite was poured from bags. Water was introduced into the mixing chamber through 16 small jet openings in its walls. The slurry leaving the plants had a carefully controlled maximum density of 65 lb per cu ft and was pumped to the trenches through an 8-in. pipeline.

Once the trench had been filled with slurry, dragline excavation proceeded within it, casting some of the native soil into a windrow away from the trench but mixing considerable quantities into the slurry. In this case, the slurry was cleaned by passing it over moving screens, instead of through a slush tank; and the desired trench density of 80 lb per cu ft was maintained by feeding slurry from the mixing plants into the returning discharge from the screens.

At rock level, an air probe was used to sound out the condition of the rock and to force adhering soil particles up into suspension in the slurry.

With the excavation complete, and the trench filled with a slurry of uniform density, backfill was begun, employing a bulldozer to move the excavation windrow, in increments, into the trench and using the air probes for agitating the mixture. Some of the slurry slopped over and was collected for local re-use, but the effect of the slow addition of soil was to create a dense impervious mass having a density of about 150 lb per cu ft. The cellular cofferdam was then placed to contain it.

Another use of bentonite slurry is illustrated. Here, a slurried trench was used to construct a foundation wall for a 21-story office building in Montreal, using equipment and procedures patented in Europe by Rodio-Marconi and resembling the procedures used at Wanapum Dam.

The site, here, consisted of clay over silty clay, over sandy till, over rock, with sufficient water to have required steel sheet piling had the walls been built in open excavation. The building foundation was a 30-in.-perimeter concrete wall 600 ft long and from 40 to 45 ft in depth.

Without prior excavation, 30-in.-diameter drill holes were put down in a continuous line, 7 to 12 ft long, under a slurry fill. The resulting mixture of slurry and cuttings was drawn up through the 8-in. telescopic hollow drill shaft by a vacuum-type pump and discharged first, over vibrating screens, and then passed through a centrifugal desander.

Excavation was carried 3 ft into the underlying rock, reinforcing steel mats were dropped into place in the slurry, and grout pipes were set for the contingency that later sealing might be needed. The sections of slurry-filled trench were then concreted by tremie, and the displaced slurry was salvaged for subsequent use. A specially designed head on the drill rig cleaned off the concrete edge of a section before the adjoining section was poured.

7.28 TOWER AND STACK FOUNDATIONS

Tower and stack foundations come in a wide variety of sizes, shapes, and functional performances having but one basic factor in common: because of the height of the structure above them, even the slightest differential settlement at bearing level can produce visible out-of-plumb tilts; and, should the direction of maximum wind loading or wind-induced oscillation occur in line with such tilts, damage or collapse can result.

The contractor, reading a set of plans and specifications and observing that tower or stack foundations are to rest on a bearing soil capable of supporting 4000 psf, and then examining the borings, which indicate varieties of rock at bearing level, confidently assumes that since these are all in excess of 4000 psf bearing capacity, these foundations need only be of limited depth. Later, chagrinned (and out of pocket), he recognizes that nothing could have been further from the truth.

Rock outcrops, near the surface, and particularly those types which weather deeply, can present considerable variation in density within a limited area. The tower designer wishes his foundations to rest on rock of a uniform density *at the same depth or elevation.* Very frequently this means carrying 50 to 75% of the footings down through quite solid rock so that one footing, or two, may go deep enough to get below soft or weathered pockets in the rock surface.

Towers, whether for elevated water-storage tanks, for television transmission, or for other similar purposes, are generally similar to piers except that frequently they have steeply sloping faces in at least one direction; and special forming, with an eye to possible flotation, may be necessary.

Stack foundations generally begin with a heavy mass of concrete placed on the subsoil, with offsets in the exterior faces until they reach flue level, where they convert to concrete walls faced with brick, or to solid brick walls, before climbing farther.

In the case of large tower footings or the base slab of stacks, it is frequently desirable to forget about bracing the form from the exterior and, using only exterior walers, to hold the forms in alignment by anchor rods tied not across the form but diagonally down into the underlying rock. Such anchors consist of steel rods ¾ or 1 in. in diameter drilled diagonally into the rock at a 45-deg slope, or less, and grouted into place. Generally two sections of rod are used; one passes through the form, between the walers and through a plate washer with slotted opening, and then is nutted over a tapered washer; the other is grouted into the rock. The two loose, threaded ends are then drawn up on a turnbuckle to adjust form alignment. The method is practical only where bearing is on solid rock.

Construction details of the foundations for two television towers built in the 1960s are illustrated. The first of these, a triangular tower built in Milwaukee, was the tallest *self-supported* tower when built, having an overall height, to its topping-out aviation light, of 1079 ft. Each of the three legs was set on a rectangular spread footing 28 ft square and 4.5 ft thick with a tapered pier rising above this. (Such piers should be poured monolithically, requiring that the top form be hung over the base form.) The legs were secured to the foundation through a circle of twelve 2½-in.-diameter anchor bolts. The precise templating of such bolts is no great problem, but the template seriously restricts the space available for placing and working the concrete, and slapdash methods of concreting cannot be used.

The second tower, also for television broadcasting, was built at Cape Girardeau, Missouri, and, although much higher—1676 ft—was a guyed tower, having 18 guy lines supporting it, six on each of the three sides in sets of three. Because the guys took care of most of the overturning potential, the three concrete foundations were somewhat smaller in size. Each of these footings consisted of two hexagonal thick slabs (see opposite), one above the other and with considerable reinforcing, permitting each of the segments to be poured separately. Also shown are the anchor blocks for the ground end of the guy wires.

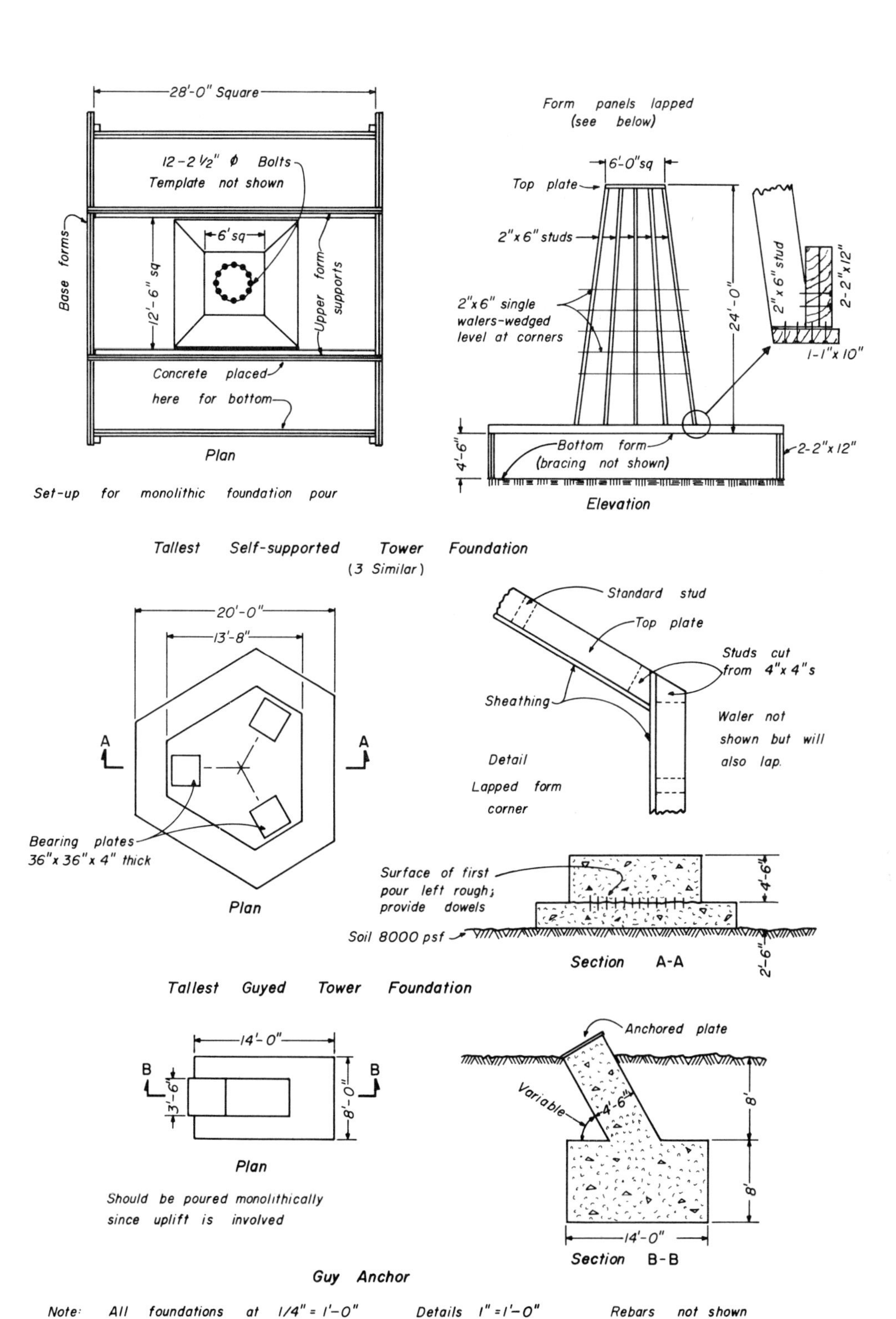

Note: All foundations at 1/4" = 1'-0" Details 1" = 1'-0" Rebars not shown

7.29 EQUIPMENT FOUNDATIONS AND VIBRATION CONTROL

The chief problem in the design of foundations for moving equipment is that of isolating the vibration induced by the machinery from the framework of the structure housing that equipment.

Light, high-speed equipment may be installed where independent slabs (see Sec. 7.8) rest on earth. Here the foundation, a single mass of concrete, is carried below floor level to solid bearing and the floor is poured up to a ½-in. pre-molded expansion filler surrounding the foundation. Larger foundations may be set on piles or caissons, carried to bearing at a greater depth, to prevent transmission of vibration through the soil. In one case, where solid clay existed below a foundation, ¾-in. steel rods were driven through the clay, the foundation being cast around the exposed tops of these rods. This furnished adequate support since, so long as the clay held the bars erect, the steel was loaded only in compression.

For heavy high-speed equipment, set on soil-supported bearing, a two-piece foundation is used. The bottom section is a concrete box with open top isolated from the floor by expansion joints. The interior of the box is lined with impregnated cork board, 2 to 3 in. thick, with the joints sealed. The bearing block is then poured on the cork, with its top generally extending above the lip of the bottom box.

For heavy low-speed equipment a more substantial two-box arrangement is used; but here the bearing, or inertia block, is carried on vibration isolation springs which may be set inside the box or on its upper edge. The basic box is isolated from the surrounding slab as noted above.

Where raft-type floor foundations are involved, it is not desirable to break the continuity of the floor, and equipment foundations must be set on the slab. In some cases the slab is thickened under the foundation, while in other cases dowels are set into the unthickened slab and a concrete foundation mass, poured later, on top. Regardless of size or type of vibration produced by the machinery, it must be isolated, by rubber pads or coiled springs, from the mat foundation.

Where moving equipment such as trains is involved, buildings subject to vibration from this source are themselves isolated. The new Pan-American building built over a portion of Grand Central Terminal in New York City has all its columns resting on lead pads to reduce vibration, and this technique was used 30 years ago when the Philadelphia Post Office Building was constructed.

The "new" Post Office Building, as it was called then, was built at 30th and Market Streets, Philadelphia, over the main railroad tracks of the Pennsylvania's Baltimore-Washington line. The building was constructed on Gow caissons carried 40 to 50 ft into sandstone. Although of varying diameters, shapes, and sizes, all were to be finished off on the surface to receive the lead pads. The construction superintendent, accustomed to using dry-pack to resolve minor variations at the surface (under base plates), poured the caissons rapidly and continuously, striking them off smoothly enough, but before the concrete had acquired any set. The concrete, in setting up, also shrank, forming a depression in the center of each caisson top.

It then became necessary for the contractor to grind off the major portion of the top of each caisson to provide a smooth base, level within 1/100 ft, as a bearing surface for the lead pad. Inevitably, since no filling was permitted, the result was to lower the finished pad grade and lengthen the steel columns above the pads correspondingly.

Details for forming simple equipment foundations are shown opposite. The top of the form is generally set to the future grade of the underside of the equipment base plate, and a chamfer strip is placed around the inside top of the form at this elevation.

Depending upon size, and hence the span across the form, timbers are then placed across the top and spiked, or toenailed, to its upper stud (now placed horizontally). Bolt centers are laid out and bolt holes drilled in the timbers.

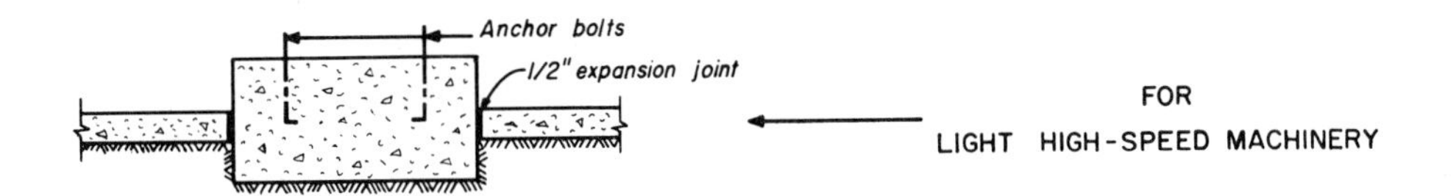

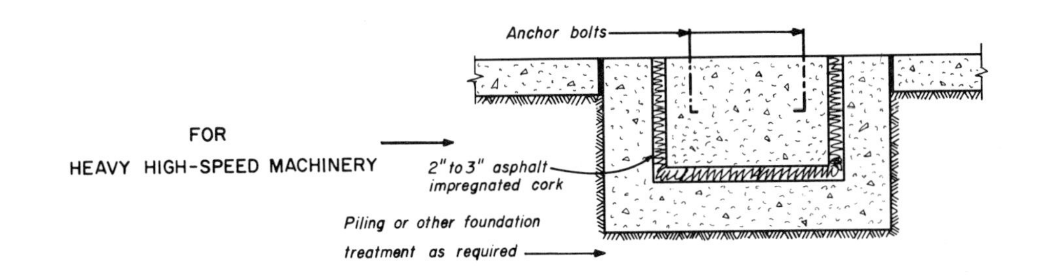

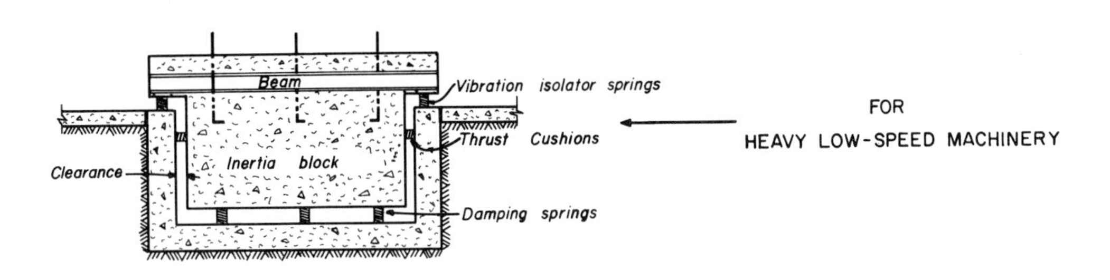

ISOLATED FOUNDATIONS

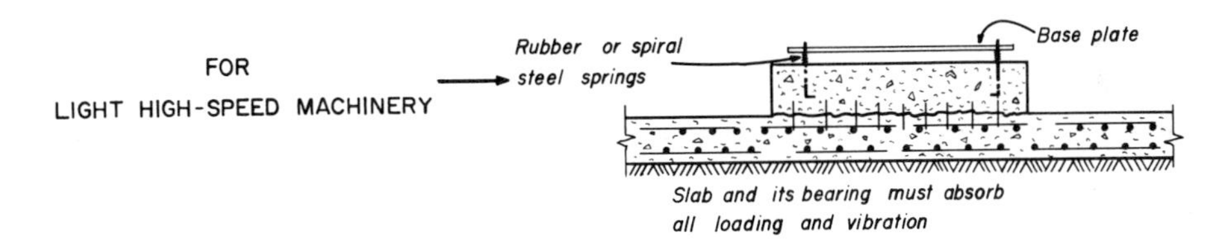

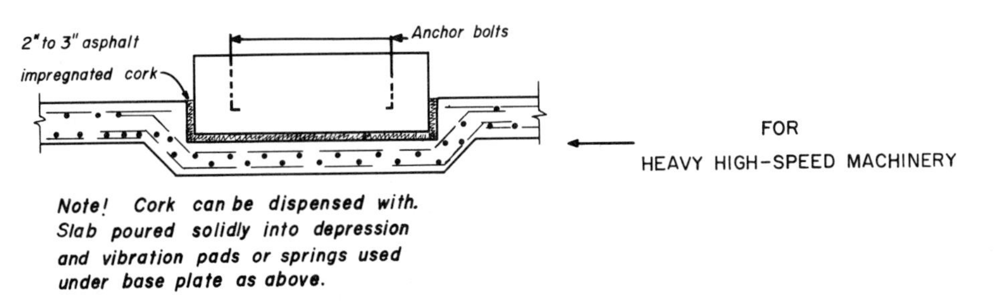

MACHINERY PADS ON RAFT FOUNDATIONS

7.30 A FORGING-PRESS FOUNDATION

Section 7.29 covered the entire range of equipment foundation possibilities in a very broad way, necessary since the possibilities are as numerous as the pieces of equipment available and are subject, as well, to wide variations in soil and site conditions. The general discussion is here carried further to a specific case involving a large pre-stressed concrete foundation for a forging press.

Following World War II, the United States Air Force set up a "Heavy Press Facilities" program to speed production of special pressed aluminum shapes for aircraft. One of the earliest presses for this program, a 14,000-ton Schloemann extrusion press, had been brought from Europe and set up in the Lafayette, Indiana, plant of the Aluminum Company of America with considerable publicity. We are here concerned with an 8000-ton forging press manufactured in the United States and set up in the Cleveland, Ohio, plant of the same company at a later date (1954).

The press was to be installed within an existing building, and the first foundation limitation was the space available between the spread footings of the structural columns, found to be 28 ft.

The subsoil consisted of layers of silt, silty clay, and sandy silt to a depth of 130 ft, where hardpan was to be found. Tests had indicated that the soil at foundation level would sustain a loading of 3400 psf, and a design load of 3200 psf was used. There was no water problem.

The press itself weighed 2300 tons, and were a solid mat footing to be used, the concrete necessary would add 3900 tons to the total weight. At 3200 psf this would require a bearing area 28 by 134 ft. Because the press loading would be concentrated within a small area at the center, such a mat size would provide poor distribution of the deflection which would occur at full load moment. It was decided therefore to use a pre-stressed concrete foundation, as shown opposite.

The basic strength was carried in two pre-stressed beams, 6 ft 9 in. by 9 ft 6 in., placed on either side of the foundation and connected by a 2-ft-thick slab. To reduce the dead weight of concrete, the bearing surface was sloped upward from the center area in two directions, lengthwise.

The area to be occupied by the foundation was first enclosed in MZ-38 steel sheet piling driven to the exact dimensions of the footing exterior. The concrete would be poured directly against the face of the sheathing; and the resulting ribs, reinforced, would act as vertical beams, in tension. Excavation was carried down to a sub-grade 3 in. below foundation grade, and a 3-in. working slab of concrete was placed to protect the bearing surface.

The first step was the construction of the two sill girders; and since jacking space for pre-stressing would be needed, these were cast 70 ft long, leaving an 8-ft pit at each end. Forms were set up and a two-way mat of reinforcing steel was supported from the working slab. The pre-stressing strands, pre-encased in flexible metal tubing, were now positioned accurately in the end forms and wired to metal supports placed at intervals on the working slab. The tubings, which stopped at the end forms, were caulked to prevent mortar from leaking into them. Additional reinforcing and end dowels were set before the top row of pre-stressing strands was wired in place.

The sill girders were poured in a continuous operation using a 4000-psi concrete with a 2½-in. slump. After 9 days, the concrete had acquired a strength of 3000 psi, end forms were stripped, and tensioning was begun. Tensioning was done in four passes, first loading every fourth bottom strand and every third top strand and starting from the centers of each row of strands to avoid setting up tensile stresses in the concrete faces of the sill girders. As can be seen, the top strands were straight, whereas the bottom strands were curved at two points. Since the curved strands required a higher gage load, these were stressed before the top strands.

After tensioning was complete, the 8-ft pits at the ends were concreted, and the remainder of the foundation concrete followed standard procedures.

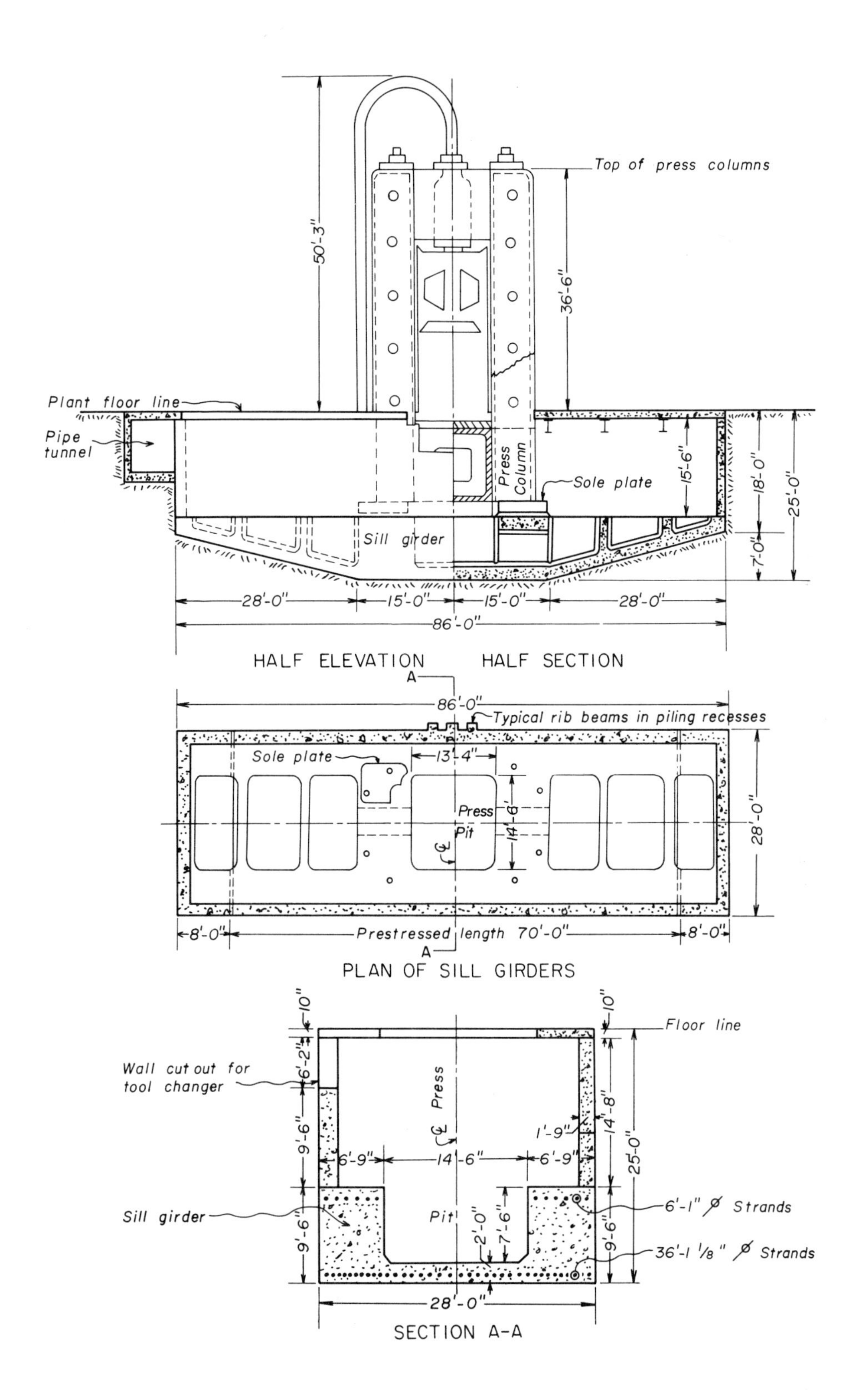
Top of press columns
50'-3"
36'-6"
Plant floor line
Pipe tunnel
Press Column
Sole plate
15'-6"
18'-0"
25'-0"
7'-0"
Sill girder
28'-0"
15'-0"
15'-0"
28'-0"
86'-0"
HALF ELEVATION
HALF SECTION
A
86'-0"
Typical rib beams in piling recesses
Sole plate
13'-4"
Press
Pit
14'-6"
28'-0"
8'-0"
Prestressed length 70'-0"
8'-0"
A
PLAN OF SILL GIRDERS
10"
10"
Floor line
Wall cut out for tool changer
6'-2"
Press
9'-6"
1'-9"
14'-8"
25'-0"
6'-9"
14'-6"
6'-9"
Sill girder
Pit
2'-0"
7'-6"
9'-6"
9'-6"
6'-1" ⌀ Strands
36'-1 1/8" ⌀ Strands
28'-0"
SECTION A-A

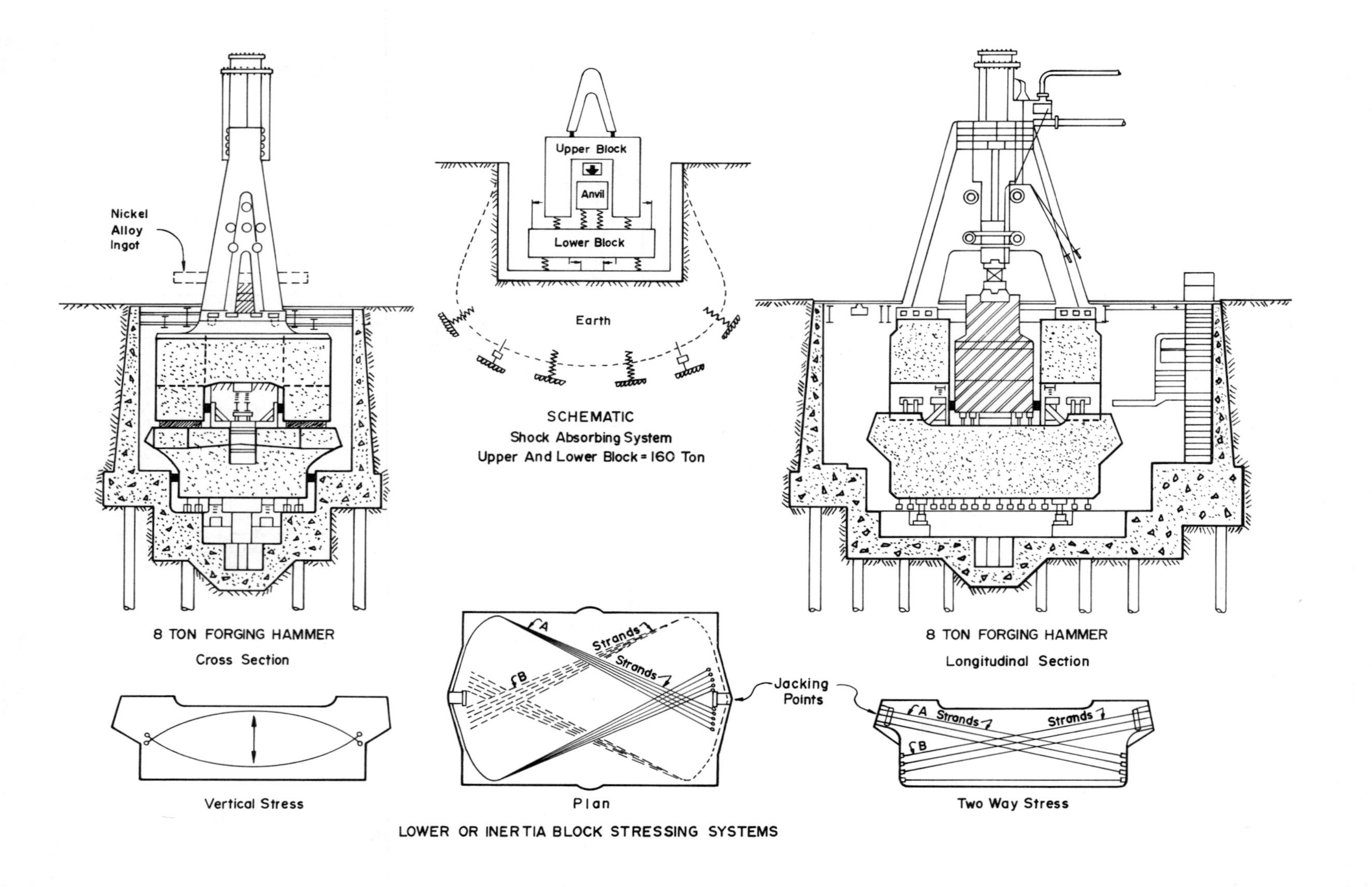
Nickel Alloy Ingot
Upper Block
Anvil
Lower Block
Earth
SCHEMATIC
Shock Absorbing System
Upper And Lower Block = 160 Ton
8 TON FORGING HAMMER
Cross Section
8 TON FORGING HAMMER
Longitudinal Section
A
B
Strands
Strands
Jacking Points
A
B
Strands
Strands
Vertical Stress
Plan
Two Way Stress
LOWER OR INERTIA BLOCK STRESSING SYSTEMS

7.31 A FORGING-HAMMER FOUNDATION

In Sec. 7.30, a forging-*press* foundation was discussed, where heavy loading rather than vibration was the principal problem. Now let us look at the foundation for a forging *hammer,* pounding out 4 × 11 in. blooms from ingots of hard nickel alloys at the rate of 45 blows per minute. This 8-ton hammer, installed in the plant of the International Nickel Company in Huntingdon, West Virginia, is to be operated by superheated steam at high pressures and sets up oscillations in the hammer, the anvil, and the foundations that must be dampened between blows.

At each hammer blow a shock wave travels through the foundation system, causing each part to vibrate at its own natural frequency. Should any two parts of this system have approximately the same natural frequency, they will be in resonance and can build up a destructive amplitude. (This is also true of the soil on which the foundation rests. Soft clay will oscillate at about 10 cps, while compacted dry sand has a frequency of some 30 cps.) To absorb the shocks developed in this instance, it was necessary to set both the anvil and the upper block on independent rubber pads resting on the lower block and to support the lower block, in turn, on helical springs thrust against the bottom and sides of the foundation pit.

As in the case of the forging-press foundation, design was influenced by site conditions. In this case an existing concrete-lined pit was to be used and the lower block was to be of pre-stressed concrete, once again limiting the jacking space to the ends. However, in this case, to control the distribution of shock waves it was desired to pre-stress the block in three directions; and, as shown in the detail, an elaborate system of pre-stressing strands, jackable from the ends, was required.

The first step in construction was to remove the pit bottom, excavate, and cut off the tops of the existing piles. A new concrete bottom was then poured, at a lower level, including a concrete frame to resist the inward thrust of the earth and the outward thrust of the side springs of the lower block. Forms were now set for the lower block.

It had been decided to use Intrusion-Prepakt concrete in this block because of the extensive number of inserts, cables, and reinforcing bars involved, as well as because the base plates were to be poured in place and would cover 25% of the surface area.

The lower mat of reinforcing was first set, supported on vertical bars, and then pre-stressing cones and helical spring housings were secured to the form surfaces. The bottom layer of pre-stressing cables, in their tubes, was now placed and tack-welded to the vertical bars supporting the reinforcing steel. At this point, a 30-in.-thick layer of coarse aggregate was spread within the form.

With the bottom layer of aggregate in place, the remainder of the cables, reinforcing steel, and wall-surface inserts were set and the level of coarse aggregate was carried up to within 12 in. of the top of the form. Supports for the base plates, bolts, and other surface insertions were now installed and the remainder of the aggregate placed.

Three-quarter-inch grout pipes, rising above the form and terminating at various depths in the aggregate, had been set and tacked to the reinforcing; now it was necessary to provide a pressure top for the form before starting the grout intrusion. The surface of the aggregate was first covered with muslin and then with a layer of fine wire mesh, both held down by a third layer of expanded metal lath. These layers, designed to permit the escape of air from the form but contain the grout, were held in place by wood battens, with half-inch spaces between them, nailed to the top of the form and weighted at their centers to prevent upward deflection.

The intrusion grout had been designed to provide a 4000-psi concrete in 7 days; and, to provide temperature control, 5½ tons of ice was spread over the surface of the pressure mat the day before grouting and was melted down by sprinkling to bring the aggregate temperature to 50°F. Cooling water was circulated through drain lines, incorporated in the block, during grouting.

7.32 FOUNDATIONS IN PERMAFROST

It has been noted in Sec. 3.14 that the principal problem in foundation construction in permafrost is that of maintaining the thermal regime or, in other words, that of preventing the permafrost from thawing. As illustrated in Sec. 3.14, permafrost is covered by an active zone which freezes in winter and thaws in summer and is generally of the same soil material as the permafrost itself.

The factors controlling the depth of active zone will be expressed by an air-thawing index and an air-freezing index, in degree-days, and by a surface-thawing index, also in degree-days. The effect of the surface-thawing index may be naturally conditioned by a heavy layer of tundra over the soil. Further conditioning the result will be the classification, the density, and the moisture content of the soil.

The effect of placing a structure near the ground surface (stripping off the tundra, perhaps) in permafrost areas will be to raise the thawing index because of the heat transmitted through the floor of the building.

The simplest protection that can be provided to resist subsurface thawing is the use of a non-frost-susceptible material, such as a free-draining gravel, placed under the building. Such a gravel pad is placed under all structures to be constructed over permafrost unless piling is used or a satisfactory rock surface is available. The thickness of the gravel pad will be the thickness of the active zone, and the pad will extend 10 to 15 ft beyond the foundation lines.

The simplest type of construction, suitable for one-story buildings 30 to 40 ft wide and up to 200 ft long, is a crib-type foundation constructed on sills placed in the gravel (see opposite). This type of construction is suitable where low temperatures can be maintained in the building and where some differential settlement is not objectionable. The longest dimension is placed parallel to the wind direction so that snow accumulations build up on the short, lee side.

An alternative procedure for the smaller structure, particularly suitable where snowfall is light, but where higher interior temperatures must be maintained and where differential settlement is not desirable, is the pan-type foundation. In this instance a 4-in. concrete slab is laid down over the gravel, and metal floor pans are positioned on this slab. These pans provide ducts 12 in. high on 32-in. centers; and over them, another 4 in. of concrete is poured. Cellular glass insulation is now laid down over the pan slab; and another slab, 6 in. thick and reinforced, tops the insulation (carried on wood divider strips placed between mats of insulation). Since drifting snow can clog the ducts and prevent passage of air, the ducts are turned at right angles to the prevailing wind direction.

The most satisfactory type of foundation for larger structures or for extreme temperature and snowfall conditions is the fully ventilated pad type of construction. In this case a series of 12-in. corrugated metal pipes are set in the gravel under the building, connected at either end to plenum chambers set outside the building and extending the full length of the structure on either side. The pipes are placed parallel to the wind direction, and stacks are provided at intervals in both plenum chambers. On the gravel, over the 12-in. pipe, a sandwich slab of 4 in. of concrete, insulation, and a 6-in. reinforced slab are laid down.

The stacks on the windward side of a fully ventilated building are carried up to a height which will prevent snow from drifting into them and are frequently provided with revolving vent cowls so that they move their open face into the wind as it changes direction. On the leeward side, stacks are carried somewhat higher, at least to eave height, and above any possible snow accumulations. This system of ducts permits cold air to flow through and maintain back freeze during the cold season, and dampers are provided on all stacks to shut off air flow during the warm season.

It is important that no moisture enter any of the floor ventilating systems shown, and this necessitates not only control of surface drainage but the proper construction and insulation of drain and supply lines where they enter the building.

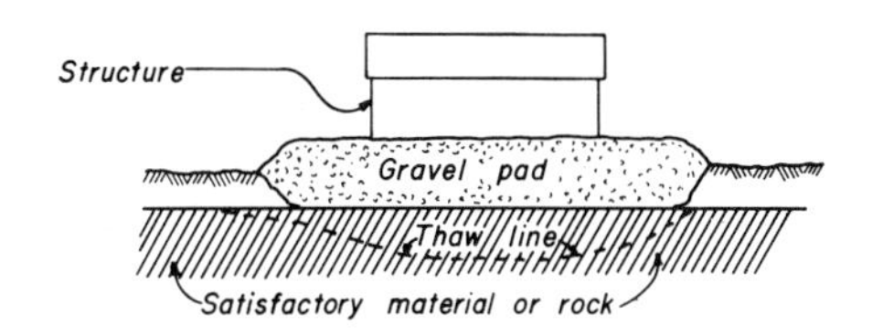

Thaw line below pad

Gravel pad
Thaw line
Unsatisfactory material

Thaw line within pad

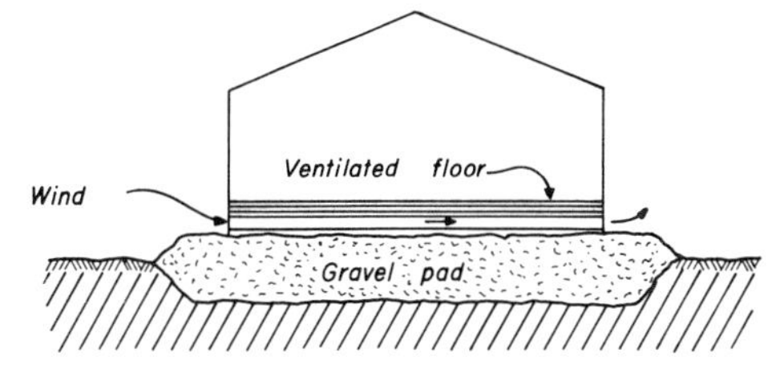

Floor slab
Insulation
Gravel
6"
4"
4"
12"
4"
32"
32"
20"

Pan type foundation

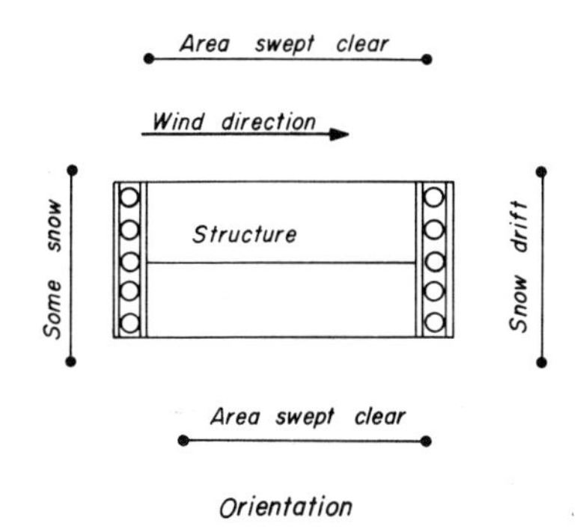

Orientation

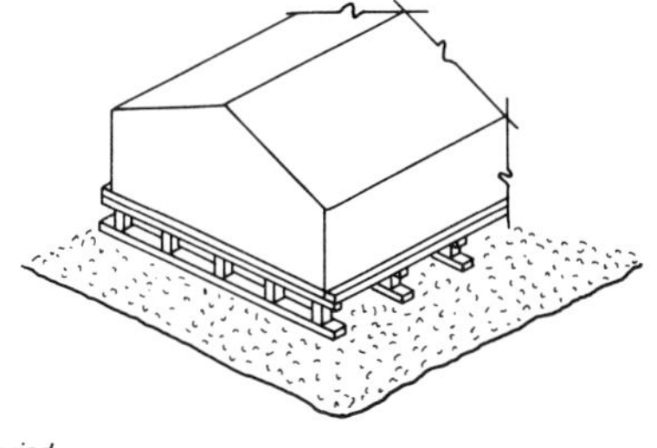

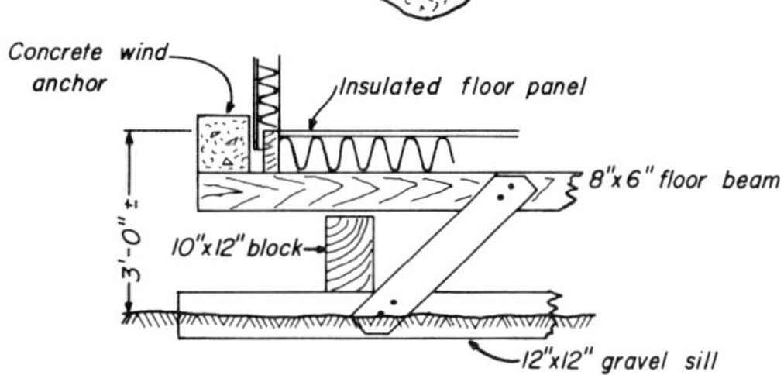

Crib type foundation

Wind direction
Under floor ventilation

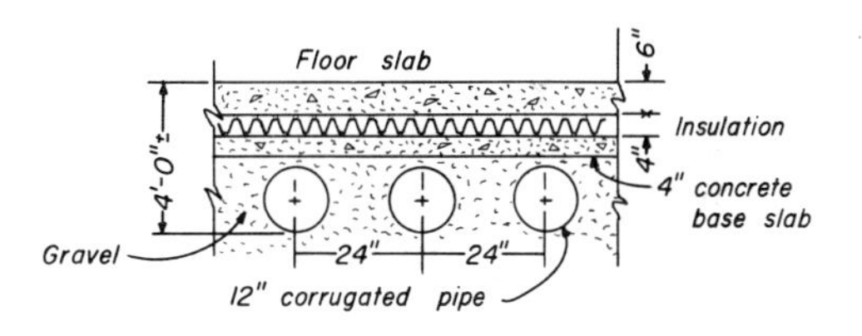

Ventilated pad foundation

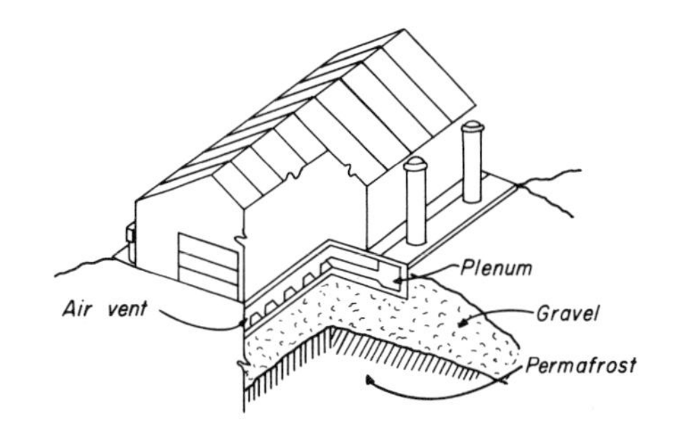

Typical installation

7.33 THE THAW-BLAST METHOD FOR PERMAFROST

The whole approach to construction over permafrost in Sec. 7.32 was that of careful coddling of the thermal regime, but suppose we say, "To hell with the thermal regime, let's change it!" This was done for the construction of foundations for a steam power plant at the Ladd Air Force Base in Alaska, to permit the use of a standard spread footing covering a 25,000-sq-ft area. The procedure, involving $50,000 worth of experimental study prior to starting and another $50,000 cost for preparing the site prior to placing the footings, was expensive, but it was done.

Experimental probing of the site revealed that permafrost lay in discontinuous masses starting at a depth from 10 to 17 ft below ground surface and extending to a depth of 60 ft. The soil in the permafrost zone was a stratified sandy gravel, river deposited, with a 25% porosity and a 6% inclusion of silt. Over this, in the active zone, was a 12-ft-thick layer of silts and sandy gravel. There were no appreciable inclusions of organic material.

A thaw test revealed that a settlement of about 0.8 in. developed upon thawing to a depth of 19 ft and thaws to a depth of 30 ft produced no further settlement.

Blasting tests, following upon the thaw tests, revealed that a combination of thawing and blasting would produce a settlement averaging 3 in.

On the basis of these tests, the structural foundation was designed for a 5000-psf bearing value and a ½-in. permissible differential settlement. It was further decided to carry thawing and blasting operations to a depth of 30 ft only.

For control and information, three temperature wells were spotted throughout the site. Each well consisted of a 6-in. pipe drilled in down to a depth of 36 ft with a 1-in. thermocouple pipe installed, the annular space packed with sand, and the 6-in. pipe casing extracted. In addition, 51 reference points were spread throughout the site. These points, consisting of a 2-in. pipe driven to a depth of 11 ft, with a 9-in.-square steel plate laid on ground surface beside it, would be used to determine the amount of settlement produced by the operation.

The first step in preparing the site was to thaw the permafrost. For thawing, a ¾-in. pipe with a chisel point, having two ¼-in. orifices set just above the point, was steam-jetted to a depth of 24 ft. These pipes were spaced at 7-ft intervals, in two directions, to encompass an area of roughly 30,000 sq ft. Once jetted to full depth, the ¾-in. pipes were fed by a steam header system supported off ground surface, with take-offs similar to those of a well-point system. Two boilers, each with a capacity of 220 hp, supplied steam at 60 psi pressure which entered the thaw pipes at pressures ranging from 12 to 35 psi. After 19 days of continuous operation, the permafrost had thawed to a depth of 30 ft below the surface.

After thawing had been completed, blasting holes were set up. For blasting, 6-in. steel pipes with drive points were driven 22 ft into the thawed ground on a pattern of 25-ft centers. After driving, the 6-in. drive pipe was withdrawn and a 5-in. pipe with a ⅜-in. plate tack-welded to its bottom was dropped into the resulting hole. Once in position (which occasionally required additional driving), the closure plate was broken free from the 5-in. pipe, the pipe was loaded for blasting, the annular space was rammed with sand, and the 5-in. casing, in its turn, was pulled out.

Blasting was done in three rounds of separate patterns. The first round consisted of 48 holes on 25-ft centers, each loaded with 20 lb of 60% gelatin dynamite, fired with primacord. The second round consisted of the same number of holes as in the first round, but each of these was offset 7 ft from those in the first round and was loaded with only 15 lb of dynamite. The third round was the same as the second.

On completion of the blasting, bearing was tested by means of a 3-ft-square test plate set at ground surface and loaded to 12,000 psf. Subsidence of the plate was 0.55 in. and, when the load was removed, rebound was 0.15 in. The required bearing value of 5000 psf was therefore considered to have been obtained.

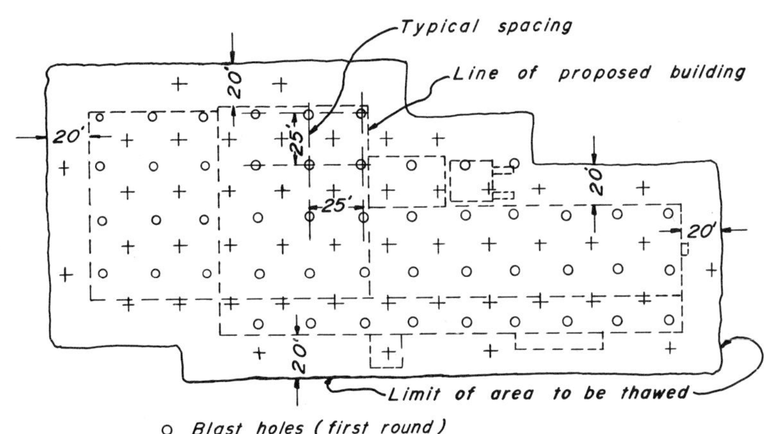

POWER PLANT SITE IN PERMAFROST

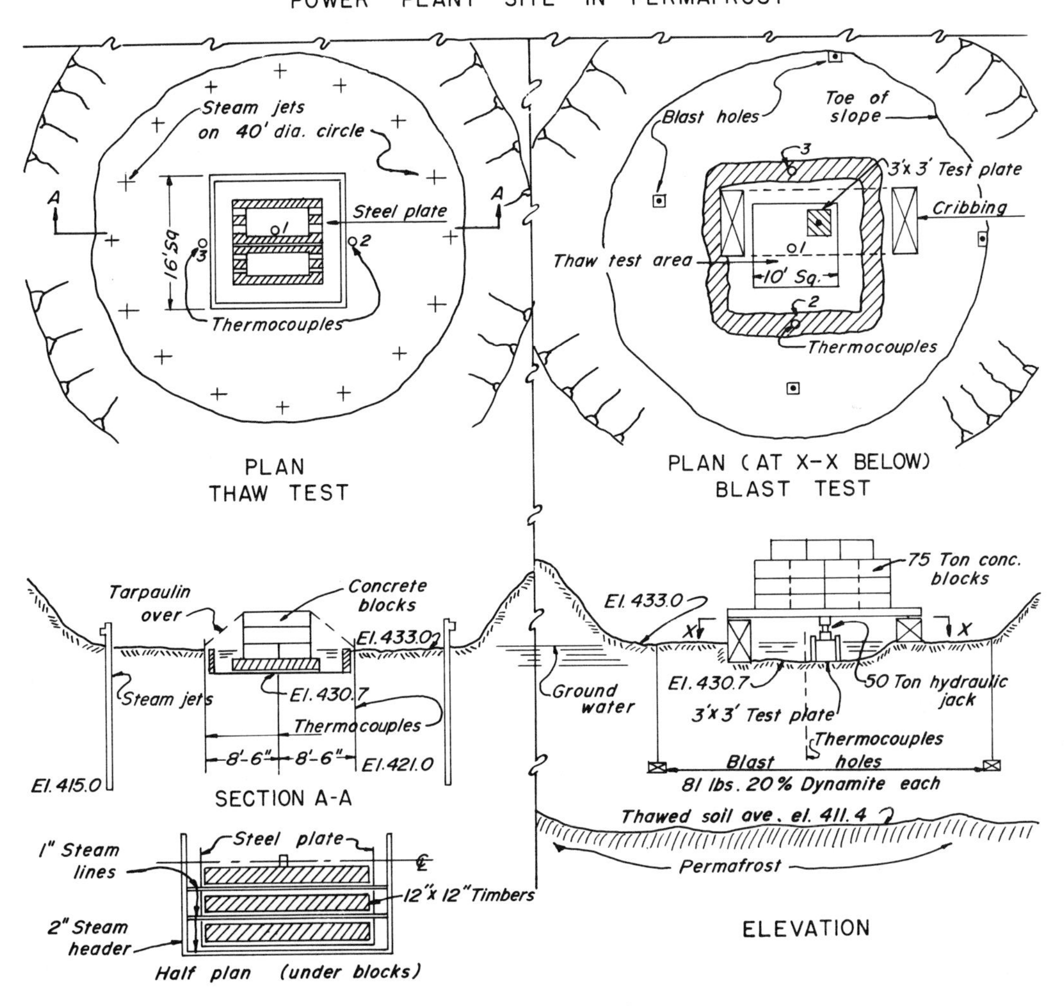

TEST PROCEDURES

7.34 STANDARD PIER CONSTRUCTIONS

To use the word "standard" in discussing piers is to introduce a misnomer since pier construction is the least susceptible to standardization of any structural item. It might be more suitable simply to say "customary" pier construction methods, yet this too is misleading. At any rate, the intention here is to describe two specific examples of pier construction, employing an assembly of techniques, each of which has previously been detailed, before taking up some special and unusual pier constructions.

Late in 1955, construction of pier No. 2 for the Morrison Street Bridge, crossing the Willamette River at Portland, Orgeon, was started. This was a bascule pier founded on a sand-and-gravel stratum at elevation −75.00 but supported on piling driven 44 ft deeper. For this construction a rectangular, steel-sheet-pile cofferdam was used.

In this case, the internal bracing system of 24-in. WF walers, with timber posts and struts, was to serve also as the template for the cofferdam. Accordingly, the two lower rings were assembled on a barge and floated into position. At the pier site, the barge was anchored, steel guide or spud piles were driven through pockets provided in the corners of the frame, and the frame was then supported by angles temporarily fastened to the spuds. The barge, somewhat smaller than the area of the frame, was then further submerged and floated out from under. The two upper framing rings were built on top of the frame previously placed, and the entire assembly was lowered, in increments, until it rested on the river bottom.

With the frame in position, Z-32 steel sheet piles 95 ft long were driven around the perimeter to a bottom elevation of −80.00. This left a 12-ft projection of the cofferdam above the mean low water elevation of +3.2. The interior of the cofferdam was excavated under water by clamshell to elevation −75.00. By using an underwater pile hammer, 154 fourteen-inch steel H-piles were driven to a point elevation of −119.00.

A 26-ft-thick tremie slab was cast over the projecting tops of the piles and allowed to sit for 7 days, and the interior of the cofferdam was then pumped out. Four electric pumps were used in the initial dewatering, but one 10-in. pump throwing 2000 gpm was able to keep the 50-ft submerged depth of sheathing dry.

A concrete slab 10 ft thick was now poured in the dry, and shaft construction was carried to elevation +25.00 well above high-water level. At this point pumping was terminated, and the sheathing was extracted.

Four years previously a Pennsylvania state highway bridge had been constructed across the Monongahela River 25 miles below Pittsburgh. This bridge was built on four piers, two of them in shallow water being constructed in rectangular cofferdams and two center piers in deeper water being constructed in cellular cofferdams. The two deep piers were carried to rock 50 ft below pool level, and their footings were carried down 25 ft into the rock stratum.

Each site was pre-dredged and a floating wooden template, held in place by four 53# H-beam spuds, was used to position six 52-ft-diameter circular cells. The cells and closure area consisted of M-112 steel sheet piling of sufficient length to provide a 10-ft projection above pool level. This height was somewhat less than potential maximum flood, and one closure cell was dropped 3 ft below the level of the main cells to permit flooding of the cofferdam before it could be overtopped. The cells, after filling with sand and gravel, were capped with a 6-in. concrete slab.

After drying up the cofferdam, excavation was carried down 25 ft into the rock by means of controlled blasting, maintaining a minimum distance of 6 ft inside the cell walls to the excavation face. At a depth of 13 ft an extensive flow of water through the rock was encountered, and excavation was temporarily halted while the surrounding rock was pressure-grouted.

These piers were relatively long and narrow, and only a pedestal at each end was carried down to the 25-ft level. Over these, at a height 20 ft above rock surface, an 8-ft-thick slab was placed; and on top of this the stone facing for the pier began.

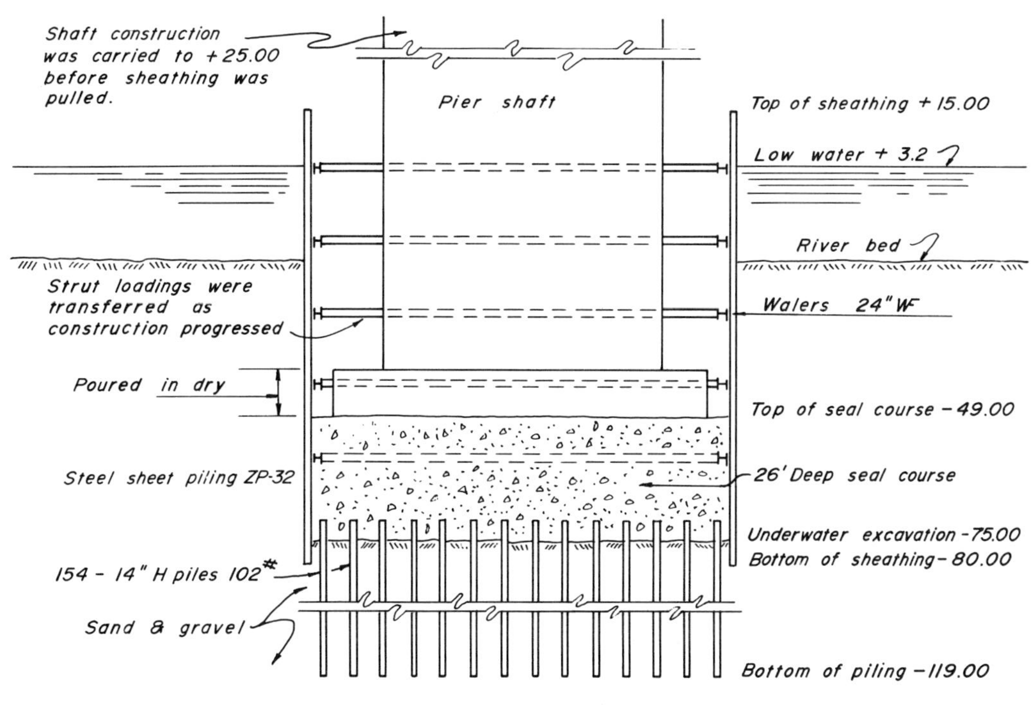

CONSTRUCTION OF PIER No. 2
MORRISON BRIDGE — WILLAMETTE RIVER
PORTLAND — OREGON

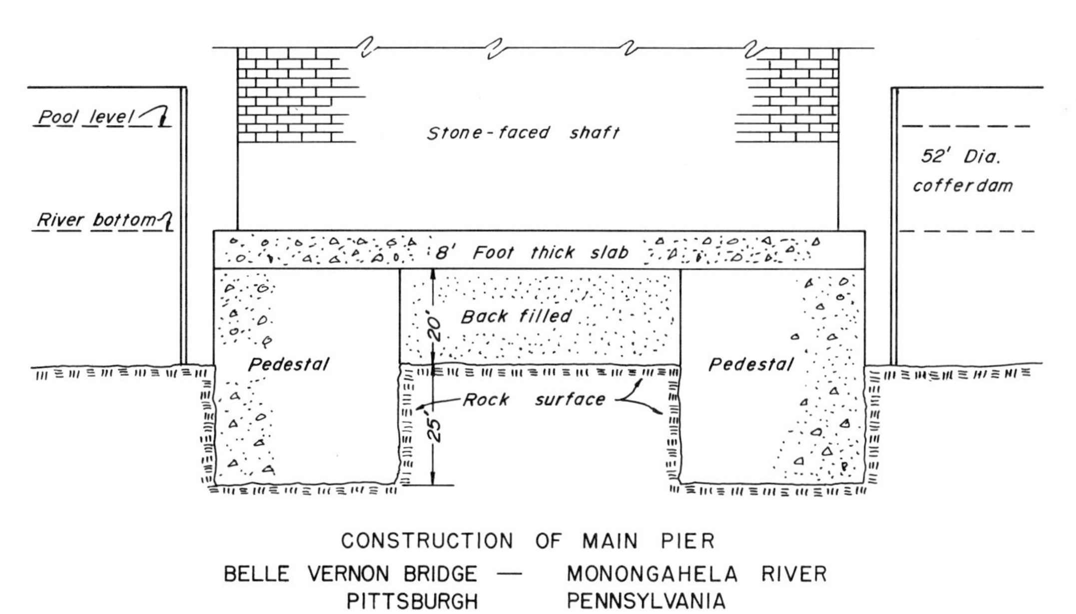

CONSTRUCTION OF MAIN PIER
BELLE VERNON BRIDGE — MONONGAHELA RIVER
PITTSBURGH PENNSYLVANIA

7.35 EXTRUDED CONCRETE PIERS

The term "extruded concrete" to describe piers concreted in moving forms was first suggested by C. M. Davis, consulting engineer of Fort Worth, Texas, in an article in *Civil Engineering* for March of 1957. Mr. Davis had long been engaged in the design and construction of concrete silos and grain elevators, where the use of slip forms had become commonplace. There was, it appeared, an essential difference between slip forms used in the construction of relatively thin walls and the application of the same process to a mass of concrete, such as that encountered in pier construction; or so he was told when he first suggested the method in 1943, during the wartime shortage of form lumber.

Mr. Davis's contention was that initial set was not the controlling factor at all, but that the initial ability of the concrete to support itself derived from the cohesion of aggregate particles after all excess water had been squeezed out of them. In the slip forming of concrete, higher than normal slumps are needed, first to facilitate placing and secondly to provide lubrication between the form and the concrete surface. Unless this water is adequately squeezed out and removed, the concrete will indeed "fall out" from below the form.

The illustration opposite, adapted from that shown by Mr. Davis, indicates that vibration forces water to the surface continuously as the form rises and that, furthermore, additional water escapes down the face of the concrete, from beneath the form, a condition which does not exist with "tight" forms. Although the sides are here shown tapered, to accentuate the squeezing action, the same thing occurs on vertically faced walls. It is true that the top layer contains a detrimental quantity of water, but this can be removed from the surface.

Whatever the theory may portend, the method was used successfully (under Mr. Davis's direction), in 1943, on piers 285 ft high for the Pecos River Bridge; in 1947, for the 240-ft piers of the Cumberland River Bridge near Burnside, Kentucky; again in 1951, for 100-ft-high piers on the Whitney Lake Bridge in Texas. These piers had vertical faces and included setbacks at two or three locations in each case.

The method was used in 1957 on the tall piers of the Carquinez Toll Bridge in California, averaging 70 ft high and varying in cross section from solid shafts, 6 by 22 ft, to boxed or hollow shafts, 20 by 76 ft. The piers were designed without batters or setbacks to permit the use of this process.

A detail of the setup for slip forming is shown opposite, here applied to a section of boxed or walled shaft. In this instance, the more customary screw jacks were replaced by hydraulic jacks similar in operation to those used in lift slab construction, manipulated through a console so that relative positioning could be minutely controlled.

The jacks ride up on 1-in. threaded high-tensile-strength bolts, set on 7-ft centers and passing through a 3-in. pipe sleeve, to permit their ultimate removal. The jack rods are first set up on the base or footing, the cage of reinforcing steel is set around them, and the 4-ft-high form is then placed. Concrete is poured in the form in 8-in. layers until it has been filled; and after 3 hr has elapsed, jacking is begun.

The form is "slipped" up at a rate of from 5 to 14 in. per hr, depending upon ambient air temperature and prevailing wind velocity. As the concrete face is exposed, concrete finishers are set to work on it at ground level; and when the form has reached a suitable height, a scaffold is hung from the underside of the form for the use of the finishers.

For curing purposes, a horizontal waterline, fitted with fog nozzles, is hung below the finishers' platform and keeps the face of the concrete continuously moist.

In this case the rate of pour averaged 20 vertical feet per 24-hr day. Where piers were over 100 ft in height, requiring more than a 5-day week, the operation was shut down over the weekend to avoid overtime labor rates, except for an operator who kept the form moving upward in tiny increments until the bottom of it stood 12 in. below the surface of pour.

EXAGGERATED DIAGRAM

(After C.M. Davis)

Showing reduction in water content due to squeezing action during slip-form concreting.

Not shown here is the additional squeezing action due to the head of concrete increasing as the forms rise.

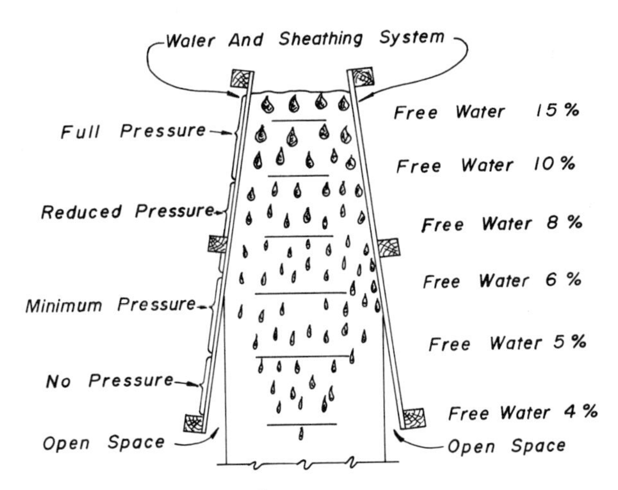

SLIP FORM SCHEMATIC

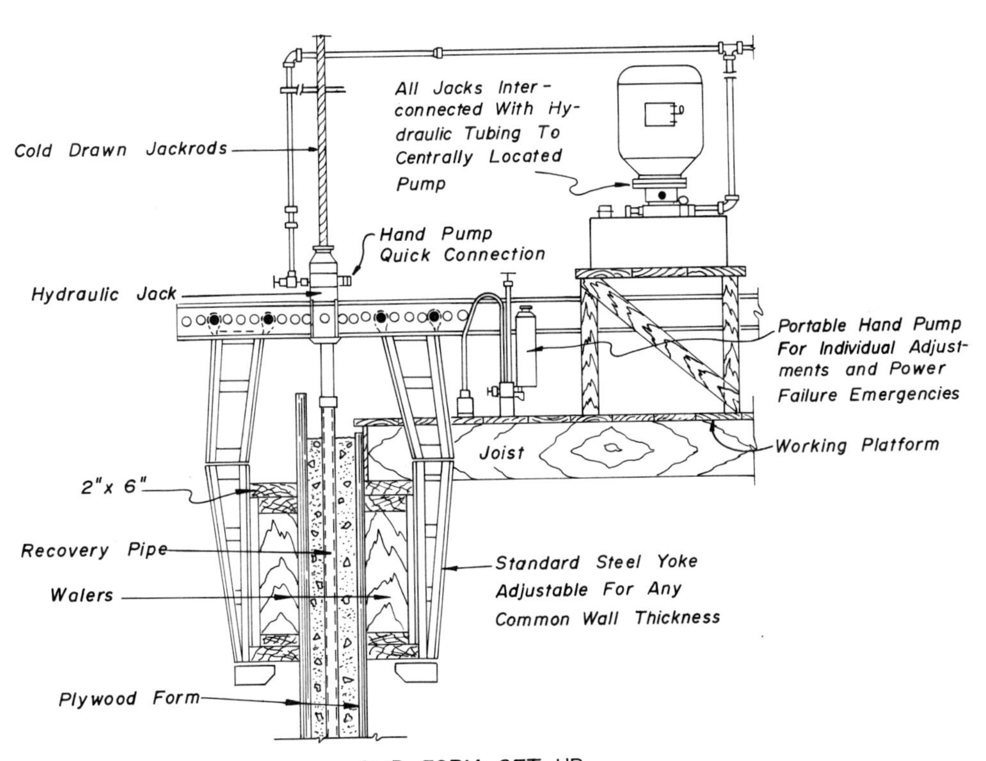

SLIP FORM SET-UP

Jacking Equipment Developed By B. M. Heade Inc.

Contractor-Peter Kiewit Sons Company

7.36 TRANSMISSION TOWER PIERS BUILT UNDER WATER

In 1952, a power transmission line was carried across the southern end of San Francisco Bay from Hayward on the east shore to San Mateo on the west side, involving 19 major towers sitting in the environs of 2 miles of salt evaporation ponds and 8 miles of bay waterway. Two of the tower foundations were set in 60 ft of water adjacent to the ship channel, where the tidal current runs as high as 7 knots per hr and the tidal range is normally 8 ft, but may run as high as 10 ft. The remaining foundations were in shallow water overlying deep deposits of mud.

The towers were spaced about 2000 ft apart, and preliminary work included the pre-dredging of all sites and the construction of a wood platform set on eight temporary wood piles to control the positioning of each permanent foundation.

The shallow-water piers started with a cluster of five 18-in.-square pre-cast concrete piles, one driven vertically in the center and the remaining four driven on opposing batters of 2 on 12, around it. Here the mud flats lay at an average elevation of −1.00 and the site had been pre-dredged to about −13.00 to −15.00. After the piling had been driven, a sand blanket 6 to 8 ft thick was placed around the cluster and topped with 6 in. of crushed stone.

The first pre-cast cylinder, 12 ft in diameter and 11 ft long, was set on end on the sand-and-gravel mat and centered on the pile cluster, using a three-point bridled frame for pickup. After the cylinder had been set, leveled, and plumbed, the interior of the shell was filled with tremied concrete to elevation −4.00, 4 ft below its top rim.

At mean tide and below, the interior of the cylinder could be dewatered and the tops of the concrete piles chipped away to expose their reinforcing steel. When this had been done, a second cylinder, only 8 ft in diameter and 12 ft long, was set upright on the tremie surface and centered in the lower cylinder. The 4-ft-deep 18-in.-wide annular space between cylinders was sealed by concreting at low tide.

The shell of the top 8-ft cylinder had been notched at the upper end, in casting, to receive the pre-cast beams. These beams were now set with end dowels projecting into the cylinder, templated anchor bolts were secured in position, and concrete was poured to encompass the pile and beam dowels and fill the cylinder to the top.

The two deep foundations, beside the ship channel, also consisted of four piers set to form a square 40 ft long on a side. The bay bottom, here, had 15 to 20 ft of soft clay overlying a firm stratum, and pre-dredging had been carried down to this stratum.

The support for these piers was a cluster of 23 wood piles driven with an underwater hammer, in submarine leads, hung from a floating barge. After driving, the piles were cut off 5 ft above the level of the pier base by a diver using an air saw. A sand mat was placed around the cluster up to the elevation of the base of the pier.

The lower shell of these piers was 19 ft in diameter, was 16 ft long, and had an 8-in. wall thickness. The cylinder was set in position with a light steel tower temporarily fastened to its upper end. This 60-ft-high tower, projecting above water level, was used to place the cylinder accurately.

In addition to wall reinforcing in the lower cylinder, a welded cage of reinforcing steel was pre-set in the interior area. To this cage of steel a dished metal ring was welded, placed 4 ft below the top of the cylinder, to serve as a seat for the upper shaft. The lower cylinder was filled with tremied concrete to the underside of this metal ring.

The shafts, 62 ft long, were octagonal on the exterior and circular on the interior, with a minimum wall thickness of 6 in. and a maximum exterior diameter of 7 ft. After being positioned and plumbed on the metal seat, the annular space between shells was sealed by tremied concrete, as was the lower 20 ft of the shaft, after another cage of reinforcing steel had been dropped into it. At this point the shaft could be pumped out and the remaining height to elevation +5.00 could be poured in the dry.

The foundation was completed by the setting of the 46-ft-long pre-cast tie beams.

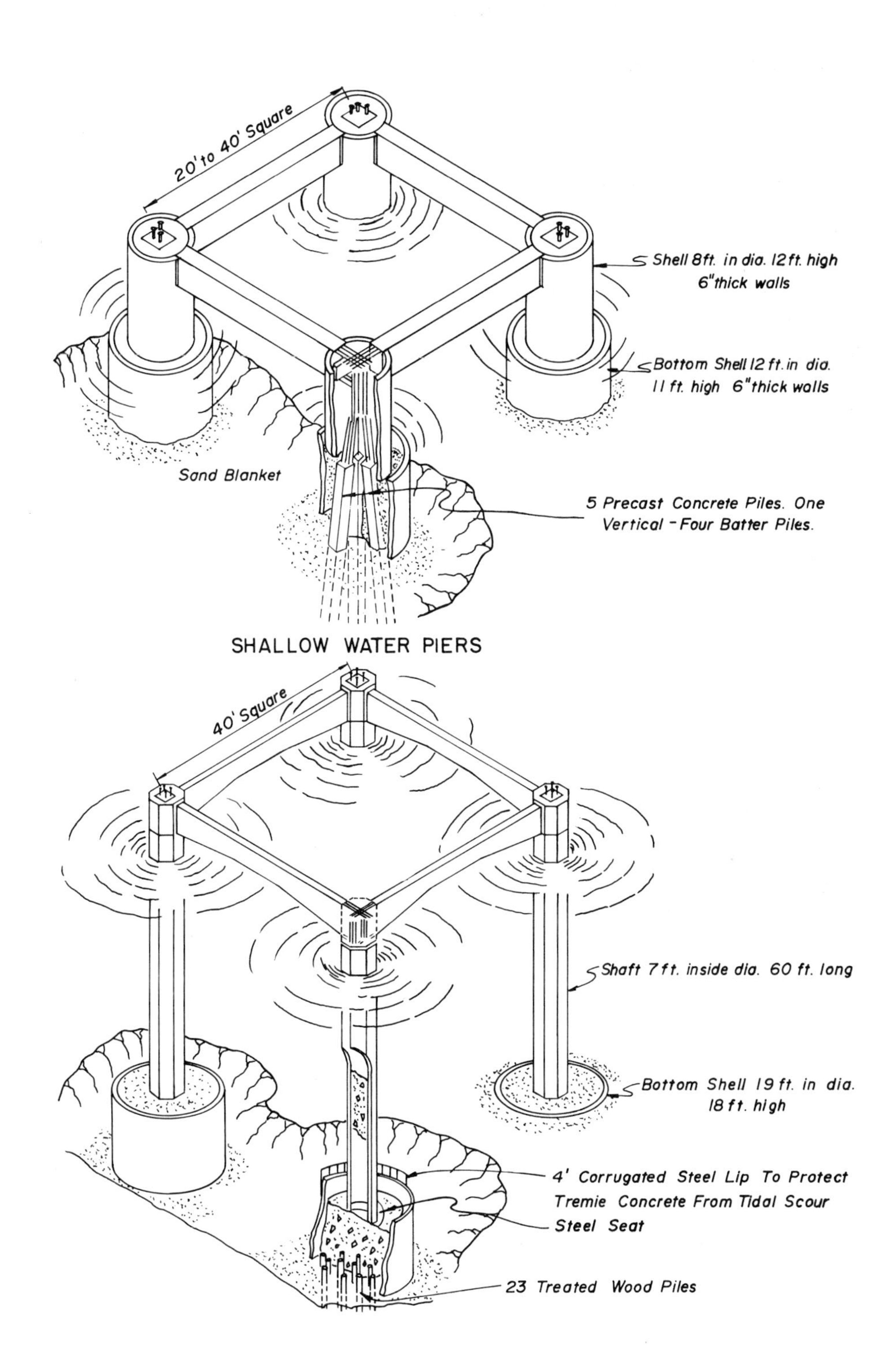
20' to 40' Square
Shell 8 ft. in dia. 12 ft. high
6" thick walls
Bottom Shell 12 ft. in dia.
11 ft. high 6" thick walls
Sand Blanket
5 Precast Concrete Piles. One
Vertical - Four Batter Piles.
SHALLOW WATER PIERS
40' Square
Shaft 7 ft. inside dia. 60 ft. long
Bottom Shell 19 ft. in dia.
18 ft. high
4' Corrugated Steel Lip To Protect
Tremie Concrete From Tidal Scour
Steel Seat
23 Treated Wood Piles
DEEP-WATER CHANNEL PIERS

7.37 BRIDGE PIERS CONSTRUCTED UNDER WATER

Once a general technique has been introduced and developed, it is not surprising that its use should be picked up and extended; but what is surprising, frequently, is the delay in the initial introduction and the incidental motives that prompted its first presentation. The contractor on the foundation construction for transmission-line towers, described in Sec. 7.36, was Ben C. Gerwick, Inc., and the introduction of this technique was largely due to the fact that this firm had a concrete casting yard operating on Petaluma Creek, a navigable arm of San Francisco Bay. It is not surprising, therefore, that only a year later the same contractor used a variation of the same method for the foundations of the Richmond–San Rafael Bridge across the northern leg of San Francisco Bay, 10 miles above the San Francisco–Oakland Bay Bridge.

The bridge in question crosses a 4-mile-wide narrows where water varies from 30 to 60 ft in depth and the bottom is covered with a thick layer of soft mud. Tidal ranges and currents are similar to those mentioned in Sec. 7.36.

At each site, a depression was dredged in the mud some 12 ft deep, and wood piling was driven under water to support a grid plate. The piles were cut off to the elevation of the bottom of the grid base and surrounded, to their tops, by a layer of sand.

Two circular concrete slabs 12 in. thick, connected by a spreader truss, were lowered to rest on the piles, with a lightweight steel sighting tower secured to them to accurately position them. Each slab (base grid or disk) was pierced with slots shaped to suit the H-piles to be driven for primary support.

H-piles in 200-ft lengths were driven to 100-ton bearing by floating pile drivers equipped with 120-ft-long telescoping leads. A diver guided the bottom end of the pile into the proper slot initially, and the piles were driven vertically or on batters.

With the pile driving completed, a pre-cast bottom shell was set over each grid, fastened down to brackets pre-cast on the plate surface, and sealed with 5 ft of tremied concrete.

On top of each pair of shells a single pre-cast unit, consisting of two cone-shaped sections connected by a diaphragm, was set. This provided precise positioning for the pre-cast shafts to be placed on top of the cones. Provisions in the shafts permitted the setting of steel forms to extend the connecting diaphragm upward. The assembly was concreted by tremie to above mean tide.

A similar set of procedures was used 2 years later for bridge piers for a river crossing between Portland, Oregon, and Vancouver, Washington. The water depth was similar, but tidal ranges were not involved.

An adaptation, necessary because the specifications limited tremie concrete to the bottom shell, was used for pier foundations for a bridge over Lake Texoma some 75 miles north of Dallas, Texas.

These piers were positioned by a steel frame template set on four 10-in. spud piles, like that for cofferdam construction. Within this frame, a separate, removable steel template was lowered to rest on the bottom and to correctly position the twelve 74# steel H-piles required. These were driven by a floating rig and employed a follower pile for driving under water.

After the piling was in place, the bottom template was removed and a second framework, in which a steel form had been assembled, was lowered through the original template frame. The forms consisted of containment for a bottom shell, a conical section, and a shaft; and they were connected, within their frame, through hydraulic rams. After the form frame was lowered, the rams would permit moving the form within the frame for precise positioning.

The bottom shell was to be left in place and was filled with 5 ft of tremie concrete. The remainder of the form was designed to be removed, under water, after concreting was complete. The form was made up in three segments, held together through push-pull hydraulic rams, the segment joints sealed by neoprene gaskets. After the forms were set and dewatered and concrete had been placed, the rams were used to thrust open the form joints.

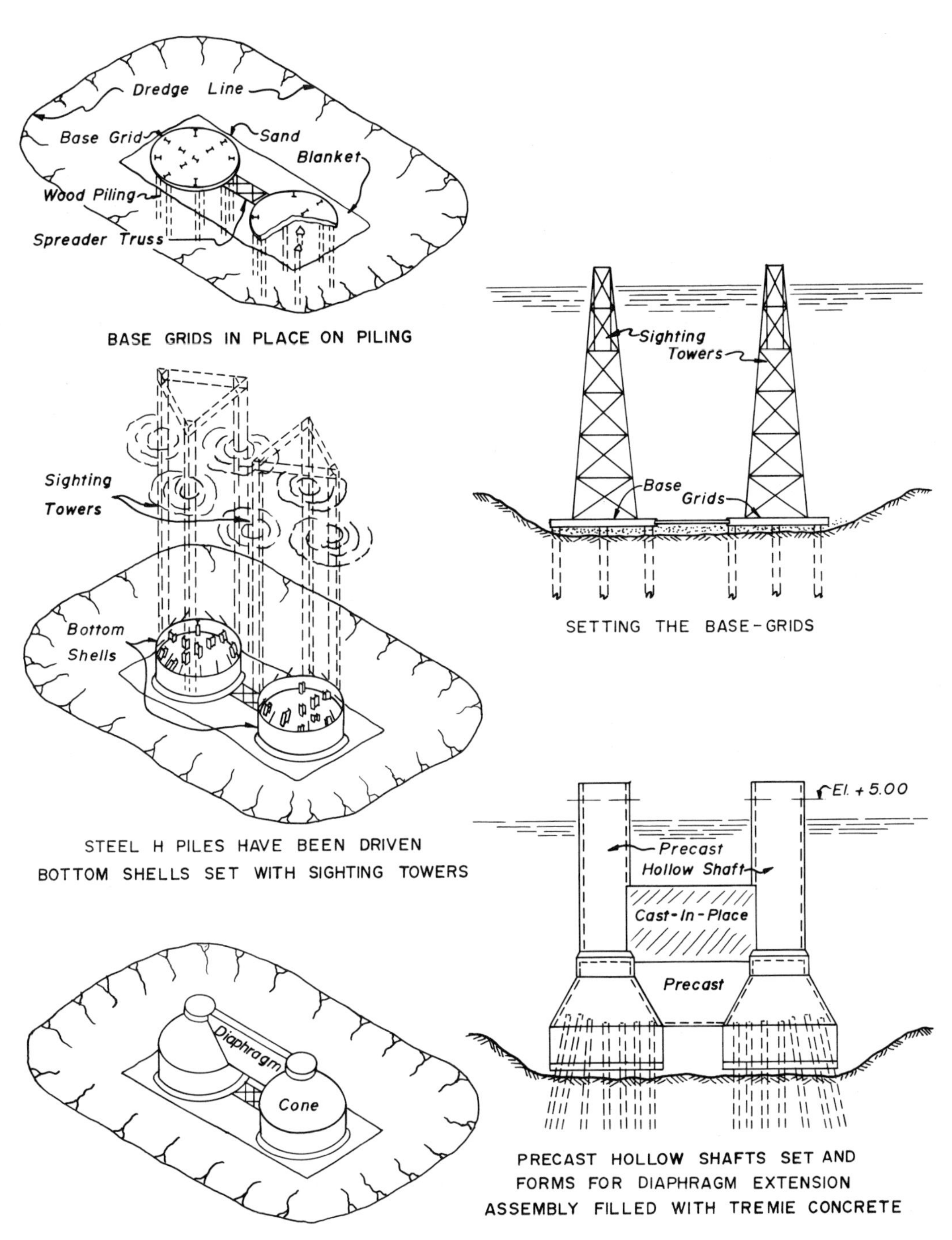

BASE GRIDS IN PLACE ON PILING

SETTING THE BASE-GRIDS

STEEL H PILES HAVE BEEN DRIVEN
BOTTOM SHELLS SET WITH SIGHTING TOWERS

PRECAST HOLLOW SHAFTS SET AND
FORMS FOR DIAPHRAGM EXTENSION
ASSEMBLY FILLED WITH TREMIE CONCRETE

BOTTOM SHELLS FILLED WITH TREMIE CONCRETE
PRECAST DUAL CONE AND DIAPHRAGM SET

7.38 A SPECIAL FOUNDATION WATER PROBLEM

If there is one premise that is basic to all construction procedures, it is that we start at the lowest point of a structure and build upward. The law of gravity being what it is, even in the space age, we can do little else. Another basic tenet is that we control groundwater to suit our foundation construction, not construct the foundation to suit the water condition, unless it be an all-pervading medium. But, in the early 1950s, in San Salvador, a bridge anchorage was constructed in such a way as to prove an exception to these fundamentals even though it was not hung from sky hooks and no gurgling aquifer burbled instructions to the field crew.

The San Marcos Bridge (officially *El Puente del Literol*) is a five-span suspension bridge crossing the Lempa River at a point 12 miles inland from the Pacific Ocean. Designed and built by John A. Roebling's Sons Corporation, it has many points of interest in its superstructure construction which need not detain us.

The east-cable anchorage is founded on a soft sedimentary rock, known as talpatate, which extends downward from close to the surface. At the west anchorage, set some 500 ft back from the edge of the river, there is no rock within 100 ft of the surface; instead, there are stratified soil deposits.

The bridge site lies in an earthquake area, equidistant from two semi-active volcanoes, and much of the soil deposit is of unresolved volcanic material. There is a tropical rainy season from May to October during which the river is high and subject to flash floods and during which the water table at the west anchorage rises from −20 ft to −3 ft.

Eighty feet down lies the top of an aquifer under artesian pressure, topped first by a 20-ft stratum of fine sand, with a 20-ft stratum of stiff brown clay over that. The bottom of the anchorage would be set deeply into the clay stratum, weakening it and possibly permitting the artesian pressure to break through.

As a result of these conditions, a decision was made to first pour half of the foundation, that portion which would normally be poured last, during the dry season from November to April.

General excavation of the site was carried down to a depth of 17 ft, and a well-point system was installed to lower the water table 16 ft. Discharge from the pumps was fed into a masonry flume 300 ft long, which discharged downstream.

With the advent of the next dry season, work was begun on the second half of the anchorage. A new well-point header was now placed in a trench 9 ft below the previous working level. From this header, pressure-relief well points 60 ft long were jetted down into the bottom of the pressure aquifer to a depth 85 ft below ground surface.

The pressure in the aquifer pushed water up in these 51 well points to a height sufficient to be picked up by the pump suction. The volume of water was tremendous. When the pumps stopped, the water would rise 10 ft in 2 min, and the recovery of this 10 ft, at maximum pump capacity, required 6 hr.

Although this depressurizing system eliminated the danger of damage to the clay stratum, further dewatering was necessary in the upper layers and even in the clay itself, which was riddled by pipes remaining from ancient organic inclusions. A second set of standard-depth well points was now installed, which brought the water table down to within 12 ft of the bottom of the foundation.

To prevent any possibility of the previously poured section of anchorage sliding down the slope into the new excavation, the final 12-ft depth was split up into five pockets, each excavated separately by hand in the order shown on the illustration, each pocket being poured before the adjoining one was excavated.

To dry up the last 12 ft, a third ring of well points, placed successively around each pocket, was used. This stabilized the clay sufficiently to permit the banks to act as forms for the concrete. There was the chance, of course, that a substantial earthquake tremor would break down the earth bank before concreting, but fortunately nothing of the kind occurred.

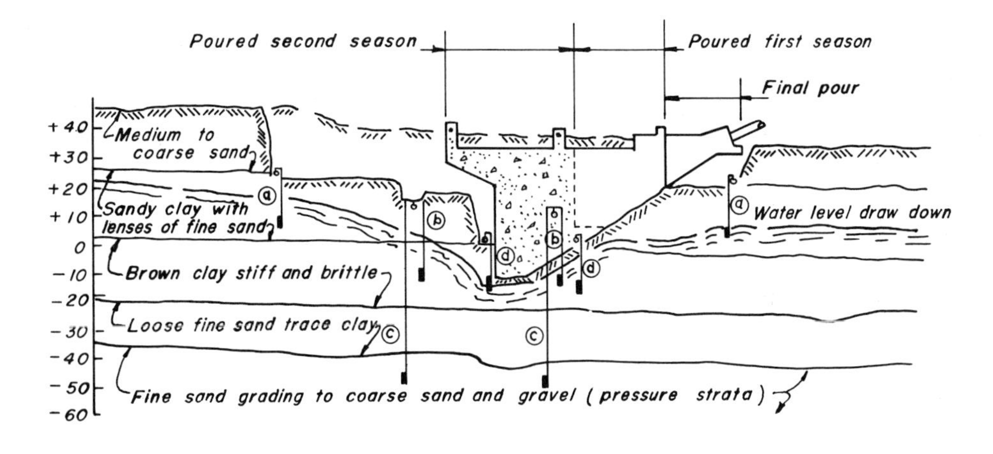

(a) Primary wellpoint system for area dewatering.
(b) Secondary wellpoint system. Same header as for (c)
(c) Depressurizing wellpoint system.
(d) Stabilizing wellpoint system.

SECTION AT WEST ANCHORAGE – SAN MARCOS BRIDGE

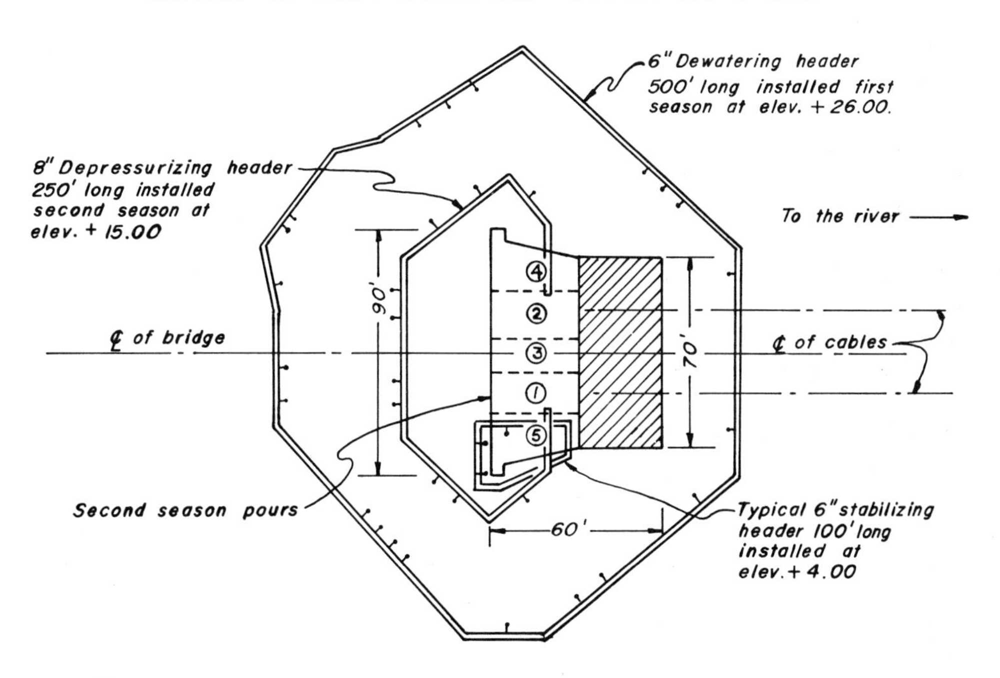

(1) Indicates order of excavation and concreting of segments

PLAN OF WEST ANCHORAGE SITE – SAN MARCOS BRIDGE

7.39 CONTROL OF SITE CONDITIONS

There is always the temptation to speak only of the successful construction operation, while quietly ignoring the flubs and dubs that occur so frequently in spite of the exercise of the greatest judgment by the best brains. We are not speaking of those out-and-out failures of structures which, collected, are made the subject of ominous volumes, but of those situations involving the control of conditions at the site where, after all rational approaches have been exhausted, sheer, bull-headed persistency gets the job done.

In Sec. 6.16, we mentioned a deflector developed by Dravo Corporation for setting cofferdam cells in swift water at the upper end of the Long Sault Rapids in the St. Lawrence River. At about the same time, at the lower end of the same rapids several miles downstream, another contractor, A. S. Wikstrom of Skaneateles, New York, was struggling with the problems of foundations for a bridge across the St. Lawrence River between New York State and Barnhart Island. It was on this island that the powerhouse would be built to utilize the water from the dam under construction by Dravo.

The substructure contract for the bridge embraced four piers, two of them onshore and involving no special problems and two of them in the river, 508 ft apart, each being about 225 ft out from the shoreline. This contract was awarded in October (of 1954), with a completion date established in June of the following year. Construction, therefore, must begin in December and be pushed through the coldest part of the year when, in this area, temperatures drop to 30° below zero and ice piles up. Although such conditions are hardly ideal for construction, they were known to the contractor at the time of bidding, and the solutions for the problems engendered could be devised.

Tales of misfortune generally start out by telling us that events began on a warm and shining day, with everyone involved "all bright-eyed and bushy-tailed" and with trouble only creeping upon the scene thereafter. But in this case trouble was standing by, leering at the contractor.

In October, when the contract was awarded, no adequate borings had been taken in the river. A separate contract for them had, indeed, been let; but the drilling contractor, working from a floating barge, had encountered trouble in moving his rig to the site and in holding it in drilling position in the swift current. Finally, in November, after three borings at the north pier and one at the south pier had been made, the drill barge broke loose from its cable moorings, one night, and was swept downstream, never to return.

The borings obtained, scanty though they were, revealed that rock lay 11 ft deeper at the north-pier site and 7 ft deeper at the south pier than had been expected. Redesign of the piers was indicated, and both of the river piers were changed from a circular section 72 ft in diameter to a rectangular section 50 by 80 ft.

The contractor, Wikstrom, even before bidding, had made careful studies of flow conditions on a prototype model of the river furnished to him by the Ontario Hydropower Commission. No one was then aware that accurate soundings had never been taken in the vicinity of the piers because of the swift water and that, in consequence, the model did not reflect the true conditions at the site. Knowledge of this deficiency would come later—the hard way.

On the basis of his model studies, the contractor had decided to construct a stone dike 250 ft long at a distance of 140 ft upstream from the south-pier site, in order to provide stilled water in which to build his pier. It was expected that this dike would raise the upstream water level about 6 ft and divert the swiftest part of the flow to a point some 40 or 50 ft beyond the end of the dike.

Construction of the dike was begun early in December, using stone from a quarry opened up by the contractor; and the first 135 ft was placed by rear-end dumps and crane, without difficulty. Here, the depth of water was 30 ft, and the stone washed away as fast as it was placed.

Believing that the solution lay in using larger stone masses, in the 5- to 7-ton range, the contractor, who could not produce these sizes in his own quarry, turned to a commercial quarry whose blasting operations could. With this stone the dike was extended another 50 ft into a water depth of 38 ft. At this depth the stone, even of this weight, took an appreciable time to settle and, during the interim, the current carried them downstream. Then, before this difficulty could be resolved, 14 ft of the dike broke loose and was washed away.

A basket was now devised consisting of horizontal steel H-beams tied together with wire mesh, 35 ft long, narrow at the top and 40 ft wide at the bottom to match the configuration of the dike. Lowered into place by an 80-ton crane, it was guided into position by cable lines to bulldozers on the shore and, after positioning, was anchored by 1⅛-in. cable, top and bottom, to two 10-ton concrete deadmen set upstream.

The basket was filled with rock and appeared to have provided the answer except that the dike now fell short of the length needed to furnish quiet water downstream. At this juncture, the engineers agreed to move the bridge piers 25 ft south, thus bringing this pier 25 ft closer to shore.

The contractor was now prepared to begin pier construction when, at this inauspicious moment, the river ice began to jam up on Lake Francis, 3 miles downstream. This raised the water level at the construction site some 7 ft, although "the best available information" had never previously indicated a greater height than 3 ft. Construction operations were suspended.

In the early part of March, the ice jam broke up, the river level started down, and operations were about to resume when ice, floating down the river, built up on the wire mesh of the baskets, tore the cables loose from their anchorages, and washed the basket and its rock contents downstream. Another, heavier basket was now built and set, but it too washed away.

Finally, in the latter part of March, new baskets were set up. These were only 12 ft in length and were anchored by driving steel H-piles into the underlying rock, and the spaces between were closed with steel sheet piling to assist in retaining the stone. This system of basketry held until completion.

Site construction was begun by sinking an "icebreaker" at a point 30 ft upstream from the pier by filling it with rock. The icebreaker was essentially a heavy timber box, 36 by 40 ft in plan and 40 ft high, which had been built on shore and floated into position. It not only served as a base for a long-boomed crane but was used to hold the cofferdam template.

Rock overburden at the pier site was only 2 ft deep, insufficient to hold the spud piles for the cofferdam template. Accordingly, two 33-in. WF 200# steel beams 76 ft long were fastened to the sides of the icebreaker, extending outward like the prongs of a two-tined fork. Between these the cofferdam template was bolted.

Before the driving of steel sheet piling could begin, it was necessary to clean up the rock masses that had been swept downstream each time the dike had failed. This took time, since a diver working in the icy water had to rig many of the stones for a crane lift.

After all that had gone before, the setting of the cofferdam and construction of the pier were anti-climactic.

At the pier off the north shore these difficulties were not repeated. A 200-ft-long dike was placed, and held, without incident.

The contractor chose to concrete the two piers by intrusion grouting and, to avoid setting up two grouting plants, had planned to pipe grout across the river from the plant on the American side. It was planned to lay two 2½-in. pipelines on the bottom of the river, to supply grout to the north pier. Experiments with cable, however, indicated that the current would belly such lines and probably result in their breaking. It therefore became necessary to build towers on either side of the 950-ft span and to hang the pipelines from suspended cables.

8 Open Caissons

Remember the old nursery rhyme, "London Bridge Is Falling Down"? This referred to Old London Bridge, built in the years between 1176 and 1209, which for 500 years was the only bridge over the Thames at London. Although portions of it spectacularly collapsed from time to time, the solid tradesmen of London fought any new construction. Indeed, in 1664, when the prospect of a new bridge seemed imminent, they raised the sum of 100,000 pounds and offered it to King Charles the Second for his services in *opposing* any new construction. Needless to say, he accepted the offer and a new bridge was further delayed almost a century. Finally, in 1738, a new structure spanning the Thames between Lambeth and Westminster at the site of an existing ferry was begun, financed in part by funds from a public lottery. The engineer selected, Charles Labelye, was a Swiss who had been trained by the French Corps Royaux des Ponts et Chaussées. For his foundations, he employed caissons, whose name is derived from the French word *caisse* meaning a box or casing.

Caissons were not a new device even in 1716, in France, when Labelye studied with the Royal Corps of Bridges and Roads. The Romans had used them in throwing bridges across the Tiber in the heyday of the Empire, but the more usual method even then, as noted by Vitruvius, was to construct bridge foundations on piling within a shallow cofferdam. The Dutch also had used caissons to found control structures in dike walls, but these again were crude and shallow.

Labelye's caissons for the Westminster Bridge were wooden boxes 80 ft long, 30 ft wide, and 16 ft deep, framed in 9 × 12 in. fir timbers and covered on two sides of the framing with 3-in. plank fastened to the frames with oak pegs or "treenails." The corners were reinforced with wrought-iron bands, and the joints between planking were caulked with pitch.

The river bottom at the bridge pier sites was leveled by dredging, and the caisson, constructed on the shore, was towed into position and anchored. The box was sunk by flooding through a sluice gate provided for the purpose to test its bearing and then, firmly anchored, was re-floated to permit three courses of masonry to be laid in the dry. When the box was re-sunk in its final

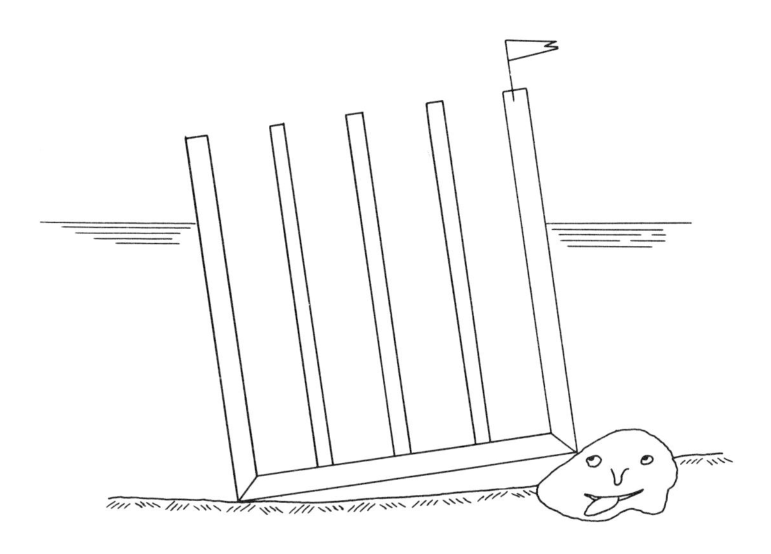

position, its depth was such that its top was above water at low tide. As the tide went out, the box was pumped down and masonry work then proceeded until the next rising tide halted it. The only piling used was around the two midstream piers to prevent their movement in the swift-flowing waters.

The method was used 10 years later on a third bridge over the Thames by Robert Milne and again in 1816 by James Walker, who built the first iron bridge over the same river; but for the most part bridge piers continued to be set on piling.

The difficulty with the caisson method, as then used, was that it did not carry the foundations deep enough to protect them from scour. Westminster Bridge survived for 100 years but was finally pulled down after scour had seriously undermined the piers.

Methods of carrying caissons to greater depths did not appear until the 1850s when the pneumatic caisson was developed. By this time, the introduction of iron bridges, followed by the longer-spanning suspension bridges, so concentrated pier loadings that greater depths were imperative on other scores besides that of scour.

Even when the pneumatic caisson was introduced, the idea of a wooden box was not abandoned. Eads used a six-sided timber structure 82 by 72 ft in outside dimensions for his caisson in 1869, but sheathed it in ⅜-in. iron plate. Roebling, for his caissons for the Brooklyn Bridge, used no steel plate but contented himself with 9-ft-thick timber walls.

This pattern prevailed for many years, particularly in the field of railroad bridge construction. The report of the Chief Engineer to the president of the Chicago, St. Louis and New Orleans Railroad Company in 1891 contained the details for the construction of 10 pneumatic caissons to be sunk to depths of from 77 to 94 ft for a bridge over the Ohio River at Cairo, Illinois. These caissons were 26 by 60 ft in size, constructed of 12 × 12 in. oak timbers, with the exterior walls faced with ⅜-in. iron plate. The roof of the pneumatic working chamber also contained 12 × 12 in. oak timbers in a double layer, covered with pine planking and stiffened with 14 × 14 in. bracing, the timbering tied together with 2-in. bolts. The unit was made up on shore and floated into position.

8-1 EARLY PNEUMATIC CAISSONS

Although Lord Cochrane has generally been credited with the idea of using air for excavation under water, and he certainly filed a patent on it in 1830, it may not have been in connection with caissons at all but with the use of air in tunnels, the idea itself having been suggested 2 years previously by Dr. (Professor) Daniel Colladon. In 1828, Marc Isambard Brunel, a French engineer, was engaged in driving a tunnel under the Thames River when the river broke in and flooded the project. Methods of salvaging the operation poured in, like the river, from all sides, including that of using compressed air, suggested by Dr. Colladon. Dr. Colladon had been studying the possibilities of compressed air for some time and was later to act in an advisory capacity when compressed air was used for rock excavation in driving the first tunnel through the Alps, but Lord Cochrane was not one to let a good idea lie idle.

Thomas Cochrane, a British naval officer, too early had advocated the use of steel hulls for the British fleet and went on to recommend that the steam engine be adapted to marine use. Whether the British Admiralty was laggard in its thinking or Thomas Cochrane's insistence was unbearable is not certain, but he presently found himself under a cloud and left England, for a time, to manage the navy of Chile.

On Cochrane's return to England, undaunted by his previous experiences, he now presented a "secret war plan" to the Admiralty, more than a hundred years before Hiroshima, that he claimed to be capable of destroying any fleet or fortress in the world. Apparently considered too frightful, his plan was rejected and the details of it were never made public; but he seems to have made his peace with the British Navy, for he ended as an Admiral and acquired the title of Lord Cochrane by succession, as the 10th Earl of Dundonald.

Twenty-one years after Cochrane's date of patent, the first pneumatic caissons were sunk by William Cubitt and John Wright to a depth of 61 ft for the foundations of a bridge over the Medway at Rochester, England; and Brunel's son again used them 4 years later in his foundations for the Royal Albert Bridge.

Whether compressed air had been used before this time for any major construction of caissons in France is not known, but by 1867 its technique was well advanced there; and it was there, in that year, that John B. Eads studied the method. It was Eads who first used the pneumatic caisson in the United States for the east pier of his bridge over the Mississippi River in 1869.

The earliest type of pneumatic caisson consisted of a wrought-iron bell, or working chamber, with a wrought-iron tubular extension providing the air lock. These caissons were floated into position and initially sunk by flooding the compartments provided for that purpose with water (see illustration). Further sinking was accomplished by loading the structure with cast-iron weights. The masonry pier, for which this caisson was devised, was constructed within the bell. When the pier had attained sufficient height, the bell was floated off for re-use.

With the development of structural steel and steel plate, the use of wood for pneumatic caissons, of any size, gradually gave way to the use of steel shells. The use of steel shells continues but has been augmented by the use of concrete shells under certain conditions. On the other hand, the use of air pressure in sinking caissons has declined in favor of the "open" caisson.

Perhaps the pneumatic method reached an apogee in 1932 when it was used for sinking one of the caissons required for the San Francisco–Oakland Bay Bridge. Here, a series of 28 shells were fabricated from steel, each being capped and filled with air under pressure. The lower portion of the shell walls was filled with concrete, but the air-filled shells provided the buoyancy required to tow the caisson into position for anchoring. From that point on, water level in the caisson cells was controlled by the application of air pressure, cell by cell or in groups, as required. This caisson was, in fact, a combination of the pneumatic and the open caisson method.

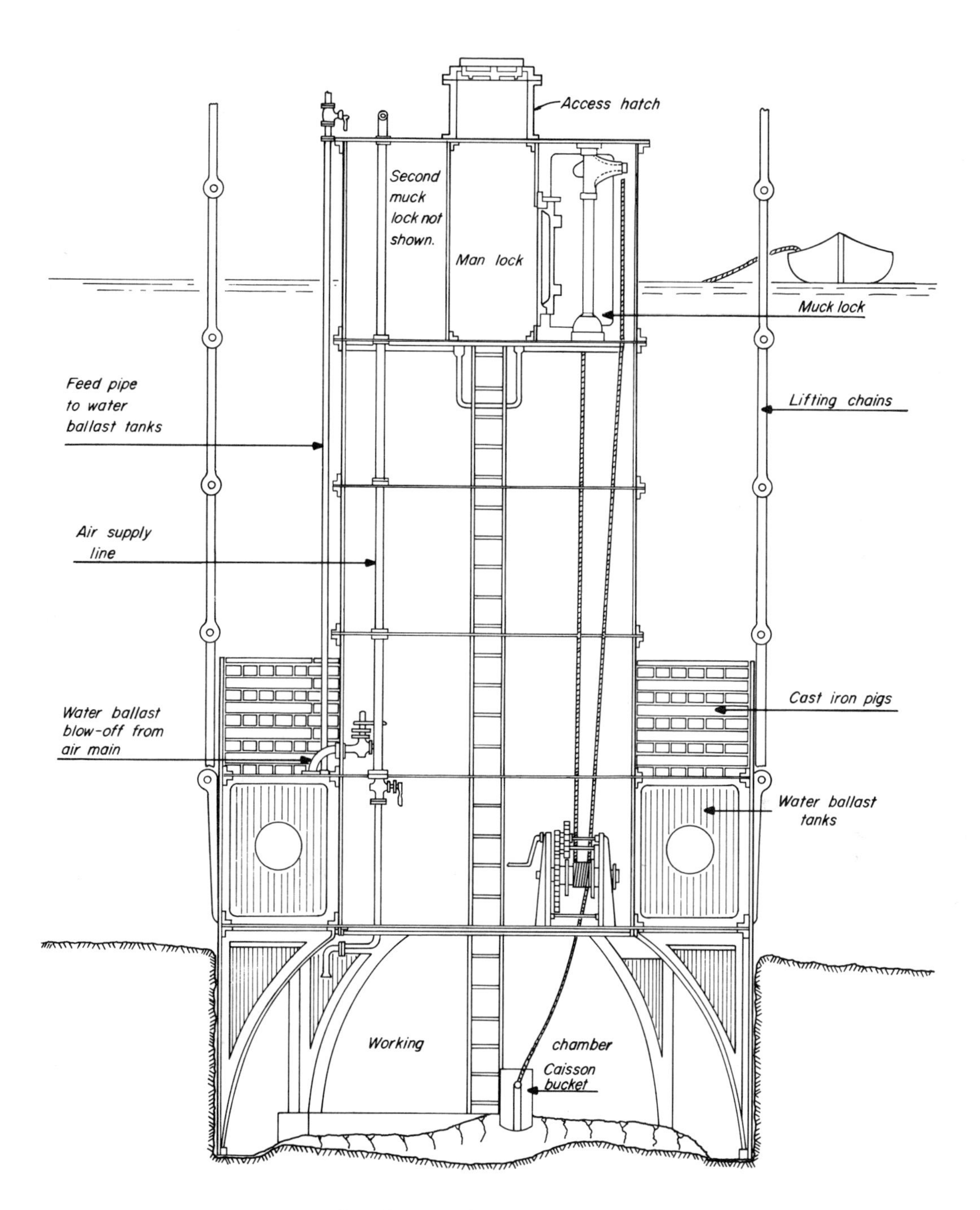

A PNEUMATIC CAISSON OF THE 1860'S

8.2 OPEN CAISSONS

The original caisson, as we have seen in the case of that for the Westminster Bridge, was a box closed at the bottom and open at the top. The next development, the pneumatic caisson, was, essentially, a box open at the bottom and closed at a point higher up. The open caisson, now generally preferred, is a box open at both the top and the bottom. The use of the open caisson was, in the beginning, approached cautiously, and provisions were made in the earliest open caissons for converting them to pneumatic operations should the need arise. Techniques developed and improved incident to World War II, however, have removed the problems that seemed to make the sinking of open caissons uncertain.

Developments in diving gear now permit divers to work at greater depths without loss of mobility. Underwater blasting techniques arising from lessons learned in underwater demolition now permit the certain removal of underwater obstacles. Underwater photography and closed-circuit television now permit the study of underwater conditions without dewatering the site. Improvements in test-boring investigations, particularly under water, and advances in the interpretation of such data due to the growth of the science of soil mechanics have removed many of the former uncertainties.

It had been long recognized that the air in a pneumatic caisson, though seldom enough to float it, still tended to provide some buoyancy and so to conflict with the sinking operation. Furthermore, it was realized that since the majority of caissons are sunk in water, the water itself, acting as a lubricant, might improve the sinking operation.

There are two principal types of open caissons, distinguishable chiefly by their respective methods of sinking. The first of these, sunk by the sand-island method, is essentially a concrete structure; the second of these, the floating caisson, returns to the concept of a steel shell within which the concrete pier is constructed. The method of sinking employed for the floating caisson is known as the "corral" method.

Generally speaking, the choice of method will depend on the depth of water at the site, as may be seen in the illustration, but this is not an infallible guide. The tower piers of the Walt Whitman Bridge spanning the Delaware River between Philadelphia and Gloucester, New Jersey, were intended to be sunk by the corral method using floating caissons and were so designed, since the 50-ft depth of water at their sites seemed to warrant it. However, the successful contractor chose to use the sand-island method; and, since he submitted an acceptable alteration in design and had it approved, the piers here were constructed by that method.

It could be considered that a depth of 50 ft is marginal, at which depth either method is practical, but the decision cannot be reached so simply, for other factors intrude an influence.

The sand-island method of open-caisson sinking employs land-based equipment, for the most part. Floating or barge-mounted equipment is required for the corral method. Primary experience in the use of one or the other type of equipment as well as the possession or availability of either type will influence the contractor's decision, where he has a choice. (This seems to have been the selective factor in the Walt Whitman Bridge caissons.) The fact that different unions may have at least partial jurisdiction in either case may weight the answer.

The type of waterway in which the construction is to be performed will have an influence. Where the caisson is to be built adjacent to shipping channels, the space required for floating equipment may make its use impractical. The extent of tidal fluctuations or the possibility of excessive floodwaters may dictate a choice. Land-based approaches can be elevated, but the relative levels between barge deck and the caisson working plateau will vary with tide, and all equipment may require complete removal in the event of floods.

The nature of the soils in the waterway bottom at the site may have some effect on the selection of method, as will be pointed out later, although this is closely allied to the depth consideration.

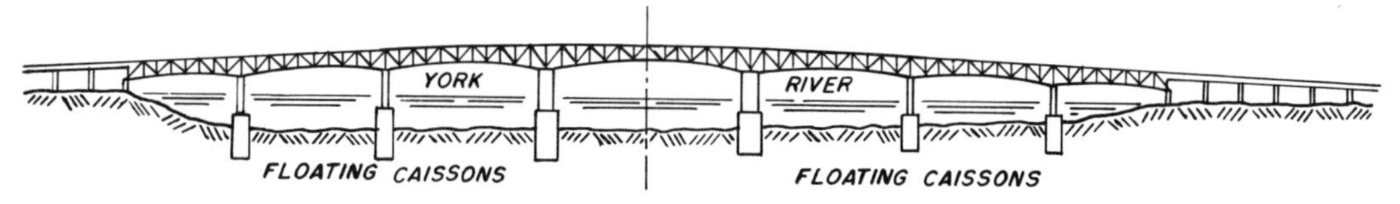

GEORGE P. COLEMAN MEMORIAL BRIDGE

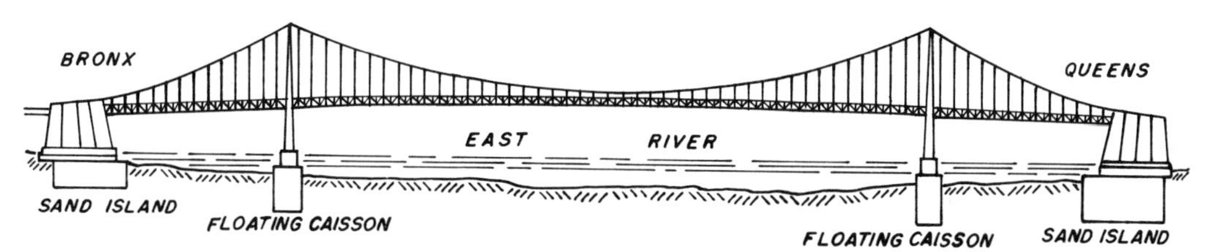

THROGS NECK BRIDGE

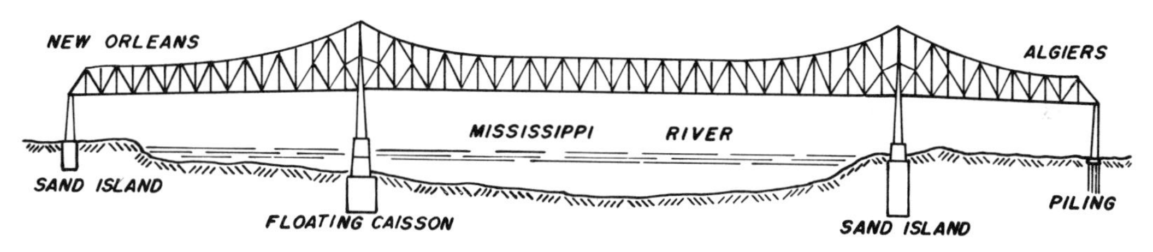

NEW ORLEANS BRIDGE

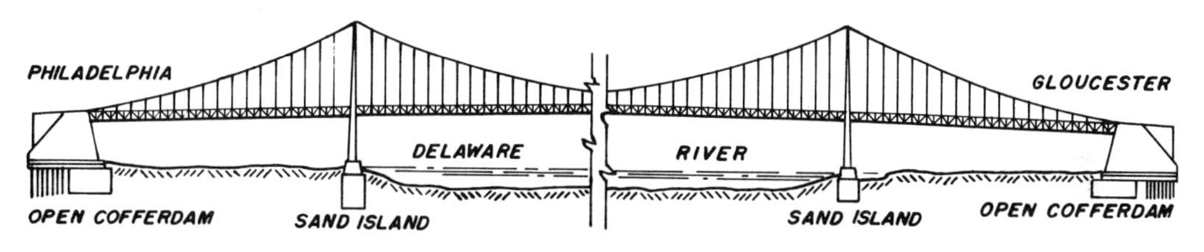

WALT WHITMAN BRIDGE

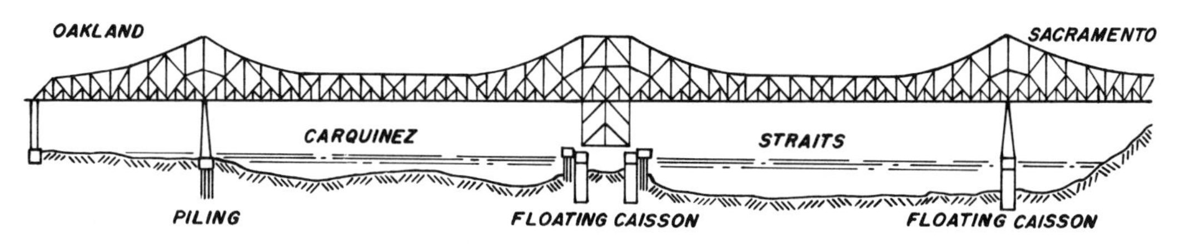

CARQUINEZ TOLL BRIDGE

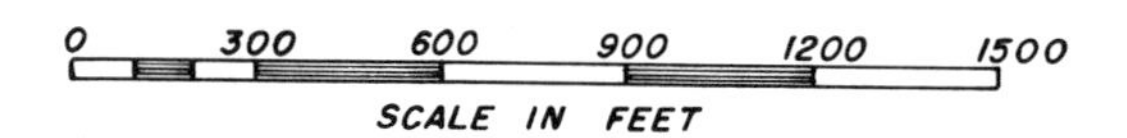

A BEVY OF BRIDGES

8.3 CAISSON DESIGN VERSUS CONSTRUCTION METHODS

It is to be expected that the elements of design will be influenced by the selection of construction methods, and this is particularly true of caissons. Indeed, with caissons it can almost be said that the construction method is first selected and the design follows.

A caisson is the lower portion of a structure and, in the case where it is most generally used, that of bridge pier construction, is the bottom of the pier; consequently the entire design of the superstructure will influence the nature of the caisson. The illustration shows some typical caisson layouts drawn to a single scale, reflecting the wide variation in size as well as shape. Part of this variation depends upon the type of bridge span, whether it be suspended, trussed, or cantilevered, and upon the length and width of span. As can be seen, the *shape* as well as the size of tower caissons may be quite different from that required for an anchorage pier, and other superstructure factors will develop the vertical and thrust loadings which will shape the caisson. These factors are determinable and detailable elements arising chiefly from command decisions and following well-established design techniques. The minute we reach the caisson stage, however, the path is not so clear.

Caisson construction is used where the foundation is to be placed under water (although some exceptions to this will be noted later); and in Sec. 8.2 the effect of depth in choosing between the sand-island and the corral method of construction has been noted. The influence of tidal range and flood flows will influence not only construction operations but caisson design as well, as will be amplified later. Beyond this are the factors of the soil conditions beneath the waterway.

It would be fine were it possible to found all footings on rock with the proper characteristics, but not only is economics opposed to this but there are other factors as well. With the best of information, rock-surface determinations can be deceptive, as in the case described in Sec. 4.13. Another factor limiting the depth of caissons is the sheer weight of the caisson and its loading. A solid concrete pier could exert a pressure of 15 tons per sq ft at a depth of 200 ft. This loading is high for many types of rock and for most soil combinations. As a result, caissons are now generally constructed with hollow cores or compartments, with these spaces filled with water rather than concrete. This substitution reduces the weight in the compartments by 58%, the difference between the weight per cubic foot of concrete and water.

It can be said of caissons that the design is final when the caisson is completed. When the caissons for the Mackinac Straits Bridge were layed out at the site, it was found that one of these was too close to the edge of an underwater gorge. There was danger of the pier's sliding along a seam toward the gorge. To move this pier alone would mean redesigning the bridge spans on either side. Accordingly, *all* the caissons were moved 30 ft to what was considered a safer position for that single pier, and the resultant discrepancies were absorbed in the approach spans at either end. This, of course, varied the founding depths and consequent heights of each of the caissons.

Not all the conditions encountered are necessarily retrogressive in nature. When one of the caissons for the Verrazano Narrows Bridge, connecting Brooklyn with Staten Island, had been partially sunk, it was found that subsurface conditions, at the level which had been reached, were adequate for founding the caisson, and construction was terminated at that elevation.

While these overages and shortages are generally provided for in the contract, the contractor must keep their possibility in mind in preparing his bid. In one instance it was necessary to increase the depth of two caissons 45 ft each, at a quoted price of $10,000 per vertical foot. This added the not inconsiderable sum of $900,000 to the contract.

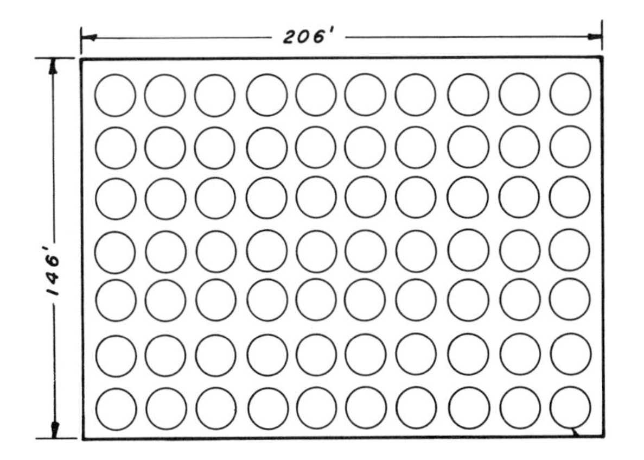

BRONX ANCHORAGE
THROGS NECK BRIDGE
BRONX TO QUEENS
NEW YORK CITY
70-16' DIA. DREDGING WELLS
DEPTH 68'

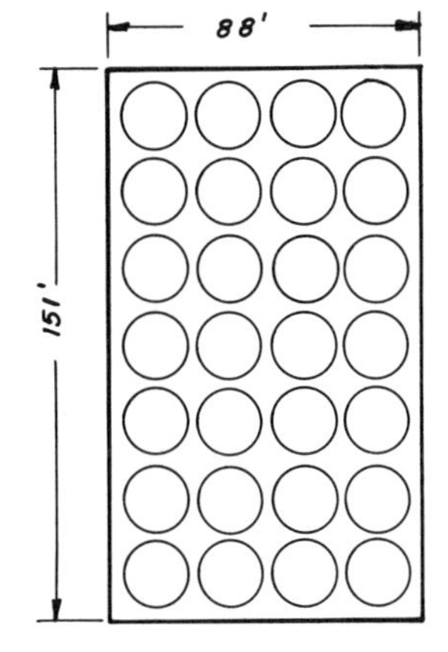

PIER NO. 2
NEW ORLEANS BRIDGE
NEW ORLEANS TO ALGIERS
OVER MISSISSIPPI RIVER
28-15'-4" DREDGING WELLS
DEPTH 180'

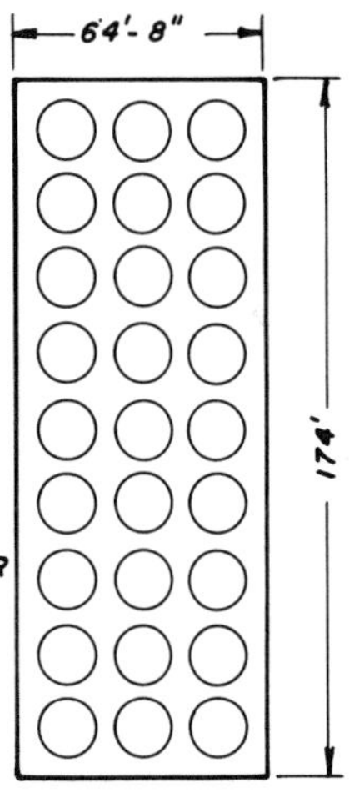

TOWER PIERS
WALT WHITMAN BRIDGE
PHILADELPHIA TO GLOUCESTER
OVER DELAWARE RIVER
27-18' DREDGING WELLS
DEPTH 78' AND 108'

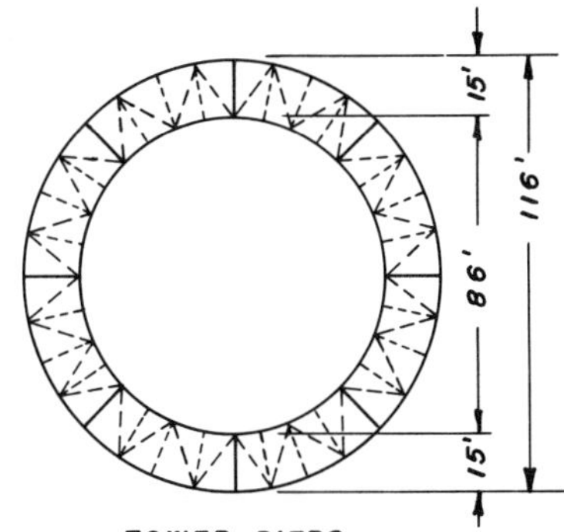

TOWER PIERS
MACKINAC STRAITS BRIDGE
COMPARTMENTED SHELL
SINGLE DREDGING WELL
DEPTHS 206' AND 211'

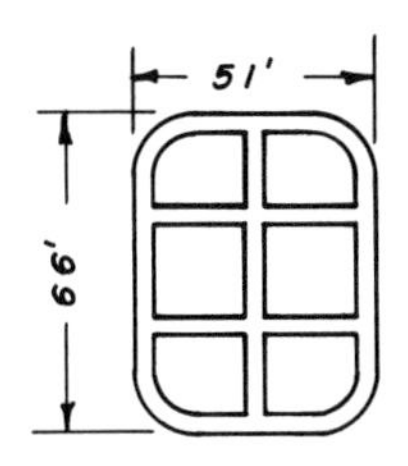

6 RIVER PIERS
GEORGE P. COLEMAN MEMORIAL BRIDGE
OVER YORK RIVER
6 VARIABLE DREDGING WELLS
DEPTHS 135' TO 150'

CAISSONS OF THE 1950'S

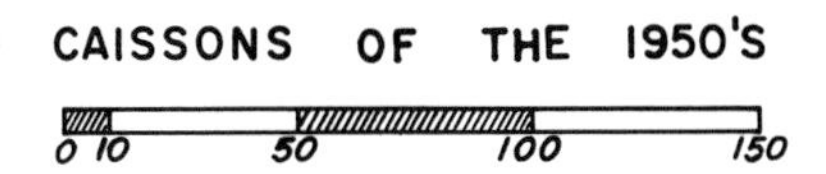

8.4 THE SAND ISLAND

The sand-island method of open-caisson construction is essentially a method by which the initial operations can be started on a dry surface with arrangements for flooding the site after a portion of the caisson is laid down. As has been noted, it is most generally used close to the shore in relatively shallow water, but there is actually no limit to the waterway depth in which it can be used except that of economics.

The sand in the sand island will require containment, partly because of the nature of the operation and partly because of space limitations. Close to the shoreline the sand island may be constructed with a single row of steel sheet piling designed as a bulkhead; but as the depth of water increases, containment must be by circular cofferdam. Because the caisson will occupy virtually all the space within the contained sand island, no cross bracing is practical, and the circular cofferdam provides the most stable condition.

In a few cases, sand islands have been built within waterway areas without containment, but the method is not generally practical. Without containment the stability of the sand will depend on its slope of repose, and under water this will be about 1 ft vertical to 3 ft horizontal. A top berm will be required for working space around the caisson some 30 to 40 ft wide, where containment is used, and this must be increased where the edge of the berm is not restrained. The resultant sand island will offer a considerable obstacle in the waterway and may interfere with shipping lanes or, under flood conditions, may dangerously restrict the channel width. The large volume of sand required may not readily be available and will require removal when the caisson has been sunk. River flows or tidal movements may tend to produce scour along its faces, washing away sand and endangering the operation. Although coarse gravel can be placed along the faces to reduce scour, this is an additional operation involving not only the availability of suitable gravel but the cost of placing.

In providing a contained sand island for caisson sinking, we start with the known size of the caisson. The minimum width of clearance between the face of the caisson and the containment should be 10 ft. (The reason for this will become apparent as the method is developed.) On one side it will generally be desirable to provide as much as 20 ft of clearance for dewatering operations, storage, and other uses. On all sides a mean must be struck between adequate working space and the added cost of increasing the area of the sand island.

Proposed access to the sand island will influence the method of constructing it, and this in turn will depend upon the type of plant available to the contractor as well as the nature of the waterway and the position of the island in relation to the shore. Where the caisson is close to or inside the shoreline, a single row of steel sheet piling may suffice, since the primary access will be from the land. As we move farther out into the waterway area, circular cofferdams will be required, not only for stability but for working space around the caisson.

Close to shore a solid causeway to the island may be possible, constructed either by progressively filling from the shore or by such expedients as sinking barge hulks loaded with stone to form a base for the roadway. However, at many river sites this may restrict the area of the waterway—dangerously. This can be overcome by constructing a trestle on piling, although there is always the possibility that this also will collect debris and thus act to dam the waterway.

At a further remove from the shore, circular cofferdams are almost mandatory, and the use of floating equipment as well. In these cases, where most of the operation is conducted independently of the shore, the cell diameter of the cofferdam must be sized for the working operations involved. They must be large enough for crawler equipment to operate on the top of the cofferdam, unless barge-mounted equipment can be used. The possible use of barge-mounted equipment will also depend on the depth of waterway surrounding the island.

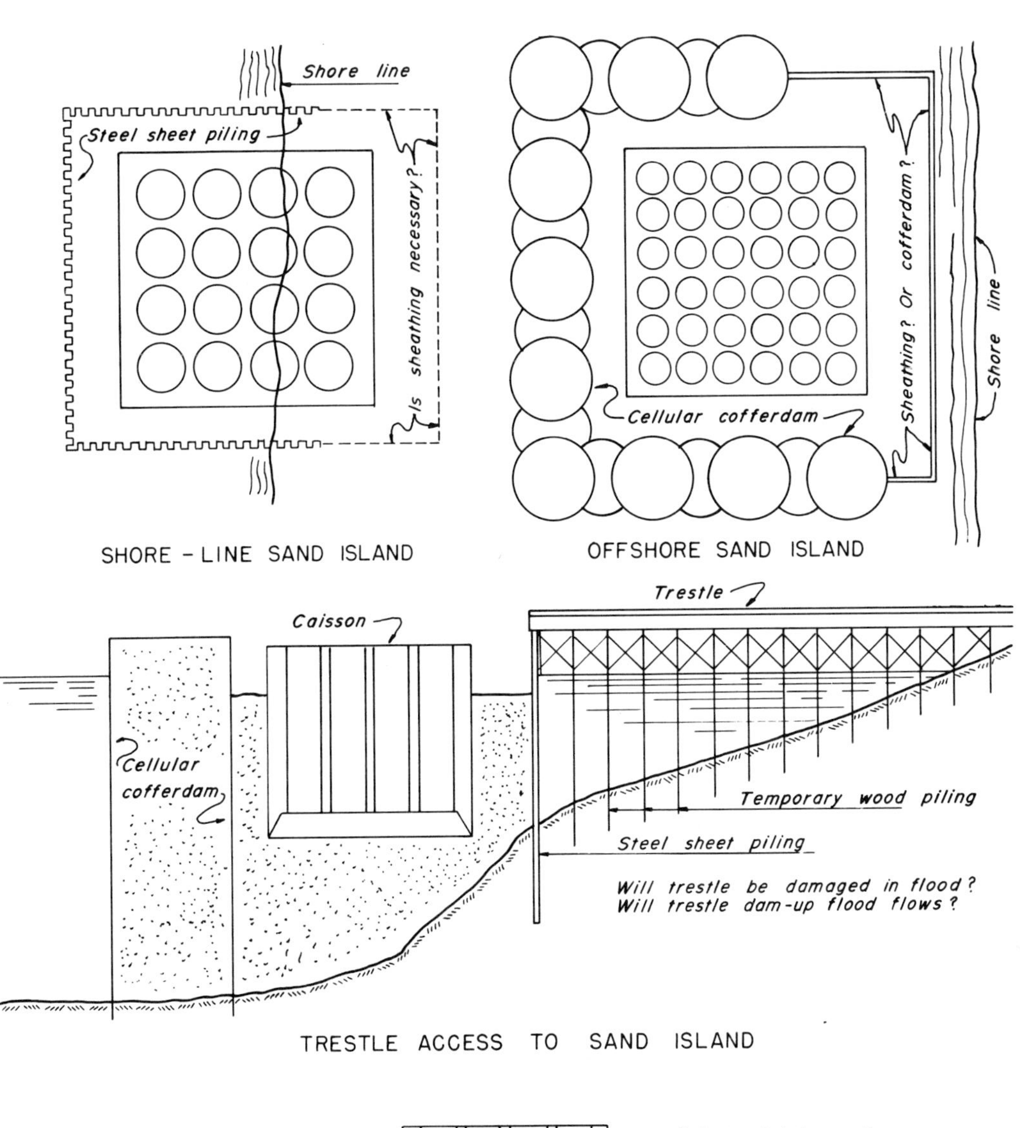

SHORE - LINE SAND ISLAND

OFFSHORE SAND ISLAND

TRESTLE ACCESS TO SAND ISLAND

Substantial berm for working space

Riprap or 1:3 Slope

coarse sand or pea gravel

Coarse gravel 1:3

Original river bed

Dredged

Caisson seating level

UNCONFINED SAND ISLAND

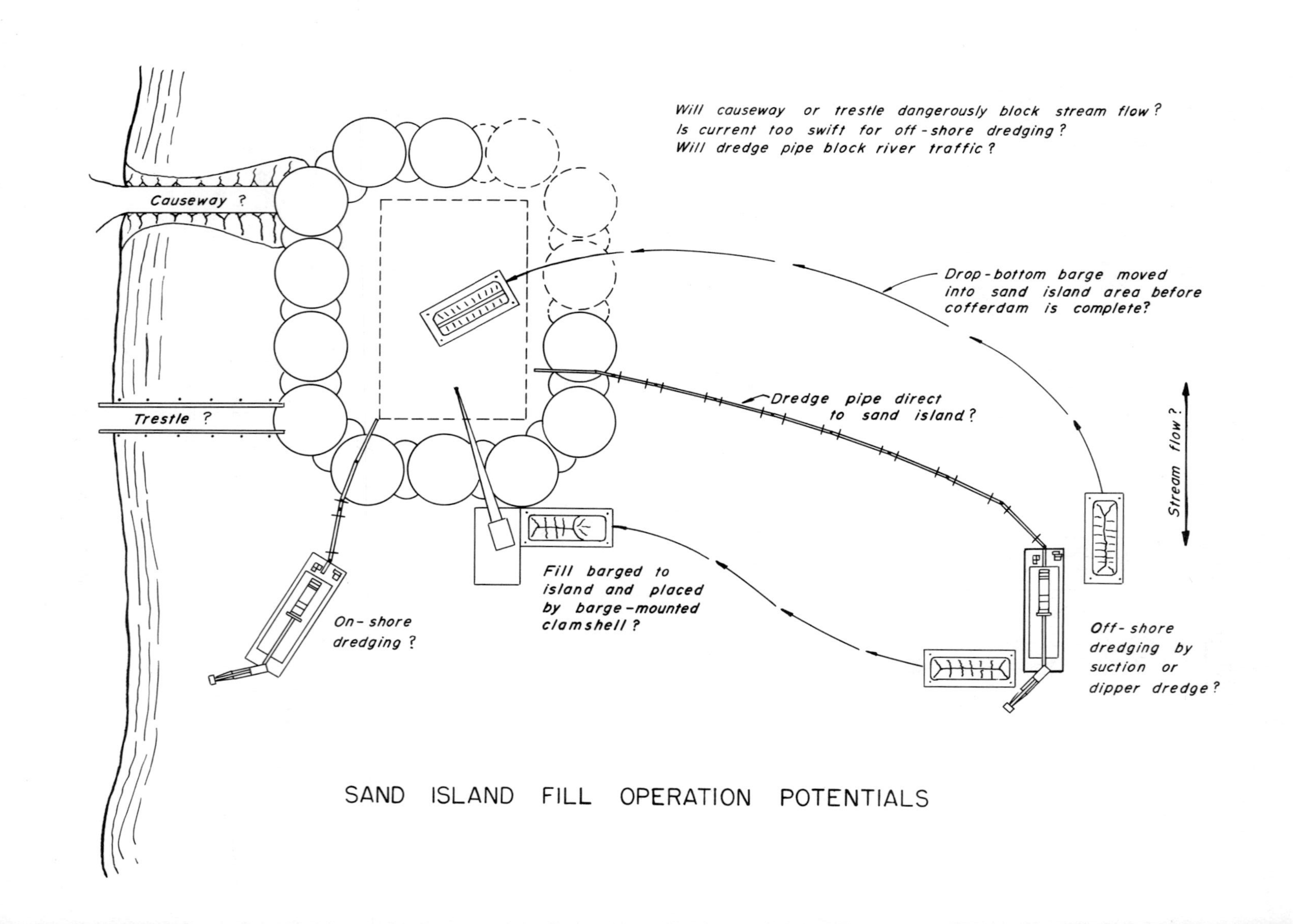
Will causeway or trestle dangerously block stream flow?
Is current too swift for off-shore dredging?
Will dredge pipe block river traffic?
Causeway ?
Trestle ?
Drop-bottom barge moved into sand island area before cofferdam is complete?
Dredge pipe direct to sand island?
Stream flow ?
Fill barged to island and placed by barge-mounted clamshell ?
On-shore dredging ?
Off-shore dredging by suction or dipper dredge ?
SAND ISLAND FILL OPERATION POTENTIALS

8.5 FILLING THE SAND ISLAND

Sand is the critical element involved in the sand-island method of sinking caissons. The use of silt or clay on the one hand or of heavy gravel on the other is wholly unsuitable. Silts and clays cannot readily be dried up to provide a basis for starting the caisson and, lacking any considerable porosity, cannot transport the water required for lubrication. Gravel, on the other hand, has too high a porosity, making it difficult to dry up, and, at the same time, does not flow readily enough to provide workability.

There is no doubt that sands within certain limits of particle size may be more ideal than others; but from an economic standpoint any sand that is not too fine, or that does not contain a high percentage of gravel in sizes over 1 in., will be satisfactory. In considering the use of the sand-island method, therefore, the availability of the sand at a reasonable cost will be a decisive factor. Fortunately, along watercourses, sand deposits from which the fines have been well leached are not rare.

The quantity of sand required within an enclosed island is readily calculated; and while the quantity of sand can be approximated for an unconfined island, the disadvantages of this type have already been noted. One factor, the condition of the waterway bottom, must be considered so far as quantity is concerned. Many waterway bottoms, especially in delta areas where the velocity of flow is low, have thick layers of light flocculent soils lying over them. If the sand is dumped on this water-laden soil, the tendency will be to force it up through the sand—often in huge gobs. In other cases the island becomes filled with a mixture of the two. Such a condition prevents adequate dewatering, fails to provide a dry working plateau, and can occasionally result in erratic caisson-sinking conditions. For this reason it is often desirable to pre-dredge the site down to a solid stratum before placing the sand fill.

Sand requirements for islands are usually inseparable from the requirements for the circular cofferdams enclosing them. Although finer sands or a higher percentage of coarse gravel may not be a disadvantage in the cofferdams, to provide two different sources of sand will generally offer no economic advantages.

Where the quantity of sand required for the island and the cofferdam is substantial, placing by dredging methods would generally appear to be the cheapest. However, the sequence of operations and the time interval required between them will have some bearing on cost. Pre-dredging is a single operation which can be performed continuously until completed. Since it will generally be desirable to fill each cofferdam cell as it is constructed, there will be delays, and dredging for this purpose will be intermittent. With the cofferdam completed, the island can be filled continuously by dredging.

Where a suitable supply of sand lies under water on the same shore as the sand island and within a few thousand feet, direct dredging would be indicated, providing that intervening bank conditions permit placing of a discharge line. Where the supply lies on the opposite shore or near stream center, navigation on the waterway or high velocities of flow may make direct dredging impractical. In these cases, or where sand must be brought in from a distant source, barge-loaded material will be required.

Sands unloaded from barges into cofferdams can be handled by clamshell or can be unloaded from hopper dredges by pumping or dumping. Sands required for the island may require the same methods, but can also be placed from self-dumping barges once a portion of the cofferdam is placed (see illustration), leaving only a limited amount to be placed by clamshell after the cofferdam is completely closed.

Islands constructed close to shore where overland access is possible can be filled progressively by dump truck. Worked in conjunction with excavation for other inshore pier footings, the costs of hauled fill may be quite reasonable, and sand islands built close to the shoreline will generally require a smaller quantity of such fill.

8.6 DEWATERING THE SAND ISLAND

After the sand island has been completed, the first step in its use will be dewatering. This operation will also determine the height to which the sand is filled inside the containment. Dewatering will generally be done in two steps: first, the water lying over the sand will be lowered to sand level; and second, a system of well points will then be installed to lower the water table to the final depth required for placing the cutting edge.

Figure *a* shows a typical case in tidal waters. Here the mean low tide is −3.0, mean high tide is +5.0, and mean tide is +0.0. The maximum high tide is +9.0, and the top of the cofferdam has been set at elevation +12.0 to provide 3 ft additional of freeboard to take care of wave action at high tide. In this case the sand island lies in shallow water close to shore, and the final sand level is to be +0.0.

Figure *b* shows a typical case where the tidal ranges are the same but where the water depth varies from 30 to 45 ft. Here the elevation of the surface of the sand has been set at −18.00 and the top of the cofferdam is set at elevation +8.00. Here a calculated risk has been taken that maximum high tide of +9.0 will not occur during the initial stage of construction and that overtopping of the cofferdam will not have serious results during the sinking operation. The saving involves the cost of placing and removing 18 ft of sand in the island.

In both cases a cutting edge 7 ft high is to be constructed in the dry, the top of the edge to be approximately at sand level. This will involve lowering the water table at least 7 ft; and to provide a margin of safety, at least 10 ft should be anticipated. The lowering of the water table 10 ft below sand level will require the use of well points.

The water to be handled comes from two sources: either through the interlocks of the cofferdam in a relatively limited quantity or up through the river bottom on which the sand island rests. The quantity of water flowing upward will depend on the nature of the underwater soils and their porosity.

In the first case, if it is assumed that the water level within the island is close to elevation +0.0, the well-point header can be blocked up off the sand and the pumps can be set on a platform resting on the sand. The center line of the header might be set at elevation +2.0 and the pump suction at elevation +4.0, providing a total suction head of 14 ft. If the pumps were to be set on top of the cofferdam, the suction would be at elevation +14.0 and the total suction head would be 24 ft. This is excessive for foolproof operation, 18 ft generally being considered the maximum.

In the second case there are two alternatives. Because of the additional 18-ft head of water to be considered, greater flows into the island may be expected, regardless of the nature of the bottom. Because of the limited freeboard allowance for high tide, the likelihood of submerging the pumps is greater.

A two-stage well-point system could be used here by sloping the outer edges of the sand up from −18.00 to +0.0 so that the first stage is set at +0.0 and the second stage is set at elevation −18.00. The added space for sloping the sand within the island seems inadvisable, however, since an additional 18 ft, at the least, will be required (providing a 1:1 slope).

In the second case the more advisable expedient would appear to be that of pulling the water level down by using open suction pumps placed at the four corners, or as required (a number of smaller-sized pumps are generally to be preferred over a few large pumps), by which an initial lowering is accomplished. Since the pump suctions will be 28 ft above the final drawdown level, complete drawdown can not be effected, and a 6-ft berm on which to rest a well-point header must be provided.

Another alternative, in the second case, is to set the open suction pumps on brackets welded to the face of the sheathing somewhere near the elevation of +0.0, providing a single drawdown to sand level. Once the well-point system has been installed at the lower level, the open suction pumps are lifted off the brackets and used as standby units in case of flooding.

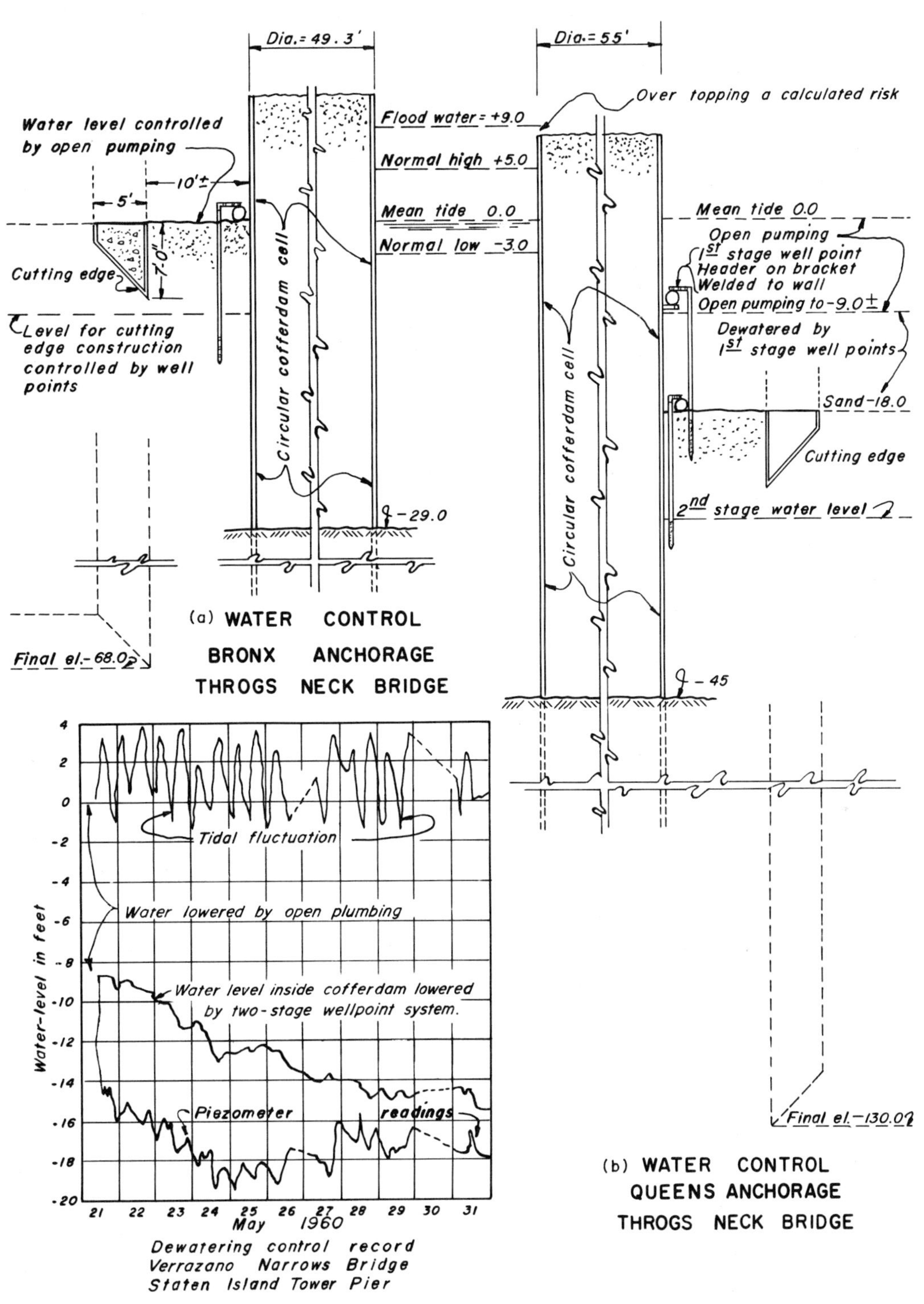

CAISSON DEWATERING

8.7 THE CUTTING EDGE

The cutting edge of an open caisson will generally consist of a steel shell filled with concrete, although there are occasional exceptions in which the cutting edge is constructed wholly of concrete. For adequate protection of this key item, however, the steel shell seems desirable.

The illustration shows several examples of cutting edges. Since the purpose is implicit in its name, those with the sharpest edges would seem the most desirable; but the long thin edge is subject to being turned over when the caisson is momentarily hung up on an obstruction, throwing all the weight on that part of the edge. Although it is termed a cutting edge, very little actual cutting is expected of it, its purpose being principally to ease the caisson into an excavation provided ahead of it.

The height of the steel shell for the cutting edge should be the full height of the working chamber. In the previous discussion a height of 7 ft was mentioned. This would be a minimum height permitting suitable working space, but heights of 8 and 9 ft are frequently used. The cutting edge surrounds the entire caisson and, where the working chamber is compartmented, interior cutting edges often are provided on these walls also.

From the top of the working chamber up, the caisson mass is pierced with dredging wells from 14 to 18 ft in diameter. Their walls provide stiffening for the exterior caisson walls and shorten the open span. In the working chamber there is no such stiffening factor, and the steel shell should be designed to reinforce the chamber walls in their full span.

Working-chamber walls are from 3 to 5 ft thick. The angle of slope at the bottom is generally 45 deg. For a 5-ft wall, the sloped face will therefore have a 5-ft vertical component. In a 7-ft chamber the steel shell should extend vertically upward the additional 2 ft to the bottom of the dredging-well concrete (the ceiling of the working chamber).

On exterior walls the outside of the cutting edge is maintained vertically and flush with the exterior concrete wall above it. Cutting edges under interior chamber walls are sloped up from a common center.

Cutting-edge walls will generally be constructed of ⅜- or ½-in. plate laid over shaped stiffeners of the same thickness and welded. As will be discussed, provisions for jetting will be constructed into the shell. Additional reinforcing steel will generally be desirable to provide doweling for the concrete above.

The best procedure is to have the cutting edge pre-fabricated in the shop in sections limited in length to the capacities of transportation facilities and, in weight, to the capacities of the lifting equipment to be used. Corners or intersections of interior and exterior walls should be included in the prefabrication, leaving only straight intermediate welds to be made in the field.

In Sec. 8.6 a height for sand in the sand island was discussed. This would be about the top elevation of the cutting edge. In placing the sand, this height should be kept down to provide for the volume of sand displaced by the cutting edge. Once dewatering is completed, a small front-end loader can be slung into the sand island to grade it off. A small backhoe can follow to dig a trench for the cutting edge. Excavated sand should be piled on both sides of the trench so that backfill can be pushed against the completed cutting edge with a minimum of rehandling.

Match-marked sections of cutting edge are slung into place and set on wooden chocks braced against the trench bank or on split wooden cradles, for easy removal. With sand islands, line and grade can be controlled from stations on the cofferdam or on shore; but in some cases, survey islands established in the waterway may be needed.

When the cutting edge has been completely welded, the trench in which it rests is partially backfilled between chocks and, with the cutting edge thus secured, the chocks are removed and backfill completed. The sand area around the cutting edge is now graded to an elevation 2 or 3 in. below the top level of the steel for a working area, and preparations for filling the cutting edge are begun.

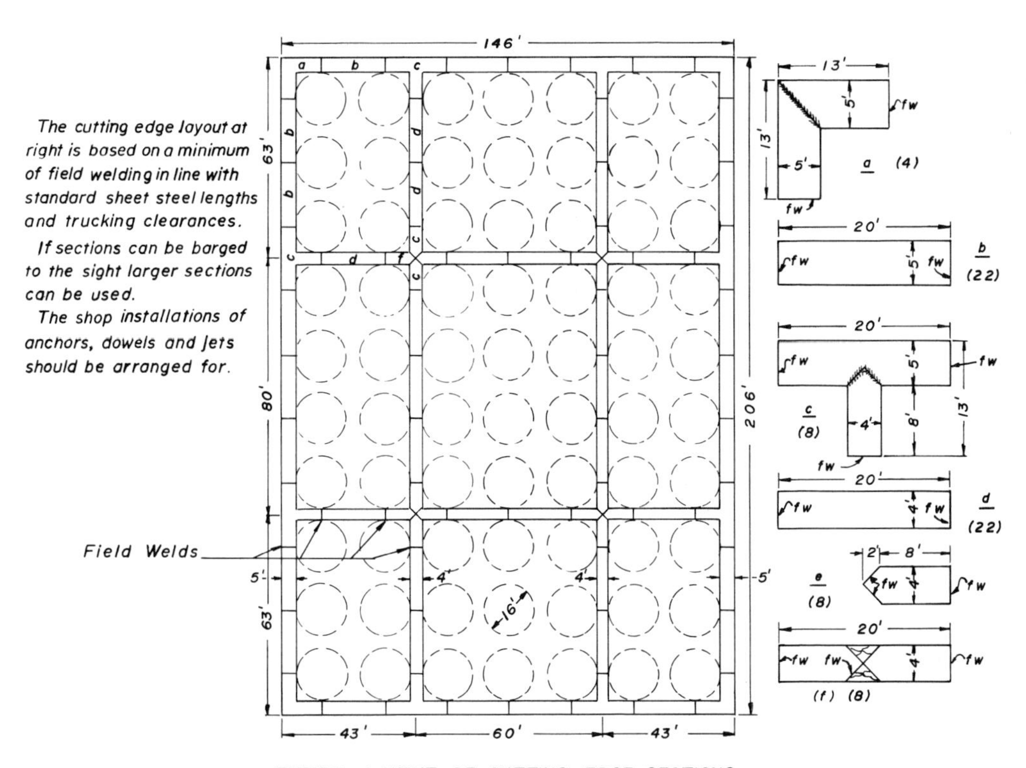

The cutting edge layout at right is based on a minimum of field welding in line with standard sheet steel lengths and trucking clearances.
If sections can be barged to the sight larger sections can be used.
The shop installations of anchors, dowels and jets should be arranged for.
146'
206'
63'
80'
63'
43'
60'
43'
5'
4'
16'
Field Welds
13'
5'
fw
a (4)
20'
b (22)
c (8)
8'
4'
d (22)
2'
e (8)
(f) (8)

TYPICAL LAYOUT OF CUTTING EDGE SECTIONS

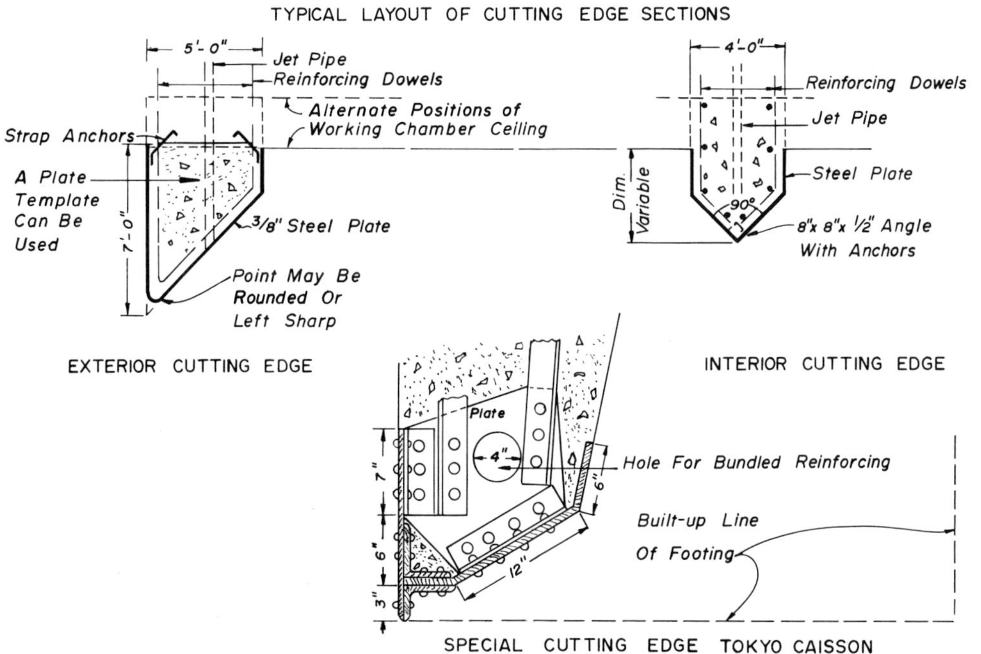

EXTERIOR CUTTING EDGE

INTERIOR CUTTING EDGE

SPECIAL CUTTING EDGE TOKYO CAISSON

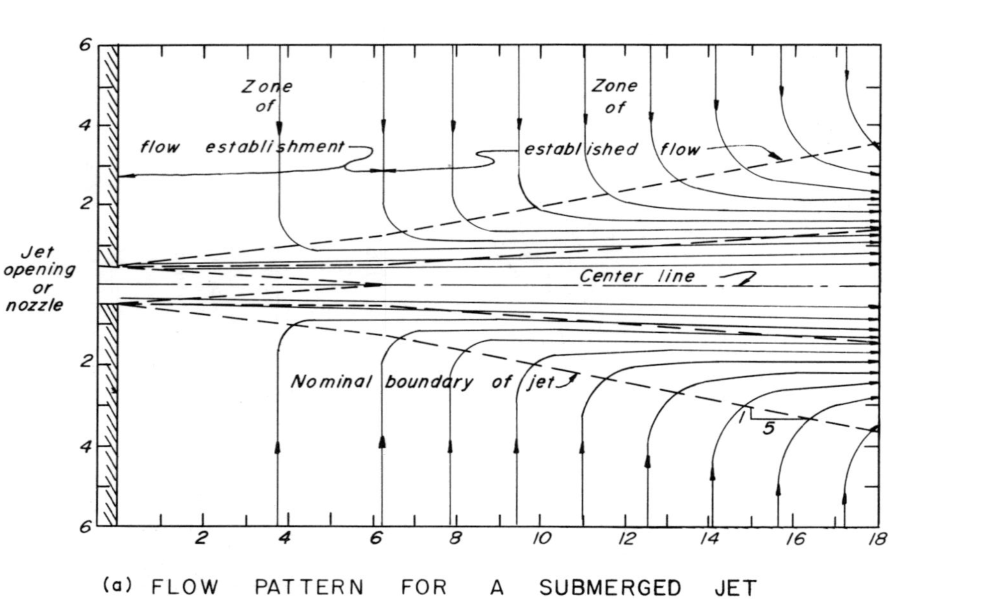

(a) FLOW PATTERN FOR A SUBMERGED JET

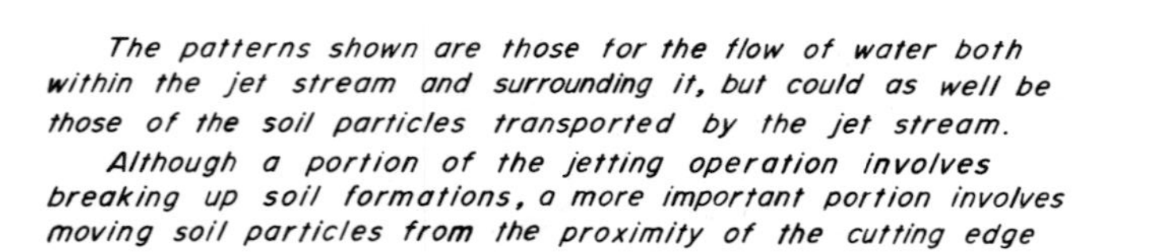

The patterns shown are those for the flow of water both within the jet stream and surrounding it, but could as well be those of the soil particles transported by the jet stream.

Although a portion of the jetting operation involves breaking up soil formations, a more important portion involves moving soil particles from the proximity of the cutting edge toward the center of the dredge wells.

It would therefore appear that portable jets lowered through sounding wells, which can be turned in varying directions, are more desirable than fixed jets along the cutting edge.

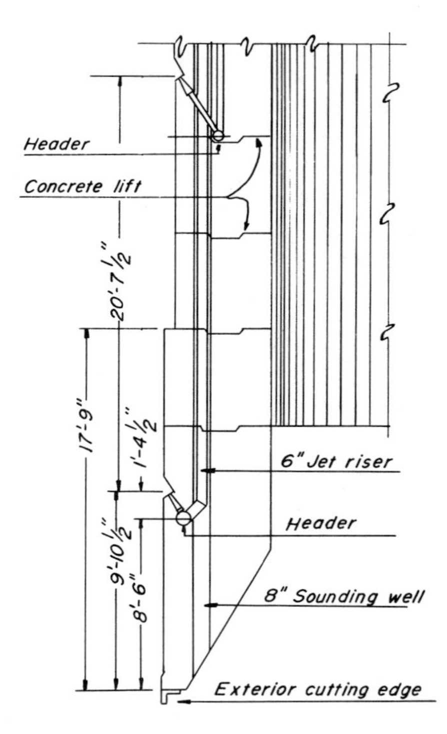

(b) JETTING ARRANGEMENT
WALT WHITMAN BRIDGE

1" Jet nozzles 12" long
2 1/2" Inlet pipe

The process of sinking an open caisson under water, that is, without the intermediate process of drying it up, is accomplished by the use of jets. A high-pressure stream of water is directed against the soil under and adjacent to the cutting edge, loosening it and tending to heap it inside the chamber, where it can be picked up by various means. Since any jets secured to the face of the cutting edge would be subject to damage during the sinking operation, jet outlets are provided in the cutting edge before it is filled with concrete.

Various arrangements of jets have been used at the cutting edge. Some have been pushed through and angled down the slope. Others have dropped vertically downward through the plate, while still others have been turned inward at an angle to the sloping surface.

Jet outlets may be straight runs of pipe 2 in. or less in diameter, or they may terminate in jet nozzles 1 in. in diameter fed by a 2-in. pipe. Jets are usually set on 5- to 7-ft spacings, tapped into a header laid down in the core of the cutting edge. The header, a 4- or 6-in. pipe, will be fed by a vertical riser, customarily an 8-in.-diameter pipe extended up to the surface of the caisson and provided with a connection to the high-pressure pumps. The outlets of these systems are *submerged* jets or nozzles, and a brief glance at their characteristics might be of interest.

Figure *a* indicates the nature of the flow from a submerged jet. It is apparent that the greatest force tending to move soil particles occurs at the center, gradually decreasing therefrom. It is also apparent that if the force is sufficient to move the particle at all, it will carry it in suspension some distance until, as the diameter of the cone of force and the distance from the orifice increases, the velocity will decline and the particle settle out. At the outer fringes of the force cone, however, the particles will tend to be hurled from the main path into the surrounding medium and then may be partially drawn back into the stream, tending to rotate the particle as it is alternately tossed from the stream and drawn back into it.

With a single jet, there is the well-known effect of producing a mound of soil around the force cone. However, if a second jet were set at an angle to the basic jet to provide a new force cone for the rotating particles, the soils could be transported to a further remove from the main cutting edge. So far as is known, such dual jetting has not been attempted.

The jetting operation is discontinuous, being shut down to excavate or to provide another lift on the height of the caisson. It has been found that when the pressure, generally about 300 psi, is cut off, a certain amount of back pressure momentarily develops, forcing soil particles substantial distances into the jet nozzle, and often plugging it entirely. As a result, it is becoming more common practice to set 8-in. pipe risers or "sounding wells" in the caisson walls with outlets through the cutting edge. These have a dual use; individual jets can be lowered through them and turned to play in any direction, and, as their name suggests, they can be used to "sound" the bottom under the cutting edge to determine soil depth.

A separate system of jets is frequently introduced into the *outside* face of the caisson, sometimes puncturing the cutting edge but also placed at higher levels. The purpose of this series of jets, piped similarly to those at the cutting edge, is to provide a lubricating sheet of water around the caisson. These outlets are turned to jet diagonally upward, washing soils away from the face and preventing binding during the sinking operation.

In order to control either jetting operation, piping must be valved so that jets in one area, where stiff resistance is being encountered, can be operated, while those in other, perhaps softer-bottomed areas, are shut off. The control of jetting operations is an important aspect of the entire sinking operation and has led to a more extensive use of independent jets lowered through sounding wells in place of the fixed jets described.

8.9 CONCRETING THE FIRST LIFT

When the cutting edge has been completed with all jet piping and accessories installed, it is filled with concrete. The methods involved in placing this concrete are similar to those used in placing the concrete in the first lift and need not be separately delineated. We are left with a concrete surface from which reinforcing-steel dowels project and in which a key has been provided. In some cases, as when there is the possibility of conversion to a pneumatic caisson, a water stop is also desirable.

The basic principle of sinking an open caisson is very simple and has already been indicated; as we increase the load on it by the addition of poured-concrete lifts, we remove the support from beneath it, permitting the increased weight to force it into the created void. With structures the size of caissons, however, the stresses that can be induced in the concrete and transferred to the reinforcing steel by a difference of a few inches in level from end to end can cause damage. In the initial stages of sinking, in a sand island, the extent of such deflections is completely unpredictable, although at later stages some support can be expected from the surrounding soils.

The solution to controlling these initially induced stresses is to develop a substantial mass of the structure, and the height of the first lift of concrete placed before the island is flooded and sinking procedures begin is determined by this factor. Consequently, depending on the overall size of the caisson, the first lift placed will be from 20 to 30 ft in height.

It is not uncommon for a caisson to have 500 cu yd of concrete per vertical foot, so that a 20-ft lift would represent 10,000 cu yd. Lifts of this magnitude cannot economically be poured in a single continuous operation.

The two economic factors involved are the design and characteristics of the formwork and the capacity of the concrete plant. The concrete plant is composed of three elements: the storage and mixing provisions, the transporting methods, and the placing equipment. Let us consider these in reverse order.

Except in special cases, concrete will be placed by buckets hung from crane hooks. Cranes will be needed for handling forms and, with clamshell buckets attached, for excavation. Special additional equipment for placing concrete is therefore economically unjustified. The working space available will limit the number of cranes that can be used. In the case illustrated, to reach the interior walls could require a crane operating radius of 70 ft. With the boom turned along the cofferdam, this could require as much as 100 ft. Since only 250 ft is available along a side, only two cranes per side could be effectively used.

A 2-cu-yd laydown concrete bucket will represent a 5-ton lift, which at a 70-ft operating radius may determine the size of crane to be used. It will require a minimum of 4 min to load, lift, swing, lower, and unload this bucket. This is equivalent to 15 lifts per hour or a rate of pour, per crane, of 30 cu yd per hr. For four cranes the rate of concrete required from the concrete plant will be 120 cu yd per hr.

With sand islands, transportation may be by mixer truck over causeways to the cranes from an on-shore mixing plant, by lighter from the same plant, or by lifting directly from the discharge hopper of a floating mixing plant. Depending upon the relative location of the island, a combination of these methods may be required.

It is possible to beef up the size of buckets, cranes, or mixing plants; but the bottleneck will generally occur in the transportation stage, more particularly since, even on the first lift, concrete must be placed with some uniformity so that loading on the caisson is evenly distributed.

It is standard practice to pour concrete in 1-ft layers over a wide area, but in caisson pours it has been found more practical to pour the whole depth at one time. This factor, together with those previously discussed, has established a pattern of individual pours 4 to 5 ft in depth.

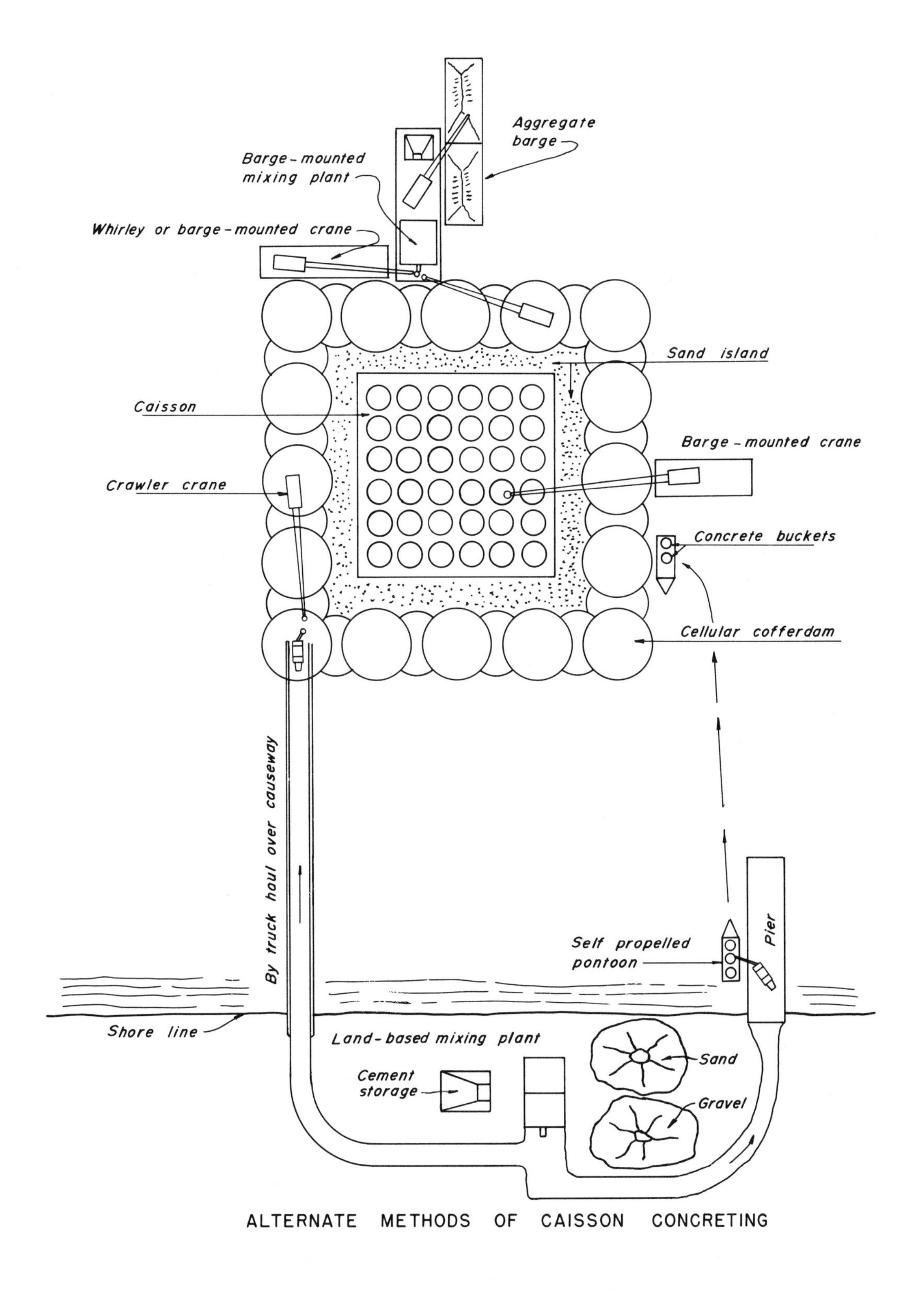

ALTERNATE METHODS OF CAISSON CONCRETING

8.10 CAISSON FORMWORK

The design and construction of caisson formwork will not be substantially different from that discussed under previous sections dealing with mass concrete. However, the controlling factors will be those which will prevail during the later stages of construction, while concrete is being placed to sink the structure, rather than those dictated by the first lift.

The bottom of the cellular structure, which will be the top of the working chamber, can be poured directly on the sand within the cutting edge. The sand is leveled off to the underside of the working-chamber ceiling, and a vapor barrier is laid down over it. Since the interior forms will rest on the vapor barrier, a reinforced paper or the heaviest plastic sheet should be used. This barrier is required to prevent draining water or grout into the sand from the concrete.

There are two elements of formwork to be considered: the circular forms for the inside of the dredging wells and the straight-wall forms for the exterior caisson walls. The design and construction of interior circular wall forms has already been discussed. Since they will be re-used a number of times, they should be well made and easily strippable. Although in a few cases what are essentially metal cans have been used and left in place, economically a re-usable form seems more desirable.

The exterior wall forms, since they are a one-sided form, require special bracing design. Since there can be movement of the caisson while the forms are in place, bracing the exterior forms against the sheathing containment of the sand island will not be practical. Interior bracing from wall to wall is not satisfactory, being too great a distance and introducing too much interference to other operations. Limited bracing at the point of tangent to the circular forms is insufficient for the entire wall. This leaves only the recourse of diagonal interior bracing down to the mass of concrete previously poured. These provisions must start with the fill placed in the cutting edge.

The first lift of concrete will require special bracing for the circular dredging-well forms; but for succeeding lifts the forms can be hung on, and pulled into place on, inserts left in the preceding pour. Intermediately, and at the top, bracing can be effected by tie-rods at the four tangent points to adjoining circular forms or, in the single case, to adjacent straight forms. The tops can be braced from unit to unit on the upper surface.

Although we are dealing, as has been said, with mass concrete, considerable precision in form setting is desirable, so that, as the forms are raised, the bottoms of the forms will accurately fit the concrete surface left by the tops of the forms. Numerous adjustments are costly, whether in damage to the forms or in the labor required for these adjustments.

The exterior wall forms can be ganged and lifted vertically as they are stripped and can be immediately reset. The circular forms, however, will be removed completely together with their bracing when concrete has reached their upper surface, to permit excavation in the dredging wells to proceed. Since they will be stored outside the caisson area, temporarily, they must be rigidly constructed to prevent distortion during this step.

Both steel and wood forms have been used for caisson formwork, and in one case pre-cast plank has been used on the exterior walls, fastened to lightweight steel trusses imbedded in the concrete. Weight is not an important problem since all handling can readily be done by crane; and here, consequently, a well-constructed metal form, providing the smoothest face to the concrete, seems the most desirable.

Although the first cost of metal forms is considerable and a substantial number of re-uses are required to balance the lower cost of wood forms, the labor and equipment costs of setting and stripping are also important. The use of such devices as the hydraulic rams mentioned in Sec. 7.37 for expanding and contracting a set of forms, and practical only where metal is used, could be adapted to dredging-well forming.

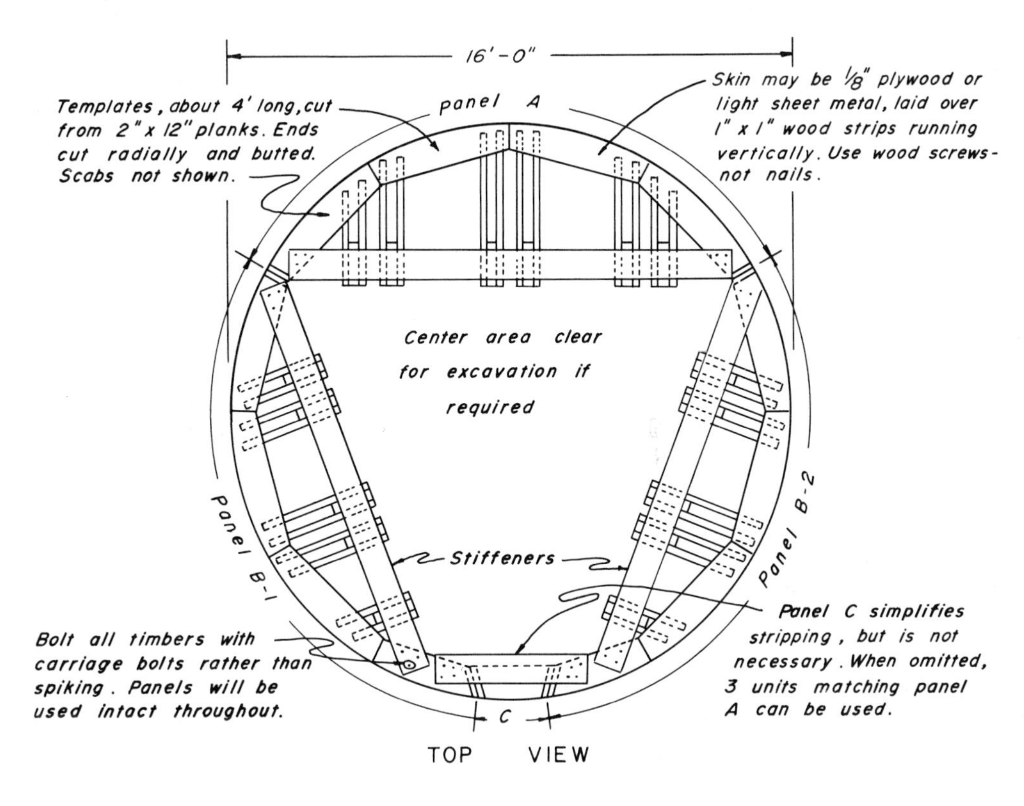

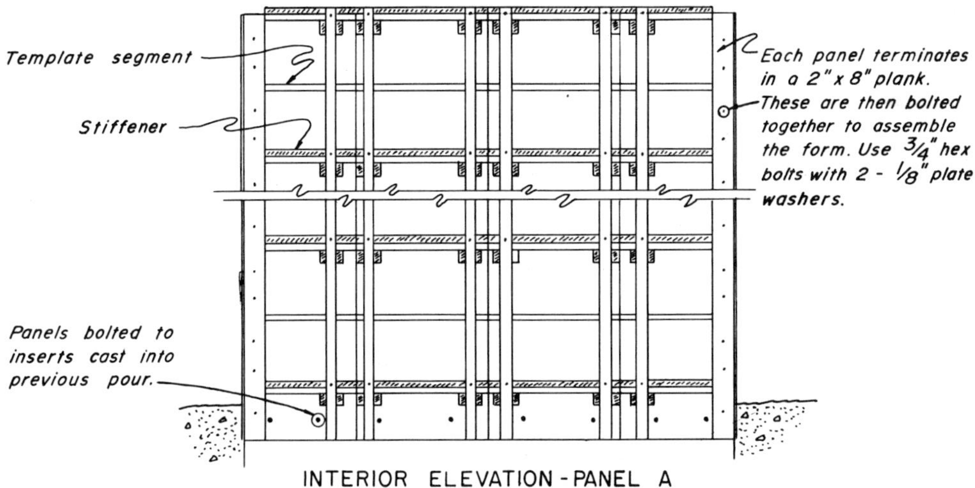

A SMALL RADIUS CIRCULAR WOOD FORM

SCALE 3/16" = 1'-0"

The wood form shown represents only one of several approaches to dredging well forms. Metal forms with bent channel, horizontal studs or straight, vertical, channel studs, covered with 11 gauge, or lighter, steel plate have been used. Liner plates can also be used.

When the first lift of concrete has been completed, the well-point pumps will be shut down to permit the water to rise. Since, as previously discussed, these may be at or below the unpumped water level in the sand island, they will generally require removal. The question of removing the well points themselves must be considered at this time. There is a slight possibility that their use may be required again. If, during the sinking operation, the caisson edge should encounter an unanticipated soft stratum and plunge downward to the extent of submerging the top of the concrete, then the island would require drying up again to add another lift of concrete. Careful control, however, should make this possibility remote. The alternative, that of maintaining the sand island during sinking through 7 to 10 ft of dry sand, will generally involve too much frictional resistance.

One possibility, controlling the water table by throttling the well-point pumps so that during excavation periods the water rises to the level of the sand only and during concreting operations the water is lowered sufficiently to provide a dry surface, has been tried. In any case, only the sand around the exterior perimeter of the caisson, between it and the cofferdam, is involved.

There are three methods of excavating beneath or within a caisson. The first of these, employing the clamshell bucket, will be used initially to excavate in the dredging wells, in the dry, before removal of the well points, to avoid handling more water than necessary. At later stages the clamshell may also be used when required, to break up compacted layers or to remove coarse gravels, rock fragments, or boulders. Two- or three-cubic-yard heavy-duty buckets are generally used. These will be lowered with the jaws open, and in this position the limiting size will be determined by the diameter of the dredging wells.

A second method of excavation is suction dredging. Eads had a dredge designed for excavation of his Mississippi River Bridge caissons, and the method has been used extensively since. The name for the openings in the caisson block—dredging wells—derives from this source. Under the best conditions, the ratio of soils to water is never more than 1:5; and as depth increases, yield decreases. Because the suction-line length must be constantly increased vertically downward, the cutterhead used in open dredging is impractical. As a result, a third method is now more commonly used.

The third method of caisson excavation employs an air lift, two varieties of which are shown in the illustration. In one type, a mass of air under pressure is forced into the riser pipe, lifting everything above it and drawing soils and water from the area below it. In the other case a mass of air bubbles is forced into the riser, aerating the soils-water mixture and making it lighter so that it rises to the surface irrespective of the pressure applied.

One advantage of the air lift is that the entrained air is dissipated to the atmosphere when released and does not have to be handled. With expert manipulation, therefore, the ratio of soils to water is higher than that with dredging. As a consequence, the size of riser can be smaller and the unit is more readily maneuvered. The air lift can be dropped down not only through the sounding well but down the dredging well and, in the latter, can be pinpointed to the precise spot desired.

The disposition of the excavated soils will depend on conditions. In sand-island construction, performed contiguous to the shore, the discharge from dredge pumps or air lifts can be carried ashore and deposited as fill. Clamshell excavation can be loaded on trucks for hauling ashore. In either case the excavated materials can be loaded onto barges, having provisions for draining off the water, and towed off to final disposition. To deposit this material on the outside of the sand island, under water, will seldom be practical, since in most cases it must later be re-handled.

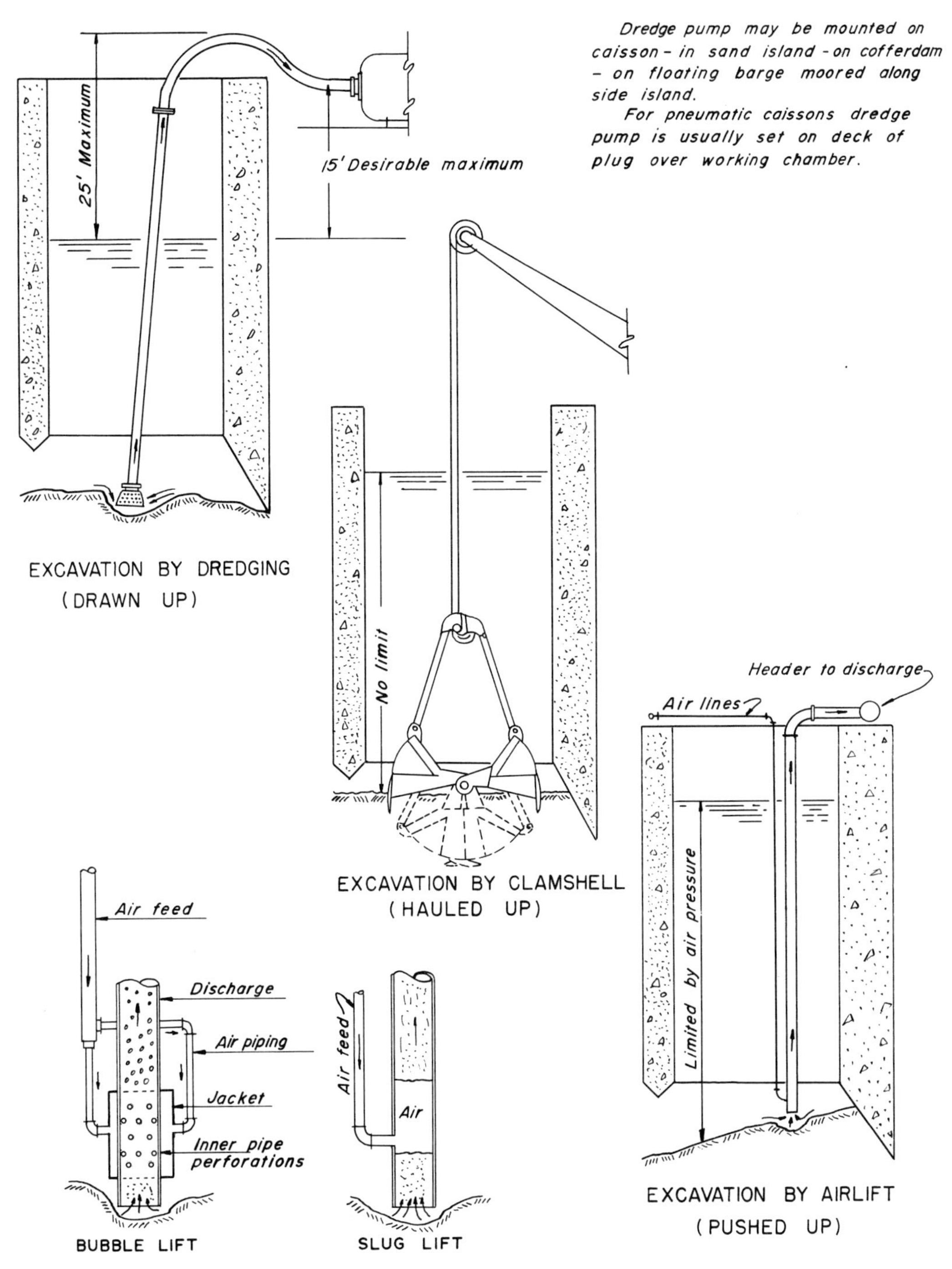

CAISSON EXCAVATION METHODS

8.12 SINKING THE CAISSON

At one time there was serious danger of tilting a caisson during sinking and, in the early 1930s, in several cases, tilts up to 30 deg from the vertical occurred. To right these tilted caissons required herculean efforts; and frequently when they had finally been tilted back to the vertical they were found to be no longer in the designed position. The sand-island method, now being discussed, and the corral method, covered in the following pages, have largely eliminated the dangers of excessive misalignment, but some dimensional tolerance must still be allowed.

The three possibilities of misalignment are tilt, twist, and shift. The specified tolerance can be checked only when the caisson has been finally seated, and careful control must be exercised from the start of sinking. Since, particularly in the sand-island method, the initial positioning of the caisson can be carefully controlled, first consideration must be given to the danger of tilting the caisson as the sinking starts.

Control of tilt involves control of excavation beneath the cutting edge so that as the caisson moves down under its own weight, it comes to rest at a lower elevation that is substantially level. This means that whatever method of jetting is used, it should be possible to employ it uniformly around the whole cutting edge simultaneously. It cannot be assumed that the soils being excavated will be homogeneous, and consequently the jetting system must be set up so that greater pressure or none at all also can be applied along any particular section of the cutting edge.

Soft formations of apparent uniformity may still contain substantial obstacles; boulders or masses thereof, trees or tree stumps, or sunken craft may all be sufficient to at least temporarily support one edge of the caisson. Where excavation is proceeding at an accelerated rate on only three sides, a tilt can develop very rapidly.

Generally the central area within the cutting edge will be excavated deeper than the strip along the cutting edges, to permit these soils to be washed to the center. However, in soft formations, a reverse proceeding may be necessary to prevent too rapid dropping of the caisson.

Settlement levels of fixed elevation markings should be taken on four diagonal points on the caisson. These levels will at once indicate tilt, but they should also indicate the elevation of the cutting edge. This should be constantly checked against a plotted log of the test borings so that the superintendent knows precisely what type of formation is currently being penetrated.

Even relatively minor tilts should be immediately corrected by limited undercutting of the high side or, in some cases, by first pouring an additional lift, to provide excess weight, and then undercutting.

Emphasis has been placed on tilting since tilting can produce both twist and shift. Twisting occurs where the caisson rotates on its center axis, and shift occurs when the caisson moves horizontally in any direction. The two movements may take place simultaneously, or one may occur independently of the other. Twist and shift occur most frequently during attempts to right tilts of substantial amounts.

Although, in the sand island, it would appear that the sand and soils surrounding the caisson should provide horizontal stability, the tilting and righting operation will distort this enclosure and change its density, at least momentarily, permitting the caisson to shift toward the side of least density.

Underwater obstacles, or changes in density of the subsoils from end to end of the caisson can cause both twist and shift. This is particularly the case where these conditions exist just outside the caisson area and cannot be detected by an examination within the working chamber.

Dimensional tolerances are frequently pre-established as 12 in. for tilt and a like amount for twist or shift. Many caissons have been sunk in recent years well within such tolerances. It should be recognized, however, that caisson sinking is working in an unpredictable medium and that even with the greatest precautions exercised, the results will seldom be precise.

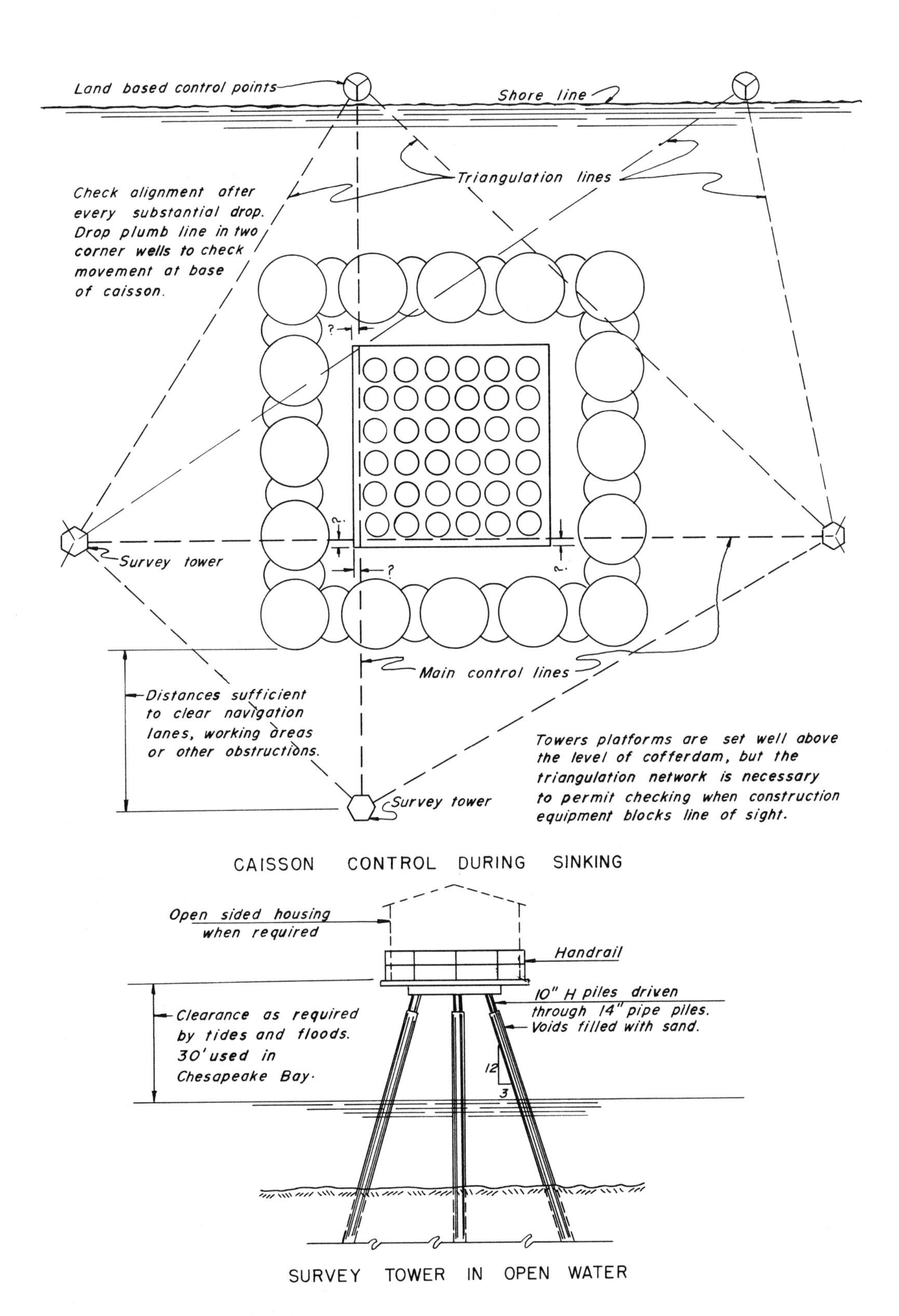

CAISSON CONTROL DURING SINKING

SURVEY TOWER IN OPEN WATER

8.13 CONTROL OF CAISSON SINKING

The use of three 4-ft-deep pours to provide a total lift of 12 ft has been previously noted. Although many other combinations are just as practical, this arrangement will be discussed here. The procedure for each lift will be essentially similar, whatever the height of the lifts.

When sinking has progressed to the depth desired, excavation in the dredging wells is terminated and suction lines, air lifts, or portable jetting units are removed. The exterior wall forms will be lifted and reset as soon after the final pour of the preceding lift as possible (24 hr should be an ample time allowance). Since they are to be braced down to the concrete previously poured, they can be set in final position at this time.

Feed lines for the jet piping system or sounding wells are now extended up to above the height of the next pour before reinforcing steel is set, to provide maximum working space for making piping connections. All reinforcing steel for the lift is then set and the forms for the dredging wells are dropped into place.

At whatever level the water may be maintained over the sand outside the caisson, there is still, below it, the sand on which scaffolding can be set. This is not the case within the dredging wells. Generally, inserts will be left in the upper part of the wall to which scaffold brackets can be bolted. These brackets can be set and the scaffold plank laid over them by a man in a boatswain's chair lowered from the crane boom or by men working from ladders hung over the side and secured to the dowel steel. Scaffold plank will generally be plank made up in sections to fit the shape of the dredging well rather than the straight plank used on continuous walls.

Once a working platform is provided, form panels are lowered and set by men working off the platform. Dredging-well form panels are conveniently fabricated in thirds, with the third unit providing for final closure after the previous two panels have been set. Continuous metal shells have been used and often left in place, but even here some provision must be made for adjusting their perimeter to the face of the concrete.

Once the dredging-well form is in place and has been braced interiorly, bracing is carried from well to well and to the outside form, at the top. Usually this top bracing will also serve as a working platform during pouring.

When the entire lift has been poured and initial set obtained, the exterior forms are again raised, the dredging-well forms removed, and the scaffold platform cleared out. A period of 3 or 4 days will generally be required to permit the concrete to reach sufficient strength to resist pressures developed in sinking. During this period and after the forms have been moved, headers feeding the jet piping can be laid down and connected to the pressure pumps, and air lifts can be set up and connected to the air-supply and discharge lines.

The weight of caisson concrete per vertical foot can amount to 1,000 tons, or 12,000 tons for a 12-ft lift. Unless the cutting edge is very solidly founded at the beginning of the pour, considerable downward movement is possible. Observations must be continued throughout the pouring period to prevent excessive sinking which might lead to tilt. It may even be necessary to stop pouring after the first or second pour in order to provide remedial measures. Since such measures would involve jetting, the dredging-well forms might have to be removed. Too rapid sinking could also engulf the forms either on the outside or in the wells, in the one case burying them below the sand and, in the other, requiring their removal under water.

In view of these possibilities, it is desirable to carry a substantial freeboard between the level of the sand (or level of the water) and the bottom of the lift. Unless predictable firm soil conditions exist at the cutting edge, it may be desirable to maintain a freeboard of a full 12-ft concrete lift above the water level at the beginning of the second lift pour. Since soil conditions will seldom be uniform vertically, it will be necessary to expect varying depths of lifts as the sinking operation proceeds.

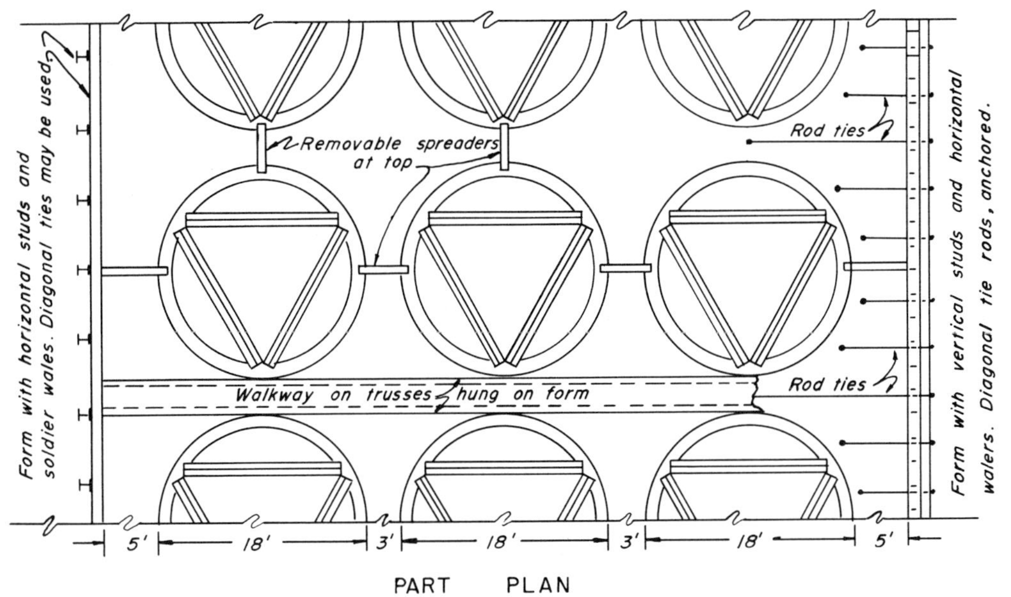

PART PLAN
CAISSON FORM ARRANGEMENT

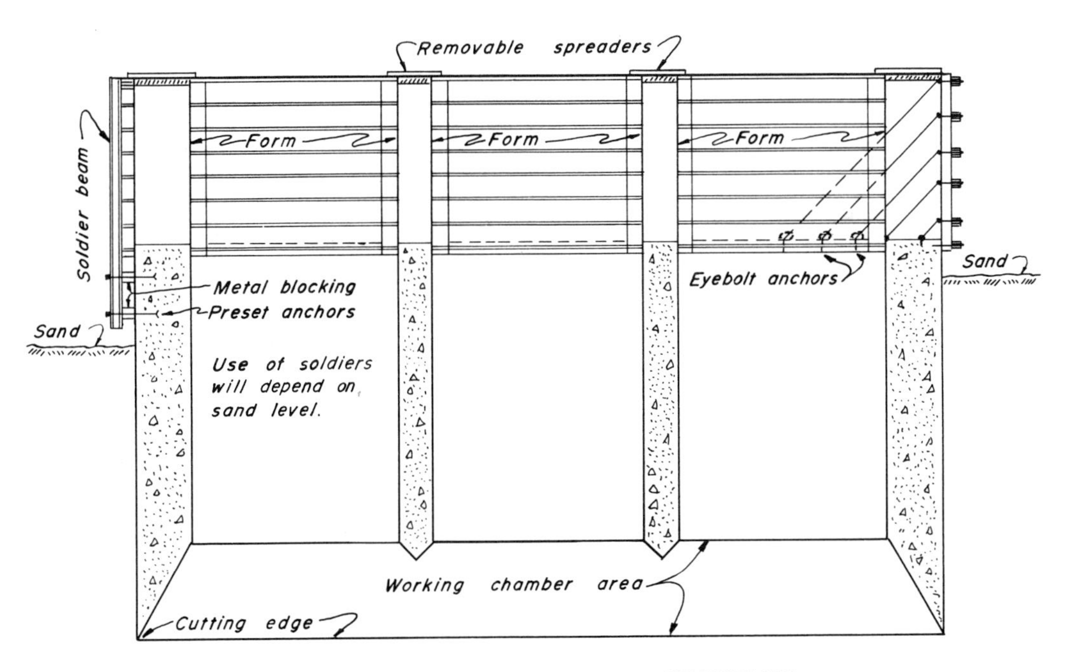

SECTION ON DREDGING WELL CENTERLINE

FORM SET-UP FOR CONCRETE LIFT

Jet piping and other details not shown

8.14 CLEARING THE CAISSON BOTTOM

In Sec. 8.13 provisions for unanticipated settling of caissons were considered. The problems of dealing with underwater conditions where settlement is not continuous must now be considered. The general aspects of underwater diving, underwater cutting and welding, and underwater blasting have been discussed.

Underwater obstacles (including hard ledges that are not substantially continuous under the whole cutting edge) will manifest themselves either by hanging the entire caisson up or by causing tilt before the superintendent is aware of their existence.

Underwater photography and closed-circuit television equipment are now available which may be lowered to observe without the use of a diver. However, since the soils on the bottom have been disturbed, time must be allowed for the water in the chamber to clear. In sands or gravels this presents no problem. In clays or silts, suspended particles may cloud the area for a considerable length of time. A submersible pump, lowered while operating, may help to clear up this condition. It is kept suspended well above the actual bottom to prevent further roiling of the soil surface. The purpose of the pump will be merely to remove the cloudy volume of water, not to lower the water level. In clays which tend to seal the bottom along the cutting edge, the water level in the dredging wells may be substantially lowered. Do not permit too great a lowering, since this can result in a blow-in, creating exactly the condition you are trying to correct.

Defogging the lower water level should be accomplished before sending a diver down. More important, the diver must have working space when he gets down there. Nothing is accomplished if he is limited to the area of a dredging well. We have said that the height of working chambers varies from 7 to 9 ft, and a helmeted diver will need a minimum of 7 ft. Excavation must therefore be carried down in the dredging wells to at least 7 ft below the chamber ceiling if the use of a diver is to be effective. The diver will require an underwater light to enable him to see, particularly along the cutting edge.

Until the source of the trouble has been located and an appraisal of the entire perimeter of the cutting edge has been made, the diver should confine himself to the area of the dredging well. There is no guarantee that a hung-up caisson will not slip downward and pin the diver under the ceiling.

Whether he is cutting away timbers with air saws or drilling for blasting or other means of attack on obstacles, these should be started, where possible, in the area of a dredging well, working outward toward the cutting edge, so that premature settling will not pin the diver.

In those cases where the caisson is to be *founded* on rock, a special problem presents itself where the rock surface is not substantially on a level plane. On a tilted stratum, one edge of the caisson may be resting on rock while the opposite edge may be several feet above it, still buried in a considerably softer stratum of soil. While at depths warranting the use of the open caisson (70 to 100 ft or more) the problem of tilt may not be serious, it still exists; and the superintendent must gage what the effect will be if the rock on the high side is shot loose.

In Sec. 4.13 there is a discussion of underpinning a caisson to equalize support around the cutting edge. This was a special case performed in a pneumatic caisson and not applicable to the open caisson unless it has provisions for converting to air operation. Although a solid concrete plug will be poured in the working chamber to develop the full area of the caisson bottom for support, the surface of rock must be cleaned off if loadings require rock-bearing values. To extend this cleaning operation over to the cutting edge may endanger the soil support under the one side of the caisson. Although several methods suggest themselves for support of the edge bearing on soil, the best practice would seem to be that of blasting out the high side before the caisson seats itself on the rock, permitting the caisson to support itself with relative uniformity when it finally reaches that level.

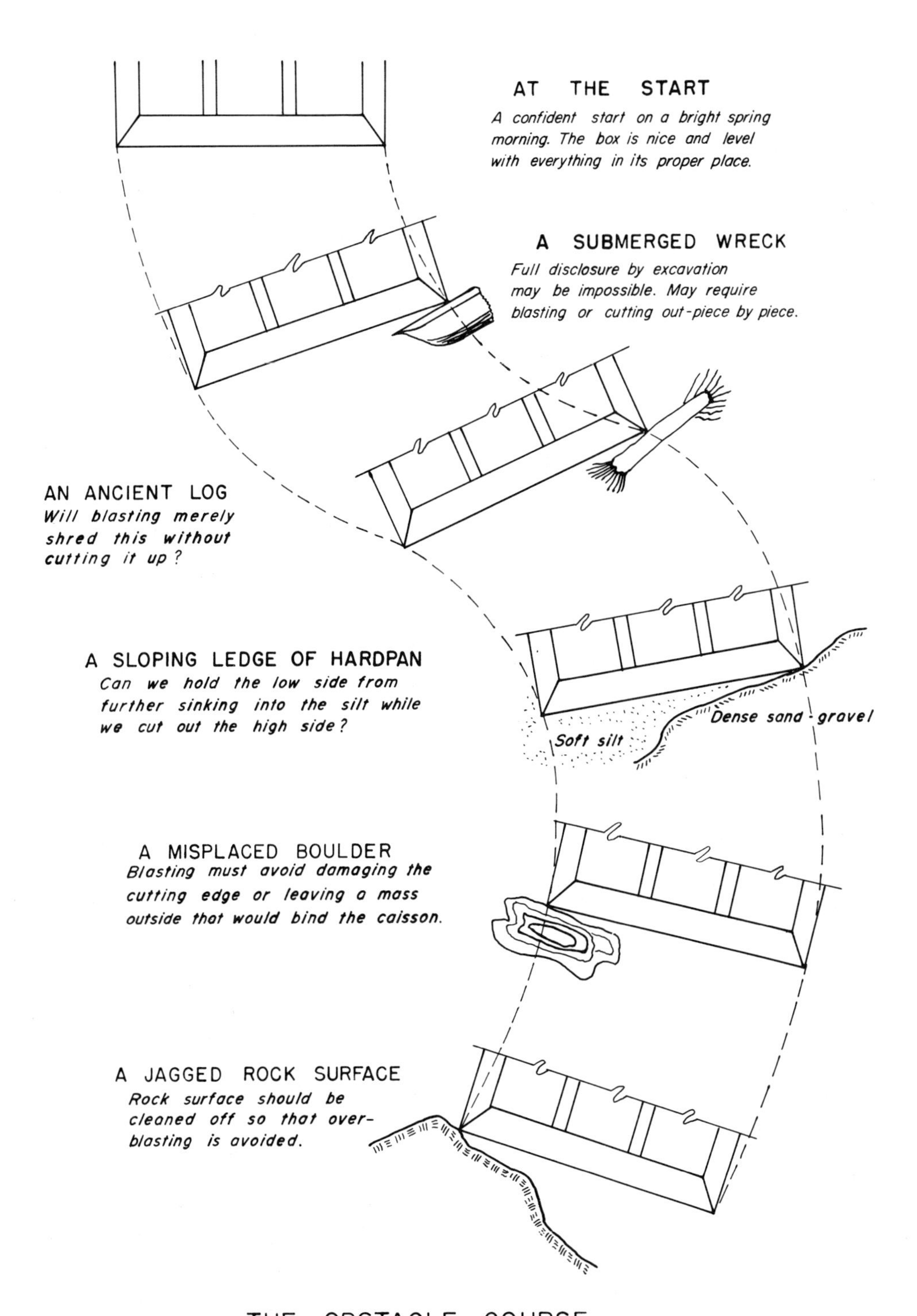

THE OBSTACLE COURSE

(None of these may have been revealed by test borings)

8.15 FLOATING CAISSONS

In Sec. 8.2, two types of open caissons were mentioned, those sunk by the sand-island method and those sunk by the corral method. The corral method is a relatively late development employing a floating caisson. Since other methods of sinking floating caissons are still pertinent, a brief discussion of methods pertaining to them, other than the corral method, would seem appropriate.

Caissons sunk in open water begin with a floating shell, once wood but now universally steel, which is moved into position and then sunk by methods generally resembling those used with the sand-island method. This shell or "can" resembles a ship in that it rides the waves, tilting and righting and heaving and twisting. There is still, however, the same need for it to be accurately positioned when it has been finally sunk.

With the mention of the word "ship" the use of anchors immediately suggests itself, and anchors have been extensively used in the past. Anchors of suitable weight, generally a mass of concrete, have been sunk at distances of from 400 to 1000 ft from the caisson position, radially away from it. Cable connections are then made from the anchor to positions close to the bottom of the caisson, where they terminate in a block and fall with the bight carried up along the side of the caisson shell so that the length of cable can be taken in or slacked off as positioning requires. This is often supplemented by barges anchored at the surface.

Anchorages have generally been used in the past in deeper waters such, for example, as the San Francisco Oakland Bay Bridge in 70 ft of water, the Carquinez Straits Bridge of 1926 in 90 ft of water, or the Tacoma Narrows Bridge in 120 ft of water; but the precise positioning of the caisson has always been difficult to control with anchors, since tides and currents will influence cable tension. When finally bottomed out, one caisson of the ill-fated Tacoma Narrows Bridge was 12 ft out of position.

In earlier days and still in the shallower waters, wood piling has been driven around a floating caisson to position it during sinking, but the method has certain disadvantages. Wood piling is flexible and, although a group of piles can be and are bound together, these tend to move under stream flow, tides, winds, or construction operations, altering the caisson position up to the time it has a very substantial penetration into the overburden of the bottom. As the length of pile increases this condition becomes worse, and the use of steel piling does not completely assuage it. Even with a circle of piling it has often been necessary to drive a line of steel sheet piling around the upstream end to provide quiet water for the caisson sinking.

To correct these conditions a mooring stall was devised for sinking the caisson for the No. 2 pier of the New Orleans Bridge over the Mississippi River in 1955, in 65 ft of water. It is shown in the illustration.

There are two problems involved in positioning. The first is to obtain an approximate location; the second, to obtain more exact positioning for actual sinking operations. The mooring stall seems to have solved these problems well.

The New Orleans stall consisted of a cross-braced frame 40 ft high with pockets provided for piling, assembled and then lowered to the floor of the river with floating derricks. In this case 36-in.-diameter pipe piling was used. Four control piles were first driven and the stall then raised on these control piles to a point where the top was about 15 ft above water level and secured there. The remainder of the piling was then driven in the sockets provided. The upstream semicircular end was enclosed in a ring of steel sheet piling to provide quiet water.

Once the piling was in place and the stall secured to each of the four piles, a rigid framework existed; and within this, vertical steel members were then accurately placed to position the caisson. The downstream end was left open, and into this the caisson shell was floated. A trussed gate was then placed across the opening, from which additional vertical members finally plumbed and leveled the shell. The can was thus tightly corralled.

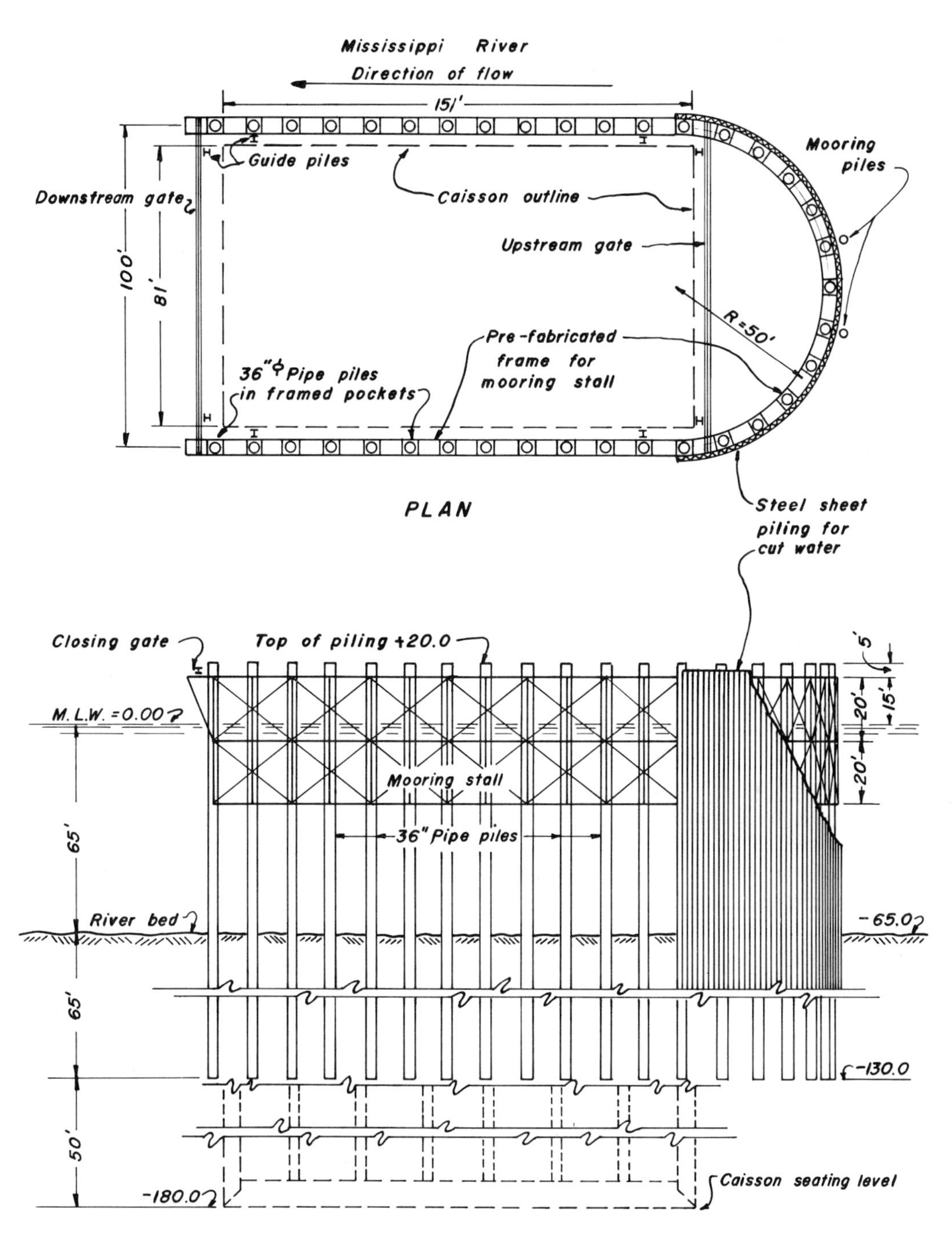

THE MOORING STALL
PIER NO 2 - NEW ORLEANS BRIDGE 1952

8.16 THE CAISSON "CORRAL"

The mooring stall described in Sec. 8.15 is, essentially, a variant of the method currently employed, using what have come to be known as Texas Towers and are familiar to many from their use first as radome tower supports in open water off the Atlantic Coast and later as structures replacing lightships, a program developed by the United States Coast Guard.

The Texas Tower was originally developed by the petroleum industry to provide drilling platforms for offshore drilling, first off the coast of Texas in the Gulf of Mexico and later world-wide, wherever oil was to be found below deep water. Drilling to depths of several thousand feet requires a stable platform. Driving piling in open water presents no great problem, but constructing a platform on these pilings in open water presents difficulties. The industry finally hit upon the device of constructing the platform ashore with hollow legs, barging it to the site, setting it upright, and driving piling through the hollow legs.

A floating derrick, or a pair of them, lifts the assembled platform, or dock, tilts it upright, and lowers it to the bottom of the seaway, where it will stand temporarily on its hollow legs while being plumbed and leveled. Since most sea bottoms contain a penetrable layer of sand or silt, the weight of the dock will partially submerge the legs and provide momentary stability. Drag anchors can be used where sufficient penetration is not available.

Once the dock is positioned, steel H or pipe piling is driven through the hollow pipe legs to a depth sufficient for permanent stability. Either the driven pile, or spud, is bolted to the leg at the surface, or the space between spud and leg is filled with grout.

For caisson sinking, four such docks have been used, placed on the four sides or, for a circular unit, at the quarter points. Three of these are set in position, and the caisson is floated into place and temporarily moored while the fourth dock is set.

The docks, which perform the function of holding the caisson in position and guiding it during sinking, are placed with a 2- to 3-ft clearance from the face of the caisson shell. Wooden bumper blocks are bolted to the inner surface of the platform to reduce the clearance to from 9 to 12 in.

The surface of these docks serves as a working platform and can be sheathed over with 3-in. plank. Additional working space is provided by connecting the docks with light trusses carried over additional independent piling placed as required by the configuration of the caisson.

The first use of such docks was for the caissons for the Yorktown Bridge in 1951 where a four-legged tower 20 ft square and 112 ft high was used. The legs were 14-in. seamless steel tubing connected and cross-braced by 8- and 10-in. pipe struts and diagonals, welded to the legs. The spuds were 10-in. pipe piles 150 ft long and, after driving, were bolted to the top of the legs.

The Yorktown caissons were originally intended to be held in place with anchors. When the docks were suggested as an alternative method, some doubt was felt as to their probable stability. Anchors were therefore also provided, pointing up and down stream, set with their leads hanging slack but arranged so that they could be quickly employed. They were never used.

By 1955 when the corral method was employed for the two tower piers of the Mackinac Straits Bridge, no qualms were expressed as to dock stability. Here the docks consisted of twelve 20-in.-diameter pipe legs tied together with box-type trusses. Eight of these legs were vertical and four were battered (see illustration). Pile spuds were 12-in. 74-lb steel H-piles driven to refusal and then grouted in place. Depths at the two sites were respectively 96 and 140 ft to the Straits' bottom.

These Mackinac tower caissons were circular shells 116 ft in diameter with four docks set at quarter points. The same design of dock was used 3 years later for the tower piers of the Throgs Neck Bridge connecting Bronx and Queens in New York City. Here the caissons were rectangular, 75 by 162 ft in outside dimensions, and the water depth was only 50 ft.

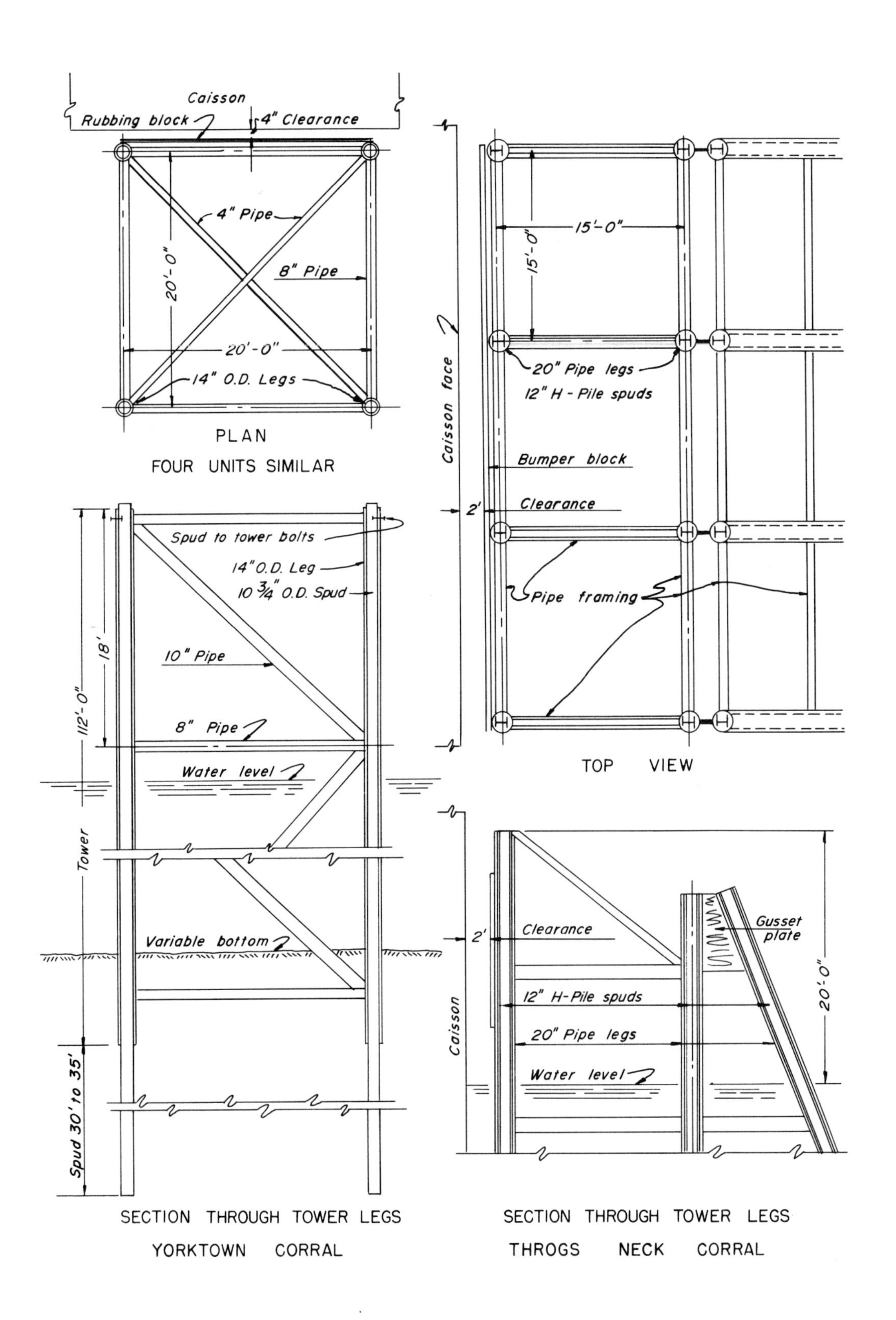
Caisson
Rubbing block
4" Clearance
4" Pipe
8" Pipe
20'-0"
20'-0"
14" O.D. Legs
PLAN
FOUR UNITS SIMILAR
Spud to tower bolts
14" O.D. Leg
10 3/4" O.D. Spud
18'
112'-0"
10" Pipe
8" Pipe
Water level
Tower
Variable bottom
Spud 30' to 35'
SECTION THROUGH TOWER LEGS
YORKTOWN CORRAL
15'-0"
15'-0"
Caisson face
20" Pipe legs
12" H - Pile spuds
Bumper block
2'
Clearance
Pipe framing
TOP VIEW
2'
Clearance
Gusset plate
20'-0"
Caisson
12" H-Pile spuds
20" Pipe legs
Water level
SECTION THROUGH TOWER LEGS
THROGS NECK CORRAL

8.17 THE "CAN"

The design and construction of the floating caisson (often referred to as the "can") and the handling of it prior to the time of corralling and sinking play an important part in the success of the caisson construction.

From a design standpoint, the floating caisson must have the general attributes of a ship; it must be, to a major extent, seaworthy; it must be properly ballasted so that it will float; it must be strong enough to resist crushing or distortion from water pressure. Though it is indeed more of a tub than a ship, it must be susceptible to launching either from shipways or from a drydock. The route of tow to the corral must provide a depth of water adequate for its passage.

Beyond the design, and within the contractor's domain, lies the decision as to whether to have the entire height of the caisson shell made up in the fabricating yard or whether to make up simply a bottom section, constructing the remainder in place.

Fabricating costs will be cheaper in a shipbuilding yard, but the taller caisson will have a deeper submergence, and a greater depth of water will be required along the tow route. There may also be greater danger of capsizing in rough water. There will be a saving not only in cost but in time in yard fabrication. Storage space for additional lifts of steel shell may be limited at the corral site and may not even be available close by on shore. A completely fabricated shell will permit higher lifts of concrete to be poured, although generally factors similar to those previously discussed for the sand-island caisson will control the rate of pour. At the corral, the placing of concrete over the side of a lower standing section will be less costly and there will be some chance to revise sinking techniques as the operation progresses.

Weeks of chewing over the factors involved in the choice can be whisked out the window by a simple pronouncement. The caissons for the Yorktown Bridge, 66 by 52 ft in plan, were 158 ft high, from the cutting edge to the top of the pier wells. The Newport News Shipbuilding and Drydock Company offered to completely fabricate, launch, and tow them 60 miles to the corral for a price within the low bidder's estimate. Further discussion was academic; the order was placed; the operation was successfully performed for four units with a towing draft of 34 ft.

Three years later at the Mackinac Straits Bridge three floating caissons were involved. One of these, a rectangle in plan, 44 by 92 ft, was shop-fabricated for the first 48 ft in height at Toledo, Ohio, and towed to the site. Two other circular cells, 116 ft in diameter, were too large to make the transit from Lake Erie to Lake Huron. Sections were barged from Gary, Indiana, via Lake Michigan, and assembled at a shipyard at Rockport, Michigan. Adequate shipways being unavailable, the first 48 ft in height were fabricated on the ground by essentially the sand-island method. They were to be launched by excavating a pit beneath them which would then be flooded through a channel dug out to the lake shore.

No one had checked ground conditions, which proved harder than expected. Blasting was tried and a hole was torn in the cutting edge. After repairs, the first of these caissons, drawing 23 ft of water, finally reached the lake, only to be hung up on an uncharted sandbar. To avoid a repetition of these contretemps the second unit was skidded over to the backfilled site of the first and launched without further mishap.

A year later, the one floating caisson required for the New Orleans Bridge was fabricated just outside Pittsburgh. This was a rectangular unit 88 by 151 ft. Only the lower 20 ft was shop-fabricated, and no concrete ballast was used. Floating low in the water, the shell was towed 1900 miles down the Mississippi River to New Orleans. Here, at an outfitting dock, it was built up to a height of 47 ft before going out to the corral.

In 1958 the floating caissons for the tower piers of the Throgs Neck Bridge were prefabricated to a height of 68 ft at a shipyard and were towed 30 miles with a 25-ft draft to the corral. (See Sec. 8.28 for discussion of another type of floating caisson.)

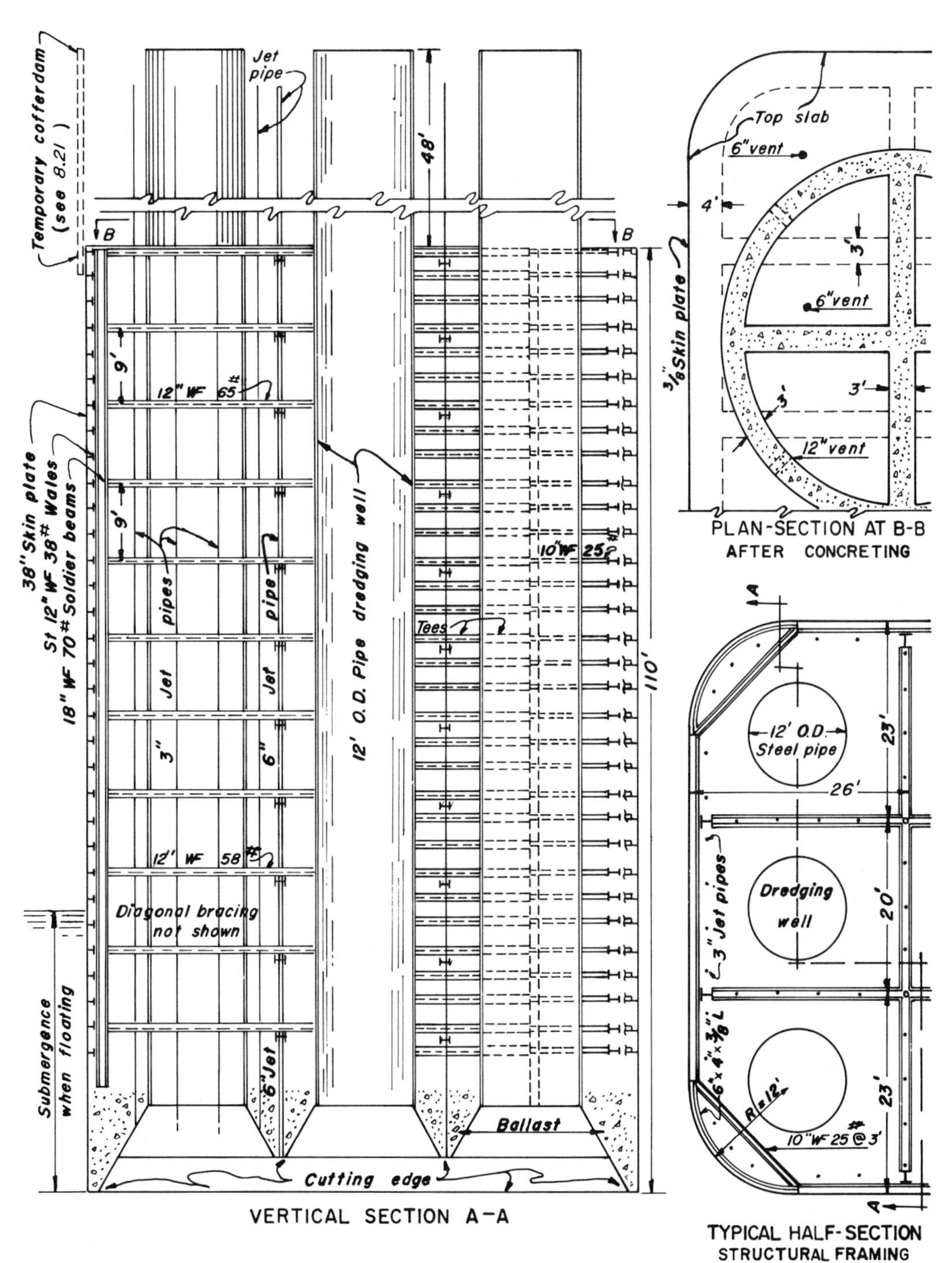

THE YORKTOWN CAN

A TYPICAL FLOATING CAISSON

SCALE 1" = 20'

8.18 SCOUR MATS

Scour occurs when swiftly flowing water picks up soil deposits in a stream bed and moves them downstream. It is a factor which must be considered in the design of bridge piers and is of equal importance in the construction period of sinking caissons in flowing water, particularly since the area of the stream bed occupied by the contractor will be considerably greater than the obstacle left by the finished pier.

The basic formula for stream flow is $Q = AV$, where Q is the quantity of water, perhaps in cubic feet per second, A is the area of the cross section of the stream in square feet, and V is the velocity of flow in feet per second. If an obstruction such as a caisson or bridge pier is inserted in the stream, the area of its cross section is reduced. Since the quantity of flow remains the same, the velocity must increase. However, the change in velocity will not be distributed throughout the whole area but will remain localized at the obstruction.

Another way of looking at this is that a portion of the flowing stream strikes an object and loses its velocity, momentarily building up a head. This head produces a new velocity greater than that in the surrounding stream, which flows down along the faces of the pier or caisson. Not only is the velocity increased, but it is no longer a smooth flow but has vertical components producing swirls and eddies. All these tend to stir up and transport the alluvium.

Completed bridge piers are generally provided with cutwaters to reduce scour, but during the construction period no such device is suitable. Completed structures can be protected from scour by having coarse rock dumped around their bases, but again this has a limited possible use during the construction phase. In watercourses such as the Mississippi River where scour has been known to cut to depths of 40 ft, it has become standard practice to lay down scour mats before construction of bridge caissons begins.

Scour mats or mattresses may be constructed of dressed lumber, but since many watercourses are abundantly lined with growths of willow, willow mats are more frequently used. Typical of this procedure was the mat used in sinking the caisson for Pier No. 2 of the New Orleans Bridge.

In this case a tract of 150 acres of river bottomland heavily overgrown with willow was acquired. Willow saplings and branches cut on this tract were bound in bundles with the stems pointing in a single direction, and tied with 16-gage soft iron wire. A launching way consisting of a sloping timber frame was constructed on the river's edge. The bundles were then assembled as though shingling a roof, the stem ends being the equivalent of the butt end of a shingle, and laced together with ⅜-in. cable to hold them in place. A mattress 300 by 500 ft was thus constructed and launched.

The construction of a scour mat is based on the premise that it will float even though partially submerged. In this case, after launching, the mat was surrounded with barges, four to a side, and the mat edges were fastened to the barges with block and tackle. Tugs towed the convoyed mat to the site and centered it. The four upstream barges were then fixed by 1000-ft-long cables to concrete anchors dropped upstream of the site. Similar shorter-cabled anchorage was provided for the side barges.

The barges had been loaded with one- and two-man stones for stability. These were now transferred from the barge holds by floating clamshells and spread over the mattress. Some care must be exercised in this procedure to provide even distribution. As the mat sank under the stone loading, the barge connections were slacked off until the mattress came to rest on the alluvial river bottom.

The mooring stall was loaded on the mat and the pipe piling was punched through it. When, later, the caisson was sunk, the cutting edge cut through the mattress and the interior portion was chopped up and removed. This left a wide band of mattress around the exterior to hold down the incidence of scour.

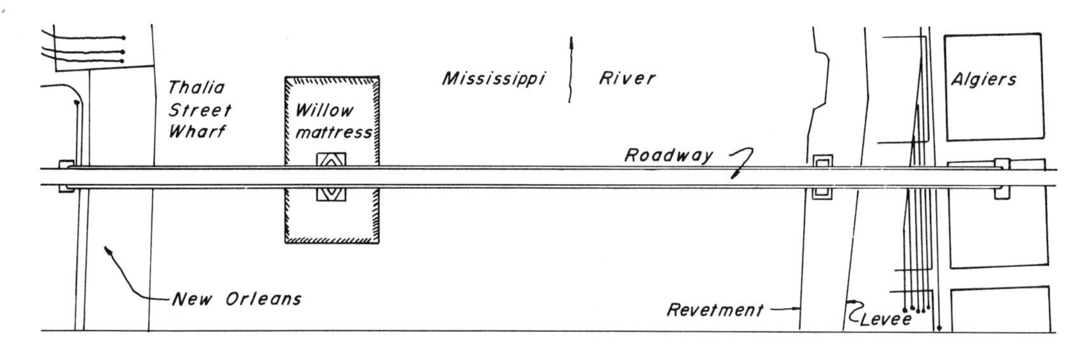

SITE PLAN

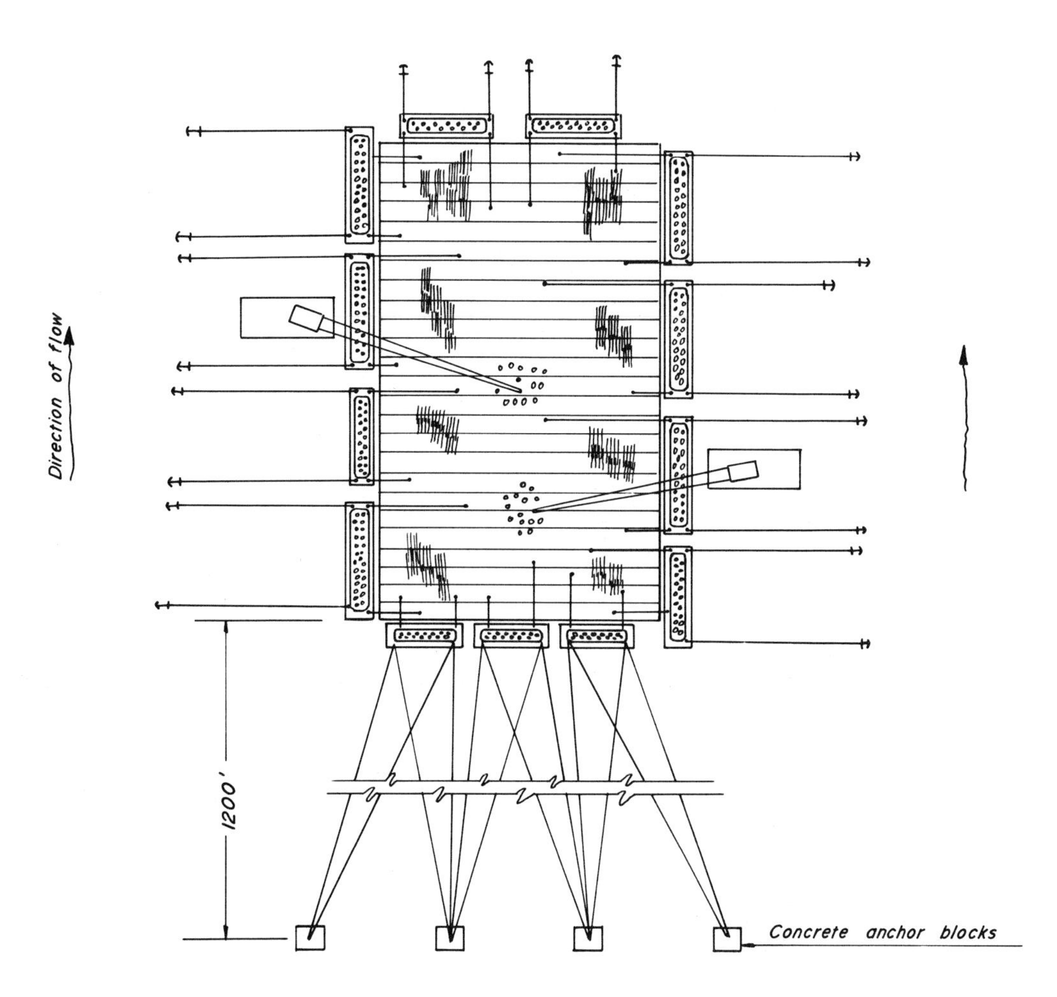

PLACING A WILLOW MATTRESS

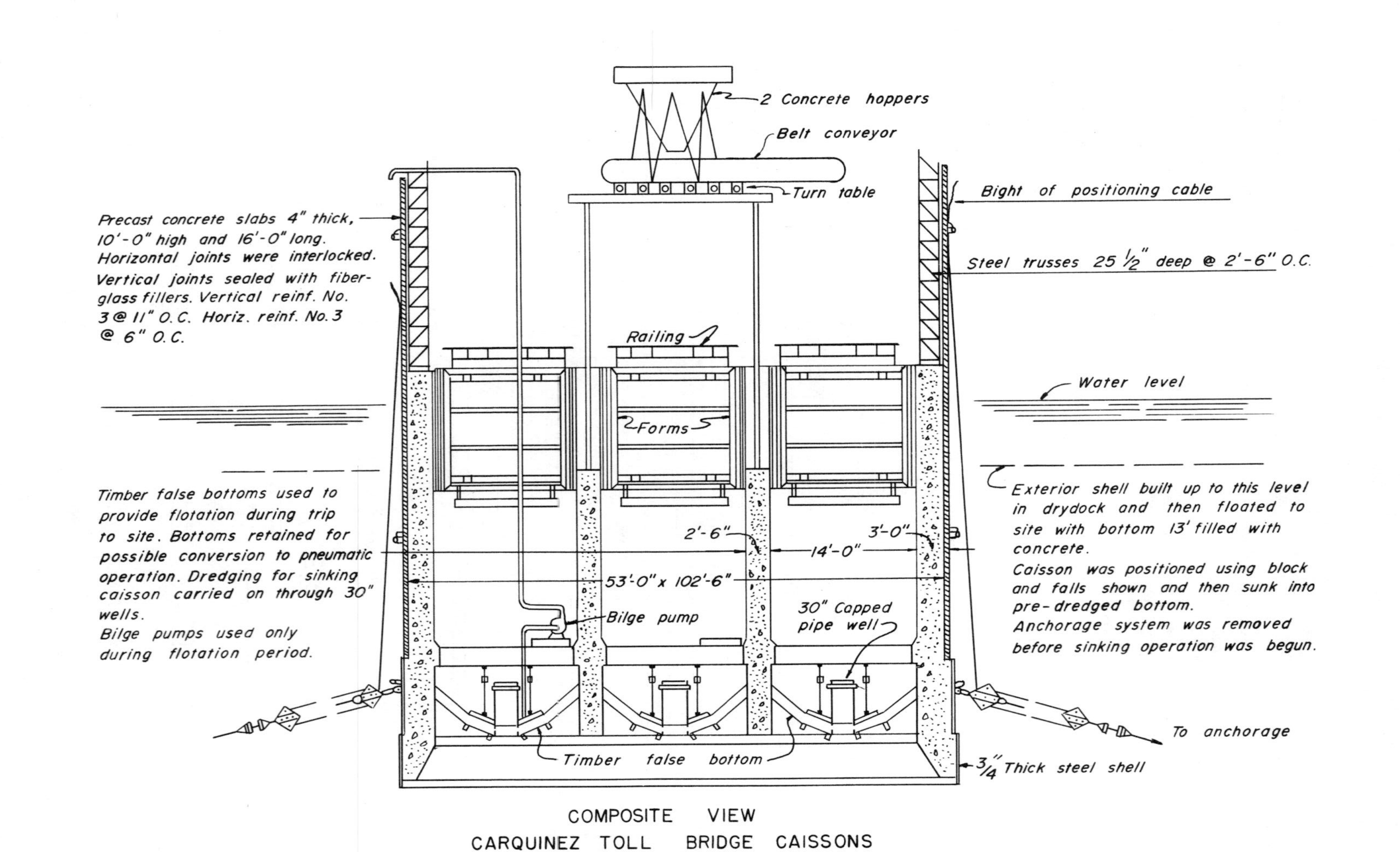

COMPOSITE VIEW
CARQUINEZ TOLL BRIDGE CAISSONS

8.19 SINKING THE FLOATING CAISSON

The procedures used in sinking floating caissons are basically the same as those used in sinking caissons in sand islands, but considerably more care must be exercised in correctly placing concrete until substantial penetration into the river bottom has been effected. So long as the can moves downward on a level keel within the clearances provided inside the corral, there is no difficulty. Once substantial tilt occurs, however, the result can be not only to throw diagonal stresses on one segment of dock but to produce *upward* pressures on the opposite side.

Although it is possible to *form* the upper portion of the floating caisson sunk in open water, it has generally seemed good practice to carry the steel plate shell to the full height whether it is built up *in situ* or prefabricated. In any case, hydrostatic pressure against the shell must be carefully and continuously gaged.

For example, on the Yorktown caissons, 5-ft lifts of concrete were poured successively to a height of 20 ft. It was found, however, that the 5-ft lift provided too much weight, sinking the caisson excessively and throwing too much loading on the shell framing. Instead of decreasing the lift, for which the interior forms were designed, it was decided to pour lifts of the exterior shell only, leaving the interior cross walls for a subsequent pour. The procedure was continued throughout the remainder of the sinking operation with the interior wall pours lagging about 20 ft behind those of the outer walls.

A continuous check must be made on buoyancy. On the Bronx tower caisson of the Throgs Neck Bridge, the buoyancy was 203.4 tons per vertical foot, requiring 100 cu yd of concrete, weighing 156 lb per cu ft, to sink it 1.03 ft. However, every vertical foot of concrete represented 245 cu yd, causing it to sink about 2.5 ft. This shell was fabricated 68 ft high with 23 ft of submergence. Sixteen feet of concrete lifts would sink it forty feet and place it on the bottom, but would leave twenty feet of shell not backed up with concrete. This, therefore, required the same limited pouring schedule as that at Yorktown.

One method of controlling buoyancy where, as in the Yorktown case, the dredging wells have been fully fabricated is to seal them at the top and force compressed air into them, displacing a part of the water. This increases buoyancy and in the cited case permitted an increase in concrete lift.

Concrete handling to caissons in open water presents more of a problem than at sand-island sites, and may reduce the rate of pouring. At the Yorktown job, concrete from an inshore plant was supplied to a dock-mounted 14-cu-yd hopper. Two-cubic-yard bottom dump buckets, carried four to a steel barge, were hopper-loaded and towed to floating cranes which lifted them into position. In this case the buckets had to be lifted initially to a height of 100 ft. Even in the smoothest water, positioning at the top was difficult; and it was found necessary to provide bucket guides, flared at the upper end, to engage and guide the buckets down into the caisson shell.

In other cases a complete floating mixing plant is moored alongside the corral. This may consist of a barge on which a mixer is mounted together with a supply crane. Barges loaded individually with cement and fine and coarse aggregate are positioned for unloading directly into the hopper by the crane. These barges should be independently anchored as well as being moored to the corral. The mixing plant on the New Orleans Bridge consisted of two 2-cu-yd mixers discharging into a common 4-yd hopper. Two-yard drop-bottom buckets were loaded at the hopper and lifted into position by a large floating whirley.

Intrusion concrete was used for the tower caissons of the Mackinac Straits Bridge. Coarse aggregate was fed in from a self-unloading barge and was subsequently intrusion-grouted from a separate self-contained barge-mounted plant. In this case it was necessary to design the shell for greater water pressure; but since it was a relatively narrow circular band, the problem was not so severe as that in rectangular structures.

8.20 SEALING THE CAISSON BOTTOM

When the caisson has been finally seated on satisfactory material, the working chamber is cleaned out and leveled off. The leveling-off elevation may be at cutting-edge height or may be held several feet higher where the exposing of the cutting edge might induce further movement. Where seating is on rock, sufficient material is removed under the cutting edge to permit uniform bearing. Within the working chamber the rock surface is cleaned off but not necessarily leveled. The working chamber and 8 or 10 ft of the dredging wells are now filled with concrete.

The tremie is the method most generally used for placing concrete under water. Methods involving closed buckets or intrusion grouting are also used.

The word "tremie," like so many terms in the field of bridgebuilding, derives from the French word meaning "hopper" and consists of a length of pipe having a hopper attached to its upper end. The pipe may be 10- or 12-in. extra heavy steel pipe with the hopper flaring up to an 8-ft square or circular opening.

The tremie is generally hung by the hopper from a crane hook (although it can be set to ride on hoist towers or on a special framework spanning the area to be poured) with the lower end of the pipe resting on the bottom of the working chamber.

The bottom of the tremie pipe can be sealed temporarily in a number of ways. Air-filled balls or bladders similar to those used for testing pipelines have been used; mechanical devices employing air-operated jacks have been tried but are not successful. The most satisfactory method seems to be the use of a wooden disk about 2 in. larger than the pipe diameter with a rubber gasket on its contact face. Lightly fastened to the bottom of the pipe, it is kept in place by water pressure once the assembly is submerged. A small cable fastened to the rim is used in the later retrieving of it.

With the pipe and plug resting on the bottom, the tremie is filled with concrete to about two-thirds its depth. The tremie is now lifted sharply a foot or two off the bottom, the weight of the column of concrete kicks the plug loose, and the concrete flows out. From this time on, the bottom of the tremie pipe must be kept a foot or more under the surface of the spreading mass of concrete. A steady flow of concrete into the hopper maintains the original level.

Tremie concrete requires no special mix; but since the surface may have the cement washed out to some extent, it is not satisfactorily used for pours less than 3 ft in depth. The slump of the concrete should be maintained between 5 and 7 in. With this slump, concrete will flow as far as 100 ft from the tremie pipe, and an area of 2500 sq ft can be placed from one position. Where the total depth to be poured is considerable, placing can be done in layers of about 8 ft in thickness.

In caisson work, interior cutting edges frequently partially compartment the working chamber. In these cases the interior compartments are conveniently poured first in order to control any uneven pressures against the exterior caisson walls.

As the pour progresses, the tremie is raised just enough to keep it well within the mass of concrete already placed. Lateral movement of the tremie is also possible, when hung from the crane hook, but generally it will be found more desirable to set tremies in their positions horizontally and keep them there. The cost of additional tremies is not great; but the difficulty of moving a 100-ft tremie pipe sideways is considerable, and the lower end may be pulled, momentarily, above the concrete.

The use of internal vibrators in tremie work is questionable. Dropping them down the tube will restrict the flow of concrete. Dropping them down on the outside is impractical unless a separate lifting hook is employed. Control of the vibrator's position from the surface may only result in mixing more water into the concrete or in unbalancing the even flow of concrete away from the tremie pipe. For slumps of 5 to 7 in. the use of a vibrator in the hopper is seldom justified.

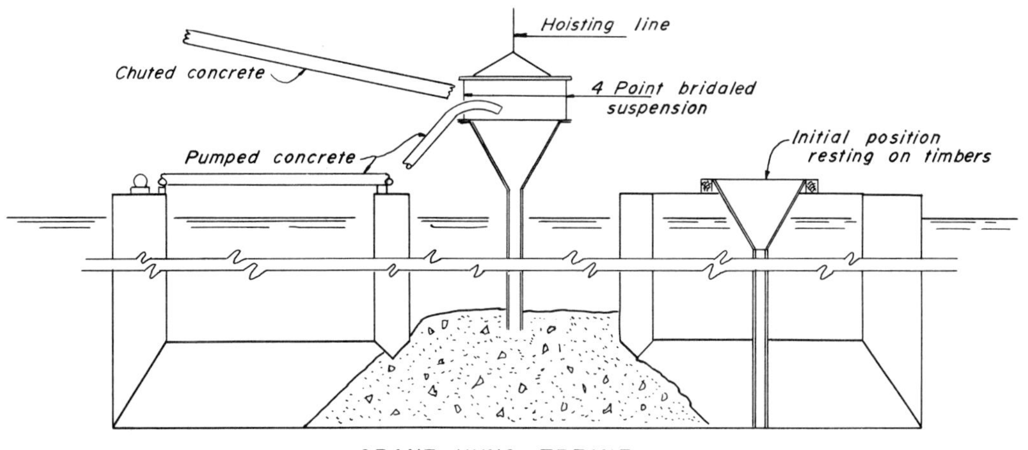

CRANE HUNG TREMIE

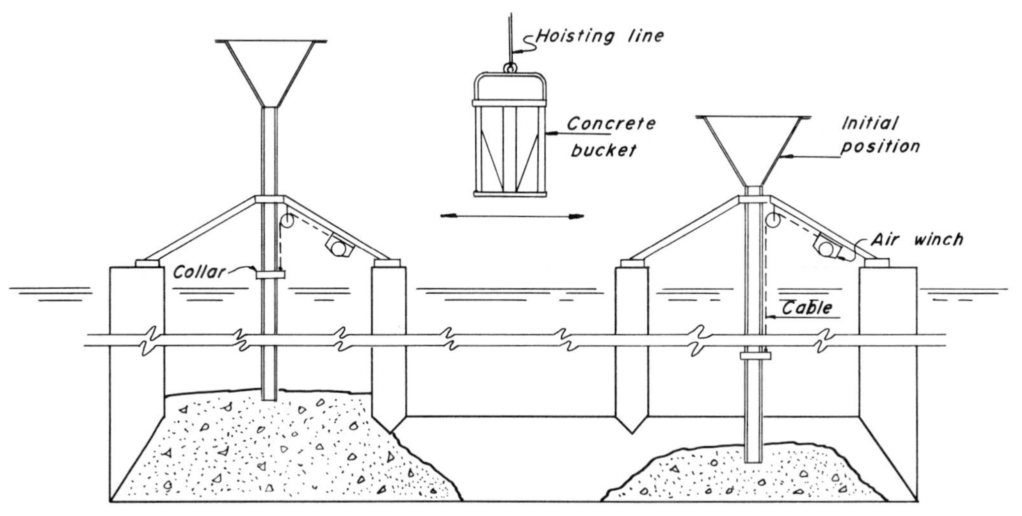

SELF - RAISING BUCKET - FILLED TREMIES

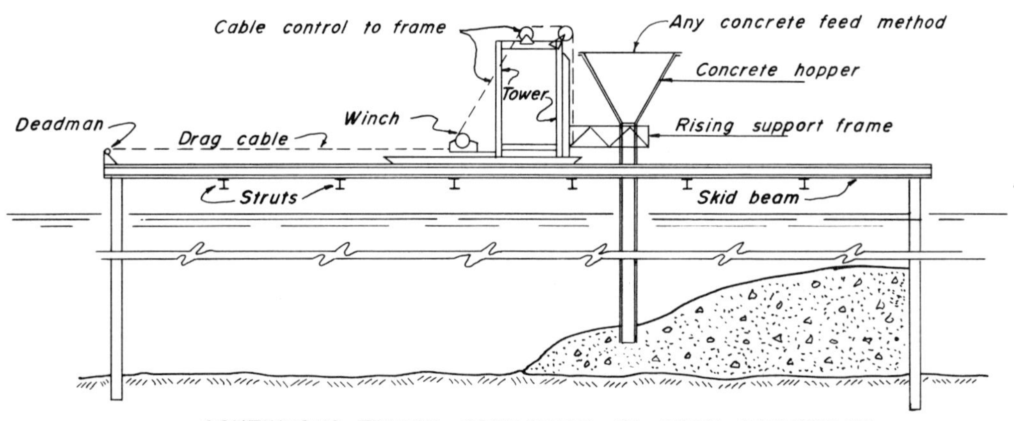

CONTINUOUS TREMIE OPERATION IN OPEN COFFERDAM

8.21 CLOSING THE TOP OF THE CAISSON

The caissons so far described are foundations on which bridge piers are to sit (although some caissons involving neither bridges nor piers will be described hereafter). The bridge pier proper will be much smaller in area than the caisson, and consideration must be given to distributing the pier load over the top of the caisson.

The dredging wells are seldom filled with concrete or other construction materials, not only to reduce the cost of construction but to reduce the load on the footing. To leave them air-filled might lead, through slow infiltration, to later changes in base loading. They are therefore left filled with water, which still provides a weight saving of 60 lb per cu ft.

After the bottom has been sealed, the next step is to close the water-filled dredging wells at the top. For this purpose a pre-cast reinforced concrete disk 6 to 12 in. thick is set into a recess left at the top of the well. The edges of the disks are sealed by grouting. Over the uniform surface thus provided, a structural slab from 4 to 10 ft thick is laid down to distribute the pier loading evenly over the tops of the caisson walls.

There is nothing particularly involved in these operations until we consider that they cannot be performed until the caisson has been seated and the bottom closed. At this time the top of the caisson will generally be from 5 to 20 ft below the water surface. It is necessary, therefore, to make provision for some kind of cofferdam around the upper perimeter of the caisson before the unit is finally seated and sealed. Such cofferdamming has its own special attributes.

Fender systems are frequently provided around bridge piers to protect them from damage by collision with vessels or floating debris. Where this is the case, the cofferdam, or portions of it, can frequently be adapted and left in place as a fender system. This occurred on the tower piers for the Walt Whitman Bridge. This aspect can influence the design of the cofferdam. Other influencing factors will be ease of removal and the condition of providing a watertight joint against the top of the caisson. Since the cofferdamming must remain in place until the bridge pier has been constructed to a height well above the water level, interior bracing passing through the pier area is undesirable.

For the tower piers cited above, 31 vertical feet of cofferdam was required. Thirty-inch WF 190# soldier beams were embedded in the concrete on 7-ft centers. These were set about a foot inside the exterior face of the caisson, which was heavily reinforced to take the toe thrust.

The soldiers were covered with horizontal timbers bolted to the outer flange, 6 by 12 in. for the lower 9 ft, and 4 by 12 in. members for the remainder of the height. One-inch shiplap was spiked to the outer surface of the timbers to improve watertightness and to reduce drag in the final stages of sinking.

Figures *c* and *d* show an alternative method using steel sag plates fastened to T-shaped wales. The wales bear against soldier beams extended upward from the framing for the caisson shell. The lapped sag plates are sealed with gaskets of ⅝-in. garden hose. This system is removed by cutting away the bolts holding the bottom angles and then burning off the soldiers.

Steel sheet piling can be fastened to the outside face of the caisson concrete. Inserts are provided in the concrete pour for drawing up bolts whose heads bear on face plates. After an initial sheet has been set, successive sheets are dropped down through the interlocks before bolting. Additional security is provided by a base angle top-bolted to the concrete. Compressible gaskets shaped to fit the section of sheeting being used will be required between the pile and the concrete, near the bottom. This does not entirely solve the leakage problem, since the interlocks can leak a considerable quantity of water. Methods of sealing interlocks, previously discussed, may be useful here, although considerable pumping may be required. Walering and bracing of the top will depend upon the height the sheeting projects above the caisson. Diagonal bracing kicked against inserts in the concrete dredge-well walls can be used.

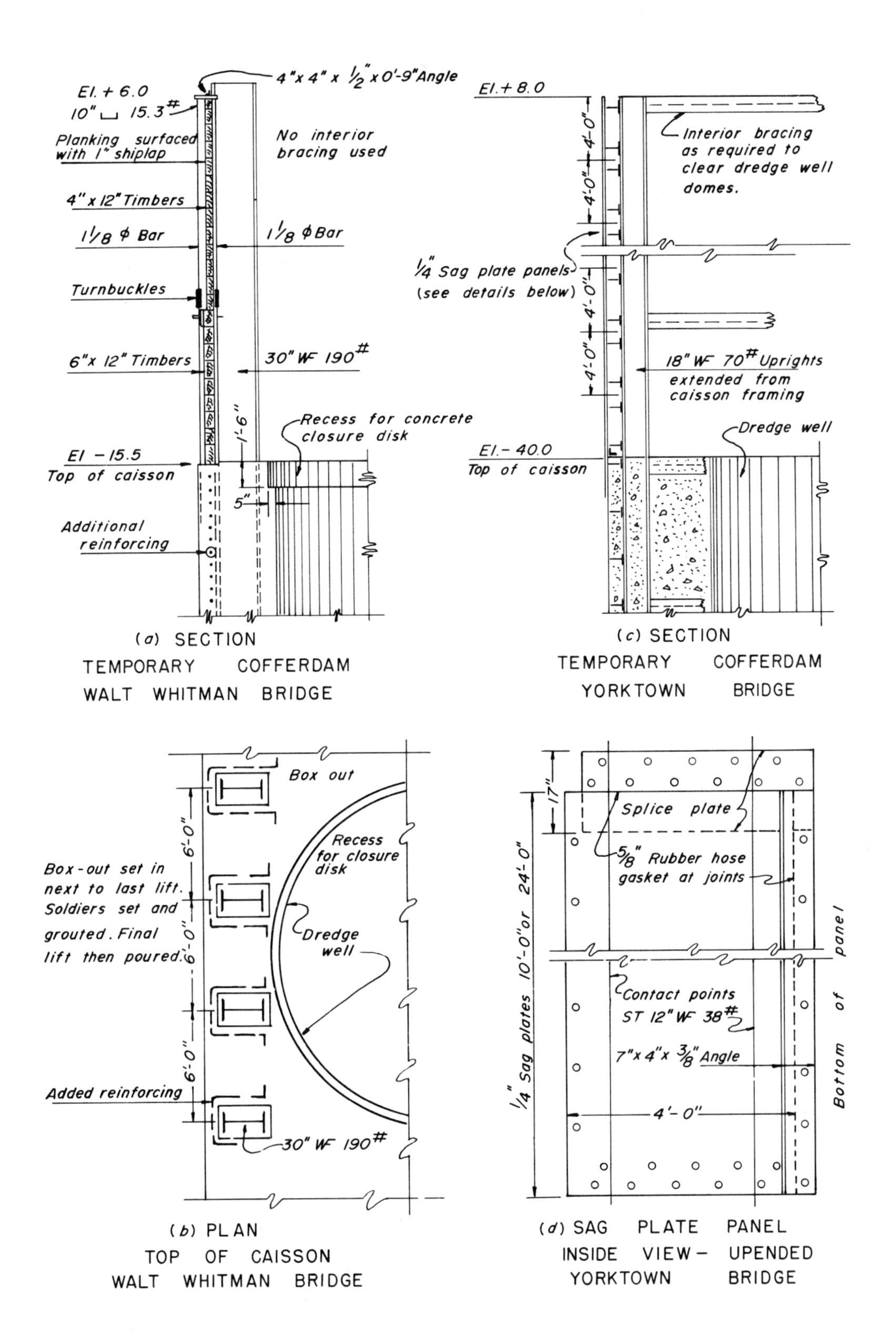

El. + 6.0
10" ⊔ 15.3#
4"x 4" x 1/2"x 0'-9" Angle
Planking surfaced with 1" shiplap
No interior bracing used
4"x 12" Timbers
1 1/8 φ Bar
1 1/8 φ Bar
Turnbuckles
6"x 12" Timbers
30" WF 190#
1'-6"
Recess for concrete closure disk
El -15.5
Top of caisson
5"
Additional reinforcing
(a) SECTION
TEMPORARY COFFERDAM
WALT WHITMAN BRIDGE
El. + 8.0
4'-0"
4'-0"
4'-0"
4'-0"
4'-0"
Interior bracing as required to clear dredge well domes.
1/4" Sag plate panels (see details below)
18" WF 70# Uprights extended from caisson framing
Dredge well
El. - 40.0
Top of caisson
(c) SECTION
TEMPORARY COFFERDAM
YORKTOWN BRIDGE
Box out
Recess for closure disk
Box-out set in next to last lift. Soldiers set and grouted. Final lift then poured.
6'-0"
6'-0"
6'-0"
Dredge well
Added reinforcing
30" WF 190#
(b) PLAN
TOP OF CAISSON
WALT WHITMAN BRIDGE
17"
Splice plate
5/8" Rubber hose gasket at joints
1/4" Sag plates 10'-0" or 24'-0"
Contact points ST 12" WF 38#
7"x 4"x 3/8" Angle
4'-0"
Bottom of panel
(d) SAG PLATE PANEL
INSIDE VIEW - UPENDED
YORKTOWN BRIDGE

8.22 CONVERSION OF OPEN TO PNEUMATIC CAISSONS

Open caissons are occasionally converted to pneumatic operation as they approach their seating level, primarily because conditions at this level indicate the necessity of working in the dry. Actually, the conversion can be made at any time during the course of sinking, but advance preparations are desirable.

The conversion procedure involves installing concrete plugs in the dredging wells in which, in turn, is cast the steel shaft extension of the air lock. To provide for the plugs, keyed sections should be cast in the wall of the dredging well, frequently by a pre-shaped metal form which is left in place. The thickness of concrete should be sufficient to withstand the air pressure to be used, and the keying arrangement should be designed on the same basis. Since the shear strength of the concrete is limited, the steel shell provides the necessary strength where the key butts the walls.

To install the concrete plug, often 10 ft thick, as close to the working chamber as possible will generally mean installation well below water level. Since to a certain extent the caisson acts as a cofferdam, it may be possible to pump down the caisson sufficiently to set forms and place the concrete.

The air-lock shaft can be suspended from the top by crane, derrick, or false work. Most codes require separate man and muck locks on all pneumatic caissons having an area greater than 150 sq ft. However, from the practical standpoint of production, both locks are desirable and, where space permits, should be provided in duplicate. The steel shafts for man locks must be a minimum of 36 in. in diameter; but these, as well as muck lock shafts, should be larger. In some cases the lock itself is sealed into a plug at the top, the lower plug is provided with a shaft and hatch, and the whole dredging-well area can be used as a communicating shaft.

One factor that determines the arrangements made is whether the caisson can be built to its entire final height or whether additional increments, particularly for more weight, must be added as sinking progresses. Where the final seating level is indeterminate, the shaft and lock should be constructed as separate units bolted together on a flanged joint, so that additional lengths of shaft can be inserted if needed to raise the air lock as the caisson concrete is raised.

Although, as noted, the final depth of a pneumatic caisson is often indeterminate and depends entirely upon conditions encountered underground near the proposed seating level, the actual depth may be greater or less. In Sec. 8.24 the proposed seating level is on rock, more or less fixing the depth. Very seldom has an attempt been made to precisely fix the seating level. In one case, using the pneumatic method, a row of caissons was sunk along the face of a quay wall in Glasgow, Scotland, to reinforce it. Completely pre-fabricated and floated into position, they were designed to be seated at a specific elevation. To effect this, timber blocking from the working face to the roof of the working chamber was installed. This supplied control within a few inches, but settlement did occur before the working chamber could be concreted.

Most industrial codes specify that the bottom of the lowest door opening of locks shall be not less than 3 ft above the water level being controlled by compressed air. This, of course, is the prevailing water level around the exterior of the caisson before air is applied, and automatically places the lock structure high—with some shafting involved.

Steel ladders are built into the man shafts, either by welding or bolting. The various services (air supply and return, tool air supply, water, recording lines from pressure gages, etc.) which are required to be carried down to the working chamber should be kept outside the man shaft. One exception might be the rigid conduit carrying the lighting system, for not only will lights be necessary at the working face but the shaft should be lighted at 10-ft intervals.

Where the whole dredging well, between plugs, is used as a shaft, a barrier should be constructed between the area used for hoisting to the muck lock and the stairs leading to the man lock, as a safety measure. The operation of the locks has been discussed.

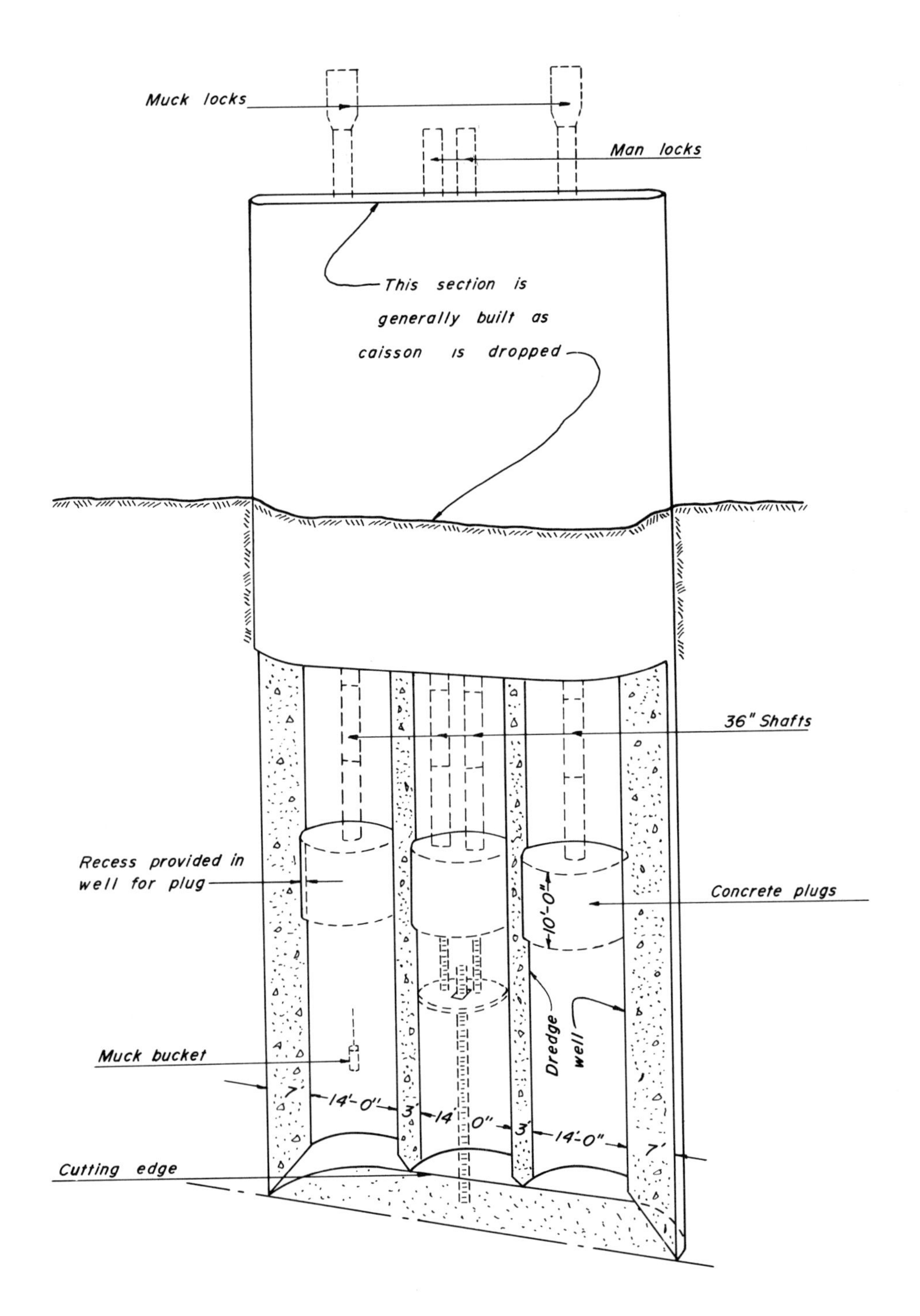

CONVERSION OF CAISSON TO PNEUMATIC OPERATION

8.23 SINKING THE PNEUMATIC CAISSON

The air pressure theoretically required to prevent water from entering the working chamber of a pneumatic caisson will be 0.433 psi per vertical foot of depth, based on clear water weighing 62.4 lb per cu ft. Waters saturated with soils, however, weigh considerably more. While it is unlikely that the whole column of liquid above a cutting edge will have a high density, it may frequently be necessary to carry higher air pressures in the caisson than those indicated by its depth in terms of water. The increase may be as much as 5 to 10 psi.

In cases where nonuniform soil conditions occur around the perimeter of the cutting edge, the excess air pressure can lead to blow-outs on the "water" side, momentarily reducing the pressure sufficiently to cause a "flow-in" on the mud side. The rapid reestablishment of air pressure in these cases is the only solution, generally to a somewhat lower pressure than before.

The air pressure in the working chamber will produce buoyancy of the caisson equal to the area of the chamber times the unit pressure. (The whole area within the cutting edge should be included, since there will be pressure on the sloping face.) In the case noted in Sec. 8.24 with a caisson area of 1500 sq ft, at a pressure of 20 psi the uplift pressure would be 2160 tons; at 50 psi this would be 5400 tons.

Before sinking of the caisson can occur, the weight of the structure must substantially exceed the uplift pressure. Even though the entire support under the cutting edge has been cut away, there is still the skin friction of the caisson barrel which must be overcome. This varies with the type of soil. (Jetting cannot be used with pneumatic sinking since it can accelerate a blow-out.) This is another reason why pneumatic operations are limited to the lower portions of the sinking after caisson construction has progressed to a point where the weight will substantially exceed the air pressure required.

It can be seen at once that if the balance of weight versus uplift is sharply altered, as would occur were the air pressure rapidly reduced, the extra loading applied to the caisson would force it down. This method of sinking a caisson has been used but can be a dangerous practice.

In 1876 a pneumatic caisson being sunk into the bed of the Neva River at what was then St. Petersburg suddenly dropped 59 ft. The ground was known to be soft, but no one was prepared for this. Nineteen men managed to scramble out ahead of the inrushing mud and water, two men were brought out alive after 28 hr of rescue efforts, and seven men lost their lives. It would appear that in this case there was a very sharp drop in buoyancy. (Work was resumed on the Neva caisson a year later. Higher air pressures were maintained since the work was being done at a greater depth. This time, the under-designed air lock blew out, hurling nine dead out of the shaft and smothering twenty others in the working chamber.)

The New York State Industrial Code of 1920 contained a provision limiting the drop of a caisson, by reducing the air pressure, to 24 in. at a time, and this is still carried in the Pennsylvania Code of 1956. However, it is apparent that such close control can be very difficult. No matter how ample the air facilities, a certain period of time, frequently not available, will be consumed in restoring air pressure.

There are other difficulties involved in the procedure of dropping caissons. Where the ground is not uniform or where unknown obstacles exist at a depth below the cutting edge, the pressure drop can cause excessive tilting, twist, or shift in position. The reduction of pressure can lead to collapse of the earth banks below the cutting edge or the upheaval of the working face. Before starting the operation, men must be pulled out and run through decompression, and tools and equipment must be removed. This procedure must be repeated after the drop (although such drops are usually scheduled for a change of shift); and, what can be more time-consuming, a reasonably dry working face must be re-established.

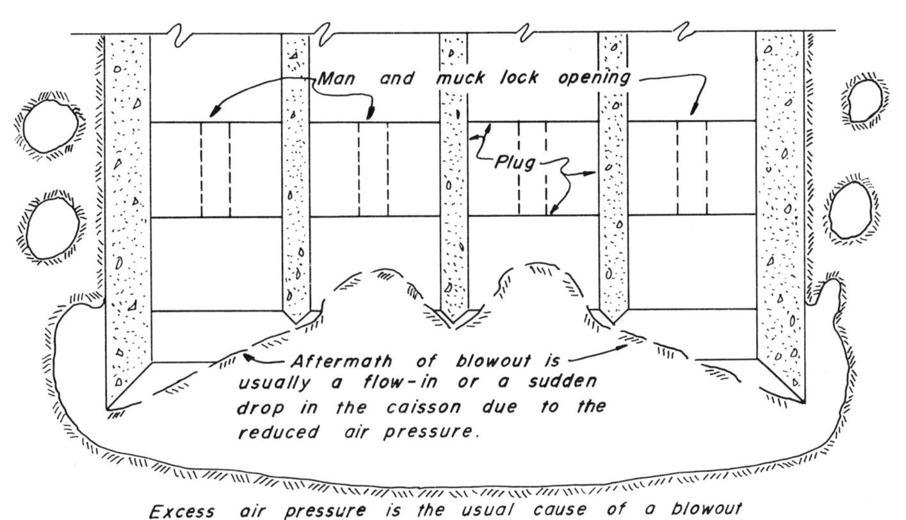

PRELUDE TO A BLOW-OUT

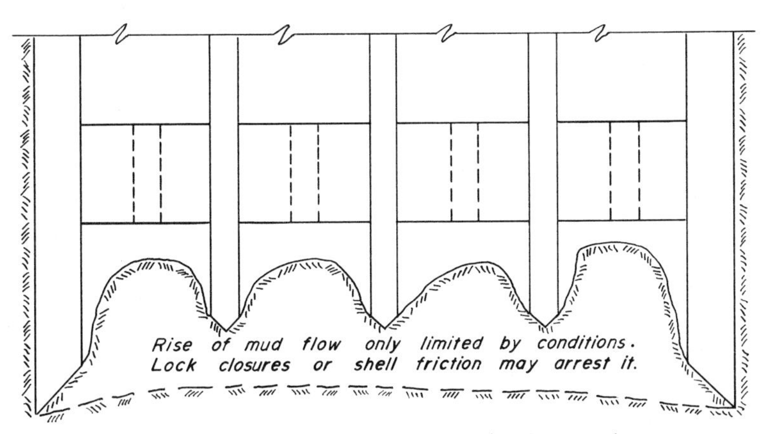

A sharp reduction in air pressure is the usual cause of a flow-in although the sudden penetration of a liquid stratum or relief of shell friction can contribute.

AFTERMATH OF A FLOW-IN

8.24 PNEUMATIC CAISSONS BELOW THE LEGAL DEPTH

It has been noted that the generally recognized legal limit of depth at which work in compressed air can be performed is 115 ft below the water level, equivalent to an air pressure of 50 psi. The technique of the open caisson has been developed to carry caissons to greater depths, but occasionally it appears necessary to prepare the seating level in the dry.

In open water, the construction of a deep caisson would require a circular cofferdam. Where conditions permit, the entire area can be dewatered and the structure built in the dry. Another alternative is that of lowering the water level sufficiently to establish a new water level at or below an elevation 115 ft above seating level.

In 1955 a land-based caisson was sunk on the Missouri side of the Mississippi River several hundred feet back of the river bank. The caisson was the foundation of one of two towers for a suspension bridge required to support two 30-in. natural-gas pipelines crossing the river at a height of 75 ft above maximum flood level and previously described in Sec. 4.13.

The normal water table lay 19 ft below the ground surface, but the rock surface on which the caisson was to be seated was down 150 ft. Moreover, this rock surface was extremely erratic, rising in projecting points as much as 40 ft above the proposed seating level. This condition indicated the desirability of sinking the final portion under air, although the open method was to be used for the first 100 ft or so. To stay within the 115-ft limit for operations under air, it was planned to lower the water table 50 ft.

A deep well was drilled some 30 ft off each of the four corners of the caisson site. Although better than 70 ft of depth was required in each well to obtain the desired water-table lowering, in one corner this was never reached, and several wells were required in the other corners before a suitable water-bearing depth was encountered. Four deep-well turbine pumps were installed with capacities of 3000 gpm which produced the desired lowering of 50 ft.

The caisson was sunk by the open method 125 ft from ground surface before being converted to air. The open sinking was conducted both with the pumps in operation and with them shut down. When the water table was lowered, it was found that the increased friction on the caisson barrel caused it to "hang up." With the pumping resumed, settling again proceeded; but at the same time it was noted that considerable soil fines were flowing into the wells.

Ultimately, this alternating drawdown and fill cycle produced a cone of subsidence around the caisson which, as it gradually widened, encompassed the wells and kicked them out of plumb. The deep-well pumps could no longer be worked in the cocked wells, and air lifts were substituted. Because of this distortion of the wells, the maximum drawdown was reduced to 30 ft, lowering the margin of safety for the air operation.

At the 125-ft depth the caisson was converted to air operation. It was then necessary to keep the air lifts operating continuously; and a tendency of the caisson to drift askew toward the most productive well, which had been apparent for some time, progressed at an accelerated rate. This drift and the subsidence noted earlier could have been due to several things.

It could have been caused by removal of fines from the silty sand, first under the caisson edge and then through the wells; flow of water through the soil could have re-oriented the particles, causing consolidation (see Sec. 3.6, Sand); or the drift alone could have been due to sliding of the caisson diagonally down the jutting rock.

Evidence of excessive pressure on the west face of the working-chamber wall appeared when the cutting edge was about 5.5 ft above the planned seating depth, when a horizontal crack appeared in the cutting-edge wall near the top of the working chamber. At this depth the cutting edge had landed on a ridge of rock extending through the center of the caisson. Because of the fear that blasting might additionally damage the wall, further sinking was abandoned and the cutting edge was underpinned (see Sec. 4.13).

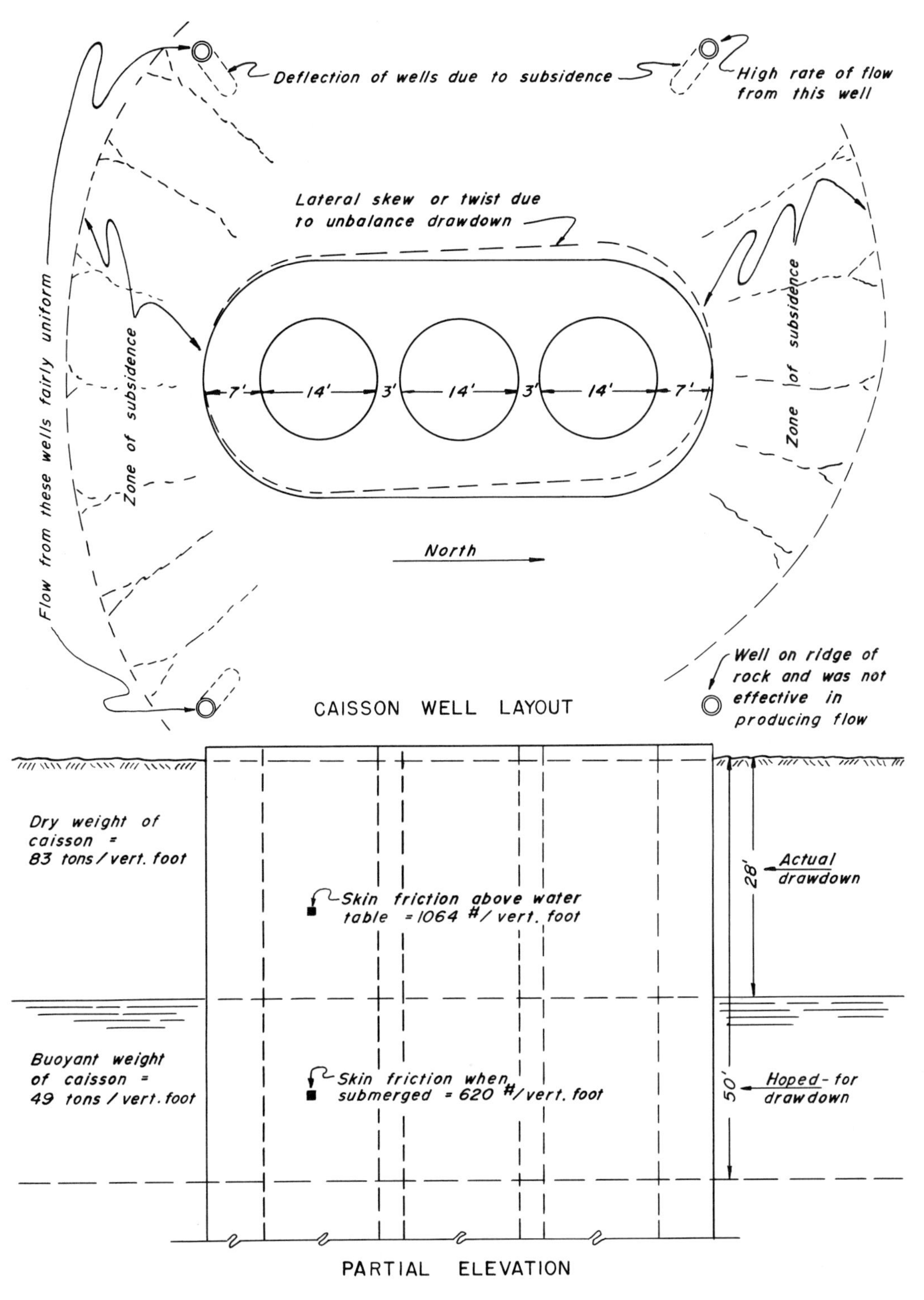

LOWERING THE WATER TABLE FOR CAISSON SINKING

The open caissons discussed so far have largely depended upon the lubricating quality of water to aid in sinking them. It is necessary now to consider the open caisson from another standpoint.

One of the more dramatic aspects of construction has been the failure of structures by excessive settlement due to inadequate bearing capacity of the soils on which they have been built. This factor alone has sped development of the science of soil mechanics. Substantial settlements, such as those in Mexico City, are not isolated events.

The caisson, basically, is an instance where an inadequate footing, resting on non-bearing soils, is overloaded, causing settlement. If this can be done in water-bearing areas, why cannot the same principle be applied to any soft soil? In other words, why not induce settlement, utilizing, controlling, and indeed accelerating it, to force a foundation into the ground? The method *has* been used successfully.

The open-caisson method of sinking foundations has been used since World War II for a number of structures in Tokyo, Japan, and certain aspects of the method were patented in that country and in the United States in 1941. In that city and in several others, the method has been used to avoid driving steel sheet piling enclosing the area of the foundation excavation where contiguous structures do not provide adequate clearance; where underpinning might otherwise be necessary; where the vibration incident to driving in clay or similar soft soils can be transmitted to adjoining foundations. The method has many advantages. Relative costs of the open-caisson method as compared with conventional methods will depend on the individual case.

Perhaps the largest foundation to be sunk by the open-caisson method on dry land was that for the Nikkatsu International Building in Tokyo. The building occupies an entire, blunted-triangular block in the heart of the city. It consists of four foundation stories, five floors of office space, and four additional floors of hotel space. It was built in 1951–1952.

Soil conditions here consisted of a 50-ft blanket of blue-grey clay, starting just below the surface soil, with a bearing value of about 1.5 tons per sq ft. Below this was a thick bed of gravel and sand which would support up to 25 tons per sq ft. The gravel was water-bearing, but little water was found in the clay stratum.

The caisson consisted of the four subsurface floors, about 60 ft in height, built on the surface and then sunk to the gravel stratum. A single cutting edge under the exterior walls was used. All interior floors, beams, and columns were constructed and braced with temporary trusses except the lower columns on the interior of the building, which were constructed just prior to seating.

The site was cleared and graded to the level on which the slab was to be poured. A trench 18 ft deep was dug for setting the cutting edge, as in the sand-island method. After the cutting edge was set and concreted, the remainder of the basement structure was built up to the first-floor hegiht, about 40 ft. The framing was of structural steel encased in concrete.

All excavation (some 87,000 cu yd) was done by hand (this being Japan), at a rate of about 2¼ cu yd per man-day, and wheeled to a skip hoist which carried it above the caisson for discharge by chute into trucks.

Excavation began in the central area and was first directed toward the corners. When the corners had been undercut, excavation was then carried along the cutting edge from the corners at each end of a side. The bearing plates of concrete, cast integrally with the structure, were provided along the interior walls between corners to reduce unit load distribution, and these were augmented by additional wood timbers as bearing plates. Excavation under the bearing plates came last, in each layered sequence, which permitted the skin friction between the exterior face of the caisson and the earth to reach a maximum value. No jetting was used, and although the weight of the caisson was adequate for sinking purposes, it was augmented further by the erection of the structural steel superstructure as the sinking progressed.

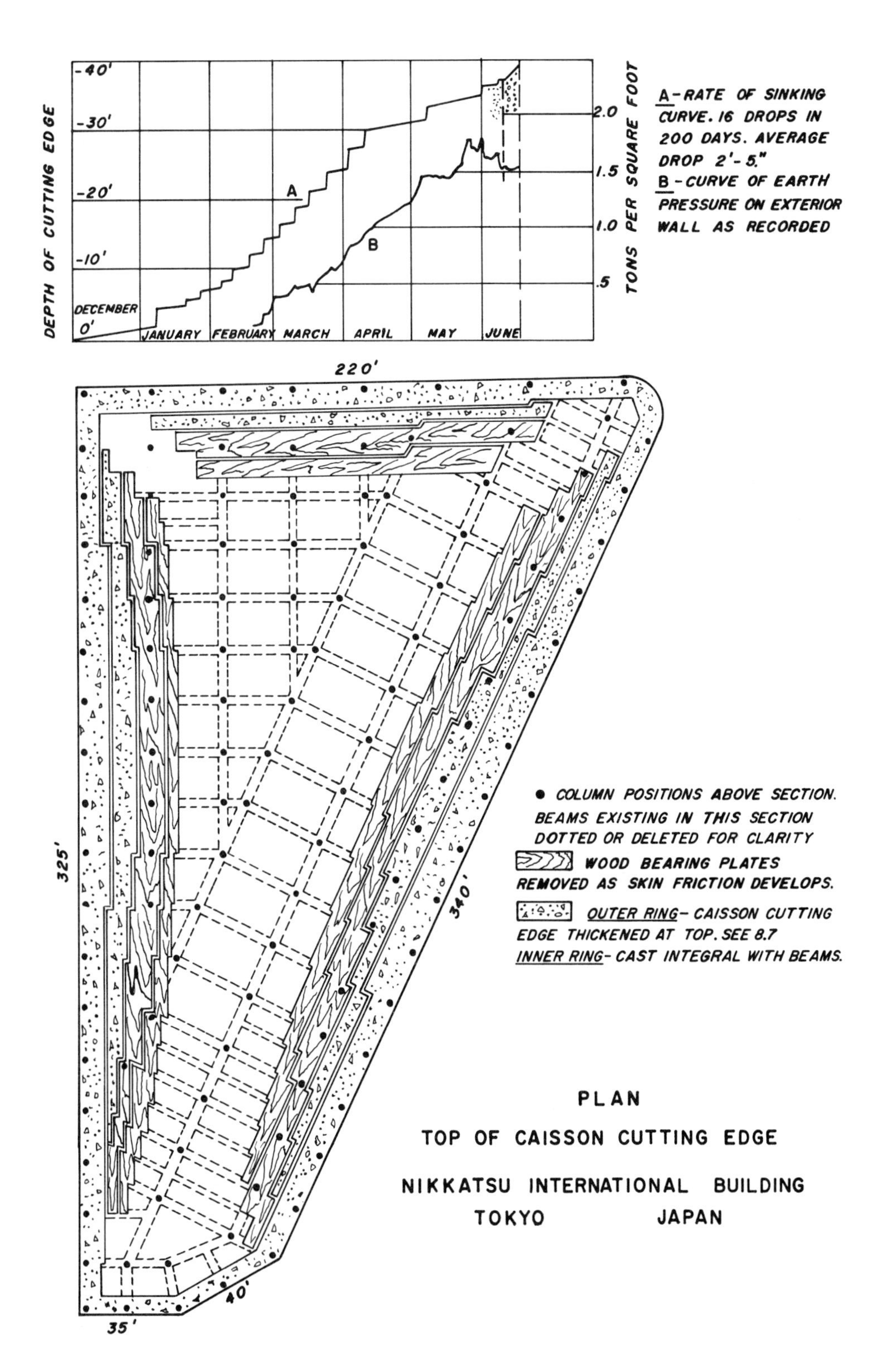

PLAN

TOP OF CAISSON CUTTING EDGE

NIKKATSU INTERNATIONAL BUILDING
TOKYO JAPAN

8.26 INSTRUMENTATION FOR OPEN-CAISSON SINKING

It is apparent that when dealing with dry-land caissons, considerably closer control of operations must be exercised. Where the caisson is sunk in a lubricating medium such as water, minor variations such as tilt can be corrected without undue stresses being applied to the structure. When, however, we are dealing with as large and as irregular a structure as the foundation of the Nikkatsu International Building, constant control is necessary; and in that particular case extensive instrumentation provided a continuous record of subsidence, tilt, strain on the temporary bracing, and earth pressures on the exterior wall.

Settlement meters recording subsidence were installed at each of the three corners, in the center of the long wall, and in the control area situated at the center of the caisson on the lower floor. Near each site a pile was driven for reference. These meters recorded settlement automatically, and the record showed that the caisson was sunk in 16 drops averaging 1.8 ft each, with a maximum single drop of 3.3 ft. On the average, an additional 6 in. of drop due to compression of the supporting soil took place between each major drop. The recordings were taken to the nearest 0.01 mm.

In the same general positions and adjacent to the settlement meters, inclinometers were installed to indicate tilt. These were electrically wired to a console at the central control area where a series of five banks of nine lights each, like a pinball machine, indicated tilt. A central blue light remained lit so long as tilt was less than 0.001 mm in any direction; four yellow lights wired to the inclinometers set near the centers of the four walls registered tilt at any of these locations between 0.001 and 0.002 mm; when tilt exceeded 0.003 mm, one of four red lights would go on. The system thus indicated not only the location of tilt but its magnitude as well.

Excavation operations were directed from the control center by means of six squawk boxes set at the three corners and in the middle of each side. As tilt was indicated by the lights, digging could be intensified or knocked off as required to correct it.

Thirty-four self-recording strain gages were spotted at various points in the temporary bracing as well as in the cross members of the bottom slab, to indicate excessive stresses that might be induced during sinking.

Pressure meters were set in the concrete bearing plates at three locations along each wall as well as in the center of each exterior wall at both the second- and third-floor levels of the caisson. These wall meters, about a yard in diameter, recorded earth pressures against the walls, of particular importance when tilt occurred. It was noted that during tilt, pressures decreased on the lower side and increased on the higher side. The pressure recorded at the second floor, with the caisson in its final position, was about 1.5 tons per sq ft.

When open caissons are sunk under dry-land conditions, some settlement of the area surrounding the caisson is to be expected, since the cutting edge will compress the soil for a short distance outside of it. In the case of the Nikkatsu Building, total exterior settlement amounted to from 6 to 7 in., and its effect tapered off over a distance of about 20 ft.

Throughout this operation the interior columns were unsupported, although cribbing has been used in other cases, less for support than to allay the fears of the workmen. When the caisson approached its final position, the interior columns were hung from the lower floor. Spread footings were then constructed to grade to support these and the columns lowered onto them. The usual grouting and bolting completed this operation.

Once the caisson was seated, the cutting edge was built up on the inside to a full width of 4 ft to act as the wall footing. The basement floor was then poured and sealed to complete the structure. The final step was the removal of the temporary trusses.

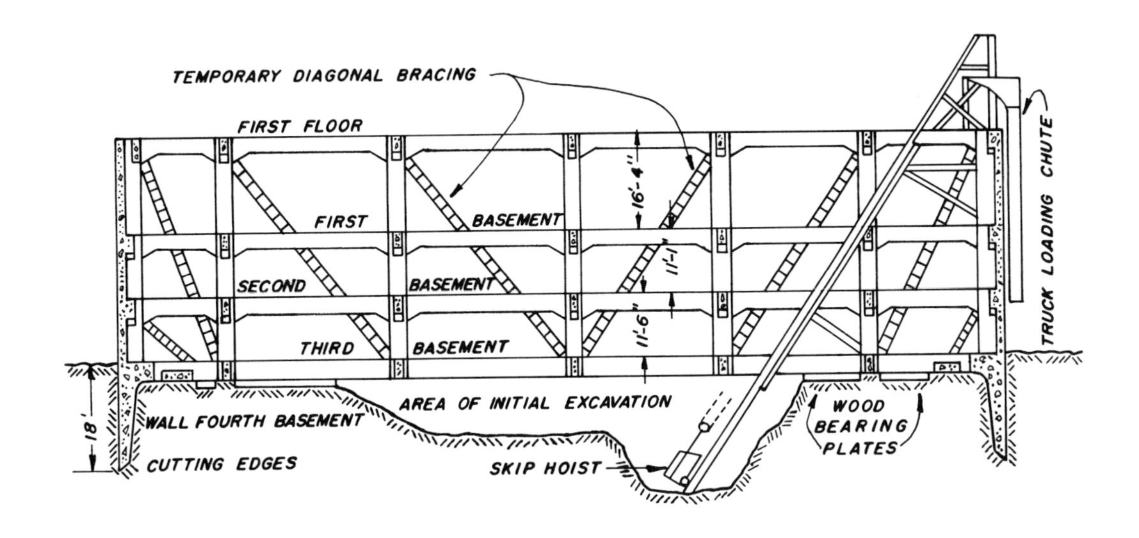

(a) CONDITION AT START OF CAISSON SINKING

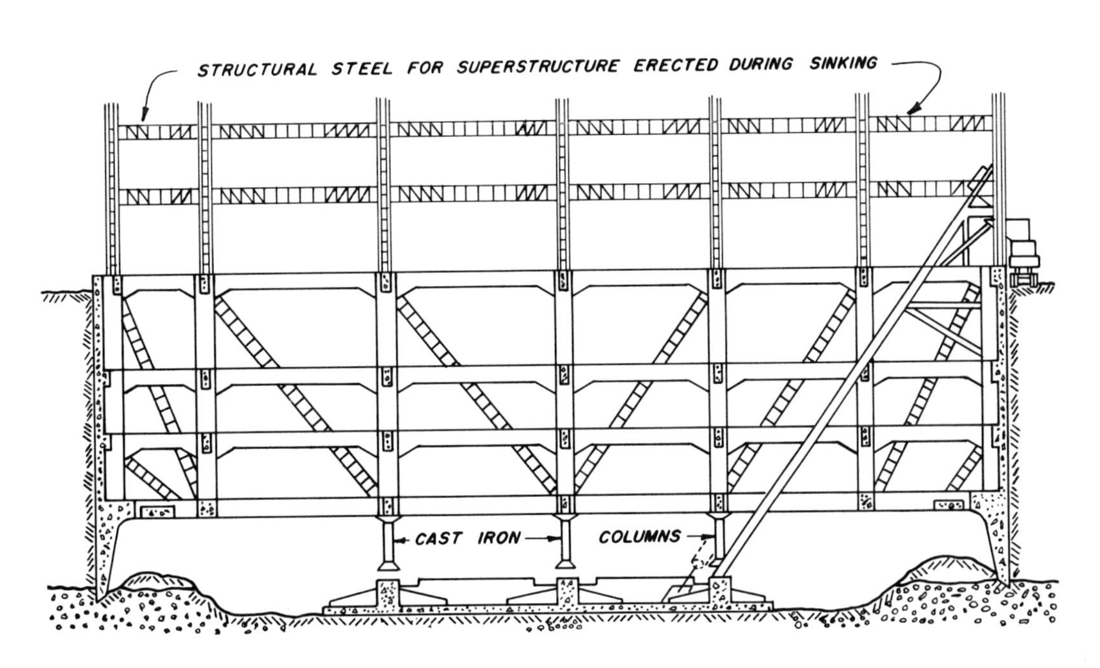

(b) CONDITION JUST BEFORE CAISSON DROPS HOME

CAISSON FOR BASEMENT STRUCTURE
NIKKATSU INTERNATIONAL BUILDING
TOKYO — JAPAN

8.27 LUBRICATED DRY-LAND CAISSONS

The use of lubricants in lieu of water when sinking open caissons on dry land is a natural thought, and this has been done. Indeed, special treatments have been applied to caissons sunk in water or water-bearing soils to reduce the skin friction. A specially designed varnish has been used on the steel plates of floating caissons; and concrete faces of sand-island caissons, whether treated or not, are constructed with as smooth an exterior face as possible (using, as noted previously, metal forms) to reduce skin friction, even where water jets are also provided. On dry land the use of a bentonite slurry has been used to provide the necessary lubricant.

In Geneva, Switzerland, a circular concrete caisson 187 ft in diameter was sunk through clay to a depth of 91 ft by the open-caisson method. The problem here resembled the case noted in Tokyo of a confined space where the use of sheeting would have occupied too much of the available area and where, moreover, the driving of steel sheet piling could seriously have disturbed adjoining buildings founded on an intervening layer of soft clay.

As shown in the illustration, the site is surrounded on three sides by streets, and a one-story building runs the length of the fourth side. The caisson foundation here (as at Tokyo) is designed to serve as a seven-story parking space, and for this purpose the circular shape was particularly suitable. (The structure above ground was made rectangular by the driving of pile clusters at the four corners of the site, after the caisson was in place.)

The upper 25 ft of soil at the site consisted of a water-bearing sand and gravel underlain by a deep layer of soft clay over a compacted clay layer of suitable bearing capacity. However, to have utilized water here for lubricating the sinking, in the heart of the business district, presented problems in the disposal of dredged material. The method used—surrounding the shell with a substantial film of bentonite slurry—not only provided lubrication but prevented water from the upper soil layer from flowing down the shell wall into the excavation.

The cutting edge, in this case, was not provided with a steel shoe, but was constructed of concrete throughout, suitably thickened, as indicated, to prevent crumbling. The shoe extended about 4 in. beyond the wall of the caisson above, to provide a 4-in. thickness of slurry.

The cutting edge, about 10 ft high, was first constructed in a trench of that depth at the surface, and then the first three levels of floors were added. These included not only the caisson wall but one full interior bay which, designed as a continuous frame, provided stiffening for the caisson wall while still leaving the center open for excavation operations.

A system of pipelines, similar to those used for jetting, was installed, with outlets between each floor level. Through these the slurry could be fed to the perimeter. The density and quantity of slurry required were determined by observations made at the surface of the ground.

Excavation under the cutting edge was performed by a chain-bucket excavator similar to a trencher, rotating on the higher ground maintained on the interior. A short conveyor carried excavated material away from the rig; and two tower cranes, operating on segments of track on two sides of the caisson, lifted the soil for loading into trucks.

While the first three floors of the caisson were being sunk, additional floors were formed up and poured by using the same tower cranes for placing concrete.

The average rate of sinking amounted to about 8 in. per day, tilt being corrected by the usual method of unbalanced excavation.

When the caisson had reached its final position, the slurry feed pipes were used to replace the bentonite with a cement grout, starting at the bottom and forcing the slurry up out of the slot. With the caisson in position, the remainder of the interior structure was completed as shown.

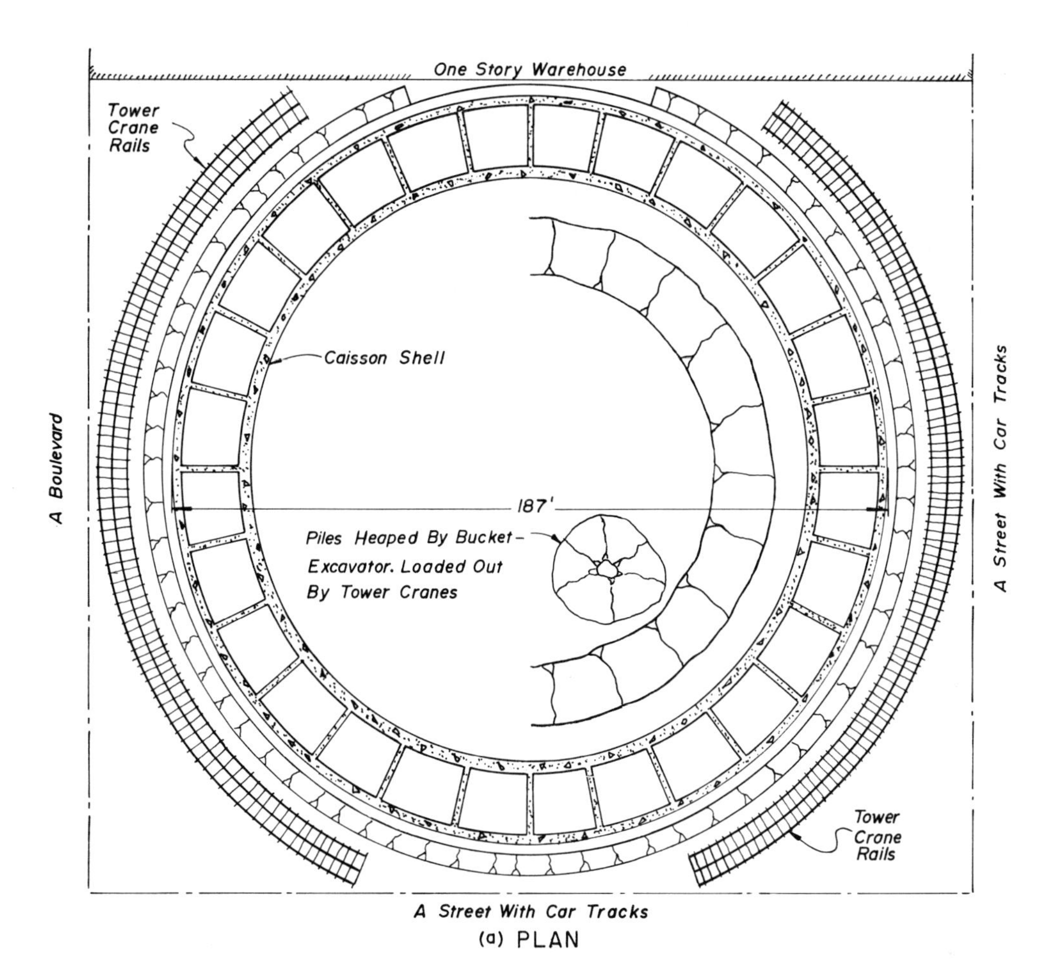

(a) PLAN

(b) FINAL POSITION

(c) THE SINKING OPERATION

LUBRICATED DRY-LAND CAISSON

8.28 TELESCOPIC FLOATING CAISSONS

Mention was made in Sec. 8.16 of the use of Texas Towers by the United States Coast Guard to establish beacon lights replacing lightships offshore along the Atlantic Coast. In Sweden, the Swedish National Board of Shipping and Navigation has also been replacing lightships, particularly in the shallow waters of the South Bothnian Sea, using pre-cast, concrete, telescopic floating caissons, the first of which they developed and placed in 1958–1959.

A "caisson-set" consists of a double-walled concrete caisson with bottom slab poured integrally, divided into six compartments with concrete bulkheads. Telescoped within the inner wall is a second caisson, of watertight construction, resting on the bottom slab, with walls projecting above the outer caisson. Towed to sea, the outer caisson is sunk by water ballasting, after which the inner caisson is floated upward by flooding the interior compartment with water.

Because shipways were remote from the position to be occupied by the first of these Swedish caissons, the unit was constructed in an available quiet cove on a platform laid down over wood piling. The piling was driven to provide adequate vertical support but without cross bracing. After the concrete slab for the caisson had been laid down, the walls were constructed by using a slipform system. When the structure was ready for launching, the piles on the seaward side were blasted away, and jacks, backed against the landside, were applied. A horizontal thrust of 150 tons applied against the shell was sufficient to cause collapse of the remaining piling, and the caisson floated out over them. Floating, the caisson-set developed a draft of about 20 ft.

The shoal site selected for this lighthouse lay in about 23 ft of water on a sloping rock shelf. The rock was leveled off by underwater blasting, and a gravel layer several feet thick and about 66 ft in diameter was placed over it. To keep the caisson plumb when seated, steel rails laid radially were set in the gravel and leveled by divers. The caisson was positioned by winch operation of cables leading to four radially set anchors and was lowered into position by controlled flooding of the exterior compartment.

When the caisson-set was in its final position, the exterior bulkheaded areas were filled with gravel. This was done by displacement without drawing down the water used to flood them. Intrusion mortar was then pumped into this gravel, displacing the remainder of the water.

In the construction of the outer caisson, capped 2-in. grout pipes were set in the floor at the time of pouring and extended up to the outer-lip level. As soon as the water fill was replaced by concrete, the stone underlying the caisson-set was intrusion-grouted. No other anchorage to the bottom was provided, the weight of the final mass of concrete developed in the caisson-set being deemed adequate for stability.

The interior compartment of the lower caisson was now flooded by pumping water into it, floating the inner caisson upward at 5 ft per hr. When it had attained its final height, about 60 ft above sea level, the inner caisson was plumbed and leveled. A temporary connection to hold it in place was made at the radial stub walls projecting inside the outer caisson.

Spaces were provided beneath the inner caisson for fresh-water tankage and for drainage. The cavity below this level was filled with gravel and intrusion-grouted. Then the water level was lowered and forms were constructed below the inner caisson by means of access through a watertight panel in the bottom. With the forms in place, the remainder of the cavity and the space between the two caisson walls was filled with gravel and grouted.

The telescopic caisson has been suggested for a number of other types of uses. It can be designed for good floating stability and relatively shallow draft, overcoming several objections to the usual floating caisson. However, the founding conditions limit its use in other directions. In the case cited, the caisson was founded on rock; and in similar cases or where the foundation can be prepared in advance, the method has some advantages.

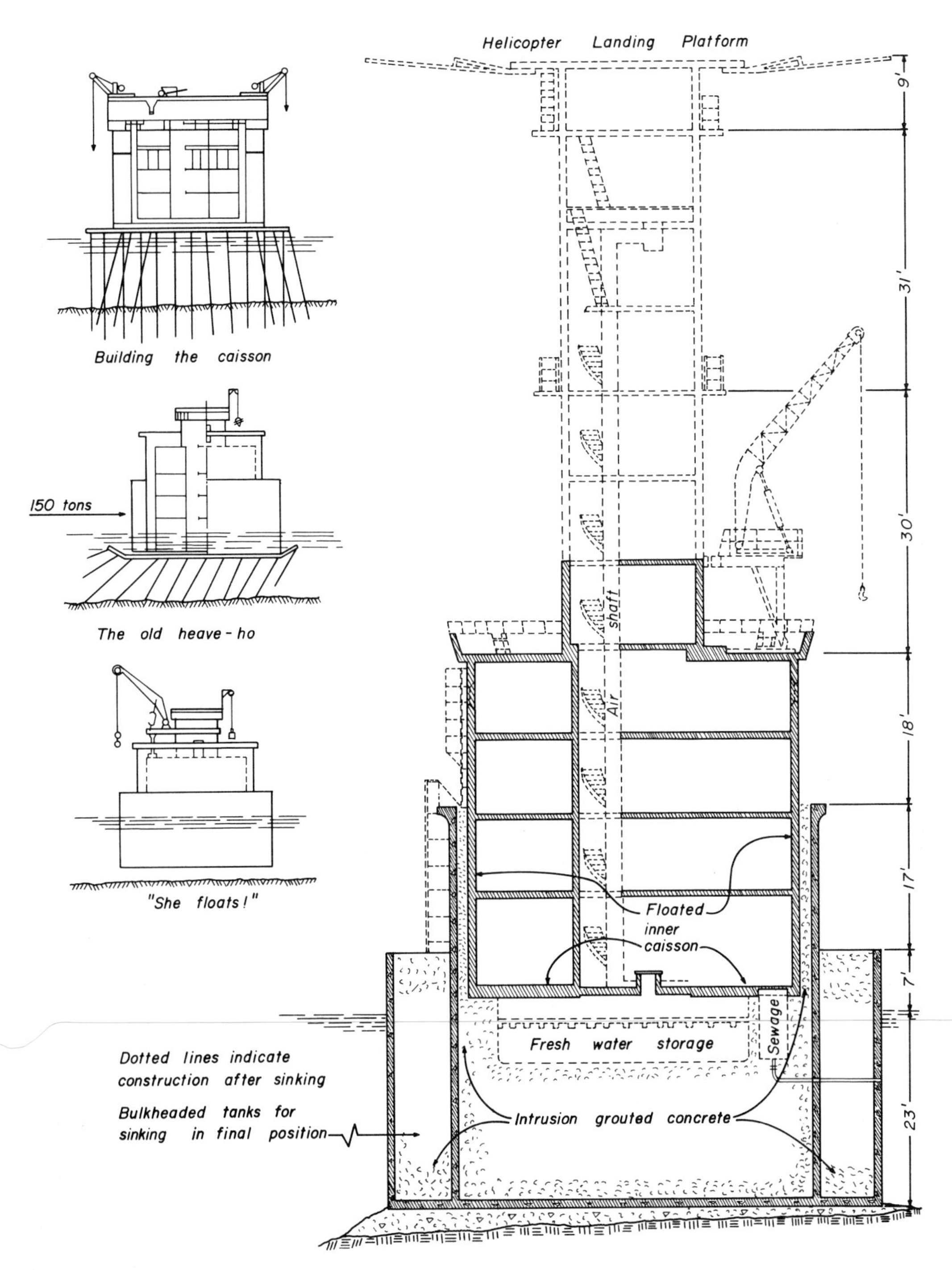

A FLOATING TELESCOPIC CAISSON

SCALE 1" = 20'

Bibliography

CHAPTER 1

1.3

Borg, Robert F.: Who Pays for the Unexpected in Subsurface Construction, *Civil Eng.,* June, 1961, pp. 64–66.

Who Pays for the Unexpected in Construction, Report of the Committee on Contract Administration of the Construction Division, *J. Construct. Div. Am. Soc. Civil Engrs.,* vol. 89, no. CO2, pp. 25–58, 1963.

1.4

Walker, Frank Norman (ed.): "Daylight Through the Mountain, Letters and Labours of Civil Engineers, Walter and Francis Shanley," Engineering Institute of Canada, Montreal, 1957.

1.5

Roberts, Palmer W.: Adverse Weather—Its Effect on Engineering Design and Construction, *Civil Eng.,* June, 1960, pp. 35–36.

1.6

Thom, H.C.S.: Distribution of Extreme Winds in the United States, *Trans. Am. Soc. Civil Engrs.,* vol. 126, part 2, pp. 450–466, 1961.

Davenport, A.G.: Rationale for Determining Design Wind Velocities, *Trans. Am. Soc. Civil Engrs.,* vol. 126, part 2, pp. 184–213, 1961.

Gentry, Robert C.: Wind Velocities during Hurricanes, *Trans. Am. Soc. Civil Engrs.,* vol. 120, pp. 169–180, 1955.

Saffir, Herbert S.: Lessons Learned from Hurricane Donna, *Civil Eng.,* March, 1961, pp. 57–59.

1.7

Roberts, Palmer W.: Adverse Weather—Humid Tropic Areas, Wet Hot Conditions, *Civil Eng.,* June, 1960, pp. 37–39.

"Tropical Engineering," U.S. Navy Civil Engineer Corps, Bureau of Yards and Docks, Navdocks P-39, 1950–1951.

Sanderson, Ivan T.: Dissecting a Jungle, in "Living Treasure,"; The Viking Press, New York, 1941, p. 126.

1.8

Roberts, Palmer W.: Adverse Weather—Desert Areas —Dry Hot Conditions, *Civil Eng.,* February, 1962, pp. 46–49.

1.9

Eyinck, Donald: Notes on Arctic Construction, *Civil Eng.,* September, 1959, pp. 62–64.

Crow, Robert L.: Cold-weather Public Works, *Civil Eng.,* September, 1959, p. 59.

Swick, E.H.: Alaska Needs Roads, *Civil Eng.,* September, 1959, pp. 54–55.

Peyton, Harold R.: Engineering Research in Arctic Alaska, *Civil Eng.,* September, 1959, pp. 82–83.

Alter, Amos J.: Civil Engineering in Alaska, *Civil Eng.,* September, 1959, pp. 49–51.

Roberts, Palmer W.: "Adverse Weather—Wet Cold Conditions," *Civil Eng.,* July, 1960, pp. 44–47.

"Cold-weather Engineering," U.S. Navy Civil Engineer

Corps, Bureau of Yards and Docks, Navdocks P-17, 1948–1949.

Building Below Zero, *Architectural Forum,* February, 1958, pp. 117–121.

1.11

Krynine, Dimitri P., and William R. Judd: "Principles of Engineering Geology and Geotechnics," McGraw-Hill Book Company, New York, 1957.

Tschebotarioff, Gregory P.: "Soil Mechanics, Foundations, and Earth Structures," McGraw-Hill Book Company, New York, 1951.

1.12

Slides, Flood Hit Mayfield Powerhouse, *Eng. News-Record,* 1959.

Barney, Keith R.: Madison Canyon Slide, *Civil Eng.,* August, 1960, pp. 72–75.

Slides Plague Scenic Pacific Coast Highway, *Eng. News-Record,* May 11, 1961, pp. 18–22.

Financial-Legal Troubles Follow Landslip, *Eng. News-Record,* Feb. 16, 1961, pp. 112–113.

Collins, Alexandre: "Landslides in Clay," translated by W.R. Schriever, University of Toronto Press, Toronto, 1956.

1.13

Terzaghi, Karl: "From Theory to Practice in Soil Mechanics," John Wiley and Sons, Inc., New York, 1960.

1.14

Wilson, S.D., and C.W. Hancock, Jr.: "Horizontal Displacement of Clay Foundations," paper presented at First Panamerican Conference on Soil Mechanics and Foundation Engineering, Mexico City, September, 1959.

"Slope Indicator," The Slope Indicator Company, 1105 N. 38 St., Seattle 3, Wash. (Brochure.)

1.15

Root, Arthur W.: Correction of Landslides and Slipouts, *Trans. Am. Soc. Civil Engrs.,* vol. 120, pp. 281–289, 1955.

Baker, R.F.: Analysis of Corrective Actions for Highway Landslides, *Trans. Am. Soc. Civil Engrs.,* vol. 119, pp. 665–689, 1954.

Schroter, G. Austin, and Ray O. Maurseth: Hillside Stability—The Modern Approach, *Civil Eng.,* June, 1960, pp. 66–69 and bibliography.

Drills Cuts to Dry Them, *Construct. Equipment,* March, 1960, pp. 104–105.

"Montana Stops Costly Slides," Case History Report 1960-1, Mobile Drilling, Inc., 960 N. Pennsylvania Ave., Indianapolis 4, Ind. (Brochure.)

1.16

Peru, Carpet of Death, *Time,* Jan. 19, 1962.

Elaborate Road Defenses Thwart Avalanches, *Eng. News-Record,* Dec. 7, 1961, p. 40.

CHAPTER 2

2.4

"Regulations for Tunnel Construction and Work in Compressed Air," Department of Labor and Industry, Commonwealth of Pennsylvania, 1956.

"The Industrial Code, Rule No. 22, Work in Compressed Air (As Amended)," Bureau of Standards and Appeals, Department of Labor, State of New York, effective Oct. 15, 1960.

2.6

"U.S. Navy Diving Manual," Navships 250-538, U.S. Govt. Printing Office, Washington 25, D.C.

Brady, Edward Martin: "Marine Salvage Operations," Cornell Maritime Press, Cambridge, Md., 1960.

2.9

"Blaster's Handbook, A Manual Describing Explosives and Practical Methods of Using Them," E.I. Du Pont de Nemours & Company, Inc., Wilmington 98, Del., 1952.

2.10

Hyzer, Peter C.: Scheduling Equipment for Great Lakes Channel Dredging, *Civil Eng.,* July, 1957, pp. 46–48.

Erickson, Ole P.: Latest Dredging Practice, *Trans. Am. Soc. Civil Engrs.,* vol. 127, part 4, pp. 1–28, 1962.

Smith, Austin B.: Southwest Pass-Mississippi River 40-ft Channel, *Trans. Am. Soc. Civil Engrs.,* vol. 126, part 4, pp. 291–310, 1961.

2.11

Thomas, William F.: Hopper Dredge Gerig, *Military Eng.,* no. 283, September–October, 1950, pp. 393–395.

2.12

Road Fill Pumped to Bridge Tight Time Spec., *Construct. Equipment,* September, 1959, pp. 71–72.

Contrasting Borrow Techniques Supply Road Fill, *Construct. Methods,* December, 1961, pp. 62–64.

2.13

Quaile, John E.: New Engineer Dredge, *Military Eng.,* no. 351, January–February, 1961, pp. 25–26.

Hayward, Henry G.A.: "Sidecast (Boom) Dredging, Foreign Experience and Local Application," paper delivered at fall meeting of American Society of Civil Engineers, New York, 1961.

Converted Tanker Opens Channel, *Construct. Methods,* March, 1960, pp. 171–174.

2.14

Dredge Builds Fill, Digs Lake—All at One Low Cost, *Construct. Equipment,* June, 1959, pp. 74–76.

Pocket-size Dredges Open New Channels of Profit, *Construct. Equipment,* May, 1962, pp. 47–60.

Brochures, Dredge Machinery Corp., P.O. Box 1002, Harrisburg, Pa.

Brochures, The Dixie-Dredge Corp., N. Miami, Fla.

Bull. 925 and 980, Ellicott Machine Corp., 1605 Bush St., Baltimore 30, Md.

Charles, Roland W.: Dredging in Newfoundland, *Military Eng.,* no. 317, May–June, 1955.

2.15

Bull. 59, Naylor Pipe Company, 1230 E. 92 St., Chicago 19, Ill.

CHAPTER 3

Introduction

Forbes, Hyde: The Geochemistry of Earthwork, *Proceedings Separate, Am. Soc. Civil Engrs.,* no. 7, March, 1950.

Physico-chemical Properties of Soils (Symposium), *Trans. Am. Soc. Civil Engrs.,* vol. 126, part 1, pp. 697–794, 1961.

3.1

"Moisture-Density Determinations for Civil Engineering Works," Testlab Corporation, 3398 Milwaukee Ave., Chicago 41, Ill. (Pamphlet.)

Nuclear Device Tests Soil Compaction within Five Minutes, *Construct. Equipment,* July, 1962, p. 60.

"d/M—Gauge System," Nuclear-Chicago Corp., 333 E. Howard Ave., Desplaines, Ill. (Pamphlet.)

"The Hydro-Densimeter," Viatec Division of Tellurometer, Inc., 206 Dupont Circle Bldg., Washington 6, D.C. (Pamphlet.)

Kuhn, S.H.: "A Comparison of Nuclear and Conventional Methods for Determination of the Moisture Content and Density of Soils," paper presented to American Association of State Highway Officials, Denver, Colo., Oct. 10, 1961.

3.4

Oneglia, Andrew: Elevator Rig Drills Holes in Face of Cliff, *Construct. Methods,* October, 1959, pp. 132–133.

Todaro, Joe: Crane-swung Drills Anchor High Rock Face Swiftly—Safely, *Construct. Equipment,* February, 1960, pp. 42–43.

Day, Ray: Drills Scale Sheer Slope to Solve Tricky Shoot, *Construct. Equipment,* March, 1961, pp. 36–39.

Allen, George W.: What You Should Know about Rock Bolting, *Eng. News-Record,* Sept. 27, 1962, pp. 32–35.

Lang, Thomas A.: Rock Bolting Speeds Snowy Mountains Project, *Civil Eng.* February, 1958, pp. 40–42.

3.5

Schmidt, A.E.: Rock Anchors Hold TV Tower on Mt. Wilson, *Civil Eng.,* January, 1956, pp. 56–58.

McKinley, Donald B.: Anchored Foundation Resists Frost Heave, *Civil Eng.,* February, 1952, pp. 42–43.

3.6

Mansur, Charles I.: Laboratory and In Situ Permeability of Sand, *Trans. Am. Soc. Civil Engrs.,* vol. 123, pp. 868–882, 1958.

Whip Troublesome Sand, *Construct. Equipment,* May, 1960, pp. 100–102.

Day, Ray: Blow Sand Loads Easy, If ——, *Construct. Equipment,* April, 1960, pp. 33–36.
Blasting Compacts Sand Fill, *Construct. Methods,* May, 1960, pp. 101–102.
Roller Gets 100% Densities in Blow Sand, *Construct. Equipment,* August, 1958, pp. 62–63.
Lambe, T. William: Capillary Phenomena in Cohesionless Soils, *Trans. Am. Soc. Civil Engrs.,* vol. 116, pp. 401–434, 1951.
Bailly, Florent H.: Controlled Porosity Pipe Dewaters Dune Sand, *Civil Eng.,* November, 1958, p. 75.

3.7

D'Appolonia, Elio, Callix E. Miller, and Thomas M. Ware: Sand Compaction by Vibroflotation, *Trans. Am. Soc. Civil Engrs.,* vol. 120, pp. 154–168, 1955.
Power, James Owen: Vibration Densifies Loose Sand, *Civil Eng.,* December, 1959, p. 50.
Coffman, Bonner S.: Estimating the Relative Density of Sands, *Civil Eng.,* October, 1960, pp. 78–79.
Holtz, W.G. and H.J. Gibbs: Research on Determining Density of Sands by Spoon Penetration Testing, *Proceedings, Fourth International Conference on Soil Mechanics and Foundation Engineering,* vol. 1, pp. 35–39, 1957.
Penetration Test and Split Barrel Sampling of Soils, *Am. Soc. Test. Materials Designation* D1586-58T.
Density of Soil in Place by the Sand Cone Method, *Am. Soc. Test. Materials Designation* D1556-58T.
"Soil Compaction by Vibroflotation," Vibroflotation Foundation Company, 930 Fort Duquesne Blvd., Pittsburgh 22, Pa. (Pamphlet.)
Missile Base Built on Sand Foundations, *Construct. Methods,* April, 1962, pp. 98–100.

3.8

Krynine, Dimitri P., and William R. Judd: The Clay Minerals, in "Principles of Engineering Geology and Geotechnics," McGraw-Hill Book Company, New York, 1957, p. 15.
Leonards, Gerald A.: Strength Characteristics of Compacted Clays, *Trans. Am. Soc. Civil Engrs.,* vol. 120, pp. 1420–1479, 1955.
Schmertmann, John H.: The Undisturbed Consolidation Behavior of Clay, *Trans. Am. Soc. Civil Engrs.,* vol. 120, pp. 1201–1233, 1955.
Seed, H. Bolton, and C.K. Chan: Compacted Clays, Symposium, *Trans. Am. Soc. Civil Engrs.,* vol. 126, part 1, pp. 1343–1425, 1961.
Wu, Tien H.: Geotechnical Properties of Glacial Lake Clays, *Trans. Am. Soc. Civil Engrs.,* vol. 125, pp. 994–1021, 1960.
Lambe, T. William: Compacted Clay: A Symposium, *Trans. Am. Soc. Civil Engrs.,* vol. 125, pp. 682–756, 1960.

3.10

Clark, Forrest D.: Experience with Expansive Clay in Peru, *Civil Eng.,* May, 1960, pp. 42–43.
Holtz, Wesley G., and Harold I. Gibbs: Engineering Properties of Expansive Clays, *Trans. Am. Soc. Civil Engrs.,* vol. 121, pp. 641–677, 1956.
Woolforton, F.L.D.: Movements in the Desiccated Alkaline Soils of Burma, *Trans. Am. Soc. Civil Engrs.,* vol. 116, pp. 443–479, 1951.

3.11

Gray, Hamilton: Field Vane Shear Tests of Sensitive Cohesive Soils, *Trans. Am. Soc. Civil Engrs.,* vol. 122, pp. 844–863, 1957.
Rubbery Shale Turns Job Upside Down, *Construct. Methods,* May, 1959, pp. 142–144.
DuBose, Lawrence A.: Compaction Control Solves Heaving Clay Problem, *Civil Eng.,* April, 1955, pp. 60–61.

3.12

Barron, Reginald A.: Consolidation of Fine-grained Soils by Drainwells, *Trans. Am. Soc. Civil Engrs.,* vol. 113, pp. 718–754, 1948.
Richart, F.E., Jr.: Review of the Theories for Sand Drains, *Trans. Am. Soc. Civil Engrs.,* vol. 124, pp. 709–739, 1959.
Carpenter, J.C., and Edward S. Berber: Vertical Sand Drains for Stabilizing Muck-peat Soils, *Trans. Am. Soc. Civil Engrs.,* vol. 121, pp. 188–204, 1956.

3.13

Clevenger, William A.: Experiences with Loess as Foundation Material, *Trans. Am. Soc. Civil Engrs.,* vol. 123, pp. 151–180, 1958.
Report of Loess Research Studies for the Ashton Pile Testing Program—Lower Platte River Area—Missouri River Basin Project, Nebraska, Earth Laboratory, *Report* EM-278, Bureau of Reclamation, Department of the Interior, Denver, Colo., Dec. 17, 1951.
Pleistocene and Recent Deposits in the Denver Area, Colorado, *Bull.* 996-c, Geological Survey, U.S. Dept. of the Interior, Denver, Colo., 1954.
Holtz, W.G., and H.J. Gibbs: Consolidation and Related Properties of Loessial Soils, *Am. Soc. Testing Materials, Special Publication* 126, 1952.
Jones, Chester W., and C.A. Lowitz: Compacted Loessial-soil Canal Linings, *Trans. Am. Soc. Civil Engrs.,* vol. 127, part 3, pp. 349–370, 1963.

3.14

Irving, Frederick F.: Engineering Research in the High Arctic, *Civil Eng.,* November, 1958, pp. 44–46.
"Cold Weather Engineering," U.S. Navy Civil Engineer Corps, Navdocks P-17, 1948–1949.

3.15

Prugh, Byron J.: Tools and Techniques for Dewatering, *Trans. Am. Soc. Civil Engrs.,* vol. 126, part 2, pp. 38–52, 1961.

3.16

Presettling Canal R-O-W by Flooding, *Eng. News-Record,* June 30, 1960, p. 37.
Golder, H.Q., and A.B. Sanderson: Bridge Foundation Preloaded to Eliminate Settlement, *Civil Eng.,* October, 1961, pp. 62–65.

Cut slope Is Prestressed with Beams and Anchors, *Eng. News-Record,* Oct. 20, 1960, pp. 52–53.

3.17

Cleaning and Grouting of Limestone Foundations, Tennessee Valley Authority, A Symposium, *Trans. Am. Soc. Civil Engrs.,* vol. 113, pp. 79–138, 1948.

Non-setting Slurry Jacks Up Slab, *Construct. Equipment,* August, 1960, p. 95.

Schutz, Raymond J.: Setting Time of Concrete Controlled by Use of Admixtures, *J. Am. Concrete Inst.,* vol. 55, January, 1959, pp. 769–781.

Machis, Alfred: Experimental Observations on Grouting Sands and Gravels, *Trans. Am. Soc. Civil Engrs.,* vol. 113, pp. 181–212, 1948.

Symposium on Grouting, *Trans. Am. Soc. Civil Engrs.,* vol. 127, part 1, pp. 1338–1403, 1962.

King, John C., and Edward G.H. Bush: Symposium on Grouting, Grouting of Granular Materials, *Trans. Am. Soc. Civil Engrs.,* vol. 128, part 1, pp. 1279–1317, 1963.

3.18

Kochis, F., and Horace A. Johnson: Faulted Foundation Complicates Construction of Folsum Dam, *Civil Eng.,* October, 1954, pp. 45–49.

Hays, James B.: Gunite Blanket Improves Foundation Grouting for Earth Dams, *Civil Eng.,* April, 1953, pp. 58–59.

Brochures, Air Placement Equipment Company, 1000 W. 25 St., Kansas City 8, Mo.

3.19

Simonds, A.W., Fred H. Lippold, and R.E. Keim: Treatment of Foundations for Large Dams by Grouting Methods, *Trans. Am. Soc. Civil Engrs.,* vol. 126, pp. 548–574, 1951.

3.20

Campbell, Chester W.: Chemicals Seal Foundation for New York Building, *Civil Eng.,* October, 1957, pp. 47–51.

Riedel, C.M.: Chemicals Stop Cofferdam Leaks, *Civil Eng.,* April, 1951, pp. 23–24.

Day, Ray: Low-cost Chemical Stops Seepage Losses, *Construct. Equipment,* August, 1960, pp. 96–100.

Three Uses of Chemical Grout Show Versatility, *Eng. News-Record,* May 31, 1962, pp. 68–69.

3.21

King, John C., and John R. Anderson: Intrusion Grouting Seals Porous Coral Rock for Vedado Cofferdam, *Civil Eng.,* July, 1953, pp. 36–39.

Grouting, *Special Report* 3, Instrusion-Prepakt Inc., Union Commerce Bldg., Cleveland 14, Ohio.

3.22

Bull. SC-62, Sika Chemical Corporation, 29-53 Gregory Ave., Passaic, N.J.

"Non-shrink Grouting of Machinery and Equipment with Embeco," The Master Builders Company, Division of American-Marietta Company, Cleveland 18, Ohio.

"Ferrolith G, Non-shrink Grout," A.I.A. File 3-0-2, Sonneborn Chemical and Refining Corp., 404 Park Ave. S., New York 16, N.Y. (Catalogue.)

CHAPTER 4

Introduction

Built-in Jacks "Float" Heavy Structure, *Construct. Equipment,* December, 1956, pp. 42–45.

4.2

Underpinning Does Triple Job—Supports Walls, Columns, Earth, *Eng. News-Record,* Sept. 18, 1961, pp. 40–42.

4.3

How to Stop Building Cracks, *Construct. Equipment,* April, 1956, p. 74.

4.4

Brentzel, R.J.: Foundation Repair Saves a Texas Sawmill, *Civil Eng.,* February, 1952, pp. 38–39.

Scruggs, E.L.: Pick-up Beams Carry 600-kip Load in Underpinning Job, *Civil Eng.,* December, 1955, pp. 52–53.

4.6

Jacks Raise Structure to Keep Concrete from Cracking, *Construct. Methods,* September, 1961, pp. 135–141.

Moroney, N.: Recharge Wells and Pit Underpinning Overcome Column Settlement Caused by Desiccation, *Civil Eng.,* January, 1961, pp. 60–61.

4.7

Gillette, D.H.: Reconstruction of the White House, *Military Eng.,* no. 303, January–February, 1953, pp. 8–13.

Dougherty, Richard E.: The White House Made Safe, *Civil Eng.,* July, 1952, pp. 46–52.

4.8

Swiger, W.F.: Prestressed Sheetpile Underpinning, *Civil Eng.,* June, 1957, pp. 66–69.

4.9

Highway Burrows beneath Subway Tracks, *Eng. News-Record,* July 14, 1960, pp. 42–44.

4.11

Schnabel, Harry V.: Pile Jacking Saves Reconstruction of Bridge Piers, *Civil Eng.,* September, 1955, pp. 56–57.

4.12

Goldfinger, Henry: Permanent Steel Used as Subway Shoring, *Civil Eng.,* March, 1960, pp. 47–49.

4.13

Newell, John M.: Pneumatic Caisson Pier for World's Longest Pipe Line Construction, *Civil Eng.,* May, 1956, pp. 51–55.

4.14

Powys, John Cowper: "Samphire," Thomas Seltzer, New York, 1922.

Selim, Mohamed A.: The High Aswan Dam, A Basic Step towards Full Utilization of the River Nile, *Civil Eng.,* August, 1958, pp. 55–59.

White, Edward E.: Saving the Temples of Abu Simbel, *Civil Eng.,* August, 1962, pp. 34–37.

4.15

Mobile Jacking Set-up Hikes Housemover's Profit, *Construct. Equipment,* May, 1960, pp. 105–107.

"Fact Sheets," Elgood Equipment Corp., 242 Randolph St., Brooklyn 37, N.Y. (Brochures.)

"Duff-Norton Jack Manual," Catalog 204, Duff-Norton Manufacturing Company, Pittsburgh 30, Pa., 1954.

"The Philadelphia Roll-Ramp," Philadelphia Gear Corp., King of Prussia, Pa. (Bulletin.)

CHAPTER 5

Introduction

East Coast Pile Needs Pile Up, *Construct. Methods,* July, 1962, p. 21

Chellis, Robert D.: "Pile Foundations," 2d ed., McGraw-Hill Book Company, New York, 1961.

5.1

Meyerhof, G.G.: Compaction of Sands and Bearing Capacity of Piles, *Trans. Am. Soc. Civil Engrs.,* vol. 126, part 1, pp. 1292–1342, 1961.

5.2

Round Timber Piles, *Am. Soc. Testing Materials,* Designation D25-52T.

"Pressure Treated Timber Foundation Piles," American Wood Preservers Institute, 1955.

5.3

"Bethlehem H. Piles," Catalog 223, Bethlehem Steel Company, Bethlehem, Pa., 1949.

5.4

"Seamless Steel Pipe Piles," Design Manual 27, National Tube Division, U.S. Steel Corp., Pittsburgh, Pa. 1956.

Follet, P.H.S.: Super-tanker Wharf on the Mississippi, *Civil Eng.,* October, 1962, pp. 42–45.

5.5

Catalogs, Raymond International Inc., 140 Cedar St., New York 6, N.Y.

"Cobi Concrete Piles," Eastern Concrete Pile Co., Inc., 80 Boylston St., Boston, Mass. (Catalogue.)

Brochures, Armco Drainage & Metal Products, Inc., 7101 Curtis St., Middletown, Ohio.

"Monotube Piles," The Union Metal Manufacturing Company, Canton 5, Ohio. (Catalogues.)

5.6

Remington, William F.: Pretensioned Piles for Navy Pier, *Civil Eng.,* August, 1957, pp. 52–53.

Lin, T.Y., and W.J. Talbot, Jr.: Pretensioned Concrete Piles—Present Knowledge Summarized, *Civil Eng.,* May, 1961, pp. 53–57.

Paulet, E.G.: Concrete Piles Designed for Marine Foundations, *Civil Eng.,* March, 1961, pp. 51–53.

Spronck, Michael A.: Concrete Piles Built to Take Heavy Side Loads, *Construct. Equipment,* April, 1955, pp. 42–47.

Van Buren, Martin.: How to Drive 192 Ft Concrete Pile, *Eng. News-Record,* Apr. 14, 1955, p. 51.

"Raymond Cylinder Piles of Prestressed Concrete," Catalog CP-3, Raymond International Inc., 140 Cedar St., New York 6, N.Y.

"Concrete Piles—Design—Manufacture—Driving," Portland Cement Association, 33 West Grand St., Chicago 10, Ill., 1951.

5.7

Liver, Norman L, Erwin C. Mardorf, and John C. King: Mixed-in-place Piles Form Jetty to Control Beach Erosion, *Civil Eng.,* March, 1954, pp. 56–57.

Thornley, J.H.: Compressed Concrete Pedestal Piles Form Foundation for Corlears Hook Apartment Project, *Civil Eng.,* December, 1954, pp. 33–38.

5.9

"Foundation Piling," National Academy of Sciences, National Research Council, Printing and Publishing Office, 2101 Constitution Ave., Washington 25, D.C., 1962.

Pipe Pile Takes 300 Ton Load, *Construct. Equipment,* April, 1959, p. 97.

40 Tons on a Wood Pile? Chicago Engineers Say "Yes," *Eng. News-Record,* Nov. 3, 1960, pp. 52–53.

Gouda, M.A.: Analysis of Load Distribution in the Piles of Piers, *Trans. Am. Soc. Civil Engrs.,* vol. 126, part 4, pp. 249–271, 1961.

Chellis, Robert D.: The Relationship between Pile Formulas and Load Tests, *Trans. Am. Soc. Civil Engrs.,* vol. 114, pp. 290–320, 1949.

5.10

Mansur, Charles I., and John A. Focht, Jr.: Pile-loading Tests, Morganza Floodway Control Structures, *Trans. Am. Soc. Civil Engrs.,* vol. 121, pp. 555–587, 1956.

Gates, Marvin: Empirical Formula for Predicting Pile Bearing Capacity, *Civil Eng.,* March, 1957, pp. 65–66.

Agerschou, Hans A.: Analysis of the Engineering News Pile Formula, *J. Soil Mech. and Found. Div. Am. Soc. Civil Engrs.,* October, 1962, pp. 1–11.

5.12

Lynch, T.J.: Pile Driving Experiences at Port Everglades, *Trans. Am. Soc. Civil Engrs.,* vol 126, part 1, pp. 216–237, 1961.

Seed, H. Bolton, and Lyman C. Reese: The Action of Soft Clay along Friction Piles, *Trans. Am. Soc. Civil Engrs.,* vol. 122, pp. 731–764, 1957.

Moore, William W.: Experiences with Predetermining Pile Lengths, *Trans. Am. Soc. Civil Engrs.,* vol. 114, pp. 351–393, 1949.

5.13

Fowler, John W.: Details for a 425 Ton Pile Test, *Civil Eng.,* December, 1960, pp. 71–72.

5.15

Floating Fleet Installs Long Cylinder Piles, *Construct. Methods,* January, 1963, pp. 96–97.
Pile Holder, Hammer Stand Help Heavy Pile Driving, *Construct. Methods,* February, 1962, pp. 98–101.
Tubular Leads on Big Crane Handle Long Piles, *Construct. Methods,* May, 1959, pp. 92–94.
How to Drive Piles on Batter without Leads, *Construct. Methods,* February, 1959, pp. 92–94.
Bull. 82, McKiernan-Terry Corp., 100 Richards Ave., Dover, N.J.

5.16

"Pile Driving, Pile-extracting Machinery," Vulcan Iron Works Inc., Riverside Dr. and Stewart St., Chattanooga, Tenn. (Bound catalogues.)
Bull. 691, 70, 68, and 721, McKiernan-Terry Corp., 100 Richards Ave., Dover, N.J.

5.17

Diesel Hammer Buy Helps Sub-contractor Keep Pile driving Crews on the Go, *Construct. Equipment,* April, 1958, pp. 62–67.
Bull. 73, 673, and 675, McKiernan-Terry Corp., 100 Richards Ave., Dover, N.J.
"Delmag, Diesel Pile Hammers," Special Construction Machines, Ltd., 166 Bentworth Ave., Toronto 19, Ont. (Bulletins.)

5.18

Advertisement of Raymond's New 5000 psi Hydraulic Hammer, *Civil Eng.,* January, 1964, inside cover.
Sonic Pile Driver Shows Great Promise, *Civil Eng.,* December, 1961, p. 52.
"Sonics" Drive a Pile 71 Ft, While Steam Drives Another 3 In., *Eng. News-Record,* November 9, 1961, pp. 24–26.
Oil Replaces Steam in New Pile Hammer, *Construct. Methods,* June, 1962, pp. 90–91.
Hydraulic Pile Hammer Works, *Eng. News-Record,* June 7, 1962, pp. 41–42.
Perez, Henry T.: Rig Drives Piles Ultra Fast—with Ultrasonic Waves, *Construct. Methods,* November, 1961, pp. 82–83.

5.19

Barnes, B.E.: Three Types of Hammers Drive Miles of Piles at Fast Clip, *Construct. Methods,* December, 1963, pp. 78–82.
Boggs, E.W.: Vibratory Driver Successful for Sheetpile Cells, *Civil Eng.,* April, 1964, pp. 58–60.

5.20

Esrig, Melvin I.: Load Test a Pile in as Little as Ten Minutes, *Eng. News-Record,* Jan. 31, 1963, pp. 38–39.
Brown, Douglas: Finding Bearing of a 24-in. Pile by a Load on a 16-in. Pile, *Civil Eng.,* February, 1959, p. 65.
Gage Tells Pile Hammer Energy, *Eng. News-Record,* Oct. 20, 1960, p. 67.

5.21

Brochures, Associated Pipe and Fitting Corp., 262 Rutherford Blvd., Clifton, N.J.
"The Caudill Drive-Point Pile," C.L. Guild Construction Company, Inc., 90 Water St., E. Providence, R.I. (Brochure.)
"The Pruyn H-pile Points," Associated Pipe and Fitting Corp., 262 Rutherford Blvd., Clifton, N.J. (Bulletin.)

5.22

"Steel H-beam Bearing Pile Splicer," Associated Pipe and Fitting Corp., 262 Rutherford Blvd., Clifton, N.J. (Bulletin.)

5.23

Piledriver on Tracks Drives Piles Diagonally, *Construct. Methods,* March, 1963.
Here's an Easy Way to Drive Batter Piles, *Eng. News-Record,* Feb. 7, 1963, pp. 39–40.
How to Drive Piles on Batter without Leads, *Construct. Methods,* February, 1959, pp. 92–94.
Floating Fleet Installs Long Cylinder Piles, *Construct. Methods,* January, 1963, pp. 96–97.
Wilhoite, Gene M.: Battered Piles for Economical Tower Foundations, *Civil Eng.,* June, 1955, pp. 48–49.
Hrennikoff, A.: Analysis of Pile Foundations with Batter Piles, *Trans. Am. Soc. Civil Engrs.,* vol. 115, pp. 351–381, 1950.

5.24

Berg, Glen V.: Efficiencies of Friction-pile Groups Found by Nomograph, *Civil Eng.,* April, 1955, p. 61.
Rau, Irving B.: Curves Give Efficiencies of Friction Pile Groups, *Civil Eng.,* July, 1954, p. 61.
Klohn, Earle J.: Pile Heave and Redriving, *Trans. Am. Soc. Civil Engrs.,* vol. 128, part 1, pp. 557–608, 1963.

5.26

Parsons, James D., and Stanley D. Wilson: Safe Loads on Dog-leg Piles, *Trans. Am. Soc. Civil Engrs.,* vol. 121, pp. 695–721, 1956.
1500 Piles Rejected, Raymond Objects, *Eng. News-Record,* Mar. 15, 1962, p. 27.
Defective Piles Halt Airport Work, *Eng. News-Record,* Nov. 9, 1961, p. 22.

5.27

Rand, Walter, and Jean Roosen: Expandable-tip Piles, *Civil Eng.,* January, 1962, pp. 25–26.
Poland, George F.: Anchoring a Tunnel in Sand, *Civil Eng.,* March, 1960, pp. 59–61.
New Cofferdam Sealing Technique Saves Excavation and Concrete, *Construct. Methods,* March, 1960, pp. 102–104.

5.28

Contractor-built Rig Cuts Timber Piles Underwater, *Construct. Methods,* March, 1963, pp. 140–141.

Fast Precasting Speeds Bridge Building, *Eng. News-Record,* Mar. 21, 1963, pp. 150–153.

5.29

Nees, L.A.: Pile Foundations for Large Towers on Permafrost, *Trans. Am. Soc. Civil Engrs.,* vol. 117, pp. 935–947, 1952.

Essoglou, Milon E.: Piling Operations in Alaska, *Military Engr.,* no. 330, July–August, 1957, pp. 282–287.

5.30

Milano, Joseph: Cathodic Protection of Marine Terminal Facilities, *Trans. Am. Soc. Civil Engrs.,* vol. 127, part 4, pp. 329–345, 1962.

Pierce, Jack W.: Fundamentals of Cathodic Protection, *Trans. Am. Soc. Civil Engrs.,* vol. 127, part 4, pp. 114–130, 1962.

Camp, Thomas R.: Corrosion and Corrosion Research, *Trans. Am.* Soc. *Civil Engrs.,* vol. 121, pp. 791–813, 1956.

5.31

Young, Edward M., Jr.: Pile Driving Gets a Broad and Intensive Probing, *Eng. News-Record,* Apr. 12, 1962, pp. 46–50.

CHAPTER 6

Introduction

Jacobus, William W., Jr.: Ohio River's New Stairway to Pittsburgh, *Eng. News-Record,* Jan. 12, 1961, pp. 28–35.

6.3

Peck, Ralph B.: Earth Pressure Measurements in Open Cuts, Chicago (Ill.) Subway, *Trans. Am. Soc. Civil Engrs.,* vol. 108, pp. 1008–1036, 1943.

White, Edward E.: Deep Foundations in Soft Chicago Clay, *Civil Eng.,* November, 1958, pp. 36–39.

6.5

"USS Steel Sheet Piling, Handbook of United States Steel," 525 William Penn Place, Pittsburgh 30, Pa., 1960.

6.7

How to Work with Sheet Piles, *Construct. Methods,* November, 1962, pp. 99–102; December, 1962, pp. 92–95; January, 1963, pp. 103–105.

6.8

Sloan, Charles L.: Cofferdam Problems Plague Harvey Tunnel Constructors, *Civil Eng.,* December, 1957, pp. 64–67.

6.9

Concrete Wall Built around Sheet Pile Bracing, *Construct. Methods,* December, 1961, pp. 65–68.

6.10

Reider, George: Power Unit Built within River-edge Cofferdam, *Civil Eng.,* March, 1962, pp. 56–57.

6.11

Grimes, Robert O.: South's Largest Ramp-type Parking Garage Completed, *Civil Eng.,* February, 1957, pp. 42–46.

Bruns, T.C.: Z-pile Cofferdam for New Orlean's Tallest Building, *Civil Eng.,* May, 1961, pp. 68–69.

Silinsh, John: Prestressed Pipe Struts Brace Sheet Pile Wall, *Construct. Methods,* April, 1961, p. 112.

6.12

Tie-back Wall Braces Building Excavation, *Construct. Methods,* November, 1962, pp. 116–119.

Jones, N.C., and G.O. Kerkhoff: Belled Caissons Anchor Walls as Michigan Remodels an Expressway, *Eng. News-Record,* May 11, 1961, pp. 28–31.

Single Wall Cofferdam Saves Time and Money, *Construct. Equipment,* November, 1959, p. 95.

Card, Richard: Dowels Anchor Piles and Piers Eliminate Rakers, *Construct. Methods,* April, 1961, pp. 98–99.

Tiebacks Remove Clutter in Excavation, *Eng. News-Record,* June 8, 1961, pp. 34–36.

6.13

Silinsh, John: Four-level Bracing Holds Twin Cofferdams, *Construct. Methods,* May, 1961, pp. 106–121.

6.14

Terzaghi, Karl: Stability and Stiffness of Cellular Cofferdams, *Trans. Am. Soc. Civil Engrs.,* vol. 110, pp. 1083–1202, 1945.

Cummings, Edward M.: Cellular Cofferdams and Docks, *Trans. Am. Soc. Civil Engrs.,* vol. 125, pp. 13–45, 1960.

White, Ardis, James A. Cheney, and C. Martin Duke: Field Study of a Cellular Bulkhead, *Trans. Am. Soc. Civil Engrs.,* vol. 128, part 1, pp. 463–508, 1963.

Irvine, John W., and Richard F. Gaston: Concrete Addition to Cellular Sheet Pile Shipway, *Trans. Am. Soc. Civil Engrs.,* vol. 126, part 2, pp. 270–279.

Richart, F.E.: Analysis for Sheet-pile Retaining Walls, *Trans. Am. Soc. Civil Engrs.,* vol. 122, pp. 1113–1138, 1957.

Boyer, Walter C., and Henry M. Lummis: Design Curves for Anchored Steel Sheet Piling, *Trans. Am. Soc. Civil Engrs.,* vol. 119, pp. 639–657, 1954.

Ayers, James R., and R.C. Stokes: The Design of Flexible Bulkheads, *Trans. Am. Soc. Civil Engrs.,* vol. 119, pp. 373–402, 1954.

Terzaghi, Karl: Anchored Bulkheads, *Trans. Am. Soc. Civil Engrs.,* vol. 119, pp. 1243–1324, 1954.

6.15

They Made Money on Floods, *Construct. Methods,* August, 1959, pp. 80–86.

6.17

Prugh, Byron J.: How to Avoid Cofferdam Boils, *Civil Eng.,* November, 1953, pp. 63.

6.18

Two Ways to Seat Cofferdams on Rock, *Construct. Methods,* November, 1960, pp. 110–114.

6.20

Dean, W.E.: Prestressed Concrete Sheet Piles for Bulkheads and Retaining Walls, *Civil Eng.,* April, 1960, pp. 68–71.

6.21

Bode, Richard L.: Telescopic Sheeting—A New Way to Shore Trenches, *Construct. Methods,* October, 1960, pp. 78–81.

"Foster Lightweight Steel Piling," Catalog 593, L.B. Foster Company, New York 7, N.Y.

"Saf-T-Jax Hydraulic Shoring System," *Bull.* 18605M, SB81611M, 41615M, Sigma Engineering Corp., San Jose, Calif.

6.22

Prugh, Byron J.: Miramar Cofferdam Dewatered by Wellpoints, *Civil Eng.,* July, 1953, pp. 34–35.

Boschen, H.C.: Havana Traffic Tunnel Built in the Dry under Almendares River, *Civil Eng.,* July, 1953, pp. 29–33.

CHAPTER 7

Introduction

Tschebotarioff, Gregory P.: "Soil Mechanics, Foundations, and Earth Structures," McGraw-Hill Book Company, New York, 1951.

Holt, Elizabeth G. (ed.): Annual Reports on the Building Operations of Milan Cathedral, in "A Documentary History of Art," vol. 1, "The Middle Ages and the Renaissance," pp. 107–114, Princeton University Press, Princeton, N.J., 1957; also Doubleday Anchor Books.

Terzaghi, Karl: The Influence of Modern Soil Studies on the Design and Construction of Foundations, in "From Theory to Practice in Soil Mechanics," John Wiley & Sons, Inc., New York, 1960.

7.1

Berggren, R. Alan: Determining Vertical Stress beneath a Footing, *Civil Eng.,* June, 1961, p. 71.

Koo, Benjamin: Design Chart for Vertical Stress under Square Footings, *Civil Eng.,* September, 1960.

Knott, Albert W.: A Graphic Solution for Soil Pressures under Eccentrically Loaded Footings, *Civil Eng.,* January, 1961, p. 70.

7.2

Willson, R.J.: The Granby Pumping Plant—A Symposium—Construction, *Trans. Am. Soc. Civil Engrs.,* vol. 119, pp. 525–541, 1954.

Bravo, Arthur C.: Variable-angle Launcher—Construction, *J. Construct. Div. Am. Soc. Civil Engrs.,* vol. 83, no. COL, p. 1343, 1957.

Smoots, Vernon A., and Philip H. Benton: Compacted Earth Fill for a Power Plant Foundation, *Civil Eng.,* August, 1961, pp. 54–57.

7.4

Pilling, Alan H., and Martin W. Boll: How to Plan Forms in Detail, *Construct. Methods,* November, 1954.

7.7

Auger and Fancy Forms Lick Tough Piers, *Construct. Methods,* February, 1961, p. 86.

7.10

Enriquez, R. Romeo, and Alejandro Fierro: A New Project for Mexico City, *Civil Eng.,* June, 1963, pp. 36–38.

Thornley, J.H., and Pedro Albin, Jr.: Earthquake Resistant Construction in Mexico City, *Civil Eng.,* November, 1957, p. 71.

Thornley, J.H., C.B. Spencer, and Pedro Albin, Jr.: Mexico's Palace of Fine Arts Settles 10 Feet. Can It Be Stopped? *Civil Eng.,* June, 1955, pp. 50–54.

Arrowhead Building Aims at Beating Shock Waves, *Eng. News-Record,* May 30, 1963, pp. 30–37.

7.11

Ballast in the Basement Balances New U.S. Embassy in Mexico City, *Eng. News-Record,* Nov. 29, 1962, p. 30.

7.12

Parsons, James D.: Foundation Installation Requiring Recharging of Ground Water, *J. Construct. Div. Am. Soc. Civil Engrs.* vol. 85, no. CO2, part 1, pp. 1–21, 1959.

7.13

Bravo, *op. cit.*

7.14

"Design and Control of Concrete Mixtures," Portland Cement Association, 10th ed., 1952.

7.15

Literature of the manufacturers of roofing materials.

7.16 and 7.17

Manning, John J.: Simple Rules for Hot or Cold Weather Concreting, *J. Construct. Div. Am. Soc. Civil Engrs.* vol. 86, no. CO3, pp. 37–43, 1960.

"Recommended Practice for Concreting Operations during Hot or Cold Weather," The Concrete Industry Board, New York, 1958.

Racey, H.J.: Lessons from Cold-weather Concrete Failures, *Civil Eng.,* November, 1957, pp. 57–59.

7.18

Brochure, Raymond Concrete Pile Company, 140 Cedar St., New York 6, N.Y.

7.19

Foundation Crews Fight Water, Lack of Space, *Construct. Methods,* September, 1960, pp. 142–144.

7.20

Brochures, Calweld, Inc., 7222 E. Slauson Ave., Los Angeles 22, Calif.

Brochures, Mobile Drilling Inc., 960 N. Pennsylvania Ave., Indianapolis, Ind.

High-speed Bucket Rig Prepares Large-diameter Drilled-in-place Footings, *Construct. Equipment,* July, 1954, pp. 30–33.

Changing a Crane to an Earth Drill, *Eng. News-Record,* May 25, 1961, p. 59.

Tough Caisson Job Calls for Special Rig, New Methods, *Construct. Methods,* August, 1961, pp. 124–125.

Crane Attachment Drills 100 Ft Caissons in 4 Hours, *Construct. Equipment,* March, 1963, pp. 73–74.

"Use of Drill Stem with Calweld Equipment," Calweld, Inc., 7222 E. Slauson Ave., Los Angeles 22, Calif. (Mimeographed.)

7.21

Switch to Caisson Buckets to Cut Concrete Over-run, *Construct. Equipment,* March, 1962, pp. 83–84.

Big Augers Go Deep So Buildings Can Go High, *Eng. News-Record,* May 25, 1961, pp. 32–33.

Juergens, Ralph E.: Drilling Experts Develop Tools to Speed Caisson Work, *Construct. Methods,* January, 1963, pp. 80–83.

7.22

"Casing Techniques," Calweld, Inc., 7222 E. Slauson Ave., Los Angeles 22, Calif. (Mimeographed.)

7.23

Krynine, Dimitri P., and William R. Judd: "Principles of Engineering Geology and Geotechnics," McGraw-Hill Book Company, New York, 1957.

"Using Reverse Circulation Principles with Calweld Equipment," Calweld, Inc., 7222 E. Slauson Ave., Los Angeles 22, Calif. (Mimeographed.)

"Using Bentonite to Control Unstable Formations in Caisson Work," Calweld, Inc., 7222 E. Slauson Ave., Los Angeles 22, Calif. (Mimeographed.)

Calcium Bentonite Slurry Seals Canals, *Eng. News-Record,* May 10, 1960, p. 53.

7.24

Gauntt, Grover C.: Marina City—Foundations, *Civil Eng.,* December, 1962, pp. 60–63.

Modified Drill Rig Sinks Big Caissons, *Contractors and Engrs.,* September, 1960.

Top and Bottom Casings Seal Deep Caissons, *Construct. Methods,* May, 1961, pp. 90–93.

7.25

Caisson Borer Drives and Mucks Casings, *Construct. Equipment,* May, 1962, p. 75.

Imported Boring Rigs Complete Tough Foundation Job, *Eng. News-Record,* Nov. 30, 1961, pp. 36–37.

Drill-drivers and T.V. Keep Caisson Job Rolling, *Construct. Methods,* May, 1960, pp. 92–95.

7.26

Drilling Sequence for Concrete Shafts, *Construct. Equipment,* September, 1956, pp. 21–22.

Giant Augers Sink 15-ft-dia. Missile Shafts, *Construct. Methods,* May, 1963, pp. 120–123.

Big Rig Drills Holes and Casts Piles, *Construct. Equipment,* August, 1955, pp. 22–26.

Save Time and Money on Bridge Work, *Construct. Equipment,* March, 1958.

7.27

Willey, C.K.: Wanapum Hydroelectric Development, *Civil Eng.,* September, 1960, pp. 65–69.

Engstrom, U.V.: Innovations at Wanapum Dam—New Ideas Pay Off, *Civil Eng.,* October, 1963, pp. 43–47.

Slurry Supports Cutoff Trench Walls, *Construct. Methods,* February, 1960, pp. 88–92.

Foundation Was Cast in Slurry, *Eng. News-Record,* Apr. 26, 1962, pp. 100–101.

7.28

Pelkey, Orville, and Austin C. Woodward: World's Tallest Structure, *Civil Eng.,* December, 1960, pp. 33–35.

Rhodes, Robert B.: Highest Self-supported Tower, *Civil Eng.,* August, 1963, pp. 48–49.

7.30

Oelschlager, Donovan E.: Prestressed Concrete Foundation Takes 8000-ton Press at Half the Cost of Conventional Design, *Civil Eng.,* January, 1955, pp. 42–45.

7.31

Klein, Alden M., and J.H.A. Crockett: Forging Hammer Foundation Built to Control Destructive Vibrations, *Civil Eng.,* January, 1953, pp. 30–36.

7.32

Williams, Roger H.: Ventilated Building Foundations in Greenland, *J. Construct. Div. Am. Soc. Civil Engrs.,* vol. 85, no. CO2, part 1, pp. 23–36, 1959.

Pihlainen, John A.: Discussion of "Ventilated Building Foundations in Greenland," *J. Construct. Div. Am. Soc. Civil Engrs.,* vol. 86, no. CO2, part 1, pp. 43–45, 1960.

7.33

Waterhouse, Robert W., and A. Nelson Sills: Thaw-blast Method Prepares Permafrost Foundation for Alaska Power House, *Civil Eng.,* February, 1952, pp. 28–31.

7.34

Bane, R.D.: Construction of Morrison Bridge, *J. Construct. Div. Am. Soc. Civil Engrs.,* vol. 86, no. CO2, part 1, pp. 1–8, 1960.

Howard, R.B.: Single Bridge Piers Built in Cellular Cofferdams, *Civil Eng.,* January, 1951, pp. 40–41.

7.35

Davis, C.M.: Extruded Concrete for Bridge Piers, *Civil Eng.,* March, 1957, pp. 58–61.

Hollister, Leonard C.: Record Approach Cut and Use of Slip Forms on 47 High Interchange Piers, *Civil Eng.,* January, 1957, pp. 54–57.

7.36

Gerwick, Ben. C., Jr.: Transmission Line Piers Built under Water without Cofferdams, *Civil Eng.,* April, 1953, pp. 33–37.

7.37

Gerwick, Ben. C., Jr.: Hollow Precast Concrete Units of Great Size Form Bridge Substructure, *Civil Eng.,* April, 1954, pp. 59–63.

Merchant, Ivan D.: Construction of the Columbia River (Portland-Vancouver) Bridges, *J. Construct. Div. Am. Soc. Civil Engrs.,* vol. 85, no. CO2, part 1, pp. 37–50, 1959.

Williams, Ira E.: Exceptional Job Engineering Pays Off on Bridge Construction, *Civil Eng.,* January, 1960, pp. 60–62.

7.38

Prugh, Byron J.: Anchorage Excavation Tests Versatility of Wellpoints, *Civil Eng.,* September, 1954, pp. 40–46, 46–48.

7.39

Wikstrom, A.S.: Contractor Finds St. Lawrence Bridge Substructure No Picnic, *Civil Eng.,* December, 1955, pp. 38–41.

CHAPTER 8

Introduction

De Maré, Eric Samuel, "The Bridges of Britain," B.T. Botsford, Ltd., London, 1954.

Smith, H. Shirley: "The World's Great Bridges," Phoenix House, Ltd., London, 1953.

"The Cairo Bridge," Report of Chief Engineer to the President of the Chicago, St. Louis, and New Orleans Railroad Company, Oct. 1, 1891.

8.1

Lloyd, Christopher: "Lord Cochrane; Seaman, Radical, Liberator, a Life of Thomas Lord Cochrane, 10th Earl of Dundonald," Longmans, Green & Co., Ltd., London, 1947.

8.2

Stone, William L., Jr.: "Sand Islands Used to Sink Caissons for Piers of Philadelphia—Gloucester Bridge," *Civil Eng.,* August, 1955, pp. 56–60.

Seely, Homer R.: Anchorages and Superstructure of Walt Whitman Bridge, *Civil Eng.,* February, 1956, pp. 52–57.

8.3

Steinman, David B., and John T. Nevill: "Miracle Bridge at Mackinac," Wm. B. Eerdmans Publishing Co., Grand Rapids, Mich., 1958.

8.4

Record Suspension Bridge Rises, *Eng. News-Record,* Sept. 22, 1960, pp. 146–147.

Seely, Homer R.: The Delaware River Memorial Bridge—Planning and Construction, *Trans. Am. Soc. Civil Engrs.,* vol. 118, pp. 398–410, 1953.

Smoot, Leonard V.: Deep Sand-island Caissons Support Cincinnati Steam Plant, *Civil Eng.,* August, 1952, pp. 42–46.

8.6

Hoffman, John F.: Wellpoints Dewater Sites for Narrows Bridge Piers, *Civil Eng.,* October, 1960, pp. 71–73.

8.8

Albertson, M.L., et al: Diffusion of Submerged Jets, *Trans. Am. Soc. Civil Engrs.,* vol. 115, pp. 639–697, 1950.

8.9

"Concrete Handling Equipment," Catalogue 300-A, Gar-Bro Manufacturing Company, 2415 E. Washington Blvd., Los Angeles 21, Calif.

8.15

Duclos, Louis: Highway Bridges on Deep Foundations, *Trans. Am. Soc. Civil Engrs.,* vol. 119, pp. 1090–1102, 1954.

Caisson Is Floated to Thruway Bridge Site, *Civil Eng.,* November, 1953, p. 83.

Jansen, Carl B.: The Toughest Part Is Under the River —Substructure for New Orleans Record Cantilever, *Civil Eng.,* February, 1956, pp. 40–44.

Steel and Concrete Caissons Support Piers for Lift Span, *Construct. Methods,* May, 1961, pp. 142–150.

Tagus Bridge Caisson Launched, *Eng. News-Record,* February 21, 1963, p. 21.

8.16

Gronquist, C.H.: Fifty Year Dream Nears Reality, *Civil Eng.,* April, 1954, pp. 53–57.

Boynton, R.M.: Mackinac Bridge—Foundations Constructed at Record Speed by Unusual Methods, *Civil Eng.,* May, 1956, pp. 46–50.

Fletcher, Gordon F.A.: Heavy Construction Goes to Sea—First Atlantic Radar Platform Installed on Georges Bank, *Civil Eng.,* January, 1956, pp. 59–64.

Four-legged Towers Replace Lightships, *Eng. News-Record,* Mar. 30, 1961, pp. 26–27.

Toler, Jack S.: Offshore Petroleum Installations, *Trans. Am. Soc. Civil Engrs.,* vol. 119, pp. 480–488, 1954.

Mathis, S.J.: Marine Foundation Construction in Oil Fields, *Trans. Am. Soc. Civil Engrs.,* vol. 127, part 2, pp. 18–38, 1962.

8.17

Quade, Maurice N.: Special Design Features of the Yorktown Bridge, *Trans. Am. Soc. Civil Engrs.*, vol. 119, pp. 109–123, 1954.

Quade, Maurice N., and George Vaccaro: Deep, Lightweight Piers for Bridge at Yorktown, Va., Built by Caisson Method, *Civil Eng.*, February, 1951, pp. 28–32.

Newell, John N.: Floating Caissons for Yorktown, Va., Bridge Sunk to 150-ft Depth in Swift-flowing Tidal Water, *Civil Eng.*, July, 1951, pp. 44–48.

8.18

Lane, E.W., and W.M. Borland: River-bed Scour during Floods, *Trans. Am. Soc. Civil Engrs.*, vol. 119, pp. 1069–1089, 1954.

Laursen, Emmett M.: Scour at Bridge Crossings, *Trans. Am. Soc. Civil Engrs.*, vol. 127, part 1, pp. 166–209, 1962.

8.19

Hollister, Leonard C.: Carquinez Toll Bridge Project—Part 2, Bridge Features Deep-bottom Caisson and High-strength Weldable Steel Superstructure, *Civil Eng.*, February, 1957, pp. 52–55.

"Bridging Mackinac on Prepakt Concrete," The Prepakt Concrete Company, Union Commerce Bldg., Cleveland 14, Ohio. (Bulletin.)

Gray, Nomer: Foundations for the Throgs Neck Bridge, *Civil Eng.*, October, 1959, pp. 50–54.

8.22

Newell, John M.: Pneumatic Caisson Pier for World's Longest Pipe Line Construction, *Civil Eng.*, May, 1956, pp. 51–55.

8.23

Rubins, Ralph E.: Caissons and Rigid Frames Combined in Unique Pier, *Civil Eng.*, February, 1956, pp. 58–60.

Holt, J.A.: Australia Completes its Longest Plate-girder Bridge, *Civil Eng.*, August, 1953, pp. 50–54.

8.25

Mason, Arnold C.: Open-caisson Method Used to Erect Tokyo Office Building, *Civil Eng.*, November, 1952, pp. 46–49.

Caissons Dig Out a Seven-story Basement, *Eng. News-Record,* July 6, 1961, pp. 42–43.

Five-story Basement Sinks Below Grade, *Eng. News-Record,* Dec. 6, 1952, p. 43.

8.28

Gellerstad, Robert V.: Telescopic Caisson for a Lighthouse Base, *Civil Eng.*, September, 1960, pp. 58–61.

INDEX